Communications in Computer and Information Science 772

Commenced Publication in 2007
Founding and Former Series Editors:
Alfredo Cuzzocrea, Xiaoyong Du, Orhun Kara, Ting Liu, Dominik Ślęzak,
and Xiaokang Yang

More information about this series at http://www.springer.com/series/7899

Jinfeng Yang · Qinghua Hu
Ming-Ming Cheng · Liang Wang
Qingshan Liu · Xiang Bai
Deyu Meng (Eds.)

Computer Vision

Second CCF Chinese Conference, CCCV 2017
Tianjin, China, October 11–14, 2017
Proceedings, Part II

 Springer

Editors

Jinfeng Yang
Civil Aviation University of China
Tianjin
China

Qinghua Hu
Tianjin University
Tianjin
China

Ming-Ming Cheng
Nankai University
Tianjin
China

Liang Wang
Institute of Automation
Chinese Academy of Sciences
Beijing
China

Qingshan Liu
Nanjing University of Information
    Science and Technology
Nanjing
China

Xiang Bai
Huazhong University of Science
    and Technology
Wuhan
China

Deyu Meng
Xi'an Jiaotong University
Xi'an
China

ISSN 1865-0929   ISSN 1865-0937   (electronic)
Communications in Computer and Information Science
ISBN 978-981-10-7301-4   ISBN 978-981-10-7302-1   (eBook)
https://doi.org/10.1007/978-981-10-7302-1

Library of Congress Control Number: 2017960863

Printed on acid-free paper

This Springer imprint is published by Springer Nature
The registered company is Springer Nature Singapore Pte Ltd.
The registered company address is: 152 Beach Road, #21-01/04 Gateway East, Singapore 189721, Singapore

Preface

Welcome to the proceedings of the Second Chinese Conference on Computer Vision (CCCV 2017) held in Tianjin!

CCCV, hosted by the China Computer Federation (CCF) and co-organized by Computer Vision Committee of CCF, is a national conference in the field of computer vision. It aims at providing an interactive communication platform for students, faculties, and researchers from industry. It promotes not only academic exchange, but also communication between academia and industry. CCCV is one of the most important local academic activities in computer vision. In order to keep track of the frontier of academic trends and share the latest research achievements, innovative ideas, and scientific methods in the field of computer vision, international and local leading experts and professors are invited to deliver keynote speeches, introducing the latest theories and methods in the field of computer vision.

The CCCV 2017 received 465 full submissions. Each submission was reviewed by at least two reviewers selected from the Program Committee and other qualified researchers. Based on the reviewers' reports, 174 papers were finally accepted for presentation at the conference. The acceptance rate is 37.4%. The proceedings of the CCCV 2017 are published by Springer.

We are grateful to the keynote speakers, Prof. Katsushi Ikeuchi from University of Tokyo, Prof. Sven Dickinson from University of Toronto, Prof. Gérard Medioni from University of Southern California, Prof. Marc Pollefeys from ETH Zurich, Prof. Xilin Chen from Institute of Computing Technology, Chinese Academy of Science.

Thanks go to the authors of all submitted papers, the Program Committee members and the reviewers, and the Organizing Committee. Without their contributions, this conference would not be a success. Special thanks go to all of the sponsors and the organizers of the five special forums; their support made the conference a success. We are also grateful to Springer for publishing the proceedings and especially to Ms. Celine (Lanlan) Chang of Springer Asia for her efforts in coordinating the publication.

We hope you find CCCV 2017 an enjoyable and fruitful conference.

Tieniu Tan
Hongbin Zha
Jinfeng Yang
Qinhua Hu
Ming-Ming Cheng
Liang Wang
Qingshan Liu

Organization

CCCV 2017 (CCF Chinese Conference on Computer Vision 2017) was hosted by the China Computer Federation (CCF) and co-organized by the Civil Aviation University of China, Tianjin University, Nankai University, and CCF Technical Committee on Computer Vision.

Organizing Committee

International Advisory Board

Bill Freeman	Massachusetts Institute of Technology, USA
Gerard Medioni	University of Southern California, USA
Ramin Zabih	Cornell University, USA
David Forsyth	University of Illinois at Urbana-Champaign, USA
Jitendra Malik	University of California, Berkeley, USA
Fei-Fei Li	Stanford University, USA
Ikeuchi Katsushi	The University of Tokyo, Japan
Song-chun Zhu	University of California, Los Angeles, USA
Long Quan	The Hong Kong University of Science and Technology, SAR China

Steering Committee

Tieniu Tan	Institute of Automation, Chinese Academy of Sciences, China
Hongbin Zha	Peking University, China
Xinlin Chen	Institute of Computing Technology, Chinese Academy of Sciences, China
Jianhuang Lai	Sun Yat-Sen University, China
Dewen Hu	National University of Defense Technology, China
Shengyong Chen	Tianjin University of Technology, China
Yanning Zhang	Northwestern Polytechnical University, China
Tao Wang	IQIYI Inc.

General Chairs

Tieniu Tan	Institute of Automation, Chinese Academy of Sciences, China
Hongbin Zha	Peking University, China
Jinfeng Yang	Civil Aviation University of China

Program Chairs

Qinghua Hu	Tianjin University, China
Ming-Ming Cheng	Nankai University, China
Liang Wang	Institute of Automation, Chinese Academy of Sciences, China
Qingshan Liu	Nanjing University of Information Science and Technology, China

Organizing Chairs

Jufeng Yang	Nankai University, China
Pengfei Zhu	Tianjin University, China

Publicity Chairs

Qiguang Miao	Xidian University, China
Wei Jia	Hefei University of Technology, China

International Liaison Chairs

Andy Yu	ShanghaiTech University, China
Zhanyu Ma	Beijing University of Posts and Telecommunications, China

Publication Chairs

Xiang Bai	Huazhong University of Science and Technology, China
Deyu Meng	Xi'an Jiaotong University, China

Tutorial Chairs

Zhouchen Lin	Peking University, China
Xin Geng	Southeast University, China

Workshop Chairs

Zhaoxiang Zhang	Institute of Automation, Chinese Academy of Sciences, China
Rongrong Ji	Xiamen University, China

Sponsorship Chairs

Yongzhen Huang	Watrix Technology
Huchan Lu	Dalian University of Technology, China

Demo Chairs

Liang Lin	Sun Yat-Sen University, China
Chao Xu	Tianjin University, China

Competition Chairs

Wangmeng Zuo	Harbin Institute of Technology, China
Jiwen Lu	Tsinghua University, China

Website Chairs

Changqing Zhang	Tianjin University, China
Shaocheng Han	Civil Aviation University of China

Finance Chairs

Guimin Jia	Civil Aviation University of China
Ruiping Wang	Institute of Computing Technology, Chinese Academy of Sciences, China

Coordination Chairs

Lifang Wu	Beijing University of Technology, China
Shiying Li	Hunan University, China

Program Committee

Haizhou Ai	Tsinghua University, China
Xiang Bai	Huazhong University of Science and Technology, China
Yinghao Cai	Chinese Academy of Sciences, China
Xiaochun Cao	Chinese Academy of Sciences, China
Hui Ceng	University of Science and Technology Beijing, China
Hongbin Cha	Peking University, China
Zhengjun Cha	University of Science and Technology of China, China
Hongxin Chen	HISCENE
Shengyong Chen	Zhejiang University of Technology, China
Songcan Chen	Nanjing University of Aeronautics and Astronautics, China
Xilin Chen	Chinese Academy of Science, China
Yongquan Chen	HuanJing Information Technology Inc., China
Hong Cheng	University of Electronic Science and Technology of China
Jian Cheng	Chinese Academy of Science, China
Mingming Cheng	Nankai University, China
Jun Chu	Nanchang Hangkong University, China
Yang Cong	Chinese Academy of Science, China
Chaoran Cui	Shandong University of Finance and Economics, China
Cheng Deng	Xidian University, China
Weihong Deng	Beijing University of Posts and Telecommunications, China
Xiaoming Deng	Chinese Academy of Science, China

Jing Dong   Chinese Academy of Science, China
Junyu Dong   Ocean University of China
Weisheng Dong   Xidian University, China
Fuqing Duan   Beijing Normal University, China
Bin Fan   Chinese Academy of Science, China
Xin Fan   Dalian University of Technology, China
Yuchun Fang   Shanghai University, China
Jufu Feng   Peking University, China
Jianjiang Feng   Tsinghua University, China
Xianping Fu   Dalian Maritime University, China
Shenghua Gao   ShanghaiTech University, China
Yong Gao   JieShang Visual technology Inc., China
Shiming Ge   Chinese Academy of Science, China
Xin Geng   Southeast University, China
Guanghua Gu   Yanshan University, China
Jie Gui   Chinese Academy of Science, China
Yulan Guo   National University of Defense Technology, China
Zhenhua Guo   Tsinghua University, China
Aili Han   Shandong University, China
Junwei Han   Northwestern Polytechnical University, China
Huiguang He   Chinese Academy of Science, China
Lianghua He   Tongji University, China
Ran He   Chinese Academy of Science, China
Yutao Hou   NVIDIA Semiconductor Technical Services Inc.
Zhiqiang Hou   Air Force Engineering University
Haifeng Hu   Sun Yat-sen University, China
Hua Huang   Beijing Institute of Technology, China
Di Huang   Beijing University of Aeronautics and Astronautics
Kaiqi Huang   Chinese Academy of Science, China
Qingming Huang   University of Chinese Academy of Sciences
Yongzhen Huang   Chinese Academy of Science, China
Yanli Ji   University of Electronic Science and Technology
                          of China
Rongrong Ji   Xiamen University, China
Wei Jia   Chinese Academy of Science, China
Tong Jia   Northeastern University, China
Yunde Jia   Beijing Institute of Technology, China
Muwei Jian   Ocean University of China
Yugang Jiang   Fudan University, China
Shuqiang Jiang   Chinese Academy of Science, China
Cheng Jin   Fudan University, China
Lianwen Jin   South China University of Technology, China
Xiaoyuan Jing   Wuhan University, China
Xiangwei Kong   Dalian University of Technology, China
Jianhuang Lai   Sun Yat-sen University, China
Zhen Lei   Chinese Academy of Science

Yao Lu	Beijing Institute of Technology, China
Bin Luo	Anhui University, China
Ke Lv	University of Chinese Academy of Sciences, China
Bingpeng Ma	University of Chinese Academy of Sciences, China
Huimin Ma	Tsinghua University, China
Wei Ma	Beijing University of Technology, China
Zhanyu Ma	Beijing University of Posts and Telecommunications, China
Lin Mei	Third Institute of Public Security
Deyu Meng	Xi'an Jiaotong University, China
Qiguang Miao	Xidian University, China
Rongrong Ni	Beijing Jiaotong University, China
Xiushan Nie	Shandong University of Finance and Economics
Jifeng Ning	Northwest A&F University, China
Jianquan Ouyang	Xiangtan University, China
Yuxin Peng	Peking University, China
Yu Qiao	Shenzhen Institutes of Advanced Technology, Chinese Academy of Sciences, China
Lei Qin	Chinese Academy of Sciences, China
Pinle Qin	North University of China
Jianhua Qiu	Beijing Wisdom Eye Technology, China
Chuanxian Ren	Sun Yat-sen University, China
Qiuqi Ruan	Beijing Jiaotong University, China
Nong Sang	Huazhong University of Science and Technology, China
Shiguang Shan	Chinese Academy of Science, China
Shuhan Shen	Chinese Academy of Science, China
Jianbing Shen	Beijing Institute of Technology, China
Linlin Shen	Shenzhen University, China
Peiyi Shen	Xidian University, China
Wei Shen	Shanghai University, China
Mingli Song	Zhejiang University, China
Fei Su	Beijing University of Posts and Telecommunications, China
Hang Su	Tsinghua University, China
Dongmei Sun	Beijing Jiaotong University, China
Jian Sun	Xi'an Jiaotong University, China
Taizhe Tan	Guangdong University of Technology, China
Tieniu Tan	Chinese Academy of Science, China
Xiaoyang Tan	Nanjing University of Aeronautics and Astronautics, China
Jin Tang	Anhui University, China
Jinhui Tang	Nanjing University of Science and Technology, China
Yandong Tang	Chinese Academy of Science, China
Zengfu Wang	Chinese Academy of Science, China
Hanzi Wang	Xiamen University, China

Hanli Wang	Tongji University, China
Hongyuan Wang	Changzhou University, China
Hongpeng Wang	Harbin Institute of Technology, China
Jinjia Wang	Yanshan University, China
Jingdong Wang	Microsoft Research
Liang Wang	Chinese Academy of Science, China
Qi Wang	Northwestern Polytechnical University
Qing Wang	Northwestern Polytechnical University
RuiPing Wang	Chinese Academy of Science, China
Shengke Wang	Ocean University of China
Shiquan Wang	Philips Research Institute of China
Sujing Wang	Chinese Academy of Science, China
Tao Wang	IQIYI Inc.
Wei Wang	Chinese Academy of Science, China
Yuanquan Wang	Hebei University of Technology, China
Yuehuan Wang	Huazhong University of Science and Technology, China
Yunhong Wang	Beijing University of Aeronautics and Astronautics, China
Shikui Wei	Beijing Jiaotong University, China
Gongjian Wen	National University of Defense Technology, China
Xiangqian Wu	Harbin Institute of Technology, China
Lifang Wu	Beijing University of Technology, China
Jianxin Wu	Nanjing University, China
Jun Wu	Chongqing Kaiser Technology, China
Yadong Wu	Southwest University of Science and Technology, China
Yihong Wu	Chinese Academy of Sciences, China
Yongxian Wu	South China University of Technology, China
Guisong Xia	Wuhan University, China
Shiming Xiang	Chinese Academy of Science, China
Xiaohua Xie	Sun Yat-sen University, China
Junliang Xing	Chinese Academy of Sciences, China
Hongkai Xiong	Shanghai Jiao Tong University, China
Chao Xu	Tianjin University, China
Mingliang Xu	Zhengzhou University, China
Yong Xu	Harbin Institute of Technology, China
Yong Xu	South China University of Technology, China
Xinshun Xu	Shandong University, China
Feng Xue	HeFei University of Technology, China
Jianru Xue	Xi'an Jiaotong University, China
Yan Yan	Xiamen University, China
Dongming Yan	Chinese Academy of Sciences, China
Junchi Yan	IBM China Research Institute, China
Bo Yan	Fudan University, China
Chenhui Yang	Xiamen University, China

Dong Yang   Beijing Wisdom Eye Technology Inc., China
Gongping Yang   Shandong University, China
Jian Yang   Beijing Institute of Technology, China
Jie Yang   Shanghai Jiaotong University, China
Jinfeng Yang   Civil Aviation University of China
Jufeng Yang   Nankai University, China
Lu Yang   University of Electronic Science and Technology
                          of China
Wankou Yang   Southeast University, China
Jian Yao   Wuhan University, China
Mao Ye   University of Electronic Science and Technology
                          of China
Xucheng Yin   University of Science and Technology Beijing, China
Yilong Yin   Shandong University, China
Xianghua Ying   Peking University, China
Xingang You   Beijing Institute of Electronics Technology and
                          Application, China
Xinge You   Huazhong University of Science and Technology,
                          China
Jian Yu   Beijing Jiaotong University, China
Shiqi Yu   Shenzhen University, China
Xiaoyi Yu   Peking University, China
Ye Yu   HeFei University of Technology, China
Zhiwen Yu   South China University of Technology, China
Jun Yu   Hanzhou Electronic Science and Technology
                          University, China
Jingyi Yu   ShanghaiTech University, China
Xiaotong Yuan   Nanjing University of Information Science
                          and Technology, China
Di Zang   Tongji University, China
Honggang Zhang   Beijing University of Posts and Telecommunications,
                          China
Junping Zhang   Fudan University, China
Junge Zhang   Chinese Academy of Sciences, China
Lin Zhang   Tongji University, China
Shihui Zhang   Yanshan University, China
Wei Zhang   Fudan University, China
Wei Zhang   Shandong University, China
Wenqiang Zhang   Fudan University, China
Xiaoyu Zhang   Chinese Academy of Sciences, China
Yan Zhang   Nanjing University, China
Yanning Zhang   Northwestern Polytechnical University, China
Yifan Zhang   Chinese Academy of Sciences, China
Yimin Zhang   Intel China Research Center, China
Yongdong Zhang   Chinese Academy of Science, China
Yunzhou Zhang   Northeastern University

Zhang Zhang	Chinese Academy of Science, China
Zhaoxiang Zhang	Chinese Academy of Science, China
Guofeng Zhang	Zhejiang University, China
Yao Zhao	Beijing Jiaotong University, China
Cairong Zhao	Tongji University, China
Yang Zhao	HeFei University of Technology, China
Yuqian Zhao	Central South University
Weishi Zheng	Sun Yat-sen University, China
Wenming Zheng	Southeast University
Bineng Zhong	Huaqiao University, China
Dexing Zhong	Xi'an Jiaotong University, China
Hanning Zhou	Beijing Gourd Software Technology Inc., China
Rigui Zhou	Shanghai Maritime University, China
Zhenfeng Zhu	Beijing Jiaotong University
Liansheng Zhuang	University of Science and Technology of China
Beiji Zou	Central South University
Wangmeng Zuo	Harbin Institute of Technology, China

Sponsoring Institutions

Athena Eyes
智慧眼

水滴科技
WATRIX TECHNOLOGY

EXTREME VISION
极视角

中科智谷
上海中科智谷人工智能工业研究院
Shanghai Academy Of Artificial Intelligence

云识图

SEGWAY
ROBOTICS

图漾科技
PERCIPIO.XYZ

Contents – Part II

Image Color and Texture

Image Composition

Image Quality Assessment and Analysis

Image Restoration

Image-Based Modeling

Face and Posture Analysis

Projective Representation Learning for Discriminative Face Recognition

Zuofeng Zhong, Zheng Zhang, and Yong Xu[✉]

Shenzhen Graduate School, Harbin Institute of Technology,
Shenzhen 518055, China
yongxu@ymail.com

Abstract. Face recognition is a challenging issue due to various appearances under different conditions of the face of a person. Meanwhile, conventional face representation methods always lead to high computational complexity. To overcome these shortcomings, in this paper, we propose a novel discriminative projection and representation method for face recognition. This method tries to seek a discriminative representation of the face image on a low-dimension space. Our method consists of two stages, namely face projection and face representation. In the face projection stage, a mapping matrix is produced by jointly maximizing the covariance of dissimilar samples and minimizing the covariance of similar samples. In the face representation stage, the representation result for each face image is obtained by minimizing the sum of representation results of each class. The proposed method achieves two-fold discriminative properties and provides a computational efficient algorithm. The experiments evaluated on diverse face datasets demonstrate that the proposed method has great superiority for face recognition task.

1 Introduction

Identifying a person from a large number of face images is the principal task of face recognition [1]. Up till now, numerous methods have been proposed and used for solving the identification problem. The commonly used classification methods include k-nearest neighbor (NN) [2,3], support vector machine (SVM) [4,5], and sparse representation [6,7], among which the study of the representation-based methods has been an attractive topic in the past decades due to its distinct performance in image classification such as face recognition [8], gait recognition [9], action recognition [10], and image representation [11].

The adaptation is an important aspect for evaluating the performance of the recognition method in real-world applications. Conventional methods, such as SVM and Neural Network etc., aim at training a specific classifier from a given dataset. Therefore, these classifiers are vulnerable for face variations. Recently, sparse representation classification (SRC) [12] method has provided an effective way to raise the performance of face recognition on diverse conditions. SRC aims to find the minimum residuals between a test sample and a linear representation of the training samples in different classes for classification.

© Springer Nature Singapore Pte Ltd. 2017
J. Yang et al. (Eds.): CCCV 2017, Part II, CCIS 772, pp. 3–15, 2017.
https://doi.org/10.1007/978-981-10-7302-1_1

The weighted coefficients indicate the contribution of each class and are referred to as the sparse solution obtained via l_1 regularization. However, the l_1 regularization often needs an iterative process for numerical optimization, which leads to heavy computational burden for real-time applications. Consequently, the l_2 regularization-based representation is proposed to overcome this drawback. The collaborative representation classification (CRC) [13] method is a typical method of the regularization-based representation. CRC is computational efficiency due to its closed-form solution. Current study [14] shows that the sparse property cannot be well guaranteed if native l_2 regularization is directly applied on the classification model. However, the study of CRC demonstrates that the collaborative representation plays more important role in classification compared with the sparsity. In addition to CRC, the robust regression for classification [15], two-phase test sample sparse representation [16], discriminative sparse representation [17] etc. are effective classification methods based on regularization. However, these methods still suffer from the issues such as insufficiently discriminative ability and heavy computational burden.

To seeking an efficient discriminative representation for face images, in this paper, we propose a novel discriminative projection and representation method for face representation and recognition. The proposed method can achieve twofold discriminative properties due to the special design of objective function. The proposed method consists of two stages, including face projection and face representation. On the face projection stage, our method produces a projection matrix by jointly minimizing the similar covariance and maximizing dissimilar covariance, and this matrix maps the face images into a discriminative low-dimension space which has the minimum similarity of samples. Moreover, we employ the quadruplets [18] to construct both covariances, which integrate the similarity of the samples to improve the distinctiveness of projection matrix. On the face representation stage, we produce the discriminative representation for the test sample by obtaining the minimum sum of representation results of all classes on the low-dimension space, which enables the representation results of different classes to be the lowest correlated. Moreover, the proposed method is very efficient in computation due to the closed-form solutions in both steps. The experiments are conducted for evaluating the superiority of the proposed method over other state-of-the-art methods.

The other parts of the paper are organized as follows. Section 2 introduces the related works. Section 3 describes the proposed method. Section 4 offers the experimental results, and we conclude this paper in Sect. 5.

2   Related Works

In this section, we briefly introduce the background of l_2 regularization representation and Linear Discriminant Analysis. Let set $X = \{x_i | x_i \in \mathbb{R}^n\}$ as a training sample set, which has n training samples. If there are c classes and each class has s samples, X can be denoted as matrix $X = [X_1, \ldots, X_j, \ldots, X_c] = [x_1, \ldots, x_{s(j-1)+1}, \ldots, x_{sj}, \ldots, x_n], j = 1, \ldots, c.$ vector y is denoted as test sample.

2.1 l_2 Regularization-Based Representation

We take CRC as an example to describe the classification procedure of l_2 regularization representation. CRC represent a test sample using a linear combination of all classes training samples to, which can be written as following equation [13]

$$y = XQ, Q = [q_1, \ldots, q_n], \tag{1}$$

where $q_i, i = 1, \ldots, N$ is representation coefficient.

The solution of Eq. (1) is $Q^* = (X^T X + \mu I)^{-1} X^T y$, where μ is a small positive constant and I is an identity matrix, and Q^* also is referred to as representation coefficients. Let Q_i^* be the coefficient vector of the i-th class, regularized class-specific representation residual $r_i = \frac{\|y - X_i Q_i^*\|_2}{\|Q_i^*\|_2}$ is used for classification. The label of y is obtained by $lable(y) = \arg\min_i r_i$.

2.2 Linear Discriminant Analysis

LDA aims to produce a projection by jointly minimizing within-class scatter and maximizing between-class scatter, and then project the data into low-dimension space. Therefore, LDA firstly needs to produce between-class scatter matrix S_b and within-class scatter matrix S_w respectively as follows:

$$S_b = \sum_{i=1}^{c} s(m_i - \bar{m})(m_i - \bar{m})^T, \tag{2}$$

and

$$S_w = \sum_{i=1}^{c} \sum_{j=1}^{s} (x_j^i - m_i)(x_j^i - m_i)^T, \tag{3}$$

where m_i and $\bar{m}$ are the mean vectors of the i-class and all classes samples respectively. The objective function of LDA is [18]

$$P = \arg\max \frac{tr(P^T S_b P)}{tr(P^T S_w P)}, \tag{4}$$

where P is the projection matrix, and $tr(\cdot)$ denotes the trace of matrix. Finally, we can obtain P by solving the following eigenvector problem:

$$S_b P = \lambda S_w P, \tag{5}$$

where λ is eigenvalue.

3 The Proposed Method

To simultaneously obtain distinctiveness and efficiency, our method represents a face image on a low-dimension space by integrating discriminative projection and l_2 regularization-based representation. Therefore, we exploit a projection to utilize the embedded discriminative information in the low-dimensional space under the regularization-based representation framework. The proposed objective function is represented as

$$\min_{P,Q} \left\| P^T y - P^T X Q \right\|_2^2 + \alpha \sum_{i=1}^c \left\| P^T X_i Q_i \right\|_2^2, \tag{6}$$

where α is a balance factor, and $Q = [Q_1, \ldots, Q_c]$ is the representation coefficient. P is the projection matrix.

To solve P and Q, in this work, we divide the optimization of model (6) into two independent stages, including face projection and face representation:

(1) Face projection: To compute a projection matrix that transforms face images into low-dimensional features which have the minimum similar covariance and maximum dissimilar covariance. The projection matrix has a closed-form solution.
(2) Face representation: To produce a discriminative representation for each sample on the new low-dimensional space via l_2 regularization. This discriminative representation is obtained by a special design of the regularization term. Moreover, this stage also generates a closed-form solution.

3.1 Face Projection

Original face image set $X = \{x_i \,|\, x_i \in \mathbb{R}^m, i = 1, \cdots, n\}$ lies in an m-dimension space. In order to find a representation for X in a d-dimension space, we take mapping $f : \mathbb{R}^m \rightarrow \mathbb{R}^d, (d < m)$ as projection function in pursuit of low-dimensional features. In this paper, instead of directly obtaining the projection via (6), we attempt to find the discriminative projection by solving a subspace problem which can obtain a closed-form solution. We use projection matrix P to denote the projection by f, which is obtained by jointly minimizing the similar covariance and maximizing the dissimilar covariance. To calculate these two covariances, we construct a similar set S of the pairs of samples from the same class and a dissimilar set D of the pairs of samples from different classes respectively. These two sets can be expressed as

$$S = \left\{ (x, x') \,|\, D(x, x') < \delta \right\}, \tag{7}$$

and

$$D = \left\{ (x, x') \,|\, D(x, x') > \delta \right\}, \tag{8}$$

where $D(\cdot, \cdot)$ is the distance of sample pair and δ is a margin coefficient. Consequently, projection matrix P can be achieved via minimizing the expectation of the pairs from similar set and maximizing the expectation of the pair from dissimilar set, which maps the samples to a subspace that has the minimum difference of similar samples and maximum difference of dissimilar samples simultaneously. It is expressed as the following loss function:

$$L = \mu \mathrm{E} \left\{ \left\| P^T x - P^T x' \right\|^2 \big| S \right\} - \mathrm{E} \left\{ \left\| P^T x - P^T x' \right\|^2 \big| D \right\}, \tag{9}$$

where μ is a balance parameter. x and x' are the pairs of samples.

To construct the similar and dissimilar sets, we extract the quadruplets from the sample set. A quadruplet [18] is that four samples x_i, x_j, x_k, and x_l from the sample set and act in this way:

$$D(x_i, x_j) + \delta < D(x_k, x_l). \tag{10}$$

Thus, for a sample x, within-class similar sample pair (x_i, x_j) is employed to produce similar set S and between-class dissimilar sample pair (x_k, x_l) is employed to construct the dissimilar set D respectively. As a result, Eq. (9) can be rewritten as

$$L = \mu E \left\{ \left\| P^T x_i - P^T x_j \right\|^2 | S \right\} - E \left\{ \left\| P^T x_k - P^T x_l \right\|^2 | D \right\}. \tag{11}$$

It is observed that

$$E \left\{ \left\| P^T x_i - P^T x_j \right\|^2 | S \right\} = \mathrm{tr} \left\{ P \Sigma_S P^T \right\}, \tag{12}$$

and

$$E \left\{ \left\| P^T x_k - P^T x_l \right\|^2 | D \right\} = \mathrm{tr} \left\{ P \Sigma_D P^T \right\}, \tag{13}$$

where $\Sigma_S = E\{(x_i - x_j)(x_i - x_j)^T | S.\}$ and $\Sigma_D = E\{(x_k - x_l)(x_k - x_l)^T | D.\}$ are the covariance matrices of similar pairs and dissimilar pairs respectively. This leads Eq. (11) to

$$L = \mu \mathrm{tr} \left\{ P \Sigma_S P^T \right\} - \mathrm{tr} \left\{ P \Sigma_D P^T \right\}. \tag{14}$$

Finally, we have

$$L \propto \mathrm{tr} \left\{ P \sum_S \sum_D^{-1} P^T \right\} = \mathrm{tr} \left\{ P \sum_R P^T \right\}, \tag{15}$$

where $\sum_R = \sum_S \sum_D^{-1}$ is a ratio matrix between similar and dissimilar covariance matrices. It is obvious that $\sum_R$ is a semidefinite matrix and can be implemented by singular value decomposition (SVD) for obtaining an orthogonal matrix. Thus, orthogonal matrix P maps the samples into the space spanned by the d smallest eigenvectors, which has minimum similarity of sample.

3.2   Face Representation

On this stage, we can represent face images in the low-dimensional space. As a result, given projection matrix P, training sample X and test sample y can be projected onto a low-dimensional space via $F = P^T X$ and $v = P^T y$. The objective function in Eq. (6) is rewritten as the following formula:

$$\min_Q \| v - FQ \|_2^2 + \alpha \sum_{i=1}^c \| F_i Q_i \|_2^2, \tag{16}$$

Because of the convexity and differentiability of Eq. (16), we can obtain the optimal solution by taking the derivative with respect to Q and setting it to 0. The computational procedure is presented as follows:

Let $f(Q) = \|v - FQ\|_2^2 + \alpha \sum_{i=1}^{c} \|F_i Q_i\|_2^2$. The derivative with respect to Q of the first term of $f(Q)$ is

$$\frac{d}{dQ}\|v - FQ\|^2 = -2F^T(v - FQ). \tag{17}$$

Then we need to determine the derivative of the second term $\frac{d}{dQ}\left(\alpha \sum_{i=1}^{c} \|F_i Q_i\|_2^2\right)$. Because $g(Q) = \alpha \sum_{i=1}^{c} \|F_i Q_i\|_2^2$ does not explicitly contain Q, it needs to compute partial derivative $\frac{dg}{dQ_k}$, and combine all $\frac{dg}{dQ_k}(k = 1, \ldots, c)$ to achieve $\frac{dg}{dQ}$.

$g(Q)$ is composed of c terms which are dependent of Q_k. It firstly calculates the c partial derivatives $\frac{dg}{dQ_k}$ as follows:

$$\begin{aligned}
\frac{\partial g}{\partial Q_k} &= \frac{\partial}{\partial Q_k}\left(\alpha \sum_{i=1}^{c} \|F_i Q_i\|_2^2\right) = \alpha \sum_{i=1}^{c} \frac{\partial}{\partial Q_k}\|F_k Q_k\|_2^2 \\
&= \alpha c \frac{\partial}{\partial Q_k}\|F_k Q_k\|_2^2 = 2\alpha c F_k^T (F_k Q_k).
\end{aligned} \tag{18}$$

Thus the derivative $\frac{dg}{dQ}$ is

$$\frac{dg}{dQ} = \begin{pmatrix} \frac{\partial g}{\partial Q_1} \\ \vdots \\ \frac{\partial g}{\partial Q_c} \end{pmatrix} = \begin{pmatrix} 2\alpha c F_1^T (F_1 Q_1) \\ \vdots \\ 2\alpha c F_c^T (F_c Q_c) \end{pmatrix} = 2\alpha c \begin{pmatrix} F_1^T F_1 & \cdots & O \\ \vdots & \ddots & \vdots \\ O & \cdots & F_c^T F_c \end{pmatrix} Q. \tag{19}$$

Let $M = \begin{pmatrix} F_1^T F_1 & \cdots & O \\ \vdots & \ddots & \vdots \\ O & \cdots & F_c^T F_c \end{pmatrix}$, the derivative of the second term is $\frac{dg}{dQ} = 2\alpha c M Q$.

Combining the derivatives of the first and second terms, the derivative of Eq. (16) is $\frac{df}{dB} = -2F^T(v - FQ) + 2\alpha c M Q$. Thus, setting the derivative to zero, we have

$$2F^T FQ - 2F^T v + 2\alpha c M Q = 0, \tag{20}$$

which leads to

$$Q = (F^T F + \alpha c M)^{-1} F^T v. \tag{21}$$

As a result, optimal Q is a closed-form solution.

3.3   Classification

The proposed method classifies the samples in a new low-dimensional space. Thus, training sample X and test sample y should be projected as F and v

by projection matrix P. Finally, we use Eq. (21) to compute representation coefficient Q of the projected test sample v. Then, a test sample v is classified to the k-th class according to the following procedure,

$$k = \arg\min_i \|v - F_i Q_i\|_2^2. \tag{22}$$

In summary, we describe a full classification algorithm in Algorithm 1.

Algorithm 1. Discriminative projection and representation algorithm for face recognition

---

Input: Training sample X, test sample y, α, dimension d.
Output: Label k, representation coefficient Q.
Procedure:

1. Generate quadruplets (x_i, x_j, x_k, x_l) of training sample X.
2. Construct similar set S using sample pair (x_i, x_j) and dissimilar set D using sample pair (x_k, x_l) respectively.
3. Compute similar covariance $\sum_S = E\left\{(x_i - x_j)(x_i - x_j)^T | S\right\}$ and dissimilar covariance $\sum_D = E\left\{(x_k - x_l)(x_k - x_l)^T | D\right\}$ respectively.
4. Perform SVD on $\sum_R = \sum_S \sum_D^{-1}$ and obtain P by selecting d smallest eigenvectors.
5. Project training sample X and test sample y onto the low-dimensional space for obtaining $F = P^T X$ and $v = P^T y$.
6. Calculate representation coefficient Q via (21).
7. Test sample y is classified to the k-th class via (22).

3.4 Computational Complexity Analysis

The major computational cost of the proposed method lies in the matrix operation. For $n \times n$ input matrix, the computational complexity of our problem consists of two parts, including projection matrix P computation and representation coefficient Q optimization. To obtain the projection matrix, the total cost is about $O(n^3)$. To obtain representation coefficient Q, we need to calculate Eq. (21) for each test sample, whose dimension is d. The total computational complexity of calculating k test samples is about $O(dn^2 + kdn)$. As a result, the total computational complexity of the proposed method is about $O(n^3 + dn^2 + kdn)$. It is pointed out that the projection matrix and representation coefficient matrix are computed completely only once and can be used for all test samples. Although the proposed method needs two computational steps, it is more efficient than the iterative methods.

4   Experimental Results

In this section, we conduct the experiments on face datasets to present the effectiveness of the proposed method. The tested datasets contain the FERET [19], Extended Yale B [20], and CMU Multi-PIE [27] datasets. Meanwhile, several state-of-the-art recognition methods including CRC [13], L1LS [21], FISTA [24], Homotopy [22], Dual augmented lagrangian method (DALM) [23], INNC [25], and a fusion classification method (FCM) [26] are used for experimental comparison. The test of Linear discriminant analysis [18] integrated with the nearest neighbor classifier is also involved in our experiments.

4.1   Dimension and Parameter Selections

The feature dimension is one of the important factors that affect the classification accuracy. Usually, the classification accuracy varies greatly under different dimensions. However, we find that the performance of the proposed method is not greatly affected by the variation of the dimension. Figure 1 shows the results of the classification accuracy under different dimensions on the FERET dataset. It is obvious that the proposed method achieves nearly stable classification accuracies for various dimensions under different numbers of training samples per class. In our experiments, the range of feature dimension is [200, 1000].

There is only one parameter α in the proposed algorithm. It is a factor of balancing the effect on the two terms in the object function. We choose the optimal value of α for each dataset among five candidate values, 0.01, 0.1, 1, 10, and 100. The search procedure can quickly find the optimal value which leads to the best classification accuracy. The proposed method can maintain a stable classification performance when the value of α varies. Figure 2 presents the relationship between α and the classification accuracy on the FERET dataset respectively. It is seen that almost stable classification rates are obtained when the value of α varies in a proper range for a certain dataset.

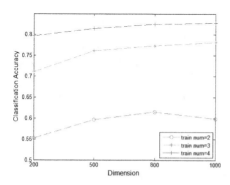

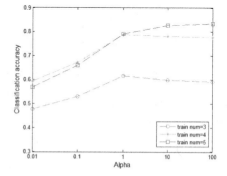

Fig. 1. Classification accuracy versus dimension on the FERET dataset.

Fig. 2. Classification accuracy versus parameter on the FERET dataset.

4.2 Experiments on the FERET Face Dataset

This experiment is conducted on the FERET face dataset [19] which contains 1400 face images from 200 subjects. Figure 3 shows some face images from this dataset. Every face image was resized to a 40 by 40 pixels. The first 3, 4, and 5 face images of each subject and the remaining face images were used as the training samples and test samples respectively. Parameter α was set to 10 and the dimension is 1000. The experimental results are presented in Table 1. From the results, we know that the proposed method achieves better recognition rate than other classification methods, which implies that the proposed method can capture more discriminative information for feature representation.

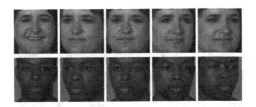

Fig. 3. Some face images from the FERET face dataset. The face images shown in the first and second rows are from three different subjects.

Table 1. Classification accuracies of different methods on the FERET dataset.

Number of training samples per subject	3	4	5
The proposed method1000	**59.88**	**78.17**	**82.75**
CRC	44.37	55.33	68.50
L1LS	59.13	76.00	82.00
FISTA	38.75	49.00	58.00
Homotopy	54.00	72.50	77.25
DALM	59.75	75.17	80.00
INNC	50.50	54.00	68.75
FCM	47.37	56.50	68.25
LDA	47.80	65.60	81.60
Pro_CRC	44.80	55.70	63.50

4.3 Experiments on the Extended Yale B Face Dataset

The Extended Yale B [20] face dataset contains 2414 single frontal facial images of 38 individuals. These images were captured under various controlled lighting conditions. The size of an image was 192168 pixels. In our experiments, all images were cropped and resized to 8496 pixels. Figure 4 shows some face images from the Extended Yale B face dataset. The first 10, 20, 30, and 40 face images of each subject were treated as original training samples and the remaining face images were viewed as testing samples. Parameter α was set to 0.001 and the dimension is 1000. The experimental results are presented in Table 2. It is obvious that the proposed method increases nearly 10 percents recognition rates under different conditions, compared with other methods. More importantly, the superior recognition rates are obtained under lower dimension (1000), compared with other methods under original dimension (8064). It means that the proposed method can capture more discriminative low-dimension feature under different conditions.

Fig. 4. Some face images from the Extended Yale B face dataset. The face images shown in the first, second, and third rows are from three different subjects.

Table 2. Classification accuracies of different methods on the Extended Yale B dataset.

Number of training samples per subject	10	20	30	40
The proposed method1000	**76.91**	**85.11**	**94.66**	**95.72**
CRC	67.54	75.30	83.28	84.10
L1LS	73.25	74.52	84.13	84.98
FISTA	42.35	40.79	43.42	41.34
Homotopy	43.86	46.41	49.38	40.79
DALM	59.45	66.21	74.77	73.14
INNC	72.47	77.21	82.74	82.35
FCM	65.20	72.19	79.88	81.58
LDA	65.51	71.98	86.74	91.67
Pro_CRC	70.70	77.80	83.80	83.80

4.4   Experiments on the CMU Multi-PIE Face Dataset

In this subsection, we evaluated the performance of our method on the CMU Multi-PIE face dataset [27]. The CMU Multi-PIE face dataset is composed of face image of 337 persons with variations of poses, expressions and illuminations. Figure 5 shows some face images from this dataset. We select a subset composed of 249 persons under 20 different illumination conditions with a frontal pose and 7 different illumination conditions with a smile expression. All images are cropped and resized to 4030 pixels. We choose face images corresponding to the first 3, 5, 7, 9 illuminations from the 20 illuminations and only one image from 7 smiling images as training samples and use remaining images as testing samples. Parameter α is set to 0.00001 and the dimension is 1000. Table 3 lists the classification accuracy on four testing sets obtained using different methods. From the results, it can be seen that our method obtains better classification accuracy than other methods. In other words, our method is more robust to variations of illuminations, poses and expressions.

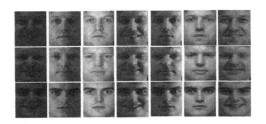

Fig. 5. Some face images from the CMU Multi-PIE face dataset. The face images shown in the first, second, and third rows are from three different subjects.

Table 3. Classification accuracies of different methods on the CMU Multi-PIE dataset.

Number of training samples per subject	4	6	8	10
The proposed method1000	**93.80**	**99.10**	**99.876**	**99.93**
CRC	90.13	96.63	98.86	99.46
L1LS	92.79	99.08	99.66	99.81
FISTA	46.74	50.01	57.26	71.16
Homotopy	55.49	56.55	63.03	71.25
DALM	68.05	67.62	72.27	84.67
INNC	82.05	95.62	98.82	99.50
FCM	90.34	95.47	97.36	98.32
LDA	89.80	95.09	98.01	99.31
Pro_CRC	90.90	96.30	98.60	99.90

5 Conclusion

Aiming at seeking an efficient discriminative representation for face images, we proposed a novel discriminative projection and representation method for face recognition. This method obtains the superiority of effective and efficient recognition by using a specific regularization term and projection matrix of the objective function. The projection produced by minimizing similar covariance and maximizing dissimilar covariance can obtain the features which have the minimum similarity of samples. The discriminative representation result is obtained by minimizing the correlation of samples. Therefore, the proposed method possesses two-fold discriminative properties, which is very helpful to improve the classification accuracy. In addition, the proposed method provides a computational efficient algorithm for face recognition tasks.

Acknowledgement. This work was supported in part by the National Natural Science Foundation of China under Grant 61332011, and partially supported by Guangdong Province high-level personnel of special support program (No. 2016TX03X164).

References

1. Turk, M.A., Pentland, A.P.: Face recognition using eigenfaces. In: Proceedings IEEE Computer Society Conference on Computer Vision and Pattern Recognition, CVPR 1991, pp. 586–591 (2002)
2. Cover, T., Hart, P.: Nearest neighbor pattern classification. IEEE Trans. Inf. Theory **13**(1), 21–27 (1967)
3. Prasad, J.R., Kulkarni, U.: Gujrati character recognition using weighted k-NN and mean 2 distance measure. Int. J. Mach. Learn. Cybernet. **6**(1), 69–82 (2015)
4. Vapnik, V.N.: The Nature of Statistical Learning Theory. Springer, New York (1995)
5. Tian, Y., Qi, Z., Ju, X., Shi, Y., Liu, X.: Nonparallel support vector machines for pattern classification. IEEE Trans. Cybern. **44**(7), 1067 (2014)
6. Zhang, Z., Xu, Y., Yang, J., Li, X., Zhang, D.: A survey of sparse representation: algorithms and applications. IEEE Access **3**, 490–530 (2017)
7. Wagner, A., Wright, J., Ganesh, A., Zhou, Z., Mobahi, H., Ma, Y.: Toward a practical face recognition system: robust alignment and illumination by sparse representation. IEEE Trans. Pattern Anal. Mach. Intell. **34**(2), 372–386 (2011)
8. Wright, J., Yang, A.Y., Ganesh, A., Sastry, S.S., Ma, Y.: Robust face recognition via sparse representation. IEEE Trans. Pattern Anal. Mach. Intell. **31**(2), 210–227 (2008)
9. Lai, Z., Xu, Y., Jin, Z., Zhang, D.: Human gait recognition via sparse discriminant projection learning. IEEE Trans. Circ. Syst. Video Technol. **24**(10), 1651–1662 (2014)
10. Liu, H., Tang, H., Xiao, W., Guo, Z.Y., Tian, L., Gao, Y.: Sequential bag-of-words model for human action classification. CAAI Trans. Intell. Technol. **1**(2), 125–136 (2016)
11. Wright, J., Ma, Y., Mairal, J., Sapiro, G., Huang, T.S., Yan, S.: Sparse representation for computer vision and pattern recognition. Proc. IEEE **98**(6), 1031–1044 (2010)

12. Zhang, L., Yang, M., Feng, X.: Sparse representation or collaborative representation: which helps face recognition? In: IEEE International Conference on Computer Vision, pp. 471–478 (2012)
13. Baraniuk, R.G.: Compressive sensing. IEEE Sig. Process. Magazine **24**(4), 118–121 (2007)
14. Naseem, I., Togneri, R., Bennamoun, M.: Robust regression for face recognition. In: International Conference on Pattern Recognition, pp. 1156–1159 (2010)
15. Xu, Y., Zhang, D., Yang, J., Yang, J.Y.: A two-phase test sample sparse representation method for use with face recognition. IEEE Trans. Circ. Syst. Video Technol. **21**(9), 1255–1262 (2011)
16. Xu, Y., Zhong, Z., Yang, J., You, J., Zhang, D.: A new discriminative sparse representation method for robust face recognition via l_2 regularization. IEEE Trans. Neural Netw. Learn. Syst. **28**(10), 2233–2242 (2017)
17. Belhumeur, P.N., Hespanha, J.P., Kriegman, D.J.: Eigenfaces vs. fisherfaces: recognition using class specific linear projection. In: Buxton, B., Cipolla, R. (eds.) ECCV 1996. LNCS, vol. 1064, pp. 43–58. Springer, Heidelberg (1996). https://doi.org/10.1007/BFb0015522
18. Law, M.T., Thome, N., Cord, M.: Learning a distance metric from relative comparisons between quadruplets of images. Int. J. Comput. Vision **121**, 1–30 (2016)
19. Phillips, P.J., Moon, H., Rizvi, S.A., Rauss, P.J.: The FERET evaluation methodology for face-recognition algorithms. IEEE Trans. Pattern Anal. Mach. Intell. **22**(10), 1090–1104 (2000)
20. Georghiades, A.S., Belhumeur, P.N., Kriegman, D.J.: From few to many: illumination cone models for face recognition under variable lighting and pose. IEEE Trans. Pattern Anal. Mach. Intell. **23**(6), 643–660 (2001)
21. Kim, S.J., Koh, K., Lustig, M., Boyd, S., Gorinevsky, D.: An interior-point method for large-scale l 1-regularized least squares. IEEE J. Sel. Top. Sig. Process. **1**(4), 606–617 (2007)
22. Yang, A.Y., Sastry, S.S., Ganesh, A., Ma, Y.: Fast l_1-minimization algorithms and an application in robust face recognition: a review. In: IEEE International Conference on Image Processing, pp. 1849–1852 (2010)
23. Yang, A.Y., Zhou, Z., Balasubramanian, A.G., Sastry, S.S., Ma, Y.: Fast l_1 minimization algorithms for robust face recognition. IEEE Trans. Image Process. **22**(8), 3234 (2013)
24. Beck, A., Teboulle, M.: A fast iterative shrinkage-thresholding algorithm for linear inverse problems. SIAM J. Imaging Sci. **2**(1), 183–202 (2009)
25. Xu, Y., Zhu, Q., Chen, Y., Pan, J.S.: An improvement to the nearest neighbor classifier and face recognition experiments. Int. J. Innov. Comput. Inf. Control **9**(2), 543–554 (2013)
26. Liu, Z., Pu, J., Huang, T., Qiu, Y.: A novel classification method for palmprint recognition based on reconstruction error and normalized distance. Appl. Intell. **39**(2), 307–314 (2013)
27. Gross, R., Matthews, I., Cohn, J., Kanade, T., Baker, S.: Multi-PIE. Image Vis. Comput. **28**(5), 807–813 (2010)

Improved Face Verification with Simple Weighted Feature Combination

Xinyu Zhang$^{(\boxtimes)}$, Jiang Zhu, and Mingyu You

College of Electronics and Information Engineering, Tongji University,
4800 Cao'an Highway, Shanghai 201804, People's Republic of China
{1510464,zhujiang,myyou}@tongji.edu.cn

Abstract. Since the appearance of deep learning, face verification (FV) has made great progress with large scale datasets, well-designed networks, new loss functions, fusion of models and metric learning methods. However, incorporating all these methods obviously takes a lot of time both at training and testing stages. In this paper, we just select training images randomly without any clean and alignment procedure. Then we propose a simple weighted average method which combines features of the last two layers with different weights on the modified VGGNet, named as *CB-VGG*. It is significantly reducing the complexity of time that one model can be treated as two models. LMNN is used as a post-processing procedure to improve the discrimination of the combined features. Our experiments show relatively competitive results on LFW, CFP, and CACD datasets.

Keywords: Face verification · Deep learning
Weighted average method · LMNN metric learning

1 Introduction

Face verification (FV), whose protocol is to classify whether a pair of faces belongs to the same person or not, is one of the most challenging face tasks with difficulty of variations on pose, age, occlusion and so on [1].

In recent years, the state-of-the-art methods based on deep learning achieved the great success on FV [2–5]. One of the most important ingredients for this success is large scale facial public datasets [6] with identity labels. However, there are often noise contaminated since images are crawled from the Internet. Some methods use various measures [4,7] to clean images, while others produce multiple patches [8–12] and synthesized images [13]. It requires a lot of extra time and human resources. Since excessive noise data usually depress the classification performance, in this paper, we only select partial training data of MsCeleb [6] randomly without any cleaning and alignment methods in order to simplify pre-processing process.

© Springer Nature Singapore Pte Ltd. 2017
J. Yang et al. (Eds.): CCCV 2017, Part II, CCIS 772, pp. 16–28, 2017.
https://doi.org/10.1007/978-981-10-7302-1_2

Some methods obtain better performance [8–11] along with concatenating features from multiple models. It needs additional time to acquire multiple models and evaluate models with high-dimensional features. Here, we propose to use both of the last two layers of a single model. Unlike concatenation, we use weighted average of the two layers without dimension increase. It can be seen that we only train one model but use it as two models by fusion of two layers, which effectively reduces the time at train and test stages.

Besides, many other strategies are effective for better performance, such as carefully-designed models [4,7] and center loss [14]. While joint Bayesian [15] has been highly successful as metric learning in many methods [9–11], Large Margin Nearest Neighbor (LMNN) [16] is used to improve accuracy. LMNN [16] is first introduced to train a matrix that maintains a large distance between imposters and constrains the k-nearest neighbors belonging to the same class. In this research, we use LMNN to further improve the discrimination of features.

Despite its simple process, our method achieves competitive results on Labeled Faces in the Wild(LFW) [17], Celebrities in Frontal-Profile in the Wild(CFP) [18] and Cross-Age Celebrity Dataset(CACD) [19]. We summarize the contributions of this paper as follows:

1. We simplify the pre-processing procedure only by selecting partial training data randomly. No any other clean measures are used to carefully select the label-corrected images.
2. We propose a new method that only requires to train one model but uses the last two layers of features via computing their weighted average, which improves performance without much time consumption.
3. We use LMNN as metric learning to strengthen the discriminative power of deeply features.
4. The proposed method performs excellent and significant results on all the three datasets of LFW, CFP and CACD.

2   Related Work

Various of methods exploited deep networks to achieve remarkable results in FV. We analyse the most critical aspects.

Data Preparation. Large scale public facial images are introduced to encourage the development of face recognition [2–5], such as CASIA-Webface [20] and MsCeleb [6]. In addition, data augmentation like multiple patches [8–12] and synthesized images [13] can make the system be more robust to various of face variation. Others like 2D/3D alignment [2,14], RGB/Gray channels [8,9] and rotated poses [13] can also contribute to performance.

Elaborately Designed Networks. Inspired by "very deep" networks, VGGNet is widely used to FV [2,13]. After that, many blocks are added to networks [4,7,11], like inception [21] and residual [22]. Besides, local connected layer(LC) [8,9,14], L2 norm [4], DeepVisage(feature normalization) [7] are designed.

Fusion of Models. Many methods show that model fusion provides additional boosts to feature presentation. DeepID series [8–11] train at least 25 models on different facial patches which are complementary with each other.

Metric Learning. Several methods use metric learning after extracting features from CNN. Principal component analysis(PCA) is mainly used [8–11] which learns a mapping matrix to reduce dimensions. Joint Bayesian is proved to be an effective way [8–11,20]. Baidu [12] uses triplets which shortens the distance of intra-class and enlarges that of inter-class.

Various Loss Functions. Softmax loss [2,12] is largely used for classification and has been used for feature embedding by removing the last classification layer. Contrastive loss [3,9–11] and triplet loss [2,5,12] are introduced to enhance intra-class binding. Then, center loss [14] is proposed to learn a center for each class.

3   The Proposed Method

We use VGGNet [2] as our baseline and modify it with a convolution-branch block(CB), called *CB-VGG*. What's more, it brilliantly exploits weighted average of the last two layers, followed by LMNN [16] before computing the cosine similarity as the verification score of a pair images. Next, we describe all the details.

3.1   Data Pre-processing

Our pre-processing process is extremely simple which saves a lot of time.

Training Data: We use MsCeleb [6] as our training data. Different from other approaches [4,7,14] cleaning the training data, we only select 29,731 identifies randomly except target datasets. For simplification, we only use Normalized Pixel Difference(NPD) [23] detector without any 2D [2,7] or 3D alignment [3]. It is inherited by [2] which found that alignment on test images instead of training data brought the best benefits. The work in [2] uses large amounts of images and the alignment step on test stage spends plenty of time. In contrast, we crop 4 corner with 1 center patches and randomly flip these images for data augmentation to promote the performance of model.

Test Data: The process is similar as training stage. Differently, LFW [17] and CACD [19] are all cropped 10 times consisting of 4 corners and 1 center with their flips. For CFP [18], we only use 1 crop without flip due to Frontal-Profile pairs.

3.2   CB-VGG Architecture

Baseline: Our baseline, VGGNet [2] comprises 11 sections with more than one linear and non-linear blocks like ReLU and max pooling. The first 8 sections use convolution as linear operator while the following 2 use fully connected layers(FC). The last FC layer is only used for classification.

CB-VGG Model: We make a small modification on the baseline model. Similar to [14], the 4th and 5th pooling units are concatenated together as the input of the 1st FC layer. In order to match the size of the two layers, we apply another convolutional operation after the 4th pooling layer, which is named as convolution-branch block(CB). It is critical for accurate face representation, because some information would be lost after successive down sampling. In this case, we can make full use of both global-abstract semantic information and high-resolution feature. Moreover, we simply use softmax loss instead of combining with other loss [14] or normalization [4,7]. The details of the model are given in Fig. 1.

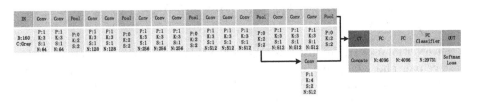

Fig. 1. Structure of CB-VGG model. **IN** is the input images. **Conv** is convolution layer. **P** represents padding. **K** is kernel size. **N** is the output number. **Pool** indicates max pooling unit. **CT** is concatenation layer which concatenates the 4th pooling(after one extra convolution) and the 5th pooling layers. **FC** and **FC Classifier** indicate fully connected layer and classification layer respectively. **OUT** is the softmax loss function.

3.3 Weighted Average Features

The deep features are taken from the last FC layer almost in all other methods. What's different in our method is that we extract features on both the last two FC layers. For an image x, we represent its feature descriptors using the 1st and the 2nd FC layer as f_{1st}^x and f_{2nd}^x. Then, the final feature descriptor f_{final}^x is expressed by taking weighted average of these two features:

$$\mathbf{f}_{final}^x = \alpha \times \frac{1}{C} \sum_{i=1}^{C} \mathbf{f}_{1st}^{x_i} + \beta \times \frac{1}{C} \sum_{i=1}^{C} \mathbf{f}_{2nd}^{x_i}. \tag{1}$$

α and β are the weights of the 1st and the 2nd FC layer's descriptors respectively which act on every elements. In order to discriminate the role of the two layers, we impose restrictions on the weights with $\beta = 1 - \alpha$. C is the number of crops which is 1 in CFP and 10 in LFW and CACD.

Model fusion is always effective. Sun et al. [9–11] concatenate 25 model's features to improve performance. Baidu [12] uses 10 embedding models to obtain high performance. However, our method only trains one model but uses two layers to extract features, which is equal to training two models but saves a lot of time and boosts the performance at the same time.

3.4 LMNN Metric Learning

Large Margin Nearest Neighbor(LMNN) [16] learns a Mahanalobis distance matrix which is good for decreasing the distance of k target neighbors, while enlarging the distance of different k nearest points. In FV task, some distances of positive pairs are larger than negative pairs, causing a bad effect on thresholds selection. Based on LMNN, we can pull the same faces closer and repel different faces further. The problem can be described as follows. i and j are same pairs, yet l is different from them. ξ_{ijl} is a slack variable. $\mathbf{M}$ is the transformable matrix.

$$min \sum_{i,j \to C_i} (\mathbf{x}_i - \mathbf{x}_j)^T \mathbf{M}(\mathbf{x}_i - \mathbf{x}_j) + \sum_{i,j \to C_i, l \to C_l} \xi_{ijl}. \tag{2}$$

$$s.t. \forall_{y_i = y_j \neq y_l} (\mathbf{x}_i - \mathbf{x}_l)^T \mathbf{M}(\mathbf{x}_i - \mathbf{x}_l) - (\mathbf{x}_i - \mathbf{x}_j)^T \mathbf{M}(\mathbf{x}_i - \mathbf{x}_j) \geq 1 - \xi_{ijl},$$
$$\xi_{ijl} \geq 0, \mathbf{M} \succeq 0. \tag{3}$$

FaceNet [5] uses triplet loss similar to LMNN based on triplets consisting of a positive pair and a negative identity. However, it is not easy to generate triplets and [5] uses 200 M images for training. Instead, we use LMNN for employing idea of triplet but avoiding triplets selection.

4 Experiments

In this section, we first introduce the details of the training stage. Then we extract features of the last two FC layers(FC6, FC7) and perform LMNN to calculate the cosine similarity in FV. In order to verify the effectiveness of our method, we evaluate it on three different datasets. All experiments we carry out are on Caffe [24].

4.1 Details of Training Stage

In this subsection, we describe the details of model training. We randomly select 2.29 M images of 29,731 identities from MsCeleb dataset [6] without including the test identities. For simplification, **no other cleaning and alignment strategies** are used to pick up images. We train our model using 95% images (2.17 M images) and other 5% images (120 K) are used for monitoring and validating the loss. Our CNN model **only uses identity information** to optimize softmax loss. **No other normalization methods** [4, 7] are used.

The baseline model and our CB-VGG model are trained from scratch with the same hyper parameters. We use stochastic gradient descent(SGD) method. Momentum and weight decay are set to 0.9 and $5e^{-4}$. We begin the training with a learning rate of 0.01 and decrease it by 10 times every $100\,K$ iterations. The training batch size is 120. Images are resized into 170×170, and then cropped into 160×160 of 4 corners and 1 center. We also apply randomly horizontal flipping to the images for data augmentation.

4.2   Results of Test Datasets

Evaluation is performed on the commonly used LFW dataset [17], frontal-profile CFP dataset [18] and cross-age CACD dataset [19]. Like training stage, we also only use NPD to detect images without any other extra alignment process. The parameters of α and β for FC6/FC7 are 0.15/0.85 respectively. The number of target neighbors k is set to 3 in LMNN.

LFW Dataset: LFW dataset [17] is one of the most popular datasets and the performance on it is close to saturation. It consists of 13,233 face images of 5,749 different identities. We evaluate our model following the most permissive protocol: unrestricted with labeled outside data. The FV task requires evaluating on 6,000 pairs in 10 folds. Each fold contains half of the genuine pairs and half of the impostor pairs.

We compare the network trained on Baseline and CB-VGG. For both FC6 and FC7, we extract features of 10 patches separately. Table 1 shows that CB-VGG is lightly better than Baseline. More over, with the weighted average method and LMNN metric learning, accuracy is further improved. The result verifies the effectiveness of weighted average of the last two layers which can be thought as fusion of two models but only trained once.

Table 1. Accuracy on LFW (%). Y is Yes (Use), N is Not (Not use)

No	Method	FC6	FC7	LMNN	Acc%
1	Baseline	Y	N	N	98.78
2	Baseline	N	Y	N	98.85
3	Baseline	Y	Y	N	98.95
4	CB-VGG	Y	N	N	98.78
5	CB-VGG	N	Y	N	98.90
6	CB-VGG	Y	Y	N	**99.02**
7	CB-VGG LMNN	Y	Y	Y	**99.07**

We also compare our method with other state-of-the-art methods. The results are given in Table 2. We observe that our method achieves relatively significant accuracy (99.07%) with relative less images and simple methods.

CFP Dataset: CFP [18] is one of the large-pose face datasets which is composed of 10 frontal and 4 profile images of each 500 individuals. Unlike LFW, CFP has two FV experiments: frontal-frontal (FF) and frontal-profile (FP). Both contains 10 folds, each with 350 same pairs and 350 different pairs.

We also compare the Baseline and CB-VGG model. However, features are extracted from both FC6 and FC7 with only 1 center crop. PCA is applied for reducing dimensions of fusional features from $4096d$ to $300d$. Table 3 shows that CB-VGG with fusional features and LMNN is more robust to pose transformation.

Following the protocal in CFP [18], we also report the EER(Equal Error Rate) and AUC(Area under the curve) values on averages of 10 splits. Table 4 and Fig. 2 provide the results and the ROC curve[1] of ours along with the state-of-the-art methods.

Table 2. Performance comparison of state-of-the-art on LFW (%). Y is Yes (Use), N is Not (Not use)

Method	Train-Images	Clean	Align	Norm	Models	Other-Loss	Metric-Learning	Acc%
DeepFace [3]	4.4 M, 4 K	N	Y	–	3	Y	Kafang	97.35
VGG Face [2]	2.6 M, 2.6 K	N	Y	Y	1	Y	Triplet	98.95
Baidu [12]	1.2 M, 1.8 K	Y	Y	–	1	Y	Triplet	99.13
Center Loss [14]	0.7 M, 17.2 K	–	Y	–	1	Y	–	99.28
DeepID2+ [9]	0.29 M, 12 K	–	Y	–	25	Y	Joint Bayesian	99.47
DeepID3 [11]	0.29 M, 12 K	–	Y	–	25	Y	Joint Bayesian	99.53
L2-softmax [4]	3.7 M, 58.2 K	Y	Y	Y	1	N	–	99.60
DeepVisage [7]	4.48 M, 62 K	Y	Y	Y	1	N	N	99.62
FaceNet [5]	200 M, 8000 K	–	Y	Y	1	Y	–	99.63
CB-VGG LMNN	2.29 M, 29.7 K	N	N	N	1	N	LMNN	**99.07**

Table 3. Accuracy on CFP (%). Y is Yes (Use), N is Not (Not use)

No.	Method	FC6	FC7	LMNN	Acc%	
					FF	FP
1	Baseline	Y	N	N	98.21	90.36
2	Baseline	N	Y	N	98.34	91.73
3	Baseline	Y	Y	N	98.43	91.99
4	CB-VGG	Y	N	N	98.11	90.41
5	CB-VGG	N	Y	N	98.39	91.94
6	CB-VGG	Y	Y	N	**98.46**	**92.06**
7	CB-VGG LMNN	Y	Y	Y	**98.59**	**92.43**

We observe that our method achieves the competitive results to the best accuracy on both FF and FP. For FF ours result is slightly lower than FV-DCNN+pool5 [25] but enjoys being simple. FV-DCNN+pool5 [25] learns the Gaussian mixture model and performs Fisher vector encoding after extracting features from models. Besides, they also use PCA, Joint Bayesian and scores fusion with pool5. For FP, we achieve the first place if not competing p-CNN [1]. p-CNN(pose-directed multi-task CNN) specially tackles pose variation by separating all poses into several groups and then jointly learn identity and PIE(pose,

---

[1] Because of lacking of data, we couldn't report the work of Sankarana *et al.* [26].

Table 4. Performance comparison of state-of-the-art on CFP (%). **ACC** is Accuracy. **EER** is Equal Error Rate. **AUC** is Area Under the Curve

Method	FF (Frontal-Frontal)			FP (Frontal-Profile)		
	ACC	EER	AUC	ACC	EER	AUC
Human [18]	96.24 ± 0.67	5.34 ± 1.79	98.19 ± 1.13	94.57 ± 1.10	5.02 ± 1.07	98.92 ± 0.46
Deep features [18]	96.40 ± 0.69	3.48 ± 0.67	99.43 ± 0.31	84.91 ± 1.82	14.97 ± 1.98	93.00 ± 1.55
Sankarana *et al.* [26]	96.93 ± 0.61	2.51 ± 0.81	99.68 ± 0.16	89.17 ± 2.35	8.85 ± 0.99	97.00 ± 0.53
FV-DCNN [25]	98.41 ± 0.45	1.54 ± 0.43	99.89 ± 0.06	91.97 ± 1.70	8.00 ± 1.68	97.70 ± 0.82
FV-DCNN+pool5 [25]	**98.67 ± 0.36**	**1.40 ± 0.37**	**99.90 ± 0.09**	89.83 ± 1.88	10.40 ± 1.85	96.37 ± 0.97
p-CNN [1]	97.79 ± 0.40	2.48 ± 0.07	99.71 ± 0.02	**94.39 ± 1.17**	**5.94 ± 0.11**	**98.36 ± 0.05**
CB-VGG LMNN	**98.59 ± 0.53**	1.47 ± 0.55	**99.90 ± 0.07**	92.43 ± 0.75	8.18 ± 0.82	97.03 ± 0.75

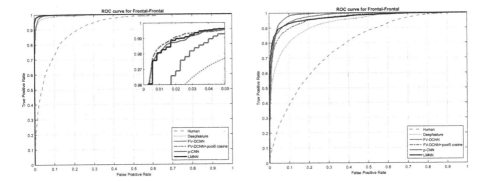

Fig. 2. ROC curve of CFP dataset. (a) is the protocol of Frontal-Frontal, while (b) is the protocol of Frontal-Profile. We only compare deep learning methods without conventional methods.

illumination and expression) for each group. Although p-CNN has great advantage on pose, it performs relatively worse than us on FF.

CACD Dataset: CACD [19] is a large cross-age celebrity dataset which consists of 163,446 images of 2,000 celebrities with age ranging from 16 to 62. For verification task, 2,000 positive image pairs and 2,000 negative pairs are selected to form CACD-VS and divided by 10 folds.

Similar to the previous process, we report comparison of baseline with ours and provide the results of other methods on CACD-VS. The results are shown in Tables 5 and 6.

MFM-CNN [28] and DeepVisage [7] are well designed that the former uses maxout activation to separate noisy signals with informative signals and the latter uses feature normalization to restrict features keeping equal contribution to cost function. LF-CNNs [27] aim at learning age-invariant features via combining latent factor layer. Note that we don't use any complex blocks or finetune measures to fit age variation.

In summary, we only use VGGNet with modification of combining FC6 with FC7 and LMNN metric learning. All the training images are randomly selected

Table 5. Accuracy on CACD (%). Y is Yes (Use), N is Not (Not use)

No	Method	FC6	FC7	LMNN	Acc%
1	Baseline	Y	N	N	97.68
2	Baseline	N	Y	N	97.38
3	Baseline	Y	Y	N	97.70
4	CB-VGG	Y	N	N	97.38
5	CB-VGG	N	Y	N	97.58
6	CB-VGG	Y	Y	N	**97.78**
7	CB-VGG LMNN	Y	Y	Y	**97.88**

Table 6. Performance comparison of state-of-the-art on CACD (%). Y is Yes (Use).

Method	Well-designed block	Aiming at age	Acc%
Human, Avg [19]	–	–	85.70
Human, Voting [19]	–	–	94.20
VGG Face [2]	–	–	96.00
LF-CNNs [27]	–	Y	98.50
MFM-CNN [28]	maxout activation	–	98.55
DeepVisage [7]	feature normalization	–	99.13
CB-VGG LMNN	–	–	**97.88**

without any clean measures. Experiments show that our method is definitely robust to pose and age variation on FV task. It means our method is robust to both frontal and pose-variable faces and weighted features of the last two layers really learns mutually complementary information. LMNN is a fast metric learning which is really able to improve discrimination.

4.3 Analysis and Discussion

For highlighting the influences of methods, we perform further analysis on (i) number and data augmentation of training datasets; (ii) different settings of weights; (iii) some examples which are corrected by our methods. All the investigations are conducted on LFW dataset.

First, we study the influence from the quantity of training data. Table 7 presents the results[2] which we observe that: (i) the larger number of images creates the better CNN performance, although the $100K$ identities decrease the results due to the large dirty images; (ii) Multi-crop augmentation is really helpful for performance, since it is equal to add more pure images for per identities

---

[2] Due to the restrictions of memory and time, we don't conduct an experiment on $100\,K$ dataset with multiple crop. The number of images is less than original images which is due to failing detection.

to increase number of training images. In this paper, we don't use any selection or alignment for training dataset. Instead, multiple crop for data augmentation helps us to extract features with location and pose invariance.

Table 7. Analysis of the influences from number and data augmentation of training dataset. $\sim$ is the approximate number which is obtained after detection.

Number	Single-crop (%)	Multi-crop (%)
1.62 M, $\sim$20 K	97.00	98.75
2.29 M, $\sim$30 K	98.83	99.02
7.66 M, $\sim$100 K	98.63	–

Next, we analyse the various weights combination of the last two layers. Figure 3 shows that the weights selection of the features is important. Since FC7 layer extracts more information than FC6, the weights of the former is must be larger than the latter. Otherwise, the performance will be pull down by the weak feature.

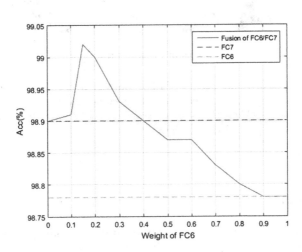

Fig. 3. Analysis of the influences from the weights of the last two layers. (Color figure online)

Finally, we provide some examples which are corrected by our methods. In Fig. 4, four misclassified pairs are presented that (a) two pairs are false rejected images and (b) another two are false accepted images. These pairs are erroneously classified by CB-CNN model with only FC7 layer. When we use weighted average features of last two layers, pairs marked with green colored rectangles are corrected. It can be seen that more discriminative features are extracted using simple weighted combination which are more robust to pose, illumination, occlusion and so on. Then, when we further use LMNN, more difficult pairs can be corrected, such as the pair with red rectangle. That's to say that our method really have ability to improve the representation of features.

(a) Examples of false rejected images.

(b) Examples of false accepted images.

Fig. 4. Some misclassified pair images of LFW dataset by CB-CNN model with only FC7 layer. Pair images with green rectangle is corrected via weighted average features of FC6 and FC7 layers. Then the pair with red rectangle is further corrected using LMNN. (a) and (b) is false rejected images and false accepted images respectively.

5   Conclusions

In this paper, we not only use extremely simple pre-processing procedure without clean or alignment, but also propose a single model named as CB-VGG. Taking care of time and accuracy, we present a simple weighted average on the last two FC layers instead of fusing several models. This can be seen as fusion of two models but only training once. After that, we use LMNN metric learning as post-processing process. Combining all these methods, we achieve competitive results and perform relatively robust to pose and age variations. These results successfully show that it may not be necessary to use complex process of selecting images or training multiple models to boost performance. Single models can be fully used to develop a good result and the time can be saved enormously. In the future, we will explore whether the cleaned images are really important for the performance.

Acknowledgements. This work was supported by Natural Science Foundation of Shanghai (No. 17ZR1431500).

References

1. Yin, X., Liu, X.: Multi-Task Convolutional Neural Network for Face Recognition. arXiv preprint arXiv:1702.04710 (2017)
2. Parkhi, O.M., Vedaldi, A., Zisserman, A.: Deep face recognition. In: Proceedings of the British Machine Vision Conference, pp 41.1-41.12 (2015)
3. Taigman, Y., Yang, M., Ranzato, M.A., et al.: Deepface: closing the gap to human-level performance in face verification. In: Proceedings of the IEEE Conference on Computer Vision and Pattern Recognition, pp 1701–1708 (2014)
4. Ranjan, R., Castillo, C.D., Chellappa, R.: L2-constrained Softmax Loss for Discriminative Face Verification. arXiv preprint arXiv:1703.09507 (2017)
5. Schroff, F., Kalenichenko, D., Philbin, J.: Facenet: a unified embedding for face recognition and clustering. In: Proceedings of the IEEE Conference on Computer Vision and Pattern Recognition, pp 815–823 (2015)
6. Guo, Y., Zhang, L., Hu, Y., He, X., Gao, J.: MS-Celeb-1M: a dataset and benchmark for large-scale face recognition. In: Leibe, B., Matas, J., Sebe, N., Welling, M. (eds.) ECCV 2016, Part III. LNCS, vol. 9907, pp. 87–102. Springer, Cham (2016). https://doi.org/10.1007/978-3-319-46487-9_6
7. Hasnat, A., Bohné, J., Gentric, S., et al.: DeepVisage: making face recognition simple yet with powerful generalization skills. arXiv preprint arXiv:1703.08388 (2017)
8. Sun, Y., Wang, X., Tang, X.: Deep learning face representation from predicting 10,000 classes. In: Proceedings of the IEEE Conference on Computer Vision and Pattern Recognition, pp 1891–1898 (2014)
9. Sun, Y., Chen, Y., Wang, X., et al.: Deep learning face representation by joint identification-verification. In: Advances in neural information processing systems, pp 1988–1996 (2014)
10. Sun, Y., Wang, X., Tang, X.: Deeply learned face representations are sparse, selective, and robust. In: Proceedings of the IEEE Conference on Computer Vision and Pattern Recognition, pp 2892–2900 (2015)
11. Sun, Y., Liang, D., Wang, X., et al.: Deepid3: Face recognition with very deep neural networks. arXiv preprint arXiv:1502.00873 (2015)
12. Liu, J., Deng, Y., Bai, T., et al.: Targeting ultimate accuracy: Face recognition via deep embedding. arXiv preprint arXiv:1506.07310 (2015)
13. Masi, I., Trần, A.T., Hassner, T., Leksut, J.T., Medioni, G.: Do we really need to collect millions of faces for effective face recognition? In: Leibe, B., Matas, J., Sebe, N., Welling, M. (eds.) ECCV 2016, Part V. LNCS, vol. 9909, pp. 579–596. Springer, Cham (2016). https://doi.org/10.1007/978-3-319-46454-1_35
14. Wen, Y., Zhang, K., Li, Z., Qiao, Y.: A discriminative feature learning approach for deep face recognition. In: Leibe, B., Matas, J., Sebe, N., Welling, M. (eds.) ECCV 2016, Part VII. LNCS, vol. 9911, pp. 499–515. Springer, Cham (2016). https://doi.org/10.1007/978-3-319-46478-7_31
15. Chen, D., Cao, X., Wang, L., Wen, F., Sun, J.: Bayesian face revisited: a joint formulation. In: Fitzgibbon, A., Lazebnik, S., Perona, P., Sato, Y., Schmid, C. (eds.) ECCV 2012, Part III. LNCS, vol. 7574, pp. 566–579. Springer, Heidelberg (2012). https://doi.org/10.1007/978-3-642-33712-3_41

16. Weinberger, K.Q., Saul, L.K.: Distance metric learning for large margin nearest neighbor classification. J. Mach. Learn. Res. **10**(1), 207–244 (2009)
17. Huang, G.B., Ramesh, M., Berg, T., et al.: Labeled faces in the wild: A database for studying face recognition in unconstrained environments. Technical report, University of Massachusetts (2007)
18. Sengupta S, Chen J C, Castillo C, et al.: Frontal to profile face verification in the wild. In: 2016 IEEE Winter Conference on Applications of Computer Vision (WACV), pp 1–9. IEEE (2016)
19. Chen, B.C., Chen, C.S., Hsu, W.H.: Face recognition and retrieval using cross-age reference coding with cross-age celebrity dataset. IEEE Trans. Multimedia **17**(6), 804–815 (2015)
20. Yi, D., Lei, Z., Liao, S., et al.: Learning face representation from scratch. arXiv preprint arXiv:1411.7923 (2014)
21. Szegedy, C., Liu, W., Jia, Y., et al.: Going deeper with convolutions. In: Proceedings of the IEEE Conference on Computer Vision and Pattern Recognition, pp 1–9 (2015)
22. He, K., Zhang, X., Ren, S., et al.: Deep residual learning for image recognition. In: Proceedings of the IEEE Conference on Computer Vision and Pattern Recognition, pp 770–778 (2016)
23. Liao, S., Jain, A.K., Li, S.Z.: A fast and accurate unconstrained face detector. IEEE Trans. Pattern Anal. Mach. Intell. **38**(2), 211–223 (2016)
24. Jia, Y., Shelhamer, E., Donahue, J., Karayev, S., Long, J., Girshick, R., Guadarrama, S., Darrell, T.: Caffe: Convolutional architecture for fast feature embedding. arXiv preprint arXiv:1408.5093 (2014)
25. Chen, J.C., Zheng, J., Patel, V.M., et al.: Fisher vector encoded deep convolutional features for unconstrained face verification. In: 2016 IEEE International Conference on Image Processing (ICIP), pp 2981–2985. IEEE (2016)
26. Sankaranarayanan, S., Alavi, A., Castillo, C.D., et al.: Triplet probabilistic embedding for face verification and clustering. In: 2016 IEEE 8th International Conference on Biometrics Theory, Applications and Systems (BTAS), pp 1–8. IEEE (2016)
27. Wen, Y., Li, Z., Qiao, Y.: Latent factor guided convolutional neural networks for age-invariant face recognition. In: Proceedings of the IEEE Conference on Computer Vision and Pattern Recognition, pp 4893–4901 (2016)
28. Wu, X., He, R., Sun, Z.: A lightened CNN for deep face representation. In: 2015 IEEE Conference on IEEE Computer Vision and Pattern Recognition (CVPR) (2015)

Pore-Scale Facial Features Matching Under 3D Morphable Model Constraint

Xianxian Zeng[1], Dong Li[1(✉)], Yun Zhang[1], and Kin-Man Lam[2]

[1] Automation, Guangdong University of Technology, Guangzhou, Guangdong, China
dong.li@gdut.edu.cn
[2] Department of Electronic and Information Engineering,
The Hong Kong Polytechnic University, Hong Kong, China

Abstract. Similar to irises and fingerprints, pore-scale facial features are effective features for distinguishing human identities. Recently, the local feature extraction based on deep network architecture has been proposed, which needs a large dataset for training. However, there are no large databases for pore-scale facial features. Actually, it is hard to set up a large pore-scale facial-feature dataset, because the images from existing high-resolution face databases are uncalibrated and nonsynchronous, and human faces are nonrigid. To solve this problem, we propose a method to establish a large pore-to-pore correspondence dataset. We adopt Pore Scale-Invariant Feature Transform (PSIFT) to extract pore-scale facial features from face images, and use 3D Dense Face Alignment (3DDFA) to obtain a fitted 3D morphable model, which is constrained by matching keypoints. From our experiments, a large pore-to-pore correspondence dataset, including 17,136 classes of matched pore-keypoint pairs, is established.

Keywords: Pore-scale facial features · Dataset · PSIFT
3D morphable model · 3DDFA

1 Introduction

Pore-scale facial features include pores, fine wrinkles, and hair, which commonly appear in the whole face region. Pore-scale facial features, which are similar to the features for irises and fingerprints, are one of the most effective features for distinguishing human identities. Recently, local feature extraction based on deep network architecture [1], namely Learned Invariant Feature Transform (LIFT), has been proposed. LIFT is a deep network architecture that implements the full feature-point handling pipeline, i.e. detection, orientation estimation, and feature description. If LIFT is trained with a large and accurate dataset, it can perform better than state-of-the-art methods for feature extraction. This inspires

---

D. Li—This work was supported by National Natural Science Foundation of China: 61503084, U1501251 and Natural Science Foundation of Guangdong Province, China: 2016A030310348, and RGC General Research Fund, Hong Kong: PolyU 152765/16E.

J. Yang et al. (Eds.): CCCV 2017, Part II, CCIS 772, pp. 29–39, 2017.
https://doi.org/10.1007/978-981-10-7302-1_3

us to believe that good pore-scale feature extraction can be achieved if LIFT is trained under a large pore-scale facial-feature dataset. However, currently, there are no large and open databases of pore-scale facial features. Therefore, in this paper, we first propose an efficient method for generating a large pore-to-pore correspondence dataset.

It is hard to set up a large pore-to-pore correspondence dataset, because the images from existing high-resolution (HR) face databases are uncalibrated and nonsynchronous. Besides, human faces are nonrigid. All these make pore-scale feature matching a great challenge. To the best of our knowledge, only a few studies have been reported in the literature that attempt to set up a pore-to-pore correspondences dataset using uncalibrated face images. Lin et al. [2] employed the SURF features [3] on facial images with viewpoints of about 45°-apart, which typically obtained no more than 10 inliers (i.e. correctly matched keypoint pairs) out of a total of 30 matched candidates in 3 poses. Li et al. [4] proposed a new framework, namely Pore Scale-Invariant Feature Transform (PSIFT), to achieve the pore-scale feature extraction, and also generate a pore-to-pore correspondence dataset, including about 4,240 classes of matched pore-keypoint pairs. PSIFT is a feature that can describe the human pore patches distinctively. However, the human face is symmetric, and PSIFT may produce some outliers. For this problem, Li [4] uses the RANSAC (Random SAmple Consensus) [14] method to discard the potential outliers, which will result in reducing the number of matched keypoints. We found that the RANSAC algorithm cannot perform satisfactorily, if the object under consideration is nonrigid. Therefore, Li's method [4] also removes many matched keypoints from facial regions. In our opinion, one of the most promising ways of establishing a larger pore-to-pore correspondence dataset is finding a new constraint, which can perform well for pore-scale feature matching.

Currently, some research solves the face-alignment problem with a 3D solution. Blanz et al. [11] proposed a standard 3D morphable model (3DMM), and Zhu et al. [10] presented a neural network structure, namely 3D Dense Face Alignment (3DDFA), to fit the 3D morphable model to a face image. Inspired by the 3DDFA algorithm, in this paper we use the fitted 3D morphable model to constrain the pore-scale keypoint matching. To the best of our knowledge, 3D-model constraint is one of the most effective constraints for keypoint matching. Our proposed framework is shown in Fig. 1. In summary, our contributions are:

1. We propose the 3D morphable model constraint, which can improve the accuracy for pore-scale keypoint matching.
2. Our proposed methods can establish a large number of correspondences between uncalibrated face images of the same person using the pore-scale features, which leads to many potential applications. Our work shows a method to merge face-based approaches with general computer-vision approaches.
3. Based on our framework, a pore-to-pore correspondences dataset containing 17,136 classes of matched pore-keypoint pairs, is established, where the same pore keypoints from 4 face images of the same subject, with different poses, are linked up.

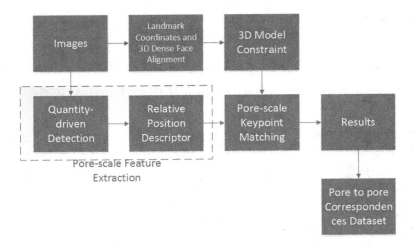

Fig. 1. The structure of the proposed overall framework.

2 Pore-Scale Invariant Feature Transform

PSIFT [4] is variant of SIFT [9], which can generate pore-scale features. The details of PSIFT will be introduced in the following sections.

2.1 Pore-Scale Feature Detection

Pore-scale facial features, such as pores and fine wrinkles, are darker than their surroundings in a skin region. Therefore, PSIFT applies the Difference-of-Gaussians (DoG) detectors for keypoint detection on multiple scales, which is shown as follows.

$$D(x,y,\sigma) = L(x,y,k\sigma) - L(x,y,\sigma) = (G(x,y,k\sigma) - G(x,y,\sigma)) * I(x,y), \quad (1)$$

where the scale space of an image $L(x,y,\sigma)$ is the convolution of the image I(x,y) and the Gaussian kernel

$$G(x,y,\sigma) = \frac{1}{2\pi\sigma^2} exp(\frac{-(x^2+y^2)}{2\sigma^2}). \quad (2)$$

PSIFT constructs the DoG in octaves, which have the σ doubled in the scale space. Li [4] found that the PSIFT detector only needs the maxima of the DoG to locate the darker pore keypoints in face regions. An example is shown in Fig. 2(c). This is because a blob-shaped pore-scale keypoint is a small, darker point due to its small concavity, where incident light is likely to be blocked. Therefore, PSIFT models the blob-shaped skin pores using a Gaussian function, as follows:

$$pore(x,y,\sigma) = 1 - 2\pi\sigma^2 G(x,y,\sigma), \quad (3)$$

where σ is the scale of the pore model. Then, the DoG response to a pore, denoted as D_{pore}, can be computed as follows:

$$D_{pore}(x, y, \sigma_1, \sigma_2) = [G(x, y, k\sigma_1) - G(x, y, \sigma_1)] * pore(x, y, \sigma_2), \qquad (4)$$

and the pore-scale keypoints are the maxima of D_{pore}.

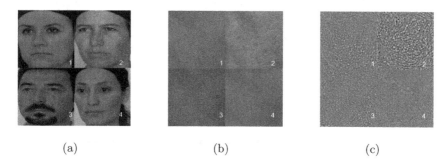

(a)   (b)   (c)

Fig. 2. (a) Four face images with different skin conditions from the Bosphorus face database, (b) local skin-texture images, and (c) the DoG of the local skin-texture image.

2.2   Pore-Scale Feature Descriptor

The local PSIFT descriptor, which is adapted from SIFT, is used to extract the relative-position information about neighboring pores. The keypoints from two facial-skin regions can be matched by using the PSIFT descriptor. Figure 2 shows some sample results of the DoG layers. The lighter points on a DoG, as shown in Fig. 2(c), represent the responses of the feature points. These points are very similar to each other: most of them are blob-shaped, and the surrounding region of the keypoints have almost the same color. However, the relative positions of the pores are unique. Therefore, the descriptor should extract not only the information around the keypoints, but also the information of a neighborhood wide enough to include the neighboring pore-scale features. Therefore, both the number of subregions and the support size of these subregions for the PSIFT descriptor should be sufficiently large. Besides, Li [4] found that the keypoints are not assigned a main orientation, because most of the keypoints do not have a coherent orientation. Some parameters of the PSIFT and SIFT descriptors are shown in Table 1.

3   Matching with the 3D Morphable Model Constraint

In order to achieve a more efficient and accurate matching, we present our method for local PSIFT feature matching by using the 3D-model constraint. The details of our method are introduced in the following sections.

Table 1. The parameters of the PSIFT and SIFT descriptors

Parameters	PSIFT	SIFT
No. of subregions	8×8	4×4
Support size of each subregion	$6 \times$ scale of keypoints	$3 \times$ scale of keypoints
Support size of total subregion	$48 \times$ scale of keypoints	$12 \times$ scale of keypoints
Dimension of the feature	512	128

3.1 3D Morphable Model

Blanz et al. [11] proposed the 3D morphable model (3DMM), which describes the 3D face space with principal component analysis (PCA), as follow:

$$S = \bar{S} + A_{id}\alpha_i d + A_{exp}\alpha_{exp}, \tag{5}$$

where S is a 3D face, $\bar{S}$ is the mean shape, A_{id} is the principal axes trained on the 3D face scans with neutral expression, α_{id} is the shape parameter, A_{exp} is the principal axes trained on the offsets between different expression scans, and α_{exp} is the expression parameter. For this, A_{id} and A_{exp} come from Basel Face Model (BFM) [12] and Face-Warehouse [13] respectively. The 3D face is then projected onto the image plane with Weak Perspective Projection, as follows:

$$V(p) = f * Pr * R * (\bar{S} + A_{id}\alpha_i d + A_{exp}\alpha_{exp}) + t_2 d, \tag{6}$$

where $V(p)$ is the constructed model and projection function, leading to the 2D positions of the model vertexes; f is the scale factor; Pr is the orthographic projection matrix $Pr = \left(\begin{smallmatrix} 1 & 0 & 0 \\ 0 & 1 & 0 \end{smallmatrix} \right)$; R is the rotation matrix constructed from rotation angles $pitch$, yaw, and $roll$; and t_{2d} is the translation vector. The collection of all the model parameters is $p = [f, pitch, yaw, roll, t_{2d}, \alpha_{id}, \alpha_{exp}]^T$.

3.2 3D Dense Face Alignment

Zhu et al. [10] presented a network structure, namely 3D Dense Face Alignment (3DDFA), to compute the model parameters p. The purpose of 3D face alignment is to estimate p from a single face image $\mathbf{I}$. 3DDFA [10] employs a unified network structure across the cascade and constructs a specially designed feature Projected Normalized Coordinate Code (PNCC). In summary, at iteration k ($k = 0, 1,..., K$), given an initial parameter set p^k, 3DDFA constructs PNCC with p^k, and trains a convolutional neutral network Net^k to predict the parameter update Δp^k:

$$\Delta p^k = Net^k(\mathbf{I}, PNCC(p^k)). \tag{7}$$

After that, a better parameter set $p^{k+1} = p^k + \Delta p^k$ becomes the input of the next network Net^{k+1}, which has the same structure as Net^k. The input is a $100 \times 100 \times 3$ color image of PNCC. The network contains four convolution layers, three pooling layers, and two fully connected layers, and the network

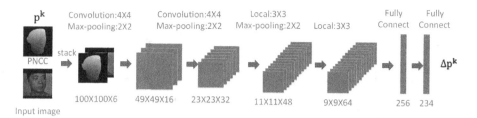

Fig. 3. An overview of 3DDFA.

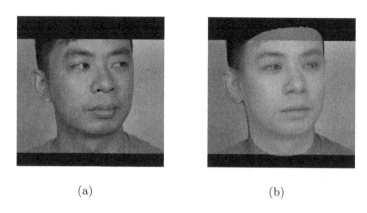

(a)   (b)

Fig. 4. (a) The original image, and (b) the image with 3D-model projection.

structure is shown in Fig. 3. The output is a 234-dimensional updated parameter set, including 6-dimensional pose parameters $[f, pitch, yaw, roll, t_{2dx}, t_{2dy}]$, 199-dimensional shape parameters α_{id}, and 29-dimensional expression parameters α_{exp}. The result, based on 3DDFA, after the 3rd iteration is shown in Fig. 4.

3.3   3D Morphable Model Constraint

A pore keypoint is a pore pointin a face image. Therefore, we can write the equations of the probe image and the gallery image from Eq. (6) as follows.

$$V_p(pore) = f_p * Pr * R_p * (\bar{S}_p(pore) + A_{id}\alpha_{id_p} + A_{exp}\alpha_{exp_p}) + t_{2d_p}, \quad (8)$$

$$V_g(pore) = f_g * Pr * R_g * (\bar{S}_g(pore) + A_{id}\alpha_{id_g} + A_{exp}\alpha_{exp_g}) + t_{2d_g}, \quad (9)$$

where $\bar{S}_p(pore)$ and $\bar{S}_g(pore)$ are the 3D location of the pores of the mean shape. From Eqs. (8) and (9), we assume that if a pore keypoint of the probe image and a pore keypoint of the gallery image are the same pore keypoint of the face, then $Err_{3d} = ||\bar{S}_g(pore) - \bar{S}_p(pore)||_2$ approximately equals 0. Then, we can compute the following:

$$V_{pg}(pore) = f_g * Pr * R_g * (\bar{S}_p(pore) + A_{id}\alpha_{id_g} + A_{exp}\alpha_{exp_g}) + t_{2d_g} \quad (10)$$

$$Err_2d = ||V_{pg}(pore) - V_g(pore)||_2 < range, \qquad (11)$$

where f_g, $R_{(g)}$, $\bar{S}_p(pore)$, α_{id_g}, α_{exp_g}, and t_{2d_g} can be computed from 3DDFA. This means that if $range$ is set correctly and the same pore patch can be detected in the probe image and the gallery image, Eq. (11) will be true. Then, we only need to compute the nearest neighbor rate of the neighboring feature of $V_{pg}(pore)$. If the rate is less than a threshold, the matched keypoint between the probe and gallery images will be found. The estimation of the keypoint positions matched based on the pore-scale facial features is summarized in Algorithm 1.

Algorithm 1. PSIFT improvement by using 3D-model constraint

1: Given two images I_1, and I_2, we assume that there are N_{k1} and N_{k2} keypoints detected in I_1 and I_2, respectively. The coordinates of the i-th keypoint in I_1 are denoted as (x_1^i, y_1^i). Similarly, the coordinates of the j-th keypoint in I_2 are denoted as (x_2^j, y_2^j);

2: Using the 3DDFA method to find the best parameters p_1 and p_2 of I_1 and I_2, respectively. Afterwards we adopt Z-Buffer to project the $\bar{S}_1$ and $\bar{S}_2$ to I_1 and I_2, and denote them as $Z(\bar{S}_1)$ and $Z(\bar{S}_2)$, respectively;

3: Without loss of generality, assume that $N_{k1} < N_{k2}$. Matching the keypoints is established from I_2 to I_1;

4: **for** the j-th keypoint in I_2 **do**

5:    Compute $V_{pg}(j) = f_g * Pr * R_g * (Z(\bar{S}_2(j)) + A_{id}\alpha_{id_g} + A_{exp}\alpha_{exp_g}) + t_{2d_g}$;

6:    Initialize a list **L**, which includes all the keypoints in I_1;

7:    **for** the i-th keypoint in I_1 **do**

8:       **if** $||V_{pg}(j) - (x_1^i, y_1^i)|| < range$ **then**

9:          **L** is not updated;

10:       **else**

11:          Remove the i-th keypoint from the list **L** of I_1;

12:       **end if**;

13:    **end for**;

14:    **if** the distance ratio based on the reduced list **L** is smaller than the threshold δ, which is a constant between 0.8 and 0.9 **then**

15:       a match is established;

16:    **end if**;

17: **end for**;

In our algorithm, we do not use RANSAC [14] to identify those inliers, because the 3D morphable model constraint can identify the inliers accurately, and detect more matched keypoints. Some examples are shown in Fig. 5, where the green point in Fig. 5(a) is one of the pore keypoints, while the red points in Fig. 5(b) are the neighbors of the green point in Fig. 5(a), by using the 3D-model constraint. Besides, the green point in Fig. 5(b) is the matched pore keypoint of the green point in Fig. 5(a).

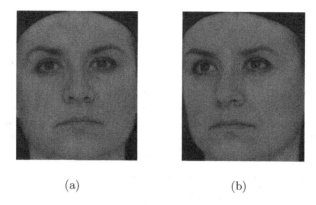

(a)   (b)

Fig. 5. (a) A face image in the neutral pose, (b) the face at a yaw rotation of $10°$. The red points in (b) represent the neighboring keypoints of $V_{pg}(pore)$, and the green point in (b) is the matched point of (a).

4   Experiment

In this section, we will evaluate the performances of our proposed method in terms of accuracy for pore matching. The face images used in the experiments are the original size from the Bosphorus database [15].

4.1   Skin Matching Based on the Bosphorus Dataset

In this section, we estimate the performance of each stage of our algorithm for facial skin matching. We use 105 skin-region pairs cropped from 420 face images, which were captured at $10°$, $20°$, $30°$, and $45°$ to the right of the frontal view in the Bosphorus database, as shown in Figs. 2 and 6. Considering the fact that the dataset is uncalibrated and unsynchronized, Li [4] set the distance threshold used in RANSAC at 0.0005, so only limited number of accurate matching results can be obtained. On the contrary, our method uses 3D-model constraint, so we can obtain more matched keypoints than Li's method [4]. Table 2 illustrates the numbers of inliers obtained by the two methods. Table 2 shows that our method can detect many more matched keypoints, so our method can be used to generate a larger pore-to-pore correspondence dataset.

Table 2. Skin matching results in terms of number of inliers detected

Method	Avg. no. of inliers	total inliers
PSIFT + RANSAC	40.4	4240
PSIFT + 3D-model constraint	163.2	17136

4.2   Pore-to-pore Correspondences Dataset

With the improvement achieved by PSIFT with the 3D-model constraint, a larger pore-to-pore correspondences dataset can be constructed, so that the learning for pore-keypoint-pair matching can be conducted. For each subject, its pore keypoints at one pose are matched to the corresponding pore keypoints at an adjacent pose. We have established three sets of matched keypoint pairs, with viewing angles at 10° and 20°, 20° and 30°, and 30° and 45°. After finding a set of matched pore keypoints between each image pair, we use the matched keypoints to form tracks. A track is a set of matched keypoints across the face images of the same subject at different poses. If a track contains more than one keypoint in the same image, it is considered to be inconsistent, and is then removed. We choose only those consistent tracks, containing 4 keypoints corresponding to the 10°, 20°, 30°, and 45° poses, as shown in Fig. 6. Finally, 17,136 tracks are established, which is much larger than the pore-to-pore correspondences dataset established in Li [4]. In addition, we have also generated another larger

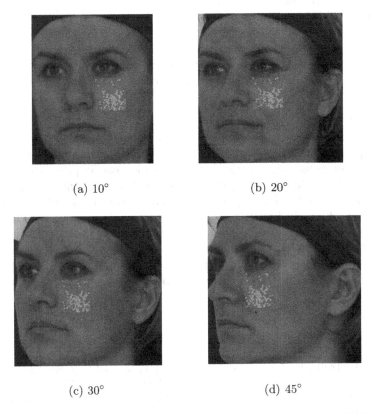

(a) 10°   (b) 20°

(c) 30°   (d) 45°

Fig. 6. Images of the same subject at different poses. The red points are the keypoints of the skin region, and the green points are the corresponding keypoints at another pose.

Fig. 7. Some patches of a subject: each row consists of the corresponding patches of the same pore keypoints of the face images of the same subject at 10°, 20°, 30°, and 45° poses.

pore-to-pore correspondences dataset, based on the whole face of the subjects in the Boshorus dataset, which contains 80, 236 tracks.

Based on our proposed method, which relies on the PSIFT features, we can match the pore-scale keypoints of the same subject from different perspectives. We extract training patches according to the scale σ of the pore keypoints detected. Patches are extracted from a $24\sigma \times 24\sigma$ support region at the keypoint locations, and then normalized to $S \times S$ pixels, where $S = 128$ in our algorithm. Some data from the pore-to-pore dataset is shown in Fig. 7.

5   Conclusion

In this paper, we have proposed using the 3D-model constraint to improve the performance of pore-scale feature matching, which can improve the matching performance when the face images to be matched have a large baseline. Using our proposed method, a larger pore-to-pore correspondences dataset, including 17,136 classes of matched pore-keypoint pairs, is established. In our future work, we will use this larger pore-to-pore correspondences dataset to train a deep neural network so as to learn a better pore-scale feature for face matching. Furthermore, we will evaluate our method under different facial expressions and different light conditions, so that we can produce a pore dataset with different conditions.

References

1. Yi, K.M., Trulls, E., Lepetit, V., Fua, P.: LIFT: learned invariant feature transform. In: Leibe, B., Matas, J., Sebe, N., Welling, M. (eds.) ECCV 2016, Part VI. LNCS, vol. 9910, pp. 467–483. Springer, Cham (2016). https://doi.org/10.1007/978-3-319-46466-4_28
2. Lin, Y., Medioni, G., Choi, J.: Accurate 3D face reconstruction from weakly calibrated wide baseline images with profile contours. In: Computer Vision and Pattern Recognition, pp. 1490–1497. IEEE (2010)

3. Bay, H., Ess, A., Tuytelaars, T., Van Gool, L.: Speeded-up robust features. Comput. Vis. Image Underst. **110**(3), 404–417 (2008)
4. Li, D., Lam, K.M.: Design and learn distinctive features from pore-scale facial keypoints. Pattern Recogn. **48**(3), 732–745 (2015)
5. Matthews, I., Baker, S.: Active appearance models revisited. Int. J. Comput. Vis. **60**(2), 135–164 (2004)
6. Tzimiropoulos, G., Zafeiriou, S., Pantic, M.: Robust and efficient parametric face alignment. In: IEEE International Conference on Computer Vision, pp. 1847–1854. IEEE (2012)
7. Spaun, N.A.: Facial comparisons by subject matter experts: their role in biometrics and their training. In: Tistarelli, M., Nixon, M.S. (eds.) ICB 2009. LNCS, vol. 5558, pp. 161–168. Springer, Heidelberg (2009). https://doi.org/10.1007/978-3-642-01793-3_17
8. Lin, D., Tang, X.: Recognize high resolution faces: from macrocosm to microcosm. In: Computer Vision and Pattern Recognition, pp. 1355–1362. IEEE (2006)
9. Lowe, D.G.: Distinctive image features from scale-invariant keypoints. Int. J. Comput. Vis. **60**(2), 91–110 (2004)
10. Zhu, X., Lei, Z., Liu, X., et al.: Face alignment across large poses: A 3D solution. In: Computer Vision and Pattern Recognition, pp. 146–155. IEEE (2016)
11. Blanz, V., Vetter, T.: Face recognition based on fitting a 3D morphable model. IEEE Trans. Pattern Anal. Mach. Intell. **25**(9), 1063–1074 (2003)
12. Paysan P., Knothe R., Amberg B., et al.: A 3D face model for pose and illumination invariant face recognition. In: International Conference on Advanced Video and Signal Based Surveillance, pp. 296–301. IEEE (2009)
13. Cao, C., Weng, Y., Zhou, S., et al.: Facewarehouse: a 3d facial expression database for visual computing. IEEE Trans. Visual Comput. Graph. **20**(3), 413–425 (2014)
14. Fischler, M.A., Bolles, R.C.: Random sample consensus: a paradigm for model fitting with applications to image analysis and automated cartography. ACM **24**, 726–740 (1981)
15. Savran, A., Alyüz, N., Dibeklioğlu, H., Çeliktutan, O., Gökberk, B., Sankur, B., Akarun, L.: Bosphorus database for 3D face analysis. In: Schouten, B., Juul, N.C., Drygajlo, A., Tistarelli, M. (eds.) BioID 2008. LNCS, vol. 5372, pp. 47–56. Springer, Heidelberg (2008). https://doi.org/10.1007/978-3-540-89991-4_6

Combination of Pyramid CNN Representation and Spatial-Temporal Representation for Facial Expression Recognition

Shulin Xu, Nan Pu, Li Qian, and Guoqiang Xiao[✉]

College of Computer and Information Science, Southwest University,
Chongqing, China
gqxiao@swu.edu.cn

Abstract. In this paper, we propose a novel framework for facial expression recognition in video sequences by combining deep convolutional feature and spatial-temporal feature. Firstly, apex frame of every sequence is selected adaptively by calculating the displacement of facial landmarks. Then, pyramid CNN-based feature is extracted on the apex frame to capture the information of global and local regions of human face. Afterwards, spatial-temporal LBP-TOP feature is generated from video sequence and integrated with pyramid CNN-based feature to represent video, which reflect dynamic and static texture information of facial expressions. Finally, the multiclass support vector machine (SVM) with one-versus-one strategy is applied to classify facial expressions. Experimental results on the extended Cohn-Kanade (CK+) and Oulu-CASIA datasets demonstrate the superiority of our proposed method.

Keywords: Facial expression recognition
Convolutional neural network · LBP-TOP · Support vector machine

1 Introduction

In recent years, human facial expression recognition (FER) emerged as an important research area, due to the facial expression can effectively represent the emotional state, cognitive activities and personality characteristics of human beings. Thus, the facial expression recognition has been widely used in the computer vision applications, such as human computer interaction (HCI) [1], psychology and cognitive science [2], access control and surveillance systems [3], driver state surveillance, and etc.

Early researches about facial expression recognition mainly focus on recognizing expressions from a static image or recognizing the video sequences by analyzing each frame. For the static image based facial expression recognition, the methods of Gabor wavel [4] and local binary pattern (LBP) [5] are usually used to explore the texture information in the face images to represent the facial expression. Moreover, active shape model (ASM) [6] and active appearance model (AAM) [7] are commonly used to extract the facial landmarks to

© Springer Nature Singapore Pte Ltd. 2017
J. Yang et al. (Eds.): CCCV 2017, Part II, CCIS 772, pp. 40–50, 2017.
https://doi.org/10.1007/978-981-10-7302-1_4

describe the changes of facial expression. The static image based methods can effectively extract texture and spatial information in the image but they cannot model the variability in morphological and contextual factors. Therefore, some studies try to capture the dynamic variation of facial physical structure by exploring the spatial-temporal in video sequences, such as 3D-HOG [8], LBP-TOP [9] and 3D-SIFT [10]. Because the video sequences of facial expression contain not only image appearance information in the spatial domain, but also the evolution details in the temporal domain, the facial appearance information, together with the expression evolution information, can further enhance recognition performance. More recently, deep learning, particularly, the convolutional neural networks (CNNs) have shown their strength in computer vision applications and also been popular used to solve the FER problem. For example, Yu et al. [11] utilize an ensemble of CNNs, and employ data augmentation at both training and testing stage in order to improve the performance of FER. Jung et al. [12] propose a small CNN architecture to capture the dynamic variations of facial appearance.

Inspired by the advantages of deep convolutional neural networks and spatial-temporal video representations, we design a static pyramid CNN-based feature generated on the apex frame and further combine it with dynamic appearance-based LBP-TOP feature extracted in the video sequences to enhance the performance of FER. The overall flow diagram of the proposed method for FER is shown in Fig. 1. First, according to prior knowledge that the video frame with the largest expression intensity plays an important role in FER, the expression intensity of each facial frame can be estimated by calculating the displacement of facial landmarks and the facial frame with maximum displacement is selected as the apex frame. Second, making use of the superior representation capbility of deep convolutional feature in the pre-trained CNN architecture, a pyramid CNN-based

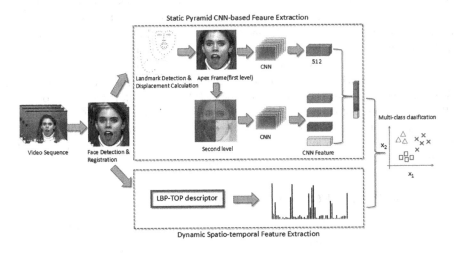

Fig. 1. The framework of our method for FER.

feature representation is generated on the apex frame to capture the information on global and local regions of human face. Afterwards, dynamic LBP-TOP feature is extracted to model the spatial-temporal information of the whole video sequences. Finally, the multi-features generated by combining the static and dynamic feature representations are fed into a classifier to accomplish the FER. The main contributions of this paper are twofold. First, in order to capture the possible slight asymmetry between left and right side of human face as well as the local subtle motion of facial local regions when facial expression changes, we propose to conduct a two level image pyramid on the apex frame, and extract the deep convolutional features on each region in the image pyramid to boost the representation capability of facial expression. Second, the static image feature representation based on the apex frame and the dynamic feature representation based on spatial-temporal information in the video sequences are combined together to further improve the performance of facial expression recognition.

The rest of paper is organized as follows. Section 2 shows detailed description of the proposed pyramid CNN-based feature generation and LBP-TOP feature. The experimental results and discussions are presented in Sect. 3. Conclusions are given in Sect. 4.

2   Methodology

The proposed FER system consists of three procedures: face preprocessing (face detection, face registration and facial expression intensity estimation), static pyramid CNN-based feature generation on apex frame, LBP-TOP feature extraction on video sequences. The details of each procedure are described as follows.

In the preprocessing stage of FER system, the region of face should be firstly detected and cropped in each frame to eliminate the interference from unnecessary noise. Following the usual protocol, the Viola-Jones face detection [13] model is used to detect the face region in the video frame. To further improve the accuracy of detected face region, the method in [14] is employed to detect the facial landmarks, and the facial landmarks at the outermost side are selected to determine the boundary of final face region.

Once we obtained the final accurate face region of each frame in the video sequences, the face registration technology [15] is then adopted to remove the influence of scale, rotation and translation changes of face region. Consequently, the difference of facial expression between the two frames is confined to the changes of facial muscle and the registered face image can also provide the optimal input for subsequent feature extraction.

2.1   Static Pyramid CNN-based Feature

As it is known to all, the frame with the largest expression intensity in video sequence contains rich discriminative expression information and it is usually termed as apex frame. Based on this factor, we can select the apex frame from video sequences and generate a feature representation for the apex frame to improve the performance of FER. In this paper, we propose to calculate the displacement of facial landmarks to estimate the facial expression intensity and select the frame with maximum displacement as apex frame. This procedure is adaptive, and it can not only apply to the expression video whose intensity changes like neutral-onset-apex in traditional datasets, but also to neutral-onset-apex-offset-neutral intensity transformation in other expression video. Inspired by the work in [14], which propose a machine learning approach called supervised descent method (SDM) to detect high accuracy positions of facial landmarks, we adopt SDM to detect the facial landmarks in our method. Under the assumption that the human facial expression transforms as neutral-onset-apex or neutral-onset-apex-offset-neutral, the first face frame in the video sequences can be treated as neutral facial expression. Therefore, for X_1^i and Y_1^i which denote the coordinate of ith landmarks in the first face frame, X_t^i and Y_t^i which are the coordinate of ith landmarks in the tth face frame, the landmarks displacement D_t between first frame and t^{th} frame can be calculated as:

$$D_t = \sum_{i=1}^{n} |X_t^i - X_1^i| + \sum_{i=1}^{n} |Y_t^i - Y_1^i| \tag{1}$$

where n denotes the number of landmarks detected by SDM and it is 66 in total as usually use. Thereby, the frame with Maximum value of D_t will be chosen as the apex frame. Figure 2 demonstrates the procedure of selecting an apex frame in the video sequences.

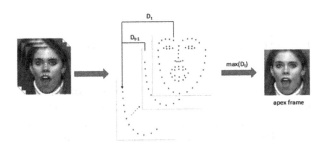

Fig. 2. The procedure of apex frame selection.

As aforementioned, along with the facial expression changes, the left and right face may be asymmetric in certain frames, especially in the eyes and mouth regions, so it is necessary to analyze the local region in the whole face from the apex frame. We propose to use the apex frame to conduct a two-level image pyramid and generate deep convolutional features on the image pyramid to represent the apex frame, such that the global and local information of the human face can be both captured to enhance the discrimination capability of the static apex frame. The proposed pyramid CNN-based face representation is established at two scale levels. The first level corresponds to the full apex face frame, and the second level consist of 4 regions by equally partitioning the full face region. Therefore, we can obtain five deep features by passing each region through a pre-trained CNN architecture: C_0 denotes the deep feature from the first level, and C_1, C_2, C_3, C_4 denote the deep features from the second level. Afterwards, we can concatenate the five deep features as: $C = [C_0, C_1, C_2, C_3, C_4]$, and the final face deep representation C is with the dimensions of 5*512. The static pyramid CNN-based features extraction process is shown in Fig. 3.

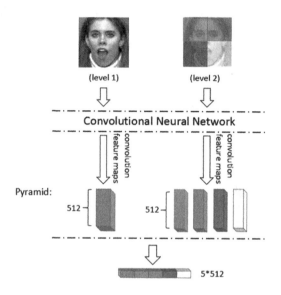

Fig. 3. Feature extraction procedure based on pyramid CNN model.

For the deep features, we propose to use the deep convolutional representation, ranther than the general outputs from the fully connected layers in CNN. Given a pre-trained CNN model with L convolutional layers and for an input image we can extract its CNN feature maps after resizing it to 224*224 for VGG [18] networks. A feature map can be denoted by $\bar{F}_i = \{F_{ij} : i = 1...L; j = 1...C_i\}$, where F_{ij} is equal to the jth feature map at i^{th} convolutional layer, and C_i equal to the number of convolutional kernels. The size of F_{ij} is $W_i \times H_i$, where W_i and H_i are the width and height of each channel. Assuming that (x, y) is the coordinate of feature map $F_{i,j}$, and $f_i(x, y)$ is the response value at the ith convolutional layer with a spatial coordinate (x, y). Then, the image representation by max-pooling can be describe as follows:

$$\dot{V}_i = [\dot{V}_{F_{i,j}} : j = 1...C_i] \tag{2}$$

$$\dot{V}_{F_{i,j}} = max(f_i(x, y)) \tag{3}$$

2.2   LBP-TOP Feature

Local binary patterns from three orthogonal planes (LBP-TOP) [9] is an extension of LBP from two-dimensional space to three-dimensional space, and LBP-TOP extracts local binary patterns features from three orthogonal planes (i.e., XY, XT and YT) of video sequences. Compared with LBP, LBP-TOP does not only contains the texture information of XY plane, but also takes into account the texture information of XT and YT, while the texture information of XT and YT record important dynamic textures. For each plane, a histogram of dynamic texture can be defined as:

$$H_{i,j} = \sum_{x,y,t} I\left\{f_j\left(x,y,t\right) = i\right\} \tag{4}$$

$$i = 0,\cdots,n_{j-1}; j = 0,1,2$$

where n_j is the number of different labels produced by the LBP operator in the jth plane (j = 0: XY, 1: XT and 2: YT), $f_j\left(x,y,t\right)$ denotes the LBP code of central pixel (x, y, t) in the jth plane and if A is true, $I\{A\} = 1$, else $I\{A\} = 0$. Afterwards, statistical histograms of three different planes will be concatenated into one histogram. Generally, considering the motion of different face regions, a block-based method is introduced which cascade histogram extracted from all block volume. In the experiment, each sequence volume is divided into 8*8 non-overlapping blocks. The procedure of LBP-TOP method extracting features from block shown in Fig. 4.

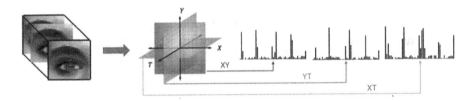

Fig. 4. Feature extraction procedure based on block-based LBP-TOP.

Furthermore, the static pyramid CNN based feature and LBP-TOP feature are cascaded as a final face video representation for training and testing, and we can note that the strength of the final representation not only contains the static feature from apex frame which has the maximum expression intensity in face frames, but also takes into account the spatio-temporal information in the video sequences.

3   Experimental Results

3.1   Dataset

The extended Cohn-Kanade dataset (CK+) dataset [16]: there are totally 593 frontal video sequences from 123 subjects. The sequences vary in duration, from

10 frames to 60 frames per video, starting from neutral to the apex of the facial expression. The CK+ dataset contains 327 expression-labeled, each of which has seven expressions, but only 309 image sequences with six basic expressions (anger, disgust, fear, happiness, sadness, and surprise) were considered in our study.

The Oulu-CASIA dataset [17]: it consists of six expressions from 80 subjects. All the image sequences were taken under three visible light conditions: normal, weak and dark. The number of video sequences is 480 (80 subjects by six expressions) for each illumination, so there are totally 2880 (4806) video sequences in the dataset. All expression sequences begin at the neutral frame and end with the apex frame. In the experiments, we evaluate our method under normal illumination condition.

3.2   Experimental Results on CK+ Dataset and Oulu-CASIA Dataset

In this part, we evaluate the proposed framework on both CK+ and Oulu-CASIA datasets. We firstly test the performance of apex frame selection based on the facial expression estimation by calculating the facial landmarks displacement. Figure 5 shows the selected apex frames from different video sequences. As the transformation of facial expression is from neutral to apex in both datasets, the estimated apex frame by our proposed method is almost the last frame of each video sequence which prove the correctness of the apex frame selection method.

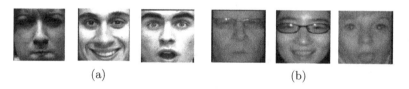

(a)   (b)

Fig. 5. Apex frames selected from (a) CK+ dataset and (b) Oulu-CASIA dataset.

We further evaluate the performance of the outputs from fully connected layers in CNN and the outputs from the last convolutional layer in CNN. Additionally, the performance of both two types of deep features on a single face image as well as the proposed two level face image pyramid is also compared. The accuracy of a specific expression recognition is measured by the ratio of correctly recognized samples over the total number of samples in the specific expression, while the total accuracy is calculated by the ratio of all correctly recognized samples over the total number of all testing samples. As illustration in Fig. 6, we can note that the deep convolutional features with only 512 dimensions show competitive or even higher accuracy than the fully connected features with 4096 dimensions on two datasets. Furthermore, we apply the deep features on the proposed two level image pyramid to evaluate the performance of pyramid CNN-based representation, and we can see that the two level image pyramid

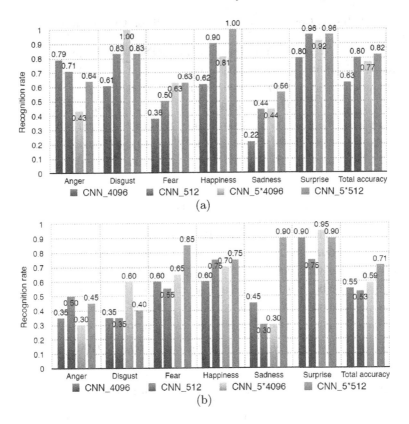

Fig. 6. Comparison recognition rate of four different dimensional CNN feature on (a) CK+ dataset and (b) Oulu-CASIA dataset.

CNN-based representation indeed improve the facial expression accuracy when comparing a single face image.

Moreover, we conduct experiments to evaluate effectiveness of combining static pyramid CNN-based feature and dynamic spatial-temporal LBP-TOP feature for facial expression recognition in video sequences. Tables 1, 2, 3 and 4 show the confusion matrices obtained by using just LBP-TOP feature as well as the combination of both features with multiclass SVM classifier. The confusion matrix includes recognition accuracy of each expression and the total classification accuracy. Based on the results on both datasets, the combination of two features achieved higher total recognition accuracy and specific expression recognition accuracy than that of just using dynamic LBP-TOP feature or static pyramid CNN-based feature. Especially for CK+ dataset, the proposed framework significantly improves the performance on the expressions of anger, disgust, happiness and surprise, meanwhile the recognition accuracy of fear expression has greatly improved compared to the LBP-TOP feature.

Table 1. Confusion matrix of LBP-TOP on CK+ dataset

	Anger	Disgust	Fear	Happiness	Sadness	Surprise	
Anger	**0.93**	0.07	0.00	0.00	0.00	0.00	
Disgust	0.00	**1.00**	0.00	0.00	0.00	0.00	
Fear	0.12	0.00	**0.38**	0.25	0.25	0.00	
Happiness	0.00	0.00	0.00	**1.00**	0.00	0.00	
Sadness	0.00	0.00	0.00	0.00	**1.00**	0.00	
Surprise	0.00	0.00	0.00	0.04	0.00	**0.96**	
Total							**0.93**

Table 2. Confusion matrix of LBP-TOP on Oulu-CASIA dataset

	Anger	Disgust	Fear	Happiness	Sadness	Surprise	
Anger	**0.65**	0.15	0.05	0.00	0.15	0.00	
Disgust	0.40	**0.55**	0.05	0.00	0.00	0.00	
Fear	0.00	0.05	**0.70**	0.00	0.20	0.05	
Happiness	0.05	0.00	0.15	**0.80**	0.00	0.00	
Sadness	0.05	0.10	0.10	0.00	**0.70**	0.05	
Surprise	0.00	0.00	0.00	0.00	0.00	**1.00**	
Total							**0.73**

Table 3. Confusion matrix of combination feature on CK+ dataset

	Anger	Disgust	Fear	Happiness	Sadness	Surprise	
Anger	**1.00**	0.00	0.00	0.00	0.00	0.00	
Disgust	0.00	**1.00**	0.00	0.00	0.00	0.00	
Fear	0.00	0.00	**0.88**	0.00	0.12	0.00	
Happiness	0.00	0.00	0.00	**1.00**	0.00	0.00	
Sadness	0.00	0.00	0.00	0.00	**1.00**	0.00	
Surprise	0.00	0.00	0.00	0.04	0.00	**0.96**	
Total							**0.98**

Table 4. Confusion matrix of combination feature on Oulu-CASIA dataset

	Anger	Disgust	Fear	Happiness	Sadness	Surprise	
Anger	**0.70**	0.10	0.05	0.00	0.15	0.00	
Disgust	0.30	**0.60**	0.00	0.00	0.10	0.00	
Fear	0.00	0.05	**0.80**	0.00	0.10	0.05	
Happiness	0.05	0.00	0.10	**0.85**	0.00	0.00	
Sadness	0.05	0.00	0.05	0.00	**0.85**	0.05	
Surprise	0.00	0.00	0.00	0.00	0.00	**1.00**	
Total							**0.80**

3.3 Comparison with State-of-the-art

In the following, we compare the proposed framework with published state-of-the-art methods on each dataset. As shown in Table 5, we can note that, for facial expression recognition, our method achieves the highest recognition accuracy which further proves the discrimination and robustness of our video sequences representation by combining static pyramid CNN-based feature and dynamic LBP-TOP feature.

Table 5. The Comparison with the state-of-the-art on both datasets

Method	Descriptor	CK+ dataset	Oulu-CASIA dataset
Klaser et al. [8]	HOG 3D	91.40%	70.63%
Liu et al. [19]	STM-ExpLet	94.19%	74.59%
Liu et al. [20]	Dis-ExpLet	95.10%	79.00%
Proposed method		**97.89%**	**80.00%**

4   Conclusions

In this paper, we presented a novel FER method, where the static and dynamic feature are integrated together to boost the FER performance. For the procedure of static feature extraction, in order to capture the global and local information of human face, a pyramid CNN model is constructed to extract features from apex frames which are selected adaptively by using the displacement information of landmarks. Moreover, spatial-temporal LBP-TOP feature is employed as the dynamic feature which is cascaded with static pyramid CNN-based feature to classify expressions by using multiclass SVM with one-versus-one strategy. The evaluation results show that our method is competitive or even superior when comparing to the state-of-the-art methods on two facial expressions datasets.

References

1. Abdat, F., Maaoui, C., Pruski, A.: Human-computer interaction using emotion recognition from facial expression. In: 2011 Fifth UKSim European Symposium on Computer Modeling and Simulation (EMS), pp. 196–201. IEEE (2011)
2. Russell, J.A.: Core affect and the psychological construction of emotion. Psychol. Rev. 110(1), 145 (2003)
3. Bettadapura, V.: Face expression recognition and analysis: the state of the art. arXiv preprint arXiv:12036722 (2012)
4. Bartlett, M.S., Littlewort, G., Frank, M., Lainscsek, C., Fasel, I., Movellan, J.: Recognizing facial expression: machine learning and application to spontaneous behavior. In: Proceedings of the IEEE Computer Society Conference on Computer Vision and Pattern Recognition, 2005, pp. 568–573, June 2005

5. Ojala, T., Pietikinen, M., Menp, T.: Multiresolution gray-scale and rotation invariant texture classification with local binary patterns. IEEE Trans. Pattern Anal.
   Mach. Intell. **24**(7), 971–987 (2002)
6. Shbib, R., Zhou, S.: Facial expression analysis using active shape model. Int. J.
   Signal Process. Image Process. Pattern Recogn. **8**(1), 9–22 (2015)
7. Lucey, S., Ashraf, A.B., Cohn, J.F.: Investigating spontaneous facial action recognition through aam representations of the face. In: Face Recognition, pp. 275–286.
   Pro Literatur Verlag, Mammendorf (2007)
8. Klaser, A., Marszaek, M., Schmid, C.: A spatio-temporal descriptor based on 3d-
   gradients. In: British Machine Vision Conference (BMVC) (2008)
9. Zhao, G., Pietikinen, M.: Dynamic texture recognition using local binary patterns
   with an application to facial expressions. IEEE Trans. Pattern Anal. Mach. Intell.
   29(6), 915–928 (2007)
10. Scovanner, P., Ali, S., Shah, M.: A 3-dimensional sift descriptor and its application
    to action recognition. In: Proceedings of the 15th International Conference on
    Multimedia. ACM (2007)
11. Yu, Z., Zhang, C.: Image based static facial expression recognition with multiple
    deep network learning. In: ACM International Conference on Multimodal Interac-
    tion (MMI), pp. 435–442 (2015)
12. Jung, H., Lee, S., Yim, J., Park, S., Kim, J.: Joint fine-tuning in deep neural
    networks for facial expression recognition. In: The IEEE International Conference
    on Computer Vision (ICCV). IEEE (2015)
13. Viola, P., Jones, M.J.: Robust real-time face detection. In: 2001 Proceedings of the
    Eighth IEEE International Conference on Computer Vision, ICCV, 2001, DBLP
    2004, p. 747 (2001)
14. Xiong, X., Torre, F.D.L.: Supervised descent method and its applications to face
    alignment. In: Computer Vision and Pattern Recognition, pp. 532–539. IEEE
    (2013)
15. Maintz, J., Viergever, M.: A survey of medical image registration. Med. Image
    Anal. **2**, 1–36 (1998)
16. Lucey, P., Cohn, J.F., Kanade, T., et al.: The extended Cohn-Kanade Dataset
    (CK+): A complete dataset for action unit and emotion-specified expression. In:
    Computer Vision and Pattern Recognition Workshops, pp. 94–101. IEEE (2010)
17. Zhao, G., Huang, X., Taini, M., Li, S.Z., Pietikinen, M.: Facial expression recog-
    nition from near-infrared videos. Image Vis. Comput. **29**(9), 607–619 (2011)
18. Simonyan, K., Zisserman, A.: Very deep convolutional networks for large-scale
    image recognition. In: Computer Science (2014)
19. Liu, M., Shan, S., Wang, R., Chen, X.: Learning expressionlets on spatio-temporal
    manifold for dynamic facial expression recognition. In: Proceedings of the IEEE
    International Conference Computer Vision and Pattern Recognition, pp. 1749–
    1756, June 2014
20. Liu, M., Shan, S., Wang, R., et al.: Learning expressionlets via universal mani-
    fold model for dynamic facial expression recognition. IEEE Trans. Image Process.
    25(12), 5920–5932 (2016). A Publication of the IEEE Signal Processing Society

Robust Face Recognition Against Eyeglasses Interference by Integrating Local and Global Facial Features

Hansheng Fang[1], Jie Wen[1], and Yong Xu[1,2,3(✉)]

[1] Bio-Computing Research Center, Shenzhen Graduate School,
Harbin Institute of Technology, Shenzhen 518055, China
yongxu@ymail.com
[2] Key Laboratory of Network Oriented Intelligent Computation,
Shenzhen 518055, China
[3] Medical Biometrics Perception and Analysis Engineering Laboratory,
Shenzhen 518055, China

Abstract. In this paper, we proposed a feature extraction method to solve a challenge problem of face recognition, i.e., recognition of faces with eyeglasses. By fusing the local and global facial features, the proposed method can extract robust facial features that can greatly reduce the negative influence of eyeglasses on face recognition. Firstly, we use the Ununiformed Local Gabor Binary Pattern Histogram Sequence (ULGBPHS) method to extract local facial features. Secondly, we apply 2D-Discrete Fourier Transform (2D-DFT) method to obtain global facial features. Finally, we use a weighted fusion strategy to combine the two kinds of facial features for face recognition. Extensive experimental results on the well-known public GT and CMU_PIE face datasets, and real scene dataset which is built by our group show that the proposed feature extraction method obtains the best performance among some state-of-the-art methods. The relevant code and data will be available at http://www.yongxu.org/lunwen.html.

Keywords: Face recognition · Eyeglasses
Local and global facial features

1 Introduction

Computer vision (CV) has become a hot research direction in recent years because of the advance of both related theories and computer hardware. There are many research areas in CV, such as the biometrics technology including fingerprint recognition, palm print recognition, face recognition etc. [3,9,24]. Among them, face recognition has received much attention in both academia and industry [2,15]. As we know, many external factors affect the actual recognition rate [5,16]. The eyeglass is a common problem in face recognition [8]. Suppose the user wears a pair of eyeglasses when he registers in a face recognition system, he may fail to pass the system when he doesn't wear eyeglasses in the

© Springer Nature Singapore Pte Ltd. 2017
J. Yang et al. (Eds.): CCCV 2017, Part II, CCIS 772, pp. 51–61, 2017.
https://doi.org/10.1007/978-981-10-7302-1_5

recognition stage. In order to reduce the negative impact caused by eyeglasses, some algorithms have been proposed in the past.

Extracting more robust facial features is one of the most effective approaches to reduce the negative influence of eyeglasses. Martinez [11] proposed to extract the local facial features to solve the occlusion problem of face recognition. He divided a face into k blocks and calculated the feature for each block. After comparing the training set and testing set, he obtained the probability of occlusion for each block. These probability values were used to adjust the weight of each block when they use Mahalanobis distances in the recognition stage. Zhang et al. [23] proposed a Local Gabor Binary Pattern Histogram Sequence (LGBPHS) method to address the occlusion problems of face recognition. They applied Gabor filters and local binary pattern (LBP) operators [1,13,14] to a face image to extract the local facial features for face recognition. Yi et al. [19] performed the sparse representation based classification method on the extracted local features for face recognition. Liu et.al. [10] proposed a novel ununiformed division strategy based on the LGBPHS method [23]. Experimental results in those papers all prove the effectiveness of the local features for eyeglass-face recognition.

In this paper, we try to solve the eyeglass-face recognition problem by extracting robust facial features [4,21]. The new feature extraction method can show outstanding performance in resisting the trouble caused by eyeglasses in a face. Firstly, we normalize the face image and extract its local facial features. We use non-uniform division strategy proposed in [10] to segment a face into several non-overlapping blocks with different sizes. For each block, Gabor filters with different scales and orientations are utilized to obtain multiple Gabor Magnitude Pictures (GMPs) [18,23]. Then we compute the histogram for each GMP and concatenate these histograms as the local facial features for each facial block. In this way, we can obtain the local facial features of each face image by integrating all histograms of these blocks. Secondly, we use the 2D-DFT method to extract the global facial features. After transforming the face image into the frequency domain via the 2D-DFT transformation, we can obtain the real component and imaginary component of a face image. In this work, we only exploit the low frequency coefficients as the global facial features since they are the intrinsic global image information [20]. Finally, the extracted local and global facial features are combined via an adaptive weighted fusion approach for face recognition.

The remainder of this paper is organized as follows: Sect. 2 introduce a brief review of the local facial feature extraction method. The details of our proposed method are described in Sect. 3. Section 4 presents the experimental evaluations. The conclusion and discussion of this paper are offered in Sect. 5.

2   A Brief Review of Local Facial Feature Extraction Method

2.1   Facial Feature Extraction Using the LGBPHS Method

The eyeglasses can be viewed as occlusion of the face. Many excellent algorithms have been proposed and achieved a good performance in solving occlusion problems of face recognition. LGBPHS is one of the representative methods. It simultaneously combines the advantages of Gabor filters and LBP operators [22]. Gabor filters have powerful ability in obtaining robust and discriminative local features. LBP is a typical visual descriptor and has been widely used in computer vision owing to its effectiveness in extracting the texture information of an image. The LGBPHS method uses Gabor filters to perform convolution operation on a normalized face image and gets plenty of GMPs in the first step. Then it uses the LBP operator to extract the LBP feature maps base on the obtained GMPs. In the second step, it divides these LBP feature maps uniformly and obtains the statistical histogram information of all blocks. Finally, it concatenates these feature histograms of all blocks into a feature histogram sequence as the final extracted local features. Experimental results show that this method is efficient for general occlusion problems.

2.2   Facial Feature Extraction Using the Ununiformed Division Strategy

Although the LGBPHS method shows good performance in general occlusion face recognition problems, it was proved to be inefficient in eyeglass-face recognition. As we know, the eyeglass-face problem is different from other general occlusion problems since eyeglasses always exists around our eyes. Human eyes are very important features in face recognition. The LGBPHS method uses a uniform way to divide a face into several blocks of the same size. This may break the integrity of eyes when we extract the facial features. In order to maintain the integrity of these facial key-points, an improved non-uniform partition strategy called Ununiformed Local Gabor Binary Pattern Histogram Sequence (ULGBPHS) was proposed in [10]. This method can be viewed as extension of the LGBPHS method which uses a non-uniform partition strategy to extract more robust local features. The main idea of this method comes from the reality that the facial keypoints in the face should keep their own completeness when we divide the face feature map. Therefore, this method proposes to partition a normalized face image into different blocks non-uniformly. Then Gabor filters are applied to each block with different sizes to obtain their GMPs. Finally, it performs LBP on each group of GMPs to obtain the LBP feature maps. It gets the histograms of the local LBP feature maps for each group and concatenated them into a final feature histogram sequence by using a weighted strategy as the final features for face recognition [7,17]. Experimental results prove that the ULGBPHS method is effective to improve the accuracy in this problem.

Although the non-uniform strategy is effective in handling eyeglass-face problems in comparison with the LGBPHS method, it doesn' t always work well since it still focuses on extraction of facial local features.

3   The Proposed Feature Fusion Method

3.1   Method Analysis

Actually each face has its own overall appearance. Rather than focusing on local feature, we care about both local and global facial features in the eyeglass-face problems. In other words, we should take overall appearance into our consideration when extracting facial features. Usually humans can recognize a person correctly no matter whether he/she wears eyeglasses or not. This is because humans can recognize a face image as a whole at first glance and confirm their appearance by recognizing the local facial keypoints. In this paper, we use the combination of local and global facial features instead of the local parts only. Specially we apply 2D-DFT method to extract the global facial feature of face image with and without eyeglasses. As a matter of fact, we can get two similar face fuzzy contours by using low frequency coefficients to reconstruct face images with and without eyeglasses. This proved the correctness of our idea.

3.2   The Procedure of the Proposed Method

Firstly, we use the ULGBPHS method to extract local facial features. We use Gabor filters of different scales and orientations to perform convolution with the facial blocks. The convolution can be expressed by the following formulation:

$$G_{\mu,\nu}(x,y) = f(x,y) * \psi_{\mu,\nu} \tag{1}$$

where $f(x,y)$ represents the pixel value of segmented face image block f and operator $*$ denotes the convolution operation. We call the $G_{\mu,\nu}(x,y)$ GMP after the convolution operation of face image block and Gabor filters. μ and ν represent the scales and orientations respectively. We use these GMPs in the next step to calculate their LBP feature maps by:

$$\cdot \ LGBP_{\mu,\nu}(x,y) = \sum_{p=0}^{7} S(G_{\mu,\nu}(x_p,y_p) - G_{\mu,\nu}(x,y))2^p \tag{2}$$

where $LGBP_{\mu,\nu}(x,y)$ denotes each feature map after using LBP operator in each obtained GMP. (x,y), (x_p,y_p) represent each pixel in a GMP and its neighbor pixels in this GMP respectively. S represents the binary pattern operation and it can be represented by the following equation:

$$S(x,y) = \begin{cases} 1, & if \quad G_{\mu,\nu}(x,y) > G_{\mu,\nu}(x_p,y_p) \\ 0, & if \quad G_{\mu,\nu}(x,y) \leq G_{\mu,\nu}(x_p,y_p) \end{cases} \tag{3}$$

The histogram of a feature map block for a face image can be represented by histogram ranging in $[0, ..., L-1]$ in the following format:

$$h = \sum_{x,y} binary\left\{LGBP_{split}(x, y) = i, i = 0, 1, ..., L-1\right\} \tag{4}$$

where h denotes the histogram for a feature map of each GMP. i refers to the gray level in this feature map and $LGBP_{split}(x, y)$ refers to the Local Gabor Binary Pattern (LGBP) feature map for each corresponding segmented face image block. The calculating procedure can be expressed by:

$$binary(A) = \begin{cases} 1, & \text{if } A \text{ is true} \\ 0, & \text{if } A \text{ is false} \end{cases} \tag{5}$$

We assume that we divide a face into m blocks, therefore the $r - th$ block histogram is:

$$H_{\mu,\nu,r} = (h_{\mu,\nu,r,0}, ..., h_{\mu,\nu,r,L-1}) \tag{6}$$

In this paper, we set μ equals to 5 and ν equals to 8 respectively. Therefore, we can obtain the local facial features by the following equation:

$$H_{local} = (H_{0,0,0}, ..., H_{0,0,m-1}, H_{0,1,1}, ..., H_{0,1,m-1}, ..., H_{7,4,m-1}) \tag{7}$$

Secondly, we use 2D-DFT to extract global facial features. We can clearly find that the current algorithms only focus on extraction of the local facial feature. Usually we can recognize a person approximately in the first sight whether he wears eye-glasses or not. We just know the faces from the overall prospective. Therefore, we proposed our method by combining global and local facial features to recognize a person. For local facial features, we use ULGBP method. For global facial features, we use DFT method. This process of extracting global facial features can be shown in the following equation:

$$F(\mu', \nu') = \frac{1}{MN} \sum_{x=0}^{M-1} \sum_{y=0}^{N-1} g(x, y) e^{-j2\pi(\frac{\mu' x}{M} + \frac{\nu' y}{N})} \tag{8}$$

where g represents a $M \times N$ size face image, μ' and ν' refers to frequency variables. The output of the above formulation can be shown in the next equation:

$$F(\mu', \nu') = R(\mu', \nu') + jI(\mu', \nu') \tag{9}$$

$R(\mu', \nu')$ and $I(\mu', \nu')$ refers to the real part and imaginary part of $F(\mu', \nu')$. Via the DFT operation, each face can be converted into a real component and imaginary component in frequency domain. And we extract the low frequency part for global information of a face (including real part and imaginary part.). We use $H_{global} = (H_R, H_I)$ to extract global facial information. H_R and H_I are feature vectors which can be calculated by:

$$H_R = \sum_{x,y} binary\left\{g_R(x, y) = i\right\}, i = 0, 1, ..., L-1 \tag{10}$$

$$H_I = \sum_{x,y} binary\left\{g_I(x,y) = i\right\}, i = 0, 1, ..., L - 1 \tag{11}$$

$g_R(x,y)$ and $g_I(x,y)$ represent the real and imaginary magnitude pictures of a face image calculated by $R(\mu^{'},\nu^{'})$ and $I(\mu^{'},\nu^{'})$ respectively. Finally, we use a weighted fusion approach to combine the two different kinds of facial features. We use $D_{local_c}^{k}$ and $D_{global_c}^{k}$ represent the Euclidean distance between the testing sample and training samples for local and global facial features respectively, where c refers to face classes and k refers to the training number of the face images in each class. We assume $C = 1, 2, ...C$, $k = 1, 2, ...K$ and the final distance between the testing sample and the training samples can be calculated by the following equation:

$$D_c^k = \omega_{local} * D_{local}^k + \omega_{global} * D_{global}^k \tag{12}$$

where ω_{local} and ω_{global} refer to the distance weight coefficients and $\omega_{local} + \omega_{global} = 1$. D_c^k refers to score level fusion of the distances calculated by two different feature vectors. The final classification result can be represented by:

$$t = arg\, min_c D_c^k \tag{13}$$

And then the testing sample is assigned to the $t - th$ class, $t \in \{1, 2, ..., C\}$. The Fig. 1 shows the whole procedure of feature extraction.

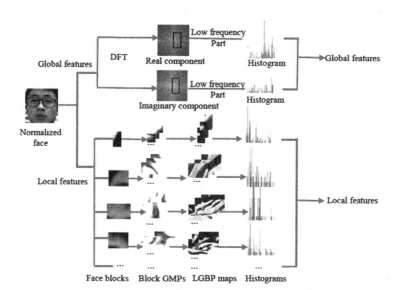

Fig. 1. Procedure of feature extraction using our proposed method

4   Experimental Evaluations

The datasets in our experiments come from two sources. Firstly, we select some typical face datasets and pick out the eyeglasses faces and non-eyeglasses faces from these original datasets. We choose the GT [12] and CMU_PIE [6] datasets for our experiments. In order to reduce the negative influence caused by some irrelevant factors, such as illumination, pose, facial expressions, we should keep these factors unchanged in the face images of each individual since they might affect the final recognition rate. We discard these unqualified face images of the original datasets before we do the experiments. After that we rename theses remaining images in order to put images with eyeglasses in the beginning half part while these without eyeglasses are set in the ending half part. Since the individuals in GT dataset do not always wear eyeglasses, we select some of individuals which contain samples with and without eyeglasses. The GT dataset we selected consists of a total 330 images of 22 individuals. Due to the different poses in the CMU_PIE dataset, we divide them into five sub-datasets, which named CMU_PIE_pose05, CMU_PIE_pose07, CMU_PIE_pose09, CMU_PIE_pose27 and CMU_PIE_pose29. The CMU_PIE_pose05 consists of a total 1,120 images of 28 individuals. And 540 face images of 27 individuals from the CMU_PIE_pose07 are used for the experiments. 25 peoples with 20 images per person from the CMU_PIE_pose09 are used for our experiments. The CMU_PIE_pose27 we use contains a total 1,120 images of 28 individuals. In the CMU_PIE_pose29, we use a total 540 images of 27 individuals. Secondly, we built a face dataset in real scene which is named BCC_Lab_Face. Our dataset consists of 50 peoples with 40 images per person. For each person, we collect four groups of face images. Two groups of face images are with eyeglasses while the other two are not. In the procedure of collection, we control the range of illumination, facial pose and facial expressions. Besides, we prepare several pairs of eyeglasses and each of them has different size and color. For a volunteer, if he wears eyeglasses, we collect the first group of face images and register another group of the face images by randomly selecting a pair of eyeglasses from our given eyeglasses. We also collect the other two groups of face images without eyeglasses which are controlled in the same outside scene. Figures 2, 3 and 4 are some samples from our experimental datasets. Actually, we carry out our experiments in two cases. In the first case, we take the first several face images with eyeglasses as training samples, and the rest of them are treated as testing samples. In the second case, we take the last several face images without eyeglasses as training samples whereas the rest of face images are set as test-ing samples. In this way, we can validate that the features extracted by using our proposed method is robust for the eyeglass-face problem. In other words, we use a kind of facial features which can resist eyeglasses problems in face recognition. The details in setting training and testing sets will be discussed in Table 1. According to the setting in Table 1, we conduct two kinds of comparative experiments based on two different experimental cases. We compare our proposed method with other two previous representative methods called LGBPHS and ULGBPHS to show the superiority of our method. Moreover, we also use single features involved in our method as facial

Fig. 2. One face sample from CMU_PIE selected dataset

Fig. 3. Two face samples from GT selected dataset

Fig. 4. Two face samples from BCC_Lab_Face

features to make comparison with our proposed method. Here, we use DFT, Gabor and LBP features as the single facial feature in comparative experiments. Figure 5 shows our experimental results on different datasets based on two different cases respectively. The different ID numbers in X axis show different datasets in our experiments and they represent CMU_PIE_pose05, CMU_PIE_pose07, CMU_PIE_pose09, CMU_PIE_pose27, CMU_PIE_pose29, GT, BCC_Lab_Face correspondingly. The Y axis shows different recognition rates of different methods. Since we mainly focus on feature extraction, we use the Euclidean distance as a common measure of standards in final recognition stage for these different methods. In this way, we can compare these different methods with our proposed methods. Figure 5 indicates that our method is the best one among these different methods in both two different testing cases. The light blue line with diamond shape shows the accuracy of our method on different datasets. We can clearly find that our method has a higher recognition rate on different datasets and it also shows stronger stability on different testing datasets.

Table 1. Training and testing sets division.

Experimental Cases	Case1: First several faces with eyeglasses as training sets		Case2: Last several face images without eyeglasses as training sets	
Dataset name	Training sets	Testing sets	Training sets	Testing sets
CMU_PIE_pose05	10	30	10	30
CMU_PIE_pose07	10	10	10	30
CMU_PIE_pose09	10	10	10	10
CMU_PIE_pose27	10	30	10	30
CMU_PIE_pose29	10	10	10	10
GT	7	8	7	8
BCC_Lab_Face	10	30	10	30

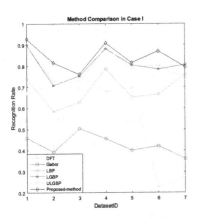 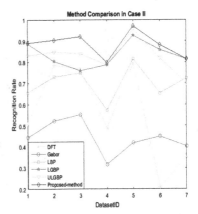

(a) Experimental results in case I   (b) Experimental results in case II

Fig. 5. Experimental results comparison in two cases

5   Conclusion and Discussion

This paper proposed a novel method to solve the challenge problem of recognition of eyeglass-faces. Compared with other methods, the proposed method simultaneously takes into account the local and global facial features. In other words, the proposed method can extract more discriminant features from the face, which enable it to obtain the best performance. Experimental results on the GT, CMU_PIE, and our collected BCC_Lab_Face datasets powerfully prove the effectives of the proposed method.

In the future, we will try to employ better classification strategies in recognition stage to improve the overall performance of the proposed method in eyeglass-face recognition problem.

Acknowledgments. This paper is partially supported by Guangdong Province high-level personnel of special support program (No. 2016TX03X164) and Shenzhen Fundamental Research fund (JCYJ20160331185006518).

References

1. Ahonen, T., Hadid, A., Pietikäinen, M.: Face recognition with local binary patterns. In: Pajdla, T., Matas, J. (eds.) ECCV 2004. LNCS, vol. 3021, pp. 469–481. Springer, Heidelberg (2004). https://doi.org/10.1007/978-3-540-24670-1_36
2. Arya, S., Pratap, N., Bhatia, K.: Future of face recognition: a review. Procedia Comput. Sci. **58**(2), 578–585 (2015)
3. Coetzee, L., Botha, E.C.: Fingerprint recognition in low quality images. Pattern Recogn. **26**(10), 1441–1460 (1993)
4. Deng, Y., Guo, Z., Chen, Y.: Fusing local patterns of gabor and non-subsampled contourlet transform for face recognition. In: IAPR Asian Conference on Pattern Recognition, pp. 481–485 (2013)
5. Givens, G., Beveridge, J.R., Draper, B.A., Grother, P., Phillips, P.J.: How features of the human face affect recognition: a statistical comparison of three face recognition algorithms. In: IEEE Computer Society Conference on Computer Vision and Pattern Recognition, pp. 381–389 (2004)
6. Gross, R., Matthews, I., Cohn, J., Kanade, T., Baker, S.: Multi-pie. Image Vis. Comput. **28**(5), 807–813 (2010)
7. Guo, K., Wu, S., Xu, Y.: Face recognition using both visible light image and near-infrared image and a deep network. CAAI Trans. Intell. Technol. **2**(1), 39–47 (2017)
8. Heo, J., Kong, S.G., Abidi, B.R., Abidi, M.A.: Fusion of visual and thermal signatures with eyeglass removal for robust face recognition. In: Conference on Computer Vision and Pattern Recognition Workshop, CVPRW 2004, p. 122 (2004)
9. Kong, A., Zhang, D., Kamel, M.: A survey of palmprint recognition. Pattern Recogn. **42**(7), 1408–1418 (2009)
10. Liu, L., Sun, Y., Yin, B., Song, C.: A novel nonuniform division strategy for wearing eyeglasses face recognition. In: Fifth International Conference on Image and Graphics, pp. 907–911 (2009)
11. Martínez, A.M.: Recognizing imprecisely localized, partially occluded, and expression variant faces from a single sample per class. IEEE Trans. Pattern Anal. Mach. Intell. **24**(6), 748–763 (2002)
12. Nefian, A.: Georgia tech face database (2013), http://www.anefian.com/research/face_reco.html
13. Ojala, T., Harwood, I.: A comparative study of texture measures with classification based on feature distributions. Pattern Recogn. **29**(1), 51–59 (1996)
14. Ojala, T., Pietikainen, M., Maenpaa, T.: Multiresolution gray-scale and rotation invariant texture classification with local binary patterns. IEEE Trans. Pattern Anal. Mach. Intell. **24**(7), 971–987 (2002)
15. Solanki, K., Pittalia, P.: Review of face recognition techniques. Int. J. Comput. Appl. **133**, 20–24 (2016)
16. Xu, Y., Fang, X., Li, X., Yang, J., You, J., Liu, H., Teng, S.: Data uncertainty in face recognition. IEEE Trans. Cybern. **44**(10), 1950–1961 (2014)
17. Xu, Y., Lu, Y.: Adaptive weighted fusion: a novel fusion approach for image classification. Neurocomputing **168**, 566–574 (2015)

18. Yang, P., Shan, S., Gao, W., Li, S.Z., Zhang, D.: Face recognition using Ada-boosted gabor features. In: Proceedings of IEEE International Conference on Automatic Face and Gesture Recognition, pp. 356–361 (2004)

19. Yi, D., Li, S.Z.: Learning sparse feature for eyeglasses problem in face recognition. In: 2011 IEEE International Conference on Automatic Face & Gesture Recognition and Workshops (FG 2011), pp. 430–435. IEEE (2011)

20. Yu, S.U., Shan, S.G., Chen, X.L., Wen, G.: Integration of global and local feature for face recognition. J. Softw. **21**(8), 1849–1862 (2010)

21. Zanchettin, C.: Face recognition based on global and local features. In: Proceedings of the 29th Annual ACM Symposium on Applied Computing, pp. 55–57. ACM (2014)

22. Zhang, W., Shan, S., Chen, X., Gao, W.: Local gabor binary patterns based on kullback-leibler divergence for partially occluded face recognition. IEEE Signal Process. Lett. **14**(11), 875–878 (2007)

23. Zhang, W., Shan, S., Gao, W., Chen, X., Zhang, H.: Local gabor binary pattern histogram sequence (LGBPHS): a novel non-statistical model for face representation and recognition. In: Tenth IEEE International Conference on Computer Vision, pp. 786–791 Vol. 1 (2005)

24. Zhao, W., Chellappa, R., Phillips, P.J., Rosenfeld, A.: Face recognition: a literature survey. ACM Comput. Surv. **35**(4), 399–458 (2003), http://doi.acm.org/10.1145/954339.954342

Face Recognition by Coarse-to-Fine Landmark Regression with Application to ATM Surveillance

Ya Li[1], Lingbo Liu[2], Liang Lin[2], and Qing Wang[2(✉)]

[1] Guangzhou University, Guangzhou 510006, China
liya@gzhu.edu.cn
[2] Sun Yat-sen University, Guangzhou 510006, China
liulingb@mail2.sysu.edu.cn, linliang@ieee.org, wangq79@mail.sysu.edu.cn

Abstract. While ATM provides us convenient banking services, it has great security risks. The authentication of only password requiring is not safe enough. With the rapid development of face recognition technology based on deep convolutional neural network (CNN), undoubtedly, applying it into ATM authentication will improve security further. In this paper, we explore a new authentication mode combine face recognition and basic password for ATM. We think that it would prevent the economic crime on ATM fundamentally. However, computational and storage costs of CNN based methods are still high. To this end, we propose a new face recognition method by landmark regression. Our pipeline integrates a landmark localization network with a light face recognition network. For landmark localization, we employ a fully convolutional neural network to produce facial landmark response maps directly from raw images in a coarse-to-fine manner. For face recognition, we train a light CNN to obtain a compact representation, where the rectified linear unit (ReLU) is replaced by max-feature-map (MFM). Our approach shows good performance on several datasets. And it is practicable due to its high speed, good accuracy, and low storage space requirement.

Keywords: Face recognition · Landmark localization
ATM surveillance · Deep CNN · Face verification

1 Introduction

ATM provides us many convenient services about banking, because most ATMs are open 24 h and their locations are spread all over a city. But at the same time, ATMs are one of the most vulnerable sites without any manual security. The criminal behaviours such as physical attack, damage on ATM, using a duplicated or stolen bank card to withdrawal or transfer money often occur. Many researches have been done to improve the security of ATM. Some works focus on suspicious wearing detection. Ray et al. [12] detected mask wearing by Viola-Jones algorithm, while Wen et al. [23] detected safety helmets wearing

© Springer Nature Singapore Pte Ltd. 2017
J. Yang et al. (Eds.): CCCV 2017, Part II, CCIS 772, pp. 62–73, 2017.
https://doi.org/10.1007/978-981-10-7302-1_6

using modified Hough Transform. Some works focused on abnormal behaviour detection. Tang et al. [22] detected peeping behavior based on Omni-Directional Vision sensor and computer vision technology. Arsic et al. [1] applied Bayes Markov chains to resolve heavy occlusions and detect robberies.

Most researchers paid their attention to the surrounding environment monitoring of ATM kiosks, and ignored the potential safety issue coming from ATM authentication mode itself. Currently, the authentication of only password requiring is not safe enough. With the rapid development of face recognition technology based on deep convolutional neural network (CNN), undoubtedly, applying it into ATM authentication will improve security further. Other biometric information like fingerprint and iris are often used for security, but in ATM system they are not very appropriate. Because we can grant privileges to other persons by upload their facial images if using face recognition technology, while the authorization is difficult to implement if using fingerprint or iris. In daily life, it would be inconvenient if only cardholders themselves are permitted to handle business, and authorization is a good resolution to guarantee both security and flexibility.

In this paper, we explore a new authentication mode combine face recognition and basic password for ATM. We think that it would prevent the economic crime on ATM fundamentally. However, real-time response, high accuracy, and low storage space requirement make this application very challenging. On one hand, there are two separated steps including face detection and alignment before recognition generally in face recognition pipeline and result in low efficiency. We simplify the general processing and present a novel landmark localization method to produce facial landmark from raw images without face detection preprocessing. On the other hand, computational and storage costs of CNN based methods are still high. To this end, we propose a new face recognition method by landmark regression. Our pipeline integrates a coarse-to-fine landmark localization network and a light face recognition network. In network architecture, we use fully convolutional network (FCN) in landmark localization and network-in-network with small convolutional kernel size in face recognition for saving storage space. Besides, considering the speed of process we obtain landmark locations in a coarse-to-fine manner. And the nine layers CNN of face recognition ensure the accuracy further.

For landmark localization, there are no fully connected layers in our network and only convolutional layers are utilized. In particular, by taking the whole image as input, the coarse locations of facial landmarks are roughly detected in global context. Then they are further refined by local regions, which taking the cropped patches containing coarse landmarks as input and produce a fine and accurate prediction. For face recognition, we train a light (CNN) where the rectified linear unit (ReLU) is replaced by max-feature-map (MFM) like paper [24]. The CNN with MFM can obtain a compact representation and shows good performance in terms of both computational cost and storage space [24].

The main contributions of this work are summarized as follows. (1) We propose a new face recognition method by coarse-to-fine landmark regression, which produces facial landmark response maps directly from raw images without relying on the result of face detection or other preprocessing done in advance. (2) Our approach is practicable due to its good accuracy and high efficiency.

The reminder of this paper is organized as follows. In Sect. 2, we briefly review some related works on landmark localization and face recognition. Then in Sect. 3, we introduce our system in detail. Finally, we present our experimental results in Sect. 4 and conclude this paper in Sect. 5.

2   Related Work

As well known, deep learning methods, especially the deep CNNs and deep auto-encoders, have made dramatic progress on many computer vision tasks including face landmark localization and face recognition. Therefore, in this section, we mainly review the works using deep learning.

2.1   Face Landmark Localization

Luo et al. [11] proposed a hierarchical face parser by combined deep belief network (DBN) with deep auto encoder (DAE). Sun et al. [16] proposed three-level cascaded CNNs to detect facial landmarks gradually, in which the first CNN produced the initial predictions and the following two CNNs refine the results. Specifically, the adjustment from third level is more subtle than the second level. Zhang et al. [30] used correlated tasks, e.g. facial expression recognition and head pose estimation to optimize facial landmark detection by deep CNN. In paper [28] Zhang Jie et al. introduced a new stacked DAEs pipeline to progressively refine the facial landmark locations by taking gradually higher resolution image version as input. Zhang Cha et al. [27] built a deep CNN to learn the face/non-face decision, the face pose estimation, and the facial landmark localization simultaneously. Lai et al. [9] proposed an end-to-end CNN architecture to learn highly discriminative shape-indexed-feature for face alignment. Recently, a deep cascaded multi-task CNN framework [29] was proposed to exploit the inherent correlation between detection and alignment. Existing methods mentioned above either require a facial bounding box or need to extract proposals by inefficient sliding window, so their performance are not good enough for applying in unconstrained settings.

2.2   Face Recognition

Currently, deep learning method has become the most common method to be used for face recognition. Earlier work proposed in paper [17] used CNN for face verification, which was composed of hybrid CNNs lower part for feature learning and restricted Boltzmann machine (RBM) top layer for classification. Lower part CNNs took twelve different face regions as input and RBM merged the twelve group outputs to give the final prediction. In 2014, DeepFace [21] and

DeepID [18] appeared almost at the same time. DeepFace employed 3D alignment for preprocessing and achieved 97.35% accuracy on LFW dataset [6]. While DeepID aligned face image based on landmarks and achieved 97.47% on LFW. Later, the improved model of DeepID named DeepID2 [15] and DeepID2+ [19] improved the verification accuracy on LFW to 99.15% and 99.47% respectively. For DeepID, the outputs of different CNNs trained on different local patches were assembled as deep feature and Joint Bayesian was applied for face verification. For DeepID2, verification loss and classification loss were further combined to increase accuracy. Compared with DeepID2, DeepID2+ was improved by increasing the dimension of patch feature and adding supervision to early convolutional layers. Recently, a triplet-based CNN model named FaceNet [13] achieved the best result 99.63% on LFW. Although these CNN based methods have achieved a practicable accuracy on LFW dataset, however, their computational costs are still high due to their deep or multi-network architectures. Hence, it is inevitable to design a light model with a smaller number of parameters for lower computation and storage cost.

3   Our Approach

3.1   Pipeline

Our model consists of two main cascaded parts: face landmark localization and face recognition, both of them are realized by using CNN. We train the whole model by stochastic gradient descent (SGD) and use the Caffe library [7]. The detailed pipeline is illustrated in Fig. 1. To begin with, the image is acquired at the moment of the card insertion. And the following operations are preprocessing, landmark localization, face alignment and face recognition. If current image is matched with card-holder or one of the authorized user's images, the process is continued. Otherwise, while asking him whether take photo again we can implicitly recognize the current user's identification to judge whether the user is a suspect, if yes, send alarm signal to monitoring center by background program. Specifically, we perform standard histogram equalization preprocessing to adjust the image contrast. And face alignment is done by horizontal rotation based on landmarks. In the next two subsections, we introduce face landmark localization and face recognition respectively.

3.2   Coarse-to-Fine Landmark Localization

Landmark locations are obtained in a coarse-to-fine way by FCNs, where convolution is used like a filter. Suppose $L_i^k = (x_i^k, y_i^k)$ is the location of i-th landmark of type k in image I, where (x_i^k, y_i^k) represents the landmark coordinates. The filtering is performed on patches. We denote $F_{\mathbf{W}^k}(P)$ as the filtering function of type k on patch P with parameters $\mathbf{W}^k$. Suppose the patch size is $w \times h$, and the filtering function is in sliding window manner with stride δ. The response map $F_{\mathbf{W}^k} * I$ whose value at location (x, y) can be computed by:

$$(F_{\mathbf{W}^k} * I)(x, y) = F_{\mathbf{W}^k}(I(x\delta : x\delta + w, y\delta : y\delta + h)), \tag{1}$$

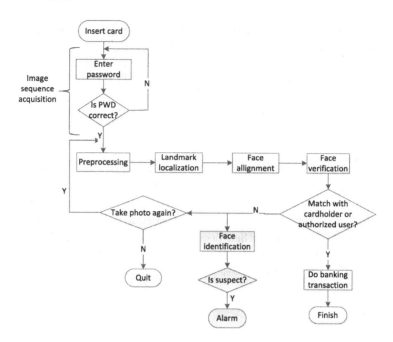

Fig. 1. The proposed pipeline of face recognition. The image is acquired at the moment of the card insertion. If current image is matched with card-holder or one of the authorized user's images, the process is continued. Otherwise, while asking him whether take photo again we can implicitly recognize the current user's identification to judge whether the user is a suspect, if yes, send alarm signal to monitoring center by background program.

where $I(x\delta : x\delta + w, y\delta : y\delta + h)$ is the image patch. At first, we take the whole image as input and obtain the response map. We wish that the response map should distinguish whether the patches containing landmarks or not. Next we only input the patches containing coarse landmarks for refining by taking the same filtering process. Thus, we can construct a multi-level networks pyramid by decreasing the path size for filtering gradually.

Our destination is learning the filtering functions which have the property: patches containing the target landmarks should have strong response, otherwise should have weak response. Suppose the threshold is θ and the landmark location is the patch center, which can be computed by:

$$Det(I) = \{(x\delta + w/2, y\delta + h/2)|(F_{\mathbf{W}^k} * I)(x,y) > \theta\}. \tag{2}$$

Let $H^k(I; \mathbf{W}_p)$ and $H_0^k(I)$ denote the predicted and ground truth response map of image I for landmark type k, where $\mathbf{W}_p$ denote the parameters of the pth level network pyramid. The value of $H^k(I; \mathbf{W}_p)$ at position (x,y) can be computed

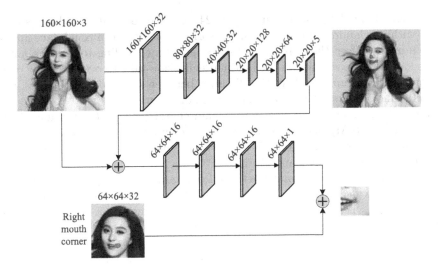

Fig. 2. The network architecture of our coarse-to-fine landmark localization. The first level network is shared by all of the landmarks. The second level sub-networks are unique to each landmark. We only illustrate "right-mouth-corner" sub-network for simplification.

by Eq.(1). The loss function used to train the network is:

$$\mathcal{L}_p(I; \mathbf{W}_p) = \sum_{k=1}^{K} ||H^k(I; \mathbf{W}_p) - H_0^k(I)||^2. \tag{3}$$

We evaluate our approach by stacking two level networks on five landmarks localization, including two eyes' centres, left mouth corners, right mouth corners and nose tip. The first level network is shared by all of the landmarks, and the second level sub-networks are unique to each landmark. The network architecture is showed in Fig. 2, where only "right-mouth-corner" sub-network is illustrated for simplification. In summary, the kernel sizes of the first level convolutional layers are 5×5, 5×5, 5×5, 9×9, 1×1, 1×1; and for the second level they are 5×5, 7×7, 9×9, 1×1 respectively.

4   Face Recognition

As well known, CNN-based face recognition systems need large amounts of memory and computational power. Although they perform well on GPU-based machines, it is often a challenge to run them on target low-power devices like ATMs due to overtaxing the compute capabilities and the limited storage. Hence, it needs us to seek a light model for the deployment on ATMs. Much research work is done to speed-up and compact CNNs. Here we utilize an off-the-shelf named light B model proposed in paper [24] for face recognition, which is cascaded with our landmark localization network. In particular, a rough face alignment is made before they are input for recognition according to the five landmarks location. We rotate two eye landmarks horizon-tally to overcome the pose variations in roll range and fix the distance between the midpoint of eyes and the midpoint of mouth for facial image normalization.

The light B model we used has two main characteristics: (1) defining Max-Feature-Map (MFM) activation function for compact representation and (2) employing net-work in network (NIN) [10] concept between convolution layers for improvements on speed and storage.

MFM is defined to obtain competitive feature maps, it outputs element-wise maximum of two convolutional feature maps. Given an input convolution layer $C \in \mathbb{R}^{h \times w \times 2n}$, where $2n$ is the channel number of convolution layer and $h \times w$ denotes the map size. The MFM can be formulated as:

$$f_{ij}^k = \max_{1 \leq k \leq n} (C_{ij}^k, C_{ij}^{k+n}), \tag{4}$$

where $1 \leq i \leq h$ and $1 \leq j \leq w$.

The light B contains 5 convolution layers, 4 NIN network layers, MFM activation function, 4 max-pooling layers and 2 fully connected layers. The small convolution kernel size with NIN can reduce the number of parameters for the model; it makes the model 20 times smaller than VGG [14] while 9 times faster on CPU time. Please refer to [24] for more academic details.

5   Experiments

In this section, we evaluate our landmark localization approach first and then show the performance of face recognition based on landmark locations using light model.

5.1   Landmark Localization

For model training, we collect 7,317 face images and 1,671 natural images from Internet. Among them 6,317 face images and 1,218 natural images for training, the rest of 1,000 face images and 453 natural images for validation. Each face is annotated with five landmarks. We employ AFLW [8] and AFW [31] for evaluation. For AFLW, we selected 3,000 faces randomly from AFLW for testing like TCDCN [30].

We report our result on mean error, which is measured by the distances between estimated landmarks and the ground truths, normalized with respect to inter-ocular distance. It is defined as:

$$err = \frac{\sqrt{(x - x')^2 + (y - y')^2}}{l}, \tag{5}$$

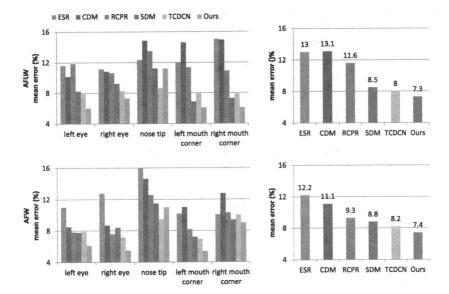

Fig. 3. Comparison with other methods on AFLW and AFW. The comparison on five landmarks is illustrated in the left sub-image, and the average mean errors of these methods are summarized in the right sub-image.

where (x, y) and (x', y') are the ground truth and predicted locations, respectively, and l is the inter-ocular distance.

We compare our method against: (1) Explicit Shape Regression (ESR) [3]; (2) A Cascaded Deformable Model (CDM) [26]; (3) Robust Cascaded Pose Regression (RCPR) [2]; (4) Supervised Descent Method (SDM) [25]; (5) Task-Constrained Deep Convolutional Network (TCDCN) [30]. The results are reported in Fig. 3. On the AFLW dataset, our method achieves 7.3% for average mean error, 8.75% improvement over TCDCN. On AFW, our average mean error is 7.4% over five parts, 9.76% improvement over TCDCN.

5.2 Face Recognition Based on Landmark Locations

We evaluate the cascaded model which integrates our landmark localization with the light B for face verification task on LFW and CACD-VS [4] datasets. LFW contains 13,233 images of 5,749 identities. For face verification, face images are divided in 10 folds and each folder contains 600 face pairs of different identities. CACD-VS dataset contains 2,000 positive pairs and 2,000 negative pairs which are collection of celebrity images on Internet.

Face verification is one-to-one matching, which identifies whether two facial images are from the same person. We compare our method against WebFace [20], VGG [14], FaceNet [13] and Light B [24] on LFW and against HFA [5], CARC [4], VGG and Light B on CACD-VS. The results are reported in Table 1.

Table 1. The accuracies of different methods on LFW and CACD-VS datasets.

LFW		CACD-VS	
Method	Accuracy	Method	Accuracy
WebFace	96.13%	HFA	84.4%
VGG	97.27%	CARC	87.6%
FaceNet	**99.63%**	VGG	96%
Light B	98.13%	Light B	97.95%
Ours	98.5%	Ours	**98.2%**

Our cascaded model shows very competitive results compared to other state-of-the-art methods on both datasets.

We also show some verification results of our approach on LFW dataset in Fig. 4. The first line illustrates some errors, where the first three pairs are errors on mismatched pairs, that is, the pairs of mismatched images that were incorrectly reported as matched pairs. And the last three pairs are errors on matched pairs, that is, the pairs of matched images that were incorrectly reported as mismatched pairs. And the second line demonstrates the robustness of our approach, where the hard pairs are verified correctly despite of variations on pose, expression, occlusion, and illumination. It is worth mentioning that our model is practicable to apply in ATM system due to the good performance on speed and storage space. The entire model is about 33 MB and it requires about 140 ms for face verification on low power PC with Xeon(R) E5-2620 v3 @ 2.40GHz CPU.

Fig. 4. Illustration of verification results of our method on LFW dataset. The first line illustrates some errors, where the first three pairs are errors on mismatched pairs, the last three pairs are errors on matched pairs. And the second line demonstrates the robustness of our approach, where the hard pairs are verified correctly despite of variations on pose, expression, occlusion, and illumination.

6   Conclusion

In this paper, we explore a new authentication mode combine face recognition and basic password for ATM. We propose a cascaded deep model which integrates a landmark localization network with a light face recognition network.

Our model shows good performance on several datasets. It is practicable due to its high speed, good accuracy, and low storage space requirement. The entire model is about 33 MB and it requires about 140 ms for face verification on low a power PC with Xeon(R) E5-2620 v3 @ 2.40 GHz CPU. The face recognition network is the bottleneck on speed and storage. The time of landmark localization on one image is only 7 ms, but for face verification is 133 ms. The landmark localization network needs less than 1 M storage space while recognition network needs 32 MB. Therefore, we will continue to optimize the recognition network further. In future, we also generalize our approach in more challenging object recognition, such as human action understanding.

Acknowledgments. This research is supported by the Research Project of Guangzhou Municipal Universities (No. 1201620302), National Undergraduate Scientific and Technological Innovation Project (No. 201711078017), the Science and Technology Planning Project of Guangdong Province (No. 2015B010128009, 2013B010406005). The authors would like to thank the reviewers for their comments and suggestions.

References

1. Arsic, D., Lyutskanov, A., Kaiser, M., Schuller, B., Rigoll, G.: Applying Bayes Markov chains for the detection of ATM related scenarios. In: Applications of Computer Vision, pp. 1–8 (2010)
2. Burgosartizzu, X.P., Perona, P., Dollar, P.: Robust face landmark estimation under occlusion. In: International Conference on Computer Vision. pp. 1513–1520. IEEE (2013)
3. Cao, X., Wei, Y., Wen, F., Sun, J.: Face alignment by explicit shape regression. Int. J. Comput. Vision **107**(2), 177–190 (2014)
4. Chen, B.C., Chen, C.S., Hsu, W.: Face recognition and retrieval using cross-age reference coding with cross-age celebrity dataset. IEEE Trans. Multimedia **17**(6), 804–815 (2015)
5. Gong, D., Li, Z., Lin, D., Liu, J., Tang, X.: Hidden factor analysis for age invariant face recognition. In: International Conference on Computer Vision, pp. 2872–2879. IEEE (2013)
6. Huang, G.B., Mattar, M., Berg, T., Learned-Miller, E.: Labeled faces in the wild: a database for studying face recognition in unconstrained environments. Technical report (2008)
7. Jia, Y., Shelhamer, E., Donahue, J., Karayev, S., Long, J., Girshick, R., Guadarrama, S., Darrell, T.: Caffe: convolutional architecture for fast feature embedding. In: ACM International Conference on Multimedia, pp. 675–678. ACM (2014)
8. Köstinger, M., Wohlhart, P., Roth, P.M., Bischof, H.: Annotated facial landmarks in the wild: a large-scale, real-world database for facial landmark localization. In: International Conference on Computer Vision Workshops, pp. 2144–2151. IEEE (2011)
9. Lai, H., Xiao, S., Cui, Z., Pan, Y., Xu, C., Yan, S.: Deep cascaded regression for face alignment. arXiv preprint arXiv:1510.09083 (2015)
10. Lin, M., Chen, Q., Yan, S.: Network in network. arXiv preprint arXiv:1502.03167 (2013)

11. Luo, P.: Hierarchical face parsing via deep learning. In: Computer Vision and Pattern Recognition, pp. 2480–2487. IEEE (2012)
12. Ray, S., Das, S., Sen, A.: An intelligent vision system for monitoring security and surveillance of ATM. In: IEEE India Conference, pp. 1–5. IEEE (2015)
13. Schroff, F., Kalenichenko, D., Philbin, J.: Facenet: a unified embedding for face recognition and clustering. In: Computer Vision and Pattern Recognition, pp. 815–823. IEEE (2015)
14. Simonyan, K., Zisserman, A.: Very deep convolutional networks for large-scale image recognition. arXiv preprint arXiv:1412.5903 (2014)
15. Sun, Y., Chen, Y., Wang, X., Tang, X.: Deep learning face representation by joint identification-verification. In: Advances in Neural Information Processing Systems, pp. 1988–1996 (2014)
16. Sun, Y., Wang, X., Tang, X.: Deep convolutional network cascade for facial point detection. In: Computer Vision and Pattern Recognition, pp. 3476–3483. IEEE (2013)
17. Sun, Y., Wang, X., Tang, X.: Hybrid deep learning for face verification. In: International Conference on Computer Vision, pp. 1489–1496. IEEE (2013)
18. Sun, Y., Wang, X., Tang, X.: Deep learning face representation from predicting 10,000 classes. In: Computer Vision and Pattern Recognition, pp. 1891–1898. IEEE (2014)
19. Sun, Y., Wang, X., Tang, X.: Deeply learned face representations are sparse, selective, and robust. In: Computer Vision and Pattern Recognition, pp. 2892–2900. IEEE (2015)
20. Taigman, Y., Yang, M., Ranzato, M., Wolf, L.: Web-scale training for face identification. In: Computer Vision and Pattern Recognition, pp. 2746–2754. IEEE (2015)
21. Taigman, Y., Yang, M., Ranzato, M., Wolf, L.: Deepface: closing the gap to human-level performance in face verification. In: Computer Vision and Pattern Recognition, pp. 1701–1708. IEEE (2014)
22. Tang, Y., He, Z., Chen, Y., Wu, J.: ATM intelligent surveillance based on omni-directional vision. In: Computer Science and Information Engineering, pp. 660–664. IEEE (2009)
23. Wen, C.Y., Chiu, S.H., Liaw, J.J., Lu, C.P.: The safety helmet detection for ATM's surveillance system via the modified hough transform. In: IEEE International Carnahan Conference on Security Technology, pp. 364–369. IEEE (2003)
24. Wu, X., He, R., Sun, Z., Tan, T.: A light CNN for deep face representation with noisy labels. arXiv preprint arXiv:1511.02683 (2015)
25. Xiong, X., De la Torre, F.: Supervised descent method and its applications to face alignment. In: Computer Vision and Pattern Recognition, pp. 532–539. IEEE (2013)
26. Yu, X., Huang, J., Zhang, S., Yan, W., Metaxas, D.N.: Pose-free facial landmark fitting via optimized part mixtures and cascaded deformable shape model. In: International Conference on Computer Vision, pp. 1944–1951. IEEE (2014)
27. Zhang, C., Zhang, Z.: Improving multiview face detection with multi-task deep convolutional neural networks. In: Winter Conference on Applications of Computer Vision, pp. 1036–1041. IEEE (2014)
28. Zhang, J., Shan, S., Kan, M., Chen, X.: Coarse-to-fine auto-encoder networks (CFAN) for real-time face alignment. In: Fleet, D., Pajdla, T., Schiele, B., Tuytelaars, T. (eds.) ECCV 2014. LNCS, vol. 8690, pp. 1–16. Springer, Cham (2014). https://doi.org/10.1007/978-3-319-10605-2_1

29. Zhang, K., Zhang, Z., Li, Z., Qiao, Y.: Joint face detection and alignment using multitask cascaded convolutional networks. IEEE Signal Process. Lett. **23**(10), 1499–1503 (2016)
30. Zhang, Z., Luo, P., Loy, C.C., Tang, X.: Facial landmark detection by deep multitask learning. In: Fleet, D., Pajdla, T., Schiele, B., Tuytelaars, T. (eds.) ECCV 2014. LNCS, vol. 8694, pp. 94–108. Springer, Cham (2014). https://doi.org/10. 1007/978-3-319-10599-4_7
31. Zhu, X., Ramanan, D.: Face detection, pose estimation, and landmark localization in the wild. In: Computer Vision and Pattern Recognition, pp. 2879–2886. IEEE (2012)

An Efficient System for Partial Occluded Face Recognition

Haoxiang Zhang[1(✉)], Peng An[1], and Dexin Zhang[2]

[1] School of Electronic and Information Engineering,
Ningbo University of Technology, 201 Fenghua Road, Ningbo 315211, Zhejiang, China
sean_public@qq.com
[2] Tianjin ISecure Technologies Co. Ltd., TEDA MSD-G1 10F, TEDA Tianjin
Economic-Technological Development Area, Tianjin, China
http://www.nbut.edu.cn

Abstract. Face recognition has attracted a lot of interest and made a wide range of applications in the real world. However the technology's efficiency is limited to well positioned, clear face images, which is not always the case in reality. In this paper, we present a software system that is resilient to various image quality degradation, such as face occlusions, illumination effects and face postures. The system exploits deep learning to build an effective model for face detection, and an Elastic Graph Matching based method to extract the key face features for comparison with the face library. The system also provides flexible functionality and extensions to adopt other biometric recognition algorithm for cross-checking and provides a higher reliability. Experiments show that the proposed system offers a good performance and is robust to face occlusions.

Keywords: Biometrics · Face recognition · Partial occlusion
Deep learning · FPCA · EGM

1 Introduction

Computer vision and image processing advances in theories and techniques boosted development of image based applications. Benefiting from it, biometric recognition technologies have been developing quickly in recent years, and face recognition has seen arguably the most wide range of applications among all image analysis and understanding technologies [1,2]. Although these machine recognition systems have brought more and more convenience and efficiency to our every day life, they still suffer from their limitations in many real application scenarios. For example, images acquired via surveillance cameras, which are very pervasive today, are often taken from an outdoor environment with variations of

H. Zhang—This work was partially supported by Chinese National Natural Science Fund 61502256.

J. Yang et al. (Eds.): CCCV 2017, Part II, CCIS 772, pp. 74–85, 2017.
https://doi.org/10.1007/978-981-10-7302-1_7

illumination conditions; cameras installed at public passages, or any uncontrolled scenarios. Such images could record faces of different poses, and faces could be occluded by glasses, hat, scarf, or even just by other people. Performing face recognition on such image signals remains a challenging problem. Under such circumstances, certain part of the face information is occluded, making it hard to extract all the important features for recognition, and resulting in degraded recognition accuracy. A lot of work has been carried out to tackle these problems [3–8]. These works are mostly theoretical research tested with experiments in labs, but rarely adopted for practical applications. In this paper, we present a face recognition software system ready for practical usage and applications, which provides robust recognition accuracy with occluded face images, while maintaining a high flexibility for users to set parameter for retrieving purpose, so that it is possible to search for certain interested individuals among the massive amount of recorded videos. It can also be easily extended to a system that combines more than one biometric recognition algorithms where cross-checking helps to satisfy a higher security requirement.

2   The Challenges for a Robust Face Recognition System

Face recognition technology is based on human facial features. A typical face recognition system works in the following procedure: 1. The system is given an input of face image, or video stream. It first analyzes the content to determine whether there exists a human face in the scene. 2. If there is a face/s, it then further locates each face, detects the size and positions of the major facial organs of each face. 3. Based on the above information, the system then further extracts each individual's unique identity characteristics. 4. The system compares with the known faces, normally from a database, to verify each person's face identity. Above is a straightforward and sensible process. But in many cases, the collected face images are not standard front face images [4,5]. And it is quite often the case that a collected face image is partly occluded by glasses, light beard, or simply distorted by reflection and other lighting effects. Such image quality would heavily degrade the system's accuracy of face recognition performance. Targeting at the above problems, the proposed system exploits deep learning algorithm to perform partial face modeling in the process and substantially improves the accuracy of face recognition. The proposed system can process the full clear recorded faces, as well as those occluded faces with glasses, beard and other obstructions, which helps to improve the searching accuracy and efficiency. In addition, the system can be used to provide an extra verification mechanism according to the needs of the security level. Cross-checking the identity using more than one biometric mechanism enables a higher level of security than the simple face recognition system [3]. The system can adapt to different security levels, with multi-mode biometric identification, such as Iris and Face recognition, to ensure an even higher security with a very brief processing time.

3   System Structure and the Sub-modules of the System

Figure 1 shows the structure of the proposed face-recognition system. As depicted in the figure, the system is organized as a series of layers. The lowest layer takes in the video or image signals, the layers in the middle level performs the face recognition function, and the highest application layer serves as an interface to support the actual security checking or individual searching job.

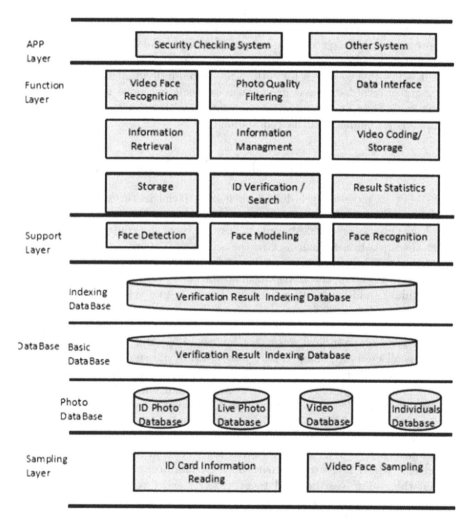

Fig. 1. System structure.

The sub-modules are described in details below in this section.

3.1 Face Detection and Locating

Here face detection refers to the process of judging whether there are human faces in an image with dynamic scene or a complex background. If there are faces, they would be then extracted. Face detection is the basic stage of face recognition, and only with the presence of a face can the recognition process be carried out. Most of the face detection techniques in the current market only works on clear and complete face images. The proposed system integrates a set of face features obtained from a large number of partial occluded face images using deep learning methods. It can process the incomplete faces and thus improve the performance under complex scenarios.

Factors Affecting Face Locating. Locating the faces in an image is not always straightforward. To perform face detection and locating, the following factors need to be considered: 1. the position, angle and the posture of the individual in the image; 2. The variable size of the face region in an image; 3. The illumination effects. In some situations the photo shooting is strictly under control and the image quality is usually guaranteed. For example, when the police takes photos of a detained suspect, the face is always aligned to the frame and locating the face in the image is simple. A typical ID photo also has very simple background color and standardized face presentation, therefore locating a face is also easy in this case. However, in many other real life cases, human faces could be recorded without regular positioning, such as video recorded by surveillance camera, or with highly dynamic background changes, such as video recorded by hand held mobile phone. Thus the faces' positions are unpredictable with these complex conditions.

Models for Face Detection. Contours and skin tones are important characters of a face. They are generally stable and enable a face to stand out of most background objects, hence, they can be used as good candidates for fast face detection in color images. The essential procedure of the feature detection method is, first, developing the skin color model, and use it to detect the pixels of skin color, then locate the possible face area according to the similarity and correlation in spatial and chromatic domains.

The picked possible face areas are then fed into the face model developed through machine learning method to go through further checking before the ultimate judgment of whether the detected area contains a face. In the process of building the database, we have included a large number of partial occluded face samples, including glasses, or beard occlusion, and the output model works effectively on such faces. Some occluded face detection results are shown in Fig. 2.

3.2 Face Alignment Module

Once the face detection is finished, the picture must go through some pre-processing stage, which is essential to keep the accuracy of the face recognition,

Fig. 2. Detection of occluded faces.

especially in a complex environment. This stage performs size and gray scale normalization, head posture correction, image segmentation etc. The purpose of performing these operations is to improve the image quality, such as reducing noise, achieving uniform image gray scale and size, which sets a good condition for feature extraction and classified identification in a later stage. Face alignment has two parts, geometric normalization and gray scale color normalization.

Geometric Normalization. Geometric normalization involves two steps: face correction and face cropping. The detected expression sub-image is transformed into a uniform size, which is conducive to the extraction of facial features.

Grayscale Normalization. The gray scale normalization is mainly to increase the contrast of the image and illumination compensation. It aims at increasing the brightness of the image, so that the details of the image are clearer, and the impact of light effects are reduced. Some resultant images of face alignment faces are shown below. Though the photos are originally taken at different postures and angles, it is clear that they are now of uniform size and presentation, and the eyes, nose mouth etc. are at similar positions of the picture, which makes it easier to extract face features from the faces. Such visually similar images are ready for further processing and comparison (Fig. 3).

3.3   Partial Occluded Face Modeling

Partial occlusion of the face often results in loss of important information of the face, thus affecting the face recognition accuracy. The proposed Partial Occluded

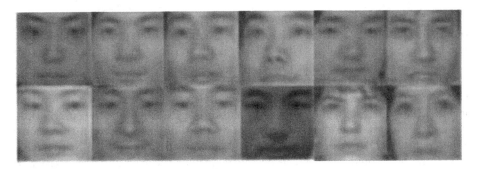

Fig. 3. The resultant face images after alignment.

Face Modeling software uses a fuzzy principal component analysis (FPCA) to carry out a separate modeling of occluded faces [9,10]. The process is as following: first, an occluded face is projected onto the eigenface space and a reconstructed face is then obtained by a linear combination of eigenfaces. The difference between the reconstructed face image and the original image is calculated, and then passed through weighted filter to calculate a probability value of the face part being occluded. This value is then used as a coefficient to combine the original image and reconstructed image to form a new face image. In the subsequent iterations, this coefficient is used in the FPCA for reconstruction, and the

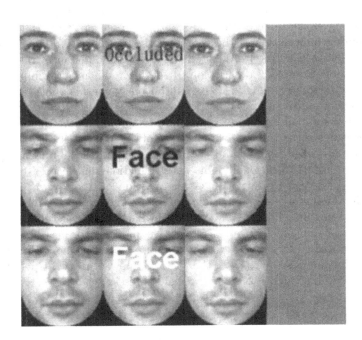

Fig. 4. Reconstruction of occluded face images.

cumulative error is used for occlusion detection. This approach can accurately locate the face occlusion area, and get a smooth and natural reconstruction of the face image. Some partially occluded faces and the corresponding reconstructed face image are shown below in Fig. 4.

The images are shown in columns side by side for comparison purpose. The first column are the original clear face images, the second column are the occluded face images, while the third column is FPCA method reconstruction of the face images. The fourth column shows the differences between the reconstructed faces and the original faces. It can be seen from the figure that the FPCA method could recover the occluded part of the face and reconstruct a natural and smooth face image. These reconstructed images are very close to the original ones, hence it can be expected that feature extraction over these images would be close to originals.

3.4   Key Face Feature Extraction

Feature extraction is the process that tries to find the clear, stable and effective face information with the presence of interference and noise in the image. The facial feature extraction method can also analyze the environment to extract the facial features with different algorithm. Face feature extraction is the core step in face recognition, which directly determines the recognition accuracy. The proposed software uses an Elastic Graph Matching method, which is based on dynamic link architecture (DLA) [11–14]. This method creates a property map for the face in a two-dimensional space, and places a topology map over the human face. Each node of the map contains an eigenvector, which records the distribution information of the face near the node [15,16]. The topological connections between nodes are denoted with geometric distance, thus forming a two-dimensional topology description of the face. When performing the face recognition with this system, we can simultaneously consider both node eigenvector matching and relative geometric position matching. In one way, we can scan the topology map structure on the face image to extract the corresponding node feature vector, and use the distance of different positions between the topological maps of the image and the face pattern in the library as the similarity measurement. Additionally, an energy function can be used to evaluate the matching between the target face vector field and the known face vector field, or, the minimum energy function matching. The method is robust to illumination and posture changes. The main drawback of this method is the high computational complexity, since calculating the model map must be performed over every individual face image, which takes up a lot of memory (Fig. 5).

In experiments over databases, the method showed a decent performance, and it adapts well to face posture and facial expression changes.

3.5   Face Recognition Software

Face recognition software works with all the above sub-modules combined. When fed in the photo of a target person, the system automatically detects and locates

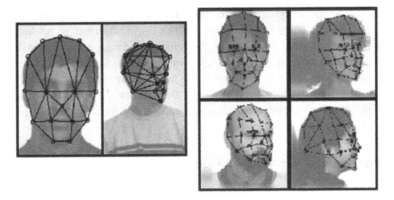

Fig. 5. Generalized elastic graph matching for face location and recognition

the face within the photo, then extracts the face image, process the image and calculates the key features of the face, then compare it with the features of recorded faces in the database. If matches are found, the detected faces are then highlighted to indicate positive match. Figure 6 shows a case of bearded face recognition, which is a typical case in practice. The individual was wearing a quite long beard and the system still accurately picks him out, and highlighted the face with a red square, as shown below.

Fig. 6. Detection of bearded face

Thanks to the earlier stages of the system, which has already reconstructed a smooth and natural face image, removed the negative effect of occlusion and other image degradation, the system recognition accuracy can be still maintained, making the system robust to such interference. In practice, the system could also enable the users to set a series of parameters, such as similarity, time of recording, and starts to search through the library. Finally, the found similar faces are displayed, and a quick confirmation by human checking is made possible.

3.6   Robustness of the Recognition

The proposed system is implemented and installed at various venues and public passages for testing. Volunteers have been walking before the cameras and the recorded videos are processed. Some results of typical occlusions are tested in these experiments, and the results are shown below in Table 1.

Table 1. Recognition Rate with Different occlusions.

Glasses	Beard	Scarves
97.13%	98.39%	99.32%

It is clearly shown that the tested typical occlusions can be effectively tackled and it achieves a good recognition rate, close to clear face images in many cases. Other tests performed include illumination affected images, and different face postures, where the system also showed substantial effectiveness. The results are not included here as it is hard to define such conditions. This part will be presented in a separate publication. In general, the system demonstrates a very good recognition performance in real life scenarios.

4   Face Recognition for Person Retrieval

Retrieving certain target person or objects is a frequent and highly desired job in practical application of video systems. Close circuit TV (CCTV) systems have shown their fastest growth in the past 2 decades. All kinds of video cameras have been installed, almost everywhere, and the number is still growing exponentially. Such system monitors a fixed area and tries to maintain a seamless records of video coverage of that area. The wide usage of such systems provides a big resource for event and human tracing. They have found many useful cases in practice, especially they provided countless examples to helping police and other public security departments to solve crime cases. However, these systems are designed to record the video signals while little has been considered about how to process these recorded files. Many systems in applications were installed in quite early years, which left a lot of room for efficiency and accuracy improvement. Some common problems have been widely seen, such as, Humans are still the major searching tools used for any specific information retrieving jobs. The CCTV systems only record the videos in case any of them could include some key information, but it is still up to manpower to search for such information. Typically, the video reviewers have to sit before the monitors for very long hours. Even when there is no active objects appearing in the scene, they still need to go through every second of the video without accelerating. Such working procedure not only result in an inefficient and low quality searching, but also harms the physical and mental health of the reviewers. Apart from the massive amount of accumulated videos, the video quality forms another challenge to make use of the files. CCTVs in real life are installed everywhere, over a long period, so the quality of recorded video varies dramatically. This could be due to different cameras, or most often, due to people's different postures when being recorded. Real life camera recording is a typical uncontrolled scenario and the resultant videos could suffer from all kinds of quality problems, such as occlusion, poor shooting angle or illumination. All these make processing the files difficult. With robust face recognition, it is then possible to search through these files effectively. The computers can save humans from manually browsing the huge amount of videos, while the algorithm ensures an effective search among the imperfect videos. Figure 7 shows the interface window of a face-recognition based people searching function in the proposed system. It can be seen in the figure that the people in the image are either of poor shooting angle, or have the face partially occluded, yet the system could still find the targeted individuals from the video. The algorithm to recognize the occluded faces advances the technology to a much larger range of applications, especially for police and public security departments, where browsing and searching through

Fig. 7. Face recognition searching module.

video are mostly performed by humans manually, which is not only tedious but also inaccurate. The proposed system provides a perfect alternative approach.

5   Conclusion

Biometrics technologies, especially face recognition, has attracted a lot of interest in both industry and academic field, When applied in practice, the algorithms often suffer from occlusion and other interference. In this paper, we propose a robust face recognition system, which tackles such interference. The system makes use of face model to reconstruct the occluded area and produces a smooth and natural face for further processing. Key features of the face can then be extracted, and the recognition accuracy is improved to be close to non-occluded faces. This robust recognition algorithm then enables effective searching through the massive amount of videos recorded by CCTV systems where video quality is not guaranteed. The system provides a searching function over video files, and it shows a reliable and efficient performance when applied in practice.

References

1. Zhao, W., et al.: Face recognition: a literature survey. ACM Comput. Surv. **35**(4), 399–458 (2003)
2. Ahonen, T., Hadid, A., Pietikinen, M.: Face description with local binary patterns: application to face recognition. IEEE Trans. Pattern Anal. Mach. Intell. **28**(12), 2037 (2006)
3. Zhang, H.: A multi-model biometric image acquisition system. Biometric Recognition. LNCS, vol. 9428, pp. 516–525. Springer, Cham (2015). https://doi.org/10.1007/978-3-319-25417-3_61
4. Tarres, F., Rama, A.: A novel method for face recognition under partial occlusion or facial expression variations. In: IEEE International Symposium on ELMAR 2005, pp. 163–166 (2005)
5. Andrés, A.M., et al.: Face recognition on partially occluded images using compressed sensing. Pattern Recogn. Lett. **36**(1), 235–242 (2014)
6. Zhang, X., Gao, Y.: Face recognition across pose: a review. Pattern Recogn. **42**(11), 2876–2896 (2009)
7. Ding, C., Tao, D.: Trunk-branch ensemble convolutional neural networks for video-based face recognition. IEEE Trans. Pattern Anal. Mach. Intell. **99**, 1 (2017)
8. Yang, A.Y, et al.: Fast 1-minimization algorithms and an application in robust face recognition: a review. In: IEEE International Conference on Image Processing, pp. 1849–1852. IEEE (2010)
9. Sharma, S.: Applied Multivariate Techniques, pp. 100–101. Wiley (1996)
10. Partridge, M., Jabri, M.: Robust principal component analysis. J. ACM (JACM) **58**(3), 1–73 (2011)
11. Wiskott, L., et al.: Face recognition by elastic bunch graph matching. IEEE Trans. Pattern Anal. Mach. Intell. **19**(7), 775–779 (1997)
12. Shin, S., Kim, S.-D., Choi, H.-C.: Generalized elastic graph matching for face recognition. Pattern Recogn. Lett. **28**(9), 1077–1082 (2007)

13. Wiskott, L., Fellous, J.M., Kruger, N., Malsburg, C.V.D.: Face recognition by elastic bunch graph matching. IEEE Trans. Pattern Anal. Mach. Intell. **19**(7), 775–779 (1997)
14. Lades, M., Vorbruggen, J.C., Buhmann, J., Lange, J., Malsburg, C.V.D., Wurtz, R.P., et al.: Distortion invariant object recognition in the dynamic link architecture. IEEE Trans. Comput. **42**(3), 300–311 (1993)
15. Yambor, W.S.: Analyzing PCA-based face recognition algorithms: eigenvector selection and distance measures, pp. 39–60. All Publications (2002)
16. Han, P.Y., Jin, A.T.B., Siong, L.H.: Eigenvector weighting function in face recognition. Discrete Dyn. Nat. Soc. 2011(2011-02-16), 2011(1026–0226), 701–716 (2011)

Image and Video Retrieval

Learning Shared and Specific Factors
for Multi-modal Data

Qiyue Yin, Yan Huang, Shu Wu, and Liang Wang[(⊠)]

Institute of Automation, Chinese Academy of Sciences,
95, Zhongguancun East Road Haidian District,
Beijing, People's Republic of China
{qyyin,yhuang,shu.wu,wangliang}@nlpr.ia.ac.cn

Abstract. In real world, it is common that an entity is represented by multiple modalities, which motivates multi-modal learning, e.g., multi-modal clustering and cross-modal retrieval. Traditional methods based on deep neural networks usually assume a joint factor or multiple similar factors are learned. However, different modalities representing the same content share both common and modality-specific characteristics, and few approaches can fully discover those features, i.e., consistency and complementarity. In this paper, we propose to learn shared and specific factors for each modality. Then the consistency can be explored through the shared factors. By combining the shared and specific factors, the complementarity will be excavated. Finally, a triadic autoencoder with deep architecture is developed for the shared and specific factors learning. Extensive experiments are conducted for cross-modal retrieval and multi-model clustering, which clearly demonstrate the effectiveness of our model.

Keywords: Multi-modal learning · Cross-modal retrieval
Multi-modal clustering

1 Introduction

Various kinds of real-world data appear in multiple modalities. For example, a web page can be described by both images and texts, and an image can be represented by either image itself or its associated tags. Since different modalities provide complementary and consistent representations of the same concept, multi-modal learning is driven to explore consistency and complementarity characteristics among multiple modalities, which has a wide range of applications, e.g., cross-modal retrieval exploring the consistency between different modalities, and multi-modal clustering discovering the complementarity among multiple modalities.

Traditional machine learning algorithms concatenate multiple modalities into a single feature set to fit multi-modal data. However, such a concatenation cannot explore the correlation between different kinds of information and ignores the incompatibility of heterogeneous feature sets. Multiple kernel learning methods

© Springer Nature Singapore Pte Ltd. 2017
J. Yang et al. (Eds.): CCCV 2017, Part II, CCIS 772, pp. 89–98, 2017.
https://doi.org/10.1007/978-981-10-7302-1_8

usually assume a kernel corresponds to a modality, and various fusion strategies are developed to combine different kernels [1]. Some other methods learn a latent space, where different kinds of modalities can be compared. Typical examples such as canonical correlation analysis [2], partial least squares [3], and bilinear model [4] obtain good results in various multi-modal learning tasks.

Recently, several deep learning methods are developed for multi-modal learning, which explore feature learning and multi-modal characteristics into a unified framework, and have obtained promising results. Some of those methods learn a joint factor among multiple modalities [5,6], and some other methods learn a factor for each modality and force the factors to be similar [7–9]. However, none of the above methods can fully excavate the consistency and complementarity among multiple modalities. For example, learning a joint factor ignores the consistency and those methods may not perform cross-modal matching tasks. Besides, learning similar factors may result in overlook of the complementarity and the data will not be fully represented.

To alleviate the above problem, a novel multi-modal learning method is developed. Firstly, higher-level factors for each modality are learned by feeding the original features into an autoencoder network. Then the learned factors are divided into shared and specific parts. Using the shared factors, the consistency can be explored based on the correspondence between the modalities with a typical triplet loss. Besides, by combining the shared and specific factors to reconstruct each modality with decoder networks, the complementarity can be excavated via a reconstruction loss. Finally, several stacked modality-friendly models are employed to learn higher-level representations of different modalities for reducing the semantic difference, and a deep architecture is accordingly developed. With the learned shared and specific factors, various multi-modal learning tasks, e.g., cross-modal retrieval and multi-modal clustering, can be performed.

The main contributions are listed as follows.

- We proposed a multi-modal learning framework that can explore consistency and complementarity characteristics simultaneously.
- We verify our model in terms of cross-modal retrieval and multi-modal clustering tasks, and the experimental results clearly demonstrate the effectiveness of our model.

2   Related Work

Multi-modal learning deals with data represented by multiple modalities, which has a wide range of applications. Various machine learning algorithms are developed to explore multi-modal characteristics based on different learning tasks. For example, cross-modal retrieval aims at discovering correlation between different modalities [2–4,10], so the modality-specific factors should be removed. As for multi-modal clustering [11,12], the main challenge lies in the mining of the complementary information among multiple modalities, so both the correlation and modality-specific factors should be considered to fully represent the data. Among various multi-modal learning methods, subspace learning based ones are popular due to their good results and ease of understanding. Those methods aim to find a low dimensional latent space, where different modalities can be compared [2,13,14]. Roughly speaking, our model also finds a space to fully explore the multi-modal characteristics.

Recently, with the resurgence of deep neural network in 2006, several deep learning methods are brought into multi-modal learning [9,15]. Ngiam et al. [15] proposed deep autoencoder models to fuse audio and video modalities for classification and retrieval tasks. Andrew et al. [5] developed a nonlinear extension of the canonical correlation analysis to obtain a higher correlation. Feng et al. [7] utilized correspondence autoencoder and deep boltzmann machine for cross-modal retrieval. Chang et al. [8] developed a highly nonlinear multi-layer embedding function to capture the complex interactions between the heterogeneous data in networks. Huang et al. [6] proposed a multi-label conditional restricted boltzmann machine to deal with modality completion, fusion and multi-label prediction. Overall, all above methods learning a joint factor or multiple similar factors cannot capture the multi-modal characteristics, i.e., consistency and complementarity, simultaneously.

It should be noted that topics of image caption [16] and image-sentence matching [17,18] are not discussed here because they go beyond of our scope.

3   Model

Taking two typical modalities, i.e., image and text, as an example, our model can be elaborated with two parts: a triadic autoencoder network for multi-modal characteristics exploring and stacked restricted boltzmann machines layers for high level representations learning.

3.1   Triadic Autoencoder

The learning architecture consists of three subnetworks with each corresponding to a basic autoencoder. Then the subnetworks are connected by a predefined triplet loss imposed on part of the code layers. More specifically, one subnetwork is fed with image representation, and the other two subnetworks sharing the same parameters are fed with similar and dissimilar text representations of the current image. With the above architecture, we can model the consistency and complementarity among multiple modalities nicely.

Formally, given an input $(\mathbf{r}, \mathbf{s}, \mathbf{t})$, which indicates $\mathbf{r}$ and $\mathbf{s}$ are a corresponding image-text pair and $\mathbf{t}$ is a randomly selected text representation, we model the consistency in the code layers. Suppose the image and text mappings are f and g respectively. Then the code layers are calculated as $f(\mathbf{r}; \mathbf{W}_f)$, $g(\mathbf{s}; \mathbf{W}_g)$ and $g(\mathbf{t}; \mathbf{W}_g)$, where $\mathbf{W}_f$ and $\mathbf{W}_g$ are the weight parameters in the subnetworks. Since multiple modalities share common and modality-specific characteristics and consistency is built based on their common parts, we use partial nodes to model the consistency:

$$L_1(r, s, t; \mathbf{W}_f, \mathbf{W}_g) = \min_{\mathbf{W}_f, \mathbf{W}_g} \max\left(0, \|\mathbf{x}\mathbf{I}_d - \mathbf{y}\mathbf{I}_d\|^2 + \gamma - \|\mathbf{x}\mathbf{I}_d - \mathbf{z}\mathbf{I}_d\|^2\right) \tag{1}$$

where L_1 is a widely used triplet loss, and γ is the size of margin. $\mathbf{x}$, $\mathbf{y}$ and $\mathbf{z}$ are code layers of r, s and t, respectively. $\mathbf{I}_d$ is a diagonal matrix with its first d elements being 1 and the others 0. Through such a matrix, we force the former d nodes to represent the shared factors.

After obtaining the code layer, we reconstruct each modality. By using all nodes in the code layer to reconstruct a modality, modality-specific factors can be represented using nodes except the shared parts. Then, the complementarity can be explored by combing shared and specific factors. The reconstruction is written as:

$$L_2(\mathbf{p}; \Theta) = \sum_{\mathbf{p}=r,s,t} \|\mathbf{p} - \tilde{\mathbf{p}}\|^2 \tag{2}$$

where L_2 is the reconstruction loss. $\mathbf{p}$ represents an image or text. Θ is parameters of the encoder and decoder networks, and $\tilde{\mathbf{p}}$ is a reconstructed image or text.

Then the final loss for an input triplet $(\mathbf{r}, \mathbf{s}, \mathbf{t})$ is:

$$L = L_1 + \alpha L_2 \tag{3}$$

where α is a parameter balancing the two terms. In summary, minimizing the loss function defined in Eq. 3 enables triadic autoencoder to explore consistency and complementarity among multiple modalities simultaneously.

3.2 Deep Architecture

Generally, data from multiple modalities consist of heterogeneous feature sets, and those descriptors may have a very big difference of semantic representation. Thus, it makes a layer of autoencoder hard to capture the consistency and complementarity characteristics. To alleviate this problem, a deep architecture is accordingly proposed. More specifically, some stacked modality-friendly models are utilized to learn higher-level representations of each modality. Then the semantic difference will be reduced for better exploring the multi-modal characteristics.

Practically, we use several restricted boltzmann machines (RBMs) to extract high-level features. To be simple, RBM is an undirected graphical model having a visible layer and a hidden layer with each layer consisting of stochastic binary units but without connections between these units. Usually, extended RBMs are utilized with the visible layer being the input image and text modalities. Here, Gaussian RBM (GRBM) [19] and replicated softmax RBM (RSM) [19] are utilized to model the real-valued feature vectors for image and the discrete sparse word count vectors for text, respectively. Then the hidden layers of GRBM and RSM will serve as the visible layers of basic RBMs. After several stacked RBMs, high level representations will be extracted for the triadic autoencoder network.

For parameters inference, all RBMs can be efficiently learned through the contrastive divergence approximation algorithm (CD). As for the triadic autoencoder network, an autoencoder can be initialized using an RBM. Then the parameters are optimized using Eq. 3 through back-propagation algorithm. Note that we can use back-propagation for the entire network, but we just finetune the triadic autoencoder network for simplicity.

4   Experiments

4.1   Datasets

Wiki Dataset: It is a widely used image-text dataset, which consists of 2,173/693 training/testing image-text pairs. In total, there are 10 categories. As for the features, the text is ten dimensional topics obtained through a topic model (Latent Dirichlet Allocation) [20], and the image is represented by 128 dimensional SIFT descriptors. Similar to [21], we split the dataset into a training set of 1,300 pairs (130 pairs per class) and a testing set of 1,566 pairs.

Pascal VOC Dataset: It is used in various multi-modal learning tasks, which consists of 5,011/4,952 training/testing image-tag pairs classified into 20 categories. As for the features, the tag feature is a 399-dimensional word frequency vector, and the image is encoded by a 512-dimensional Gist feature. For simplicity, we remove image-tag pairs with their tag features all being zero as did in [21].

4.2   Tasks and Evaluation Metrics

Since our model aims to explore consistency and complementarity among multiple modalities, we perform two kinds of tasks, i.e., cross-modal retrieval and multi-modal clustering, which depend mainly on the consistency and complementarity respectively.

Cross-modal retrieval: We map the testing images and texts into the code layers and select the former d nodes as their final embeddings. Then the two embeddings can be compared using the Euclidean distance. Finally, two cross-modal retrieval tasks, i.e., Image query vs. Text database and Text query vs. Image database, are conducted. As for the metrics, We use mean average precision (MAP) and precision-recall curve (PR) to evaluate the overall performance [22].

Multi-modal clustering: We map the testing images and texts into the code layers and concatenate all the code layers as their final embeddings. Then we cluster those embeddings through K means algorithm. As for the metrics, five widely used measures [12], i.e., the accuracy (ACC), normalized mutual information (NMI), F-measure (F1), R-Index (RI) and Entropy, are utilized for performance evaluation. As for the former four metrics, the bigger the better performance, and for the Entropy, the lower the values the better performance.

4.3   Compared Methods

PLS [3], **BLM** [4] and **CCA** [2] are three representative unsupervised methods that use pairwise information for embeddings learning. **CorrAE**: Feng et al. [7] proposed a correspondence autoencoder that learns similar factors for multiple modalities. **DCCA**: Andrew et al. [5] extended traditional canonical correlation analysis to a deep architecture.

CDFE [13], **GMLDA** [14] and **GMMFA** [14] are three typical supervised multi-modal leaning methods. They use labels to obtain relatively discriminative subspaces to enhance the performance. We compared those methods to further validate the effectiveness of our unsupervised model.

Our method is denoted as **LSSF**. Besides, LSSF without dividing the code layer into common and specific factors is denoted as **BaseF**. Comparing with this method will further validate the effectiveness of fully exploring multi-modal characteristics.

4.4   Cross-Modal Retrieval

For the Wiki dataset, all features are low dimensional and well extracted, so we do not use RBMs for feature preprocessing. As for the VOC dataset, We use two RBM layers to extract high level representations. The parameter d deciding the number of nodes being the shared factors is empirically selected to be half number of the code layer.

The MAP results of image query and text query on the Wiki and VOC datasets are shown in Tables 1 and 2 respectively. Overall, it can be seen that our method almost outperforms all the compared methods on all the datasets. Compared with BaseF, we learn shared and modality-specific factors, which is more reasonable due to the characteristics of multi-modal data.

Table 1. MAP comparison of different methods on the Wiki dataset.

Wiki dataset	Image query	Text query	Average result
PLS [3]	0.2402	0.1633	0.2032
BLM [4]	0.2562	0.2023	0.2293
CCA [2]	0.2549	0.1846	0.2198
CDFE [13]	0.2655	0.2059	0.2357
GMMFA [14]	0.2750	0.2139	0.2445
GMLDA [14]	**0.2751**	0.2098	0.2425
CorrAE [7]	0.2611	0.1748	0.2180
DCCA [5]	0.2406	0.2050	0.2228
BaseF	0.2674	0.2025	0.2350
LSSF	**0.2751**	**0.2183**	**0.2467**

Table 2. MAP comparison of different methods on the VOC dataset.

VOC dataset	Image query	Text query	Average result
PLS [3]	0.2757	0.1997	0.2377
BLM [4]	0.2667	0.2408	0.2538
CCA [2]	0.2655	0.2215	0.2435
CDFE [13]	0.2928	0.2211	0.2569
GMMFA [14]	0.3090	0.2308	0.2699
GMLDA [14]	0.3094	0.2448	0.2771
CorrAE [7]	0.2545	0.1368	0.2160
DCCA [5]	0.2303	0.2016	0.2160
BaseF	0.3205	0.2220	0.2713
LSSF	**0.3290**	**0.2556**	**0.2923**

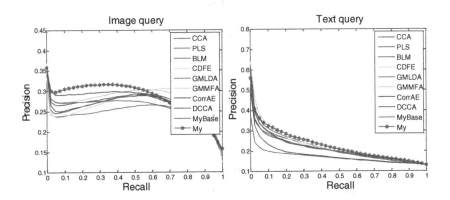

Fig. 1. Precision recall curves of different methods on the Wiki dataset.

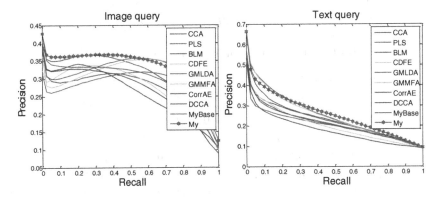

Fig. 2. Precision recall curves of different methods on the VOC dataset.

Compared with PLS, CCA and BLM, our method is much better because we use triplet loss to model the relation between images and texts, which may be more effective than the pairwise relation. More importantly, consistency and complementarity are considered simultaneously in our model. Methods CorrAE and DCCA learn similar factors. However, none of these methods can fully discover multi-modal characteristics. Thus our method is better than them.

Finally, the precision-recall curves of Image query and Text query on the Wiki and VOC datasets are shown in Figs. 1 and 2 respectively. The results are similar with that of MAP, and this further validates the effectiveness of our method.

4.5   Multi-modal Clustering

The training settings in the clustering task is the same with in the retrieval task. Then the clustering results on the Wiki and VOC datasets are shown in Tables 3 and 4 respectively. Overall, it can be seen that our method almost beats all the competing methods on the two datasets. This validates that our model can well discover the complementarity characteristic. Taking the retrieval tasks into consideration, we can draw the conclusion that considering the consistency and complementarity characteristics of multi-modal data can promote the learning performance.

Table 3. Clustering results on the Wiki datasets.

Wiki dataset	ACC	NMI	F1	RI	Entropy
PLS [3]	0.4764	0.4830	0.4141	0.3291	1.6132
BLM [4]	0.5519	**0.5493**	0.4814	0.4115	**1.3738**
CCA [2]	0.5598	0.5227	0.4816	0.4127	1.4556
CorrAE [7]	0.5791	0.5499	0.5222	0.4588	1.3681
DCCA [5]	0.5796	0.5364	0.5181	0.4524	1.4187
BaseF	0.5883	0.5426	0.5174	0.4539	1.3890
LSSF	**0.6081**	0.5440	**0.5321**	**0.4707**	1.3828

Table 4. Clustering results on the VOC datasets.

VOC dataset	ACC	NMI	F1	RI	Entropy
PLS [3]	0.4764	0.5773	0.4214	0.3772	1.5640
BLM [4]	0.5636	0.5727	0.4562	0.4113	1.6131
CCA [2]	0.5355	0.5135	0.4235	0.3783	1.8389
CorrAE [7]	0.5026	0.5120	0.4038	0.3506	1.8895
DCCA [5]	0.4857	0.4693	0.3616	0.3033	2.0787
BaseF	0.6296	0.6593	0.5758	0.5434	1.2342
LSSF	**0.6640**	**0.6891**	**0.6226**	**0.5934**	**1.1209**

5   Conclusion

In this paper, we have proposed a novel multi-modal learning method. By learning shared and modality-specific factors for each modality through a triadic autoencoder network, our model can explore consistency and complementarity characteristics among multiple modalities simultaneously. Finally, extensive experiments including cross-modal retrieval and multi-modal clustering have validated the proposed method by comparing with the state-of-the-art methods.

References

1. Gonen, M., Alpaydin, E.: Multiple kernel learning algorithms. J. Mach. Learn. Res. **12**, 2211–2268 (2011)
2. Kim, T.-K., Kittler, J., Cipolla, R.: Discriminative learning and recognition of image set classes using canonical correlations. IEEE Trans. Pattern Anal. Mach. Intell. **29**(6), 1005–1018 (2007)
3. Rosipal, R., Krämer, N.: Overview and recent advances in partial least squares. In: Saunders, C., Grobelnik, M., Gunn, S., Shawe-Taylor, J. (eds.) SLSFS 2005. LNCS, vol. 3940, pp. 34–51. Springer, Heidelberg (2006). https://doi.org/10.1007/11752790_2
4. Tenenbaum, J.B., Freeman, W.T.: Separating style and content with bilinear models. Neural Comput. **12**(6), 1247–1283 (2000)
5. Andew, G., Arora, R., Bilmes, J., Livesu, K.: Deep canonical correlation analysis. In: International Conference on Machine Learning, pp. 1247–1255 (2013)
6. Huang, Y., Wang, W., Wang, L.: Unconstrained multimodal multi-label learning. IEEE Trans. Multimedia **17**(11), 1923–1935 (2015)
7. Feng, F., Wang, X., Li, R.: Cross-modal retrieval with correspondence autoencoder. In: ACM on Multimedia, pp. 7–16 (2014)
8. Chang, S., Han, W., Tang, J., Qi, G.-J., Aggarwal, C.C., Huang, T.S.: Heterogeneous network embedding via deep architectures. In: ACM SIGKDD International Conference on Knowledge Discovery and Data Mining, pp. 119–128 (2015)
9. Wang, W., Arora, R., Livescu, K., Bilmes, J.: On deep multi-view representation learning: objectives and optimization. In: arXiv (2016)
10. Cao, Y., Long, M., Wang, J., Liu, S.: Collective deep quantization for efficient cross-modal retrieval. In: AAAI Conference on Artificial Intelligence (2017)
11. Yin, Q., Wu, S., Wang, L.: Unified subspace learning for incomplete and unlabeled multi-view data. Patern Recogn. (2017)
12. Kumar, A., Daume III, H.: A co-training approach for multi-view spectral clustering. In: International Conference on Machine Learning, pp. 393–400 (2011)
13. Lin, D., Tang, X.: Inter-modality face recognition. In: Leonardis, A., Bischof, H., Pinz, A. (eds.) ECCV 2006. LNCS, vol. 3954, pp. 13–26. Springer, Heidelberg (2006). https://doi.org/10.1007/11744085_2
14. Sharma, A., Kumar, A., Daume III, H.: Generalized multiview analysis: a discriminative latent space. In: IEEE Conference on Computer Vision and Pattern Recognition, pp. 2160–2167 (2012)
15. Ngiam, J., Khosla, A., Kim, M., Nam, J., Lee, H., Ng, A.Y.: Multimodal deep learning. In: International Conference on Machine Learning, pp. 689–696 (2011)

16. Xu, K., Ba, J., Kiros, R., Cho, K., Courville, A.C., Salakhutdinov, R., Zemel, R.S., Bengio, Y.: Show, attend and tell: neural image caption generation with visual attention. In: International Conference on Machine Learning, pp. 2048–2057 (2015)
17. Nam, H., Ha, J.-W., Kim, J.: Dual attention networks for multimodal reasoning and matching. In: arXiv (2016)
18. Huang, Y., Wang, W., Wang, L.: Instance-aware image and sentence matching with selective multimodal LSTM. In: IEEE Conference on Computer Vision and Pattern Recognition, pp. 2310–2318 (2017)
19. Srivastava, N., Salakhutdinov, R.: Multimodal learning with deep boltzmann machines. J. Mach. Learn. Res. **15**, 2949–2980 (2014)
20. Blei, D.M., Ng, A.Y., Jordan, M.I.: Multimodal learning with deep boltzmann machines. J. Mach. Learn. Res. **3**, 993–1022 (2003)
21. Wang, K., He, R., Wang, W., Wang, L., Tan, T.: Learning coupled feature spaces for cross-modal matching. In: IEEE International Conference on Computer Vision, pp. 2088–2095 (2013)
22. Rasiwasia, N., Pereira, J.C., Coviello, E., Doyle, G., Lanckriet, G.R.G., Levy, R., Vasconcelos, N.: A new approach to cross-modal multimedia retrieval. In: ACM Conference on Multimedia, pp. 251–260 (2010)

Massively Parallel Image Index for Vocabulary Tree Based Image Retrieval

Qingshan Xu[1], Kun Sun[1], Wenbing Tao[1,2(✉)], and Liman Liu[2]

[1] National Key Laboratory of Science and Technology on Multi-spectral Information Processing, School of Automation, Huazhong University of Science and Technology, Wuhan, Hubei, People's Republic of China
{qingshanxu,sunkun,wenbingtao}@hust.edu.cn
[2] Hubei Key Laboratory of Medical Information Analysis & Tumor Diagnosis and Treatment, School of Biomedical Engineering, South-Central University for Nationalities, Wuhan, Hubei, People's Republic of China
limanliu@mail.scuec.edu.cn

Abstract. Although vocabulary tree based algorithm has high efficiency for image retrieval, it still faces a dilemma when dealing with large data. In this paper, we show that image indexing is the main bottleneck of vocabulary tree based image retrieval and then propose how to exploit the GPU hardware and CUDA parallel programming model for efficiently solving the image index phase and subsequently accelerating the remaining retrieval stage. Our main contributions include tree structure transformation, image package processing and task parallelism. Our GPU-based image index is up to around thirty times faster than the original method and the whole GPU-based vocabulary tree algorithm is improved by twenty percentage in speed.

Keywords: Large-scale image retrieval · Vocabulary tree Image index · GPU-based model

1 Introduction

Image retrieval is to search for similar images in a large scale database given a query image. Bag of Features (BOF) image representation and its many variants [1,2,4–7,11,15,16] are well known for addressing the image search problem, which quantized local invariant features such as SIFT (Scale Invariant Feature Transform) [3] to visual words. Then, combining with the inverted file retrieval method, visual words can become more discriminative and be accessed faster. Generally speaking, there exists more than hundreds of thousands of visual words for each individual retrieval. In order to speed up the assignment of individual feature descriptors to its corresponding visual words, David Nistér and Henrik Stewénius [2] introduced the vocabulary tree algorithm which leverages the hierarchical structure of tree. Although this method has gained a lot efficiency in image retrieval, it still suffers from the great challenge in large-scale image database.

© Springer Nature Singapore Pte Ltd. 2017
J. Yang et al. (Eds.): CCCV 2017, Part II, CCIS 772, pp. 99–110, 2017.
https://doi.org/10.1007/978-981-10-7302-1_9

Meanwhile, with the popularity of GPU (Graphics Processing Units) hardware, CUDA parallel programming method has drawn much attentions to accelerate processing massive data, such as SiftGPU [8] and MCBA [9], which make 3D reconstruction in the large much faster. It leverages thousands of threads to execute the same instructor simultaneously. Although CUDA parallel programming method succeeds in regular applications, such as dense linear algebra, it will face a dilemma due to the irregular program of vocabulary tree, which performs unpredictable and data-driven access, and memory restrictions of GPU when compared to the amount of RAM available to CPU.

In this paper, we design and implement a GPU-based vocabulary tree algorithm for efficient large-scale image search. We first analyze the original vocabulary tree algorithm and draw the conclusion that the main bottleneck of vocabulary tree is image index. Then, we show how to transform the original tree structure to linear continuous array to fit the GPU architecture and how to organize the large-scale data to maximize the usage of memory width. Meanwhile, we explore the task parallelism (that is, asynchronism of CPU and GPU) to further boost the speedup performance.

The rest of this paper is organized as follows. We begin in Sect. 2 by analysing the original vocabulary tree algorithm and conclude some matters and challenges to be noted in GPU architecture. In Sect. 3 we describe the explicit implementation of the GPU-based vocabulary tree algorithm including the transformation of tree structure, the organization of massive data and task parallelism for SIFT descriptor search and histogram compression. In Sect. 4 we report the performance of our algorithm, and conclude with a discussion and directions for future work in Sect. 5.

2   Theoretical Background of Vocabulary Tree

Given a query image, image retrieval aims to search for similar images with common objects or scenes from a database. Denote the query images as $Q_i, i = 1, \ldots, m$ and the database images as $D_i, i = 1, \ldots, n$. SIFT feature descriptors are extracted from every query image and database image. The content of each image is represented by the Bag of Features (BOF) model, which draws an image as a point in a high dimensional feature space. However, directly searching for nearest neighbors in such a space is time consuming. In order to have a fast and high accuracy performance, vocabulary tree is introduced [2]. Different from the traditional visual words of BOF, which are learned by k-means algorithm, the visual words of vocabulary tree are built by hierarchical k-means clustering. In the vocabulary tree algorithm, k is no longer on behalf of the final number of visual words, it defines the branch factor (number of children of each node) of the tree. Before arriving at the pre-defined level L, the k-means clustering is recursively applied to each group of descriptors in the last layer, which splits each group of descriptors into finer k new parts. Eventually the internal nodes represent the centroids of each clustering, while the all leaf nodes are the final visual words.

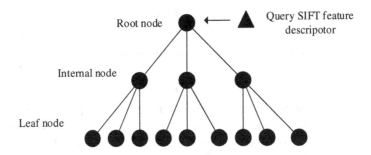

Fig. 1. Example of vocabulary tree and query SIFT feature descriptor searching. Internal nodes are centroids, while leaf nodes are visual words.

In order to assign a visual word to a feature descriptor on the query image, the algorithm will find a path from the root to a leaf node by comparing with the k candidate cluster centers at each layer. This results in at most kL comparisons for each feature descriptor, which is illustrated in Fig. 1. Just as David Nistér et al. pointed out in the literature [2], compared with the visual words defined by the non-hierarchical manner, the computational cost of the establishment of visual words in vocabulary tree is logarithm in the number of leaf nodes and the total number of feature descriptors that must be represented is $\sum_{i=1}^{L} k^i = \frac{k^{L+1}-k}{k-1} \approx k^L$.

Moreover, to characterize the relevance between a database image and the query image better, vocabulary tree borrows ideas from the inverted file index, which is called Term Frequency Inverse Document Frequency (TF-IDF). Term Frequency describes the number of feature descriptors assigned to a specific visual word, while Inverse Document Frequency reflects the significance of a visual word. Assume that for the database images, the Term Frequency of a visual word i is N_i, and the number of database images is N, then the modified weight of the visual word i is

$$w_i = \ln \frac{N}{N_i}, \tag{1}$$

which makes the visual word become more discriminative. Then both the query image vector and the database image vector can be defined as

$$q_i = n_i w_i \tag{2}$$
$$d_i = m_i w_i \tag{3}$$

where n_i and m_i are the number of feature descriptors assigned to the visual word i for the query image and the database image, respectively. In order to remove the influence of the vector bias for a better similarity measure, normalization is further utilized, which means final representative vectors of both the query image and database image are modified as

$$q' = \frac{q}{\|q\|} \tag{4}$$

$$d' = \frac{d}{\|d\|} \tag{5}$$

Subsequently, the similarity score between the query image and the database image is given by

$$s(q', d') = \|q' - d'\| \tag{6}$$

After obtaining similarity scores between the query image and the database images, sort algorithm is applied to get the final score list. And people can truncate the list to get the top j database images according to their need.

In sum, the vocabulary tree algorithm can be divided into 3 different phases, including the visual words learning, database images index and query images search. The first phase is off-line and the remaining two are online. As there exists more and more well-learned visual words, the first phase computational overhead can be ignored. However, due to the increasing of database images, the main computational overhead falls in the second image index phase, which can be verified by Tables 1, 2 and Fig. 4(a). So we need to fully exploit the modern GPU hardware to speedup the image index, which will be explicitly described in the following section.

3   GPU-Based Vocabulary Tree Algorithm

It is well-known that GPU can perform the same operation across thousands of individual data elements owing to its highly parallel architecture. In contrast to the CPU architecture where each thread execute its own set of instructions, GPU adopts the single instruction multiple threads (SIMT) model, which means that a large collection of threads can execute the same instruction simultaneously. Using this model, GPU not only achieves the massively parallel computation, but also high data throughout [10]. Based on this, we start the algorithm specially designed for vocabulary tree.

As analysed in Sect. 2, there are tremendous feature descriptors in all the database images (considering that a image of high resolution contains at least 10,000 feature descriptors), which makes the image index lower the search efficiency. An intuition is that every feature descriptor can be allocated to its respective thread to search the visual word it belongs to. Following this idea, we will further discuss the subsequent problems and solutions, which constitute the subject of our GPU-based vocabulary tree algorithm.

What we first need to consider is that how to transplant the tree structure of visual words into the GPU platform. Because the tree structure is organized by the form of non-continuous pointers, it is impossible to transfer all the visual words by just transferring the root node. Thus the tree structure of visual words should be reorganized in the form of array. In order to maintain the search efficiency of original tree structure, all nodes should also record the address offset of its first child and the number of its children except for the original SIFT descriptors and node ID. The first information is used for SIFT descriptors to search their next corresponding node until they reach the bottom level while

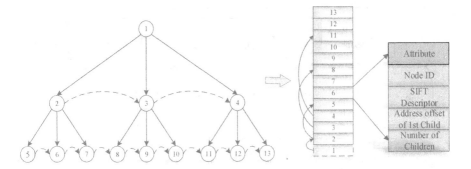

Fig. 2. Structure transformation from original irregular tree to instructive array. Except for its original tree node attribute, every structured array includes the address offset of its first child and the number of its children.

the function of the second information is to judge whether the search should be terminated or not. The above transformation can be executed by breadth first search combined with the stack operation. Once the instructive array of visual words is obtained, it can be transplanted into the GPU platform easily. See Fig. 2 for example.

Next, we will arrange the SIFT descriptors more carefully. Because the computationally intensive sections of the vocabulary tree algorithm are the image index, we need to assign the visual word attribute for each SIFT descriptor in parallel. We may first want to transfer the SIFT descriptors of an image once every time to GPU platform to execute the above mission. However, it has been shown that this scheme does not fully utilize the parallelism ability of GPU hardware. In order to further dig the GPU hardware parallelism ability, we readjust the structure of SIFT descriptor. In addition to reserve the original attribute of SIFT (scale, orientation, position and descriptor), we also mark up every SIFT descriptor with the image ID attribute so that we can organize them in a larger scale, which means that we pack several images together every time and transfer all their descriptors to the GPU platform.

When we assign the visual word attribute for every SIFT descriptor, we inevitably will face the dilemma that there exists the assignment conflicts when two or more SIFT descriptors of one image contribute to the frequency of the same visual word simultaneously. In order to overcome this contradiction, we employ the Atomic operation, which protects the current operation and make the other operation to the same variable to wait. Another subsequent problem is that the visual word histogram produced in the GPU is very sparse because we break the original tree structure for the convenience of transmission. For the efficiency of subsequent operation and the reduction of memory footprint, we need to compress the sparse histogram. Instead of handling this task serially, we exploit the hybrid execution of using both CPU and GPU resources. See Fig. 3 for example. In the sequential timeline, when the ith image patch is indexed, then the ith sparse histogram is compressed sequentially. While in the asynchronous

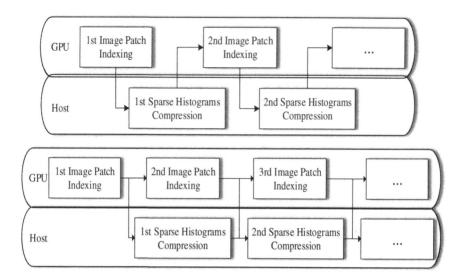

Fig. 3. The combination of data parallelism and task parallelism. The picture above shows the sequential timeline, while the picture below shows the asynchronous timeline.

timeline, when the ith image patch is indexed in GPU, the $(i + 1)$th sparse histograms is simultaneously compressed in CPU.

Furthermore, once we finish the database image query index, query image scoring can also be executed in parallel. When a query instructor is arriving, it generally compares its own histogram to all the database image histograms sequentially. It limits the capability of GPU. Thus, we make the database images simultaneously compute their similarity scores to the query image, which can be massively done in GPU.

4   Experiments and Discussions

In this section we present experimental results on the application of our GPU-based vocabulary tree algorithm to some large-scale data sets. We test our method on datasets of Noah Snavely et al. [12] including 1DSfM_Roman_Forum and 1DSfM_Vienna_Cathedral. The first dataset contains 2360 images while the second contains 6280 images. In order to compare our GPU-based vocabulary tree algorithm performance with CPU-based vocabulary tree algorithm, we will re-implement the Noah Snavley's CPU-based vocabulary tree algorithm [13], which is used to retrieval the most relevant image pairs for every image to produce compact subsets for large-scale 3D reconstruction [14,17,18].

Since there are many generic well-trained visual words for vocabulary tree, we skips the vocabulary tree training phase and test the remaining two phase. One thing that should be noted is that for most image retrieval applications, it is general to first delineate the Maximally Stale Extremal Regions (MSERs) [19] or

Table 1. Time consumed (in seconds) for 1DSfM_Roman_Forum with CPU-based and GPU-based vocabulary tree. Our model gives significant speedup in search and compression.

Stage	GPU time	CPU time	Speedup performance
Pre-processing	0.771	0	–
Search(+Compression)	7.815	298.765	**38.2**
Weighting	0.005	0.252	–
Normalization	0.183	0.613	–
Scoring	0.745	2.020	–
Data copying	2.415	0	–
Others	0.475	0.146	–
Total	12.409	301.796	**24.3**

Table 2. Time consumed (in seconds) for 1DSfM_Vienna_Cathedral with CPU-based and GPU-based vocabulary tree. Our model gives significant speedup in search and compression.

Stage	GPU time	CPU time	Speedup performance
Pre-processing	0.772	0	–
Search(+Compression)	27.467	829.172	**30.2**
Weighting	0.021	0.326	–
Normalization	0.428	1.375	–
Scoring	7.876	27.880	–
Data copying	5.967	0	–
Others	1.003	0.063	–
Total	43.534	859.942	**19.8**

Hessian-Affine interest points [20], which is proved to be benefit to the retrieval accuracy. However, here we only focus on the retrieval efficiency so that we extract all the SIFT features from all the database images for general purpose. Although we do not execute the abstraction of MSERs or Hessian-Affine interest points, this requires us to deal with more SIFT descriptor index missions, which can further demonstrate our GPU-based algorithm efficiency. Users can adaptively choose the abstraction of MSERs or Hessian-Affine interest points in practice according to their own requirements.

All the experiments are conducted on a machine with two Intel Xeon CPU E5-2630 v3 2.40 GHz, one NVIDIA GeForce GTX TitanX graphics card with 12 GB global memory and 64-bit Linux operating system. The CPU-based vocabulary tree algorithm is implemented using C++, and our GPU code is implemented with CUDA.

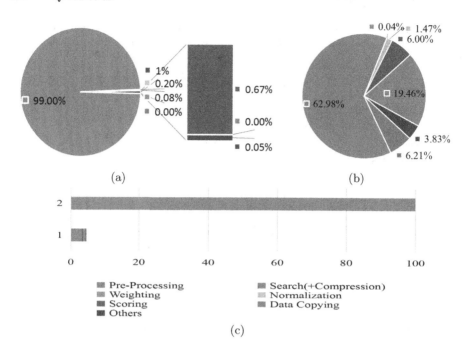

Fig. 4. (a) The runtime ratio for the CPU-based vocabulary tree. (b) The runtime ratio for GPU-based vocabulary tree. (c) The runtime comparison between GPU-based and CPU-based vocabulary tree. The runtime of GPU-based image index is up to around thirty times faster than the CPU-based method and the whole GPU-based vocabulary tree algorithm is improved by twenty percentage.

Note that here we ignore the SIFT reading time and the score sorting time because they are implemented in the same way for the compared methods. In all our experiments, we enforce the vocabulary tree with branch factor $k = 10$ and level $L = 6$ to test both the CPU-based and GPU-based algorithms. Figure 4(a) shows that the SIFT descriptor search occupies up to 99% of the runtime and illustrates that image index becomes the bottleneck in the overall runtime. In order to highlight the speedup performance of every stage, we explicitly record the detailed time in seconds for every stage in Tables 1 and 2. We can see that our GPU-based image index is around thirty times faster than CPU-based image index and the whole vocabulary tree algorithm is improved by twenty percentage, whilst the other stages (include weighting, normalization and scoring) also have a certain speedup. We also note that the data copying item that records the whole data transmission time is not so time-consuming owing to the package processing for large-scale images. Last but not least, from the 2nd row of both tables, search and compression stages are executed asynchronously, which makes the compression time hidden under the search time. Essentially, search stage and compact representation are tightly coupled in CPU-based vocabulary tree algorithm owing to the tree structure. This makes the compact representation time almost indiscriminate. Because we adopt the task parallelism execution, we eliminate the data redundancy problem from the another point of view. Figure 4(a)(b) shows the GPU-based algorithm runtime ratio and runtime comparison between CPU-based and GPU-based algorithm in which we can see SIFT descriptor search runtime ratio decreases from 99% to 63% of the overall runtime obviously. As illustrated in Fig. 4(c), the runtime is reduced greatly for both the image index phase and the overall algorithm.

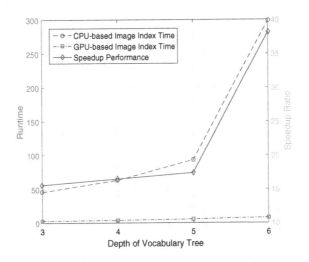

Fig. 5. The runtime analysis for the CPU-based and GPU-based vocabulary tree.

We also conduct experiments to understand how the tree depth influences the speedup performance and the accuracy robustness for both algorithms. We train the vocabulary tree with different depth progressively and test the runtime performance between CPU-based and GPU-based algorithm on the 1DSfM_Roman_Forum. As shown in Fig. 5, when depth increases, the CPU-based image index runtime is increased exponentially while the GPU-based image index runtime varies linearly with a small increment. Besides, we can see that the speedup ratio is above 15. The reason why the shallow vocabulary tree speedup performance is lower than the deep is that shallow vocabulary tree has less leaf nodes and search collision occurs more frequently. We evaluate the accuracy by extracting the SIFT descriptors after detecting the Hessian-Affine regions [22] and executing CPU-based and GPU-based vocabulary tree in the dataset UKbench [21]. The UKbench dataset contains 10200 images, consisting of 2550 groups of 4 images each. All the images are 640 × 480. We follow the performance measure of [21] to count how many of the 4 images which are top-4 when using a query image from the same group. Figure 6 shows the accuracy trend with different dataset number and different depth of vocabulary tree. The accuracy of GPU-based and CPU-based algorithm is no different so that we just

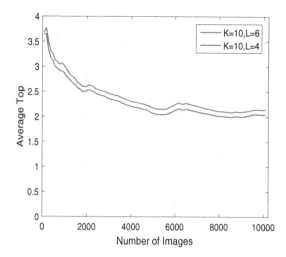

Fig. 6. The accuracy of GPU-based vocabulary tree with different depth and different image scale.

show the GPU-based algorithm's accuracy. We note that since a different vocabulary is used, our results are a bit different from [21]. When the size of image database increases, there is a small drop in the accuracy performance. This is because SIFT descriptor collision is more likely to happen when more images are used. Just as demonstrated in [2], our results also show that the most importance for the retrieval quality is to have a large vocabulary (large number of leaf nodes). Thus, the speedup performance in the vocabulary tree with branch factor $k = 10$ and level $L = 6$ is particularly import, where our GPU-based image algorithm performs well.

5   Conclusion

In this paper we present the GPU execution scheme to the problem of vocabulary tree based large-scale image retrieval. The GPU-based image indexing delivers a 30x boost in speed over the CPU-based algorithm and the whole algorithm achieves 20x faster. This is done by carefully transforming the tree structure to the array form and SIFT descriptors package processing. We also explore the task parallelism for data redundancy. Although the problem addressed in this paper is vocabulary tree image retrieval, we believe that the above speedup design here can applied to other large-scale tree-based irregular programs. In the future, we would like to further combine the GPU-based feature extraction and GPU-based sort algorithm to design an end-to-end platform to further accelerate the large scale image retrieval. We also hope to improve the retrieval accuracy by leveraging the asynchronism of GPU and CPU.

Acknowledgments. The authors would like to acknowledge Henrik Stewénius, David Nistér, Mikolajczyk, K. and Noah Snavely et al. for making their related datasets and source codes publicly available to us. This work is supported by the National Natural Science Foundation of China (Grant 61772213 and 61371140) and the Special Fund CZY17011 for Basic Scientific Research of Central Colleges, South-Central University for Nationalities, also in part by Grants 2015CFA062, 2015BAA133 and 2017010201010121.

References

1. Sivic, J., Zisserman, A.: Video Google: a text retrieval approach to object matching in videos. In: CVPR (2003)
2. Nistér, D., Stewénius, H.: Scalable recognition with a vocabulary tree. In: CVPR (2006)
3. Lowe, D.G.: Distinctive image features from scale-invariant keypoints. Int. J. Comput. Vision **60**, 91–110 (2004)
4. Lazebnik, S., Schmid, C., Ponce, J.: Beyond bags of features: spatial pyramid matching for recognizing natural scene categories. In: CVPR (2006)
5. Philbin, J., Chum, O., Isard, M., Sivic, J., Zisserman, A.: Object retrieval with large vocabularies and fast spatial matching. In: CVPR (2007)
6. Jegou, H., Douze, M., Schmid, C.: Hamming embedding and weak geometric consistency for large scale image search. In: ECCV (2008)
7. Jgou, H., Douze, M., Schmid, C.: Improving bag-of-features for large scale image search. Int. J. Comput. Vision **87**, 316–336 (2010)
8. Wu, C., SiftGPU: A GPU implementation of david lowe's scale invariant feature transform (SIFT). http://cs.unc.edu/~ccwu/siftgpu/
9. Wu, C., Agarwal, S., Curless, B., Seitz, S.M.: Multicore bundle adjustment. In: CVPR (2011)
10. Farber, R.: CUDA Application Design and Development. Morgan Kaufmann, San Francisco (2011)
11. Arandjelović, R., Zisserman, A.: DisLocation: scalable descriptor distinctiveness for location recognition. In: ACCV (2014)
12. Wilson, K., Snavely, N.: Robust global translations with 1DSfM. In: ECCV (2014)
13. Snavely, N.: A CPU implementation of David Nistér and Henrik Stewénius's vocabulary tree algorithm. https://github.com/snavely/VocabTree2
14. Agarwal, S., Snavely, N., Simon, I., Seitz, S., Szeliski, R.: Building Rome in a day. In: ICCV (2009)
15. Sattler, T., Havlena, M., Schindler, K., Pollefeys, M.: Large-scale location recognition and the geometric burstiness problem. In: CVPR (2016)
16. Koniusz, P., Yan, F., Gosselin, P.H., Mikolajczyk, K.: Higher-order occurrence pooling for bags-of-words: visual concept detection. IEEE Trans. Pattern Anal. Mach. Intell. **39**, 313–326 (2017)
17. Schönberger, J.L., Frahm, J.-M.: Structure-from-motion revisited. In: CVPR (2016)
18. Shen, T., Zhu, S., Fang, T., Zhang, R., Quan, L.: Graph-based consistent matching for structure-from-motion. In: ECCV (2016)
19. Matas, J., Chum, O., Urban, M., Pajdla, T.: Robust wide baseline stereo from maximally stable extremal regions. In: BMVC (2002)

20. Mikolajczyk, K., Schmid, C.: Scale & affine invariant interest point detectors. Int. J. Comput. Vision **60**, 63–86 (2004)
21. Stewénius, H., Nistér, D.: UKbench dataset. http://vis.uky.edu/~stewe/ukbench/
22. Mikolajczyk, K.: Binaries for affine covariant region descriptors. http://www.robots.ox.ac.uk/~vgg/research/affine/

Hierarchical Hashing for Image Retrieval

Cheng Yan[1,2,3(✉)], Xiao Bai[1,2,3], Jun Zhou[1,2,3], and Yun Liu[1,2,3]

[1] School of Computer Science and Technology, Beihang University, Beijing, China
{beihangyc,baixiao,liuyun}@buaa.edu.cn
[2] School of Information and Communication Technology,
Griffith University, Nathan, Australia
jun.zhou@griffith.edu.au
[3] School of Automation Science and Electrical Engineering,
Beihang University, Beijing, China

Abstract. Hashing has been widely used in large-scale vision problems thanks to its efficiency in both storage and speed. The quality of hashing can be boosted when supervised information is used to learn hash functions. On large-scale hierarchical datasets, hierarchical semantic information reflects the relationship between classes and their children, which however has been ignored by most supervised hashing methods. In this paper, we propose a hierarchical hashing method for image retrieval. This method models and fuses both hierarchical semantic level relationship through taxonomy structure of dataset and feature level relationship of images into an integrated learning objective, then an optimization scheme is developed to solve the learning problem. Experiments are performed on two large-scale datasets: ImageNet ILSVRC 2010 and Animals with Attributes (AWA) dataset. Besides standard evaluation criteria, we also developed hierarchical evaluation criteria for image retrieval and classification tasks. The results show that the proposed method improves the accuracy of supervised hashing in both types of criteria.

1 Introduction

Nearest neighbor (NN) search has been widely adopted in image retrieval. The time complexity of the NN method on a dataset of size n is $O(n)$, which is infeasible for real-time retrieval on large dataset. Approximate nearest neighbor (ANN) search has been proposed to make NN search scalable, and becomes a preferred solution in many computer vision and machine learning applications [4,19]. The goal of ANN search is to find approximate results rather than exact ones so as to achieve high speed data processing [2,3,13,26]. Amongst various ANN search techniques, hashing is widely studied because of its efficiency in both storage and speed. By generating binary codes for image data, image retrieval on a dataset with millions of samples can be completed in a constant time using only tens of hash bits [32,34].

Classic hashing methods, such as locality sensitive hashing (LSH) [6], try to guarantee that close samples in the original space having similar binary codes.

© Springer Nature Singapore Pte Ltd. 2017
J. Yang et al. (Eds.): CCCV 2017, Part II, CCIS 772, pp. 111–125, 2017.
https://doi.org/10.1007/978-981-10-7302-1_10

These methods are data-independent hash, which do not employ a training set to learn the hash function. Besides LSH, some other data-independent hashing schemes have been proposed [1,14,27]. These approaches can be divided into unsupervised [11,22,35] and supervised [19,28,30] depending on whether they use label information. Most unsupervised hashing methods [11,22,35] retrieve neighbors based on metric distance, ignoring the semantic information. Supervised methods take full advantage of the annotation of the dataset and have higher search accuracy than unsupervised hashing methods. Some supervised methods take simple semantic relationship of data into consideration [25,28].

Though these supervised methods achieve good performance for retrieval, when dealing with multi-class data with complex taxonomy structure, most of them are not effective because they are developed on flat datasets in which the relationship of the categories of data is relatively simple. Over the past few years, there have been growing interests in building large-scale sets with hundreds or thousands of classes with hierarchical structure for computer vision tasks. ImageNet [9], as an example, is organized according to the Wordnet hierarchy. There can be thousands of nodes in the hierarchy, each of which represents a class and each node is depicted by hundreds and thousands of images. Some researchers focus on studying the semantic taxonomy to finish vision tasks [33,36]. They learn or construct specific structure for datasets to facilitate classification or recognition. For image retrieval, especially for hash retrieval, the hierarchy semantic information shall also be considered.

Suppose Figure 1 is a dataset and we have an image of E-guitar as a query. If the retrieval result is a group of images of E-guitar, it is absolutely the best result. If it returns the images of Trumpet, Saxophone, or Trombone, which are not Guitar, not to mention E-guitar, we consider the retrieve result as unsatisfactory. If it returns C-guitar, the result is quite good, since the C-guitar looks a lot like E-guitar and they belong to the same category Guitar.

In this paper, we study hierarchical hashing for image retrieval on large-scale dataset. Based on this motivation, our work aims to develop a supervised hashing

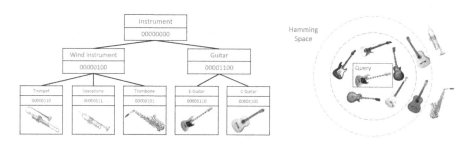

Fig. 1. An example that shows the hierarchical semantic structure of data. The images and structural relationship are extracted from the ImageNet Large Scale Visual Recognition Challenge 2010 (ILSVRC2010). Classic guitar and electric guitar are different objects. However, they belong to the same category of guitar. Their hash codes shall be different but with short Hamming distance. The right part is a sketch map to show the distances between a query and retrieval results in the Hamming space.

retrieval method that explores hierarchical semantic information on large-scale hierarchical dataset. There are three contributions of our work. First, we propose a hierarchical hashing for image retrieval on large-scale datasets by combining both semantic level relationship and feature level relationship into a hashing learning objective. Second, we develop an effective optimization solution for solving the learning objective. Third, we boost retrieval and classification performance with respect to both general evaluation criteria and specific hierarchical criteria for hierarchical datasets.

This paper is organized as follows. Section 2 introduces related work. Section 3 describes our hashing framework. The experimental results are presented in Sect. 4. The conclusions are drawn in Sect. 5.

2   Related Work

ImageNet [9] is a large visual database built for computer vision tasks such as recognition and classification. Some researchers study the semantic taxonomy to complete vision tasks on ImageNet. Verma et al. [33] proposed a method using taxonomy to learn similarity metrics. They showed that the learned metrics lead to improved classification performance. The metrics can provide effective classification amongst categories local to specific subtrees of the taxonomy. Zhao et al. [36], on the other hand, focused on using semantic taxonomy to guarantee classification accuracy and reduce dimensionality of parameter space at the same time. Nonetheless these research explored the semantic taxonomy of the hierarchical dataset, there are few works that explore the semantic taxonomy of dataset effectively for fast image retrieval.

Supervised hashing methods, such as semantic hashing (SH) [28], learning to hash with binary reconstructive embeddings (BRE) [18], minimal loss hashing for compact binary codes (MLH) [25], supervised hashing with kernels (KSH) [21], and supervised Discrete Hashing (SDH) [29], try to make the hash codes keep semantic similar or dissimilar relationship between samples. Most of them minimize the discrepancy between the data similarities and the Hamming distances of hash codes. During the training process, labels are provided to model the similarity/dissimilarity of data. For example, KSH [21] uses label 1 to specify similar pairs, label -1 to designate dissimilar pairs, and 0 for the unknown pairs. However, simply dividing the sample relationship into several specific values are not sufficient for characterizing the complex taxonomy structure in very large datasets. Our work takes such taxonomy structure into consideration, in which the hash codes of data pairs are partly based on their distances in the semantic structure.

Large-scale training sets is key to achieving good performance for many vision tasks [13,26,31]. Large datasets with taxonomy structure, e.g. ImageNet and ILSVRC [9], have been adopted by many retrieval or classification tasks [17,33,36]. Krizhevsky et al. [17] trained a deep convolutional neural network for classification on hierarchical datasets. Liong et al. [10] used CNN for binary code learning. Though having good performance, these methods have to learn a large number of model parameters. Verma et al. [33] proposed a framework based on probabilistic nearest-neighbor classification, aiming at learning similarity metrics using the class taxonomy, with focuses on using metrics for classification. Hashing was used to reduce the dimensionality of parameter space, but only for classification. Deng et al. [8] exploited prior knowledge of a semantic hierarchy for image retrieval. Its hashing retrieval method is close to LSH [6] which has advantage in efficiency but not accuracy. Our work is a hashing based solution, using hierarchy semantic information of dataset to learn projection for image retrieval. Meanwhile it also can be used for classification tasks.

3   Hierarchical Hashing Method

The basic idea of hashing is to project data onto a low-dimensional space so that each data is mapped to a binary vector. When a query arrives, the most similar items in the dataset are retrieved by comparing the mapped vectors (hash key) of the query and the dataset. Different from general hashing, our method is based on the hierarchical structure of the dataset. After training, each node in the hierarchy is also assigned with a hash key. This enables hierarchical retrieval.

3.1   Prediction Function

The purpose of hashing is to derive a group of hash functions accounting for the hash bits generation. Given a training set $\mathbf{X} = \{\mathbf{x}_1, \mathbf{x}_2, ..., \mathbf{x}_n\} \subset \mathbb{R}^d$, we use a kernel formulation to cope with linearly inseparable data in the target hash functions. We first define a prediction function

$$y = sgn(\sum_{i=1}^{m} \kappa(\mathbf{x}_i, \mathbf{x})w_i - b) \tag{1}$$

where y is a single bit hash code for the sample $\mathbf{x}$. $\mathbf{x}_1,...,\mathbf{x}_m$ are m samples uniformly and randomly selected from the training set and $m \ll n$ in which n is the size of the training set. $\mathbf{w} = [w_1, ..., w_m]^\top$ is a vector of coefficients. b is the bias which is defined as the mean of $\sum_{j=1}^{n} \sum_{i=1}^{m} \kappa(\mathbf{x}_i, \mathbf{x}_j)w_i$ [34,35]. The kernel function κ is defined as

$$\kappa(\boldsymbol{x}_i, \boldsymbol{x}_j) = exp(-(d(\boldsymbol{x}_i, \boldsymbol{x}_j))^2/2\sigma^2) \tag{2}$$

where σ is used to scale the exponential function. It derives a nonlinear model by mapping the data points to an infinite dimensional Hilbert space, which is more appropriate for the nonlinear dataset [34]. For convenience, when we have r hash bits for each sample, the hash function can be written as

$$\mathbf{y}_i = sgn(\mathbf{z}_i^\top \mathbf{W}) \tag{3}$$

where $\mathbf{y}_i = [y_1, ..., y_r]$ is the hash code of sample $\mathbf{x}_i$, and $\mathbf{z}_i$ is calculated as

$$\mathbf{z}_i = [\kappa(\mathbf{x}_1, \mathbf{x}_i) - \mu_1, \kappa(\mathbf{x}_2, \mathbf{x}_i) - \mu_2, ..., \kappa(\mathbf{x}_m, \mathbf{x}_i) - \mu_m]^\top \tag{4}$$

where

$$\mu_i = \sum_{j=1}^{n} \kappa(x_j, x_i)/n \tag{5}$$

and

$$\mathbf{W} = [\mathbf{w}_1, ..., \mathbf{w}_r] \tag{6}$$

In the prediction function, $\mathbf{W}$ defines the hash functions so that we need to learn $\mathbf{W}$ with supervised hierarchical information.

3.2 Semantic Level Learning

In many supervised hashing research [21, 25], pairwise labels are given as supervised information for training. When a pair of samples are similar, their pairwise label is 1, otherwise it is -1. The similarity between pairs may be acquired from their semantic relationship. However, most semantic information correspond to flat relationship. When structural semantic relationship is available on the dataset such as tree structure of the ImageNet, merely using -1 and 1 to reveal the pairwise relationship seems to be insufficient. Here we define a label matrix $\mathbf{M}_{ij}$ to reflect the relationship of all samples. The distance between two samples can be defined as

$$dis(\mathbf{x}_i, \mathbf{x}_j) = h(\mathbf{x}_i, \mathbf{x}_j)/h^* \tag{7}$$

where $h(\mathbf{x}_i, \mathbf{x}_j)$ is the hierarchical distance, defined as the number of levels to the lowest common ancestor of $\mathbf{x}_i$ and $\mathbf{x}_j$, e.g. $dis(E - guitar, C - guitar)$ is 1 in Fig. 1, and h^* is the height of the tree. We not only compute the height of the lowest common ancestor of two samples but also take the height of the tree they belong to into consideration. This is the normalized height that can better reflect the sample relationship. When the pairwise distances between all samples are generated, we can calculate an appropriate value to reveal the pairwise relations according to the distances. Since many works have used 1 and -1 to define the similarity, a value between -1 and 1 is preferred and can be defined as

$$g(dis(\mathbf{x}_i, \mathbf{x}_j)) = 1 - \gamma dis(\mathbf{x}_i, \mathbf{x}_j) \tag{8}$$

where γ is a parameter to control the contribution of the pairwise distance. Then the label matrix $\mathbf{M}$ can be calculated as

$$\mathbf{M}_{ij} = \begin{cases} g(dis(\mathbf{x}_i, \mathbf{x}_j)), & \text{if } (\mathbf{x}_i, \mathbf{x}_j) \in T \\ -1, & otherwisie \end{cases} \tag{9}$$

where $(\mathbf{x}_i, \mathbf{x}_j) \in T$ means $\mathbf{x}_i$ and $\mathbf{x}_j$ belong to the same tree.[1] When $\mathbf{x}_i$ and $\mathbf{x}_j$ have no common parent in the hierarchical dataset, $\mathbf{M}_{ij}$ is -1 which suggests they are not related. When they have a common parent, $\mathbf{M}_{ij}$ is based on the distance between them in the hierarchical structure.

---

[1] ImageNet is decomposed to some subtrees, and the reason is given in Sect. 4.

Since a node in the tree structure only has binary codes, the hash functions are only learned for images. We can learn hash functions by leveraging hierarchical supervised information, so that the generated hash codes of each sample reflect the hierarchical structure of the dataset as shown in Fig. 1. On one hand, the codes of nodes in different trees have larger Hamming distances in the binary space. On the other hand, in the same tree, the codes of nodes have longer common subsequence but are still different from each other. Inspired by the work of Liu et al. [21] that minimizes the code inner products and the Hamming distances, we define the objective function by incorporating the label matrix $\mathbf{M}$ as follows:

$$\mathbf{E_W} = ||\frac{1}{r}\mathbf{YY}^\top - \mathbf{M}||_F^2 \tag{10}$$

$\mathbf{M}$ is the label matrix and $\mathbf{Y}$ is defined as follows

$$\mathbf{Y} = \begin{bmatrix} \mathbf{y}_1(\mathbf{x}_1) \ ... \ \mathbf{y}_r(\mathbf{x}_1) \\ \mathbf{y}_1(\mathbf{x}_2) \ ... \ \mathbf{y}_r(\mathbf{x}_2) \\ ..... \\ \mathbf{y}_1(\mathbf{x}_n) \ ... \ \mathbf{y}_r(\mathbf{x}_n) \end{bmatrix} = sgn(\mathbf{Z}^\top\mathbf{W}) \tag{11}$$

where $\mathbf{Z} = [\mathbf{z}_1, ..., \mathbf{z}_n]$ and $\mathbf{W} = [\mathbf{w}_1, ..., \mathbf{w}_r]$. Equation (10) guarantees that the hash codes of each sample match its semantic relationship in the data hierarchy because both semantic and feature level relationships have been taken into consideration. In the next section, we analyze the feature level relationship.

3.3   Feature Level Learning

Hierarchical image retrieval requires binary codes for each node in the tree structure. Although hash codes for image samples can be calculated using hash functions, it is not straightforward to generate the hash codes for the nodes because they have category labels rather than vectorized data. Inspired by the hash clustering method proposed by Gong et al. [12], we propose to calculate each hash codes of node by binary clustering the hash codes of training samples belonging to this node. Assuming we have r bit hash codes for data and nodes, the objective function is

$$\min_{\mathbf{c}} \sum_{i=1}^{p} ||\mathbf{y}_i^\top - \mathbf{c}||_2^2 \tag{12}$$

$$s.t. \quad \mathbf{c} \in \{-1, +1\}^{1 \times r}.$$

where p is the number of training samples in the cluster, $\mathbf{y}_i$ is the binary hash codes of sample x_i, and $\mathbf{c} \subset \mathbb{R}^r$ is the center of the cluster. Then $\mathbf{c}$ is treated as the hash codes of the node. Note that p varies with the structure of the dataset. When the codes of leaf nodes are obtained, we can calculate the codes of their father node. Step by step, we can obtain the codes of all nodes. Let $\mathbf{C} = [\mathbf{c}_1, ..., \mathbf{c}_q] \subset \mathbb{R}^{r \times q}$ be the binary codes of nodes to be learned, where the number of nodes in the tree structure is q, all $\mathbf{c}$ must be restrained by

$$\mathbf{E_C} = \sum_{l=1}^{q} \sum_{i=1}^{p} ||\mathbf{y}_i^\top - \mathbf{c}_l||_2^2 \tag{13}$$

Minimizing Eq. (13) guarantees that each cluster center or node be close to its children at the feature level. Therefore, the hash codes of the father node and the hash codes of its children nodes have short Hamming distance (see the example of hash codes in Figure (1)). In this way, when we have the hash codes of training samples, we can estimate the hash codes of all nodes throughout the whole hierarchical structure by Eq. (13). Our optimization method guarantees the feature level relationship of these nodes.

3.4 Final Objective Function

Combining semantic level and feature level objective functions, the final objective function to be minimized is

$$
\begin{aligned}
\mathbf{E} &= \mathbf{E_W} + \lambda \mathbf{E_C} \\
&= ||\frac{1}{r} sgn(\mathbf{Z}^\top \mathbf{W}) sgn(\mathbf{Z}^\top \mathbf{W})^\top - \mathbf{M}||_F^2 \\
&\quad + \lambda \sum_{l=1}^{q} \sum_{i=1}^{p_c} ||sgn(\mathbf{z}_i^\top \mathbf{W}) - \mathbf{c}_l||_2^2
\end{aligned}
\tag{14}
$$

where the first term is from Eq. (10) and the second term is from Eq. (13). In the second term, we minimize the distance between every node and their children nodes in the hierarchical structure[2]. λ is a parameter to balance the contribution of two terms. When λ is small, the contribution of first term increases so that the whole function is more likely to follow the semantic relationship of the hierarchical structure since $\mathbf{M}$ models the semantic relationship. When λ is large, it is more likely to guarantee the feature level relationship since more weights are assigned to minimize the distances between samples and their cluster center. The model reduces to traditional flat hashing when λ is set to 0. The goal here is to estimate $\mathbf{W}$ by minimizing Eq. (14). In the next subsection, the optimization method is discussed.

3.5 Optimization

After some mathematical deductions, optimization of Eq. (14) can be re-written as

$$
\begin{aligned}
\min_{\mathbf{w}_k, c_{l_k}} &\ || \sum_{k=1}^{r} sgn(\mathbf{Z}^\top \mathbf{w}_k) sgn(\mathbf{Z}^\top \mathbf{w}_k)^\top - r\mathbf{M}||_F^2 \\
&+ \lambda r^2 \sum_{l=1}^{q} \sum_{i=1}^{p_c} \sum_{k=1}^{r} ||sgn(\mathbf{z}_i^\top \mathbf{w}_k) - c_{l_k}||_2^2
\end{aligned}
\tag{15}
$$

where $\mathbf{w}_k$ $(k = 1, ..., r)$ are separated from $\mathbf{W}$ in Eq. 14, and c_{l_k} is the is the k-th entry of $\mathbf{c}_l$. There are two terms to be optimized, $\mathbf{W}$ and $\mathbf{c}$. We alternatively optimize the objective function in r iterations. In each iteration, we only solve $\mathbf{w}_k$ and c_{l_k} which generates the k-th bit code of all the cluster centers given the previously solved vectors $\mathbf{w}_1^*, ..., \mathbf{w}_{k-1}^*$ and $\mathbf{c}_{l_1}^*, ..., \mathbf{c}_{l_{k-1}}^*$ $(l = 1, ..., q)$.

---

[2] The children of the leaf nodes are the training image samples.

Optimize W. In each iteration of optimization, we first optimize $\mathbf{w}_k$ with fixed c_{l_k} ($l = 1, ..., q$), which is the k-th dimension of the codes of all the cluster centers. Let

$$T_{k-1} = rM - \sum_{t=1}^{k-1} sgn(\mathbf{Z}^\top \mathbf{w}_t^*)sgn(\mathbf{Z}^\top \mathbf{w}_t^*)^\top \tag{16}$$

where $T_0 = rM$, then we can minimize the following cost function

$$||sgn(\mathbf{Z}^\top \mathbf{w}_k)sgn(\mathbf{Z}^\top \mathbf{w}_k)^\top - T_{k-1}||_F^2$$

$$+ \lambda \sum_{l=1}^{q} \sum_{i=1}^{p_c} \sum_{k=1}^{r} ||sgn(\mathbf{z}_i^\top \mathbf{w}_k) - c_{l_k}||_2^2$$

$$= (sgn(\mathbf{Z}^\top \mathbf{w}_k)(sgn(\mathbf{Z}^\top \mathbf{w}_k))^\top)^2$$

$$- 2(sgn(\mathbf{Z}^\top \mathbf{w}_k))^\top T_{k-1} sgn(\mathbf{Z}^\top \mathbf{w}_k) + tr(T_{k-1}^2)$$

$$+ \lambda r^2 \sum_{l=1}^{q} \sum_{i=1}^{p_c} \sum_{k=1}^{r} (sgn(\mathbf{z}_i^\top \mathbf{w}_k))^2 + \lambda r^2 \sum_{l=1}^{q} \sum_{i=1}^{p_c} \sum_{k=1}^{r} c_{l_k}^2 \tag{17}$$

$$- 2\lambda r^2 \sum_{l=1}^{q} \sum_{i=1}^{p_c} \sum_{k=1}^{r} sgn(\mathbf{z}_i^\top \mathbf{w}_k)c_{l_k}$$

$$= const - 2\lambda r^2 \sum_{l=1}^{q} \sum_{i=1}^{p_c} \sum_{k=1}^{r} sgn(\mathbf{z}_i^\top \mathbf{w}_k)c_{l_k}$$

$$- 2sgn(\mathbf{Z}^\top \mathbf{w}_k)^\top T_{k-1} sgn(\mathbf{Z}^\top \mathbf{w}_k)$$

Discarding the constant term, we get

$$\phi(\mathbf{w}_k) = -sgn(\mathbf{Z}^\top \mathbf{w}_k)^\top T_{k-1} sgn(\mathbf{Z}^\top \mathbf{w}_k)$$

$$- \lambda r^2 \sum_{l=1}^{q} \sum_{i=1}^{p_c} \sum_{k=1}^{r} sgn(\mathbf{z}_i^\top \mathbf{w}_k)c_{l_k} \tag{18}$$

Since $sgn(.)$ function is difficult to process, we replace it with a sigmoid-shaped function

$$\varphi(a) = \frac{2}{1 + e^{-a}} - 1 \tag{19}$$

which is sufficiently smooth and is an approximation of $sgn(.)$. Then Eq. (18) can be written as

$$\widetilde{\phi}(\mathbf{w}_k) = -\varphi(\mathbf{Z}^\top \mathbf{w}_k)^\top T_{k-1}\varphi(\mathbf{Z}^\top \mathbf{w}_k)$$

$$- \lambda r^2 \sum_{l=1}^{q} \sum_{i=1}^{p_c} \sum_{k=1}^{r} \varphi(\mathbf{z}_i^\top \mathbf{w}_k)c_{l_k} \tag{20}$$

The gradient of $\widetilde{\phi}$ with respect to w_k is

$$\nabla\widetilde{\phi} = -\mathbf{Z}((T_{k-1}\varphi(\mathbf{Z}^\top \mathbf{w}_k)) * (1 - \varphi(\mathbf{Z}^\top \mathbf{w}_k) * \varphi(\mathbf{Z}^\top \mathbf{w}_k)))$$

$$- \lambda r^2 \sum_{l=1}^{q} \sum_{i=1}^{p_c} \sum_{k=1}^{r} c_{l_k}\mathbf{z}_i * (1 - \varphi(\mathbf{z}_i) * \varphi(\mathbf{z}_i)) \tag{21}$$

where the symbol $*$ represents the Hadamard product and $\mathbf{1}$ is all one vector. Then $\mathbf{w}$ can be optimized by the gradient descent method [24].

Optimize c. We fix $\mathbf{w}_k$ to optimize c_{l_k}. Discarding the constant term and the terms that do not contain $\mathbf{c}$, Eq. (17) can be written as

$$\min_{c_{l_k}} -2\lambda r^2 \sum_{l=1}^{q} \sum_{i=1}^{p_c} \sum_{k=1}^{r} sgn(\mathbf{z}_i^\top \mathbf{w}_k) c_{l_k} \tag{22}$$

Since $c_{l_k} \in \{-1, 1\}$, minimizing Eq. (22) leads to

$$\psi = sgn(\lambda \sum_{l=1}^{q} \sum_{i=1}^{p_c} sgn(\mathbf{z}_i^\top \mathbf{W})^\top) \tag{23}$$

When $\mathbf{w}_k$ has been generated from Sect. 3.5, we can directly calculate the k-th code c_{l_k} for the center $\mathbf{c}_l$. With the new c_{l_k}, we can continue to optimize $\mathbf{w}_k$. This updates $\mathbf{W}$ so that c_{l_k} can be optimized again. $\mathbf{w}_k$ and c_{l_k} are optimized iteratively until the change of Eq. (15) is lower than a threshold. This completes the optimization for $\mathbf{w}_k$ and c_{l_k} $(k = 1, ..., r)$. For r bits, optimal $\mathbf{W}$ and $\mathbf{c}_l$ $(l = 1, ..., q)$ can be generated after r iterations. In the first iteration to estimate $\mathbf{w}_k$, we keep the first term of Eq. (15) unchanged and apply spectral relaxation by dropping the $sgn(.)$ term to compute $\mathbf{w}_k^0$ for initialization of $\mathbf{w}_k$

$$\min_{\mathbf{w}_k} -(\mathbf{Z}^\top \mathbf{w}_k)^\top T_{k-1}(\mathbf{Z}^\top \mathbf{w}_k) \tag{24}$$

which is similar to the method by Liu et al. [21]. This is a standard eigenvalue problem, i.e., $\mathbf{Z}T_{k-1}\mathbf{Z}^\top\mathbf{w}_k = \eta\mathbf{Z}\mathbf{Z}^\top\mathbf{w}_k$, where $\mathbf{w}_k$ is sought as the eigenvector associated with the largest eigenvalue. With $\mathbf{w}_k$ estimated, we can calculate c_{l_k} corresponding to $\mathbf{w}_k$, which is brought into the optimization function as the input to the initial round of calculation.

The time complexity for training stage is $O((nmt_w + pq)t_r r)$ where $n \gg p$, $n \gg q$, $n \gg m$, t_w is the number of iterations for $\mathbf{w}_k$, and t_r is the number of iterations for each r bits. The time complexity for testing stage is constant $O(mrd)$.

4   Experiments

4.1   Datasets

The first dataset is ImageNet ILSVRC2010 [9]. It is a subset of ImageNet which is organized according to the WordNet hierarchy. It contains about 1.2 million images belonging to 1,000 classes. A tree structure can be generated from the dataset, in which all 1,000 classes are considered as the leaf nodes whose children are images in the classes and the higher level nodes contain multiple categories in their corresponding children nodes. SIFT-based bag-of-words features were extracted to represent each image [23], so that each image was represented as a vector of 1,000 dimensions. We did not used the whole ImageNet dataset in the experiments because many image classes do not have multi-children hierarchical structures. They have only one node on each layer and only the leaf node contains images. In the experiments, we only used classes with more complex hierarchical structure (Fig. 2).

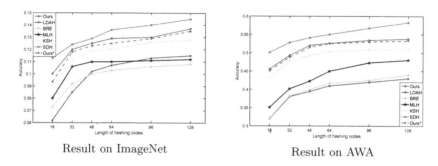

Result on ImageNet   Result on AWA

Fig. 2. The result of top-20 retrieved images using different number of bit codes with the state-of-art methods on Imagenet and AWA datasets.

The second dataset is Animals with Attributes (AWA) dataset [20] which contains 30475 images in 50 animals classes. The features adopted in our experiments were pre-trained using a seven layer CaffeNet [15]. Each image was represented by a vector of 4,096 dimensions. In the experiment, we used five subtrees of ImageNet and built a tree structure for the AWA dataset with the hierarchical relationship according to the WordNet hierarchy. The details are described in the corresponding experiments.

4.2   Hierarchical Hash Retrieval

In our method, each category in the taxonomy structure of the dataset is a node and has a hash code. To provide evaluation of different methods on the hierarchical retrieval task, we present a novel hierarchical search strategy here. In this strategy, we have multiple hash tables, the number of which equals the maximum depth of the trees. We can select different levels of retrieval results. For instance, given an E-guitar as a query, when we select the first level, i.e., the leaf node level which is the most accurate, the results are supposed to be formed by all E-guitar images. When we use the second level, the results may be formed by E-guitar and C-guitar. When we select the third level, the results may contain E-guitar, C-guitar and some other musical instruments such as Bass, and therefore, are less accurate. For each level, we build a hash table, in which we can replace the code of each image with the code of the node it belongs to in the binary space by calculating their hamming distances. Our model can provide an optional search strategy for different targets and most important of all, it requires only one time of training.

We used 502 classes with hierarchical structure for this experiment. We randomly selected 1000 samples in each class for training and the rest samples for testing. The max depth of these subtrees of ImageNet was 4, so we built 4 hash tables for retrieval. All the compared methods, including LDAhash (LDAH) [30], learning to hash with binary reconstructive embeddings (BRE) [18], minimal loss hashing for compact binary codes (MLH) [25], iterative quantization (CCA-ITQ) [11], supervised hashing with kernels (KSH) [21], and supervised discrete hashing (SDH) [29], were trained with the labels in the first level (leaf node level). With respect to retrieval, for different levels, the labels of both training and

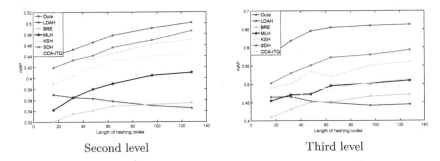

Second level   Third level

Fig. 3. The mAPs of different methods with varying code lengths for the second and third levels on the AWA dataset.

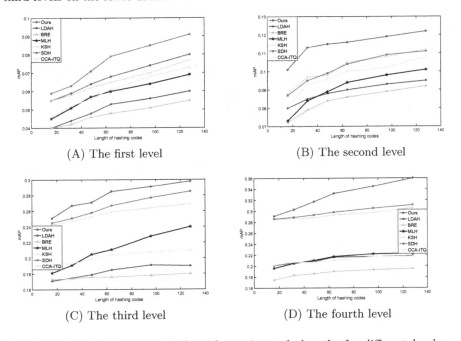

(A) The first level   (B) The second level

(C) The third level   (D) The fourth level

Fig. 4. mAPs of different methods with varying code lengths for different levels on ImageNet dataset.

testing images in the compared method were replaced by the labels of nodes in the corresponding levels. The results are displayed in Fig. 4.

With respect to the AWA dataset, we built a tree structure according to the WordNet hierarchy. The structure included 3 levels with 16 classes in the leaf nodes. The labels of all images were also replaced by the labels of cluster centers (nodes) in the corresponding level. The number of training samples in each class was 70% and the rest were used for testing. The results are shown in Fig. 3.

It is clear that our method has significantly outperformed all other approaches in high levels. Since the retrieval on the first level is equivalent to flat retrieval, the mAP of SDH is higher than ours. However, our method can better deal with hierarchical retrieval especially on high levels with one time of training. We use the structure relationship of the classes in the learning process, which guarantees that the classes belonging to the same father node have short Hamming distances than the classes that have no common father node. These results indicate that our method ensures the retrieval results by including images not only with similar features as the query but also close to the query in the dataset hierarchy.

4.3   Results on Classification

The purpose of this experiment was to evaluate the effectiveness of the intermediate nodes in the classification task. The experiments were performed on a subset of ImageNet containing 6 subtrees whose depth and the number of classes was shown in Table 1. The length of hash code was set to 256. The evaluation criterion for classification shall be able to reflect the effectiveness of hierarchical hashing.

To this end, we use both flat error and hierarchical error to compute the error rates. Given an image, if the predicted class is its true class, the flat error is 0, otherwise, the flat error is 1. The hierarchical error reports the distance from the lowest common ancestor to the level of predicted class and its true class. For example, in Fig. 1, if the query is Trombone and the predicted class

Table 1. Subtrees extracted from the ImageNet dataset for classification performance evaluation.

Subtree	Train	Test	Class	Depth
Music	2500	1250	25	6
Furniture	2300	1150	23	4
Tool	2600	1300	26	6
Amphibian	800	400	8	4
Geo	1200	600	12	5
AWA	1600	800	16	3

is E-Guitar, the hierarchical error is 2 since their lowest common ancestor is Instrument. This criterion has been used by Zhao et al. [36] to evaluate hierarchical feature hashing method.

The node classification process was as follows. We first compared the hash codes of a query with the root node of each tree, then compared it to the children of the root node, and moved to the child that had the shortest Hamming distance with the query. We repeated this process until the query image was classified into a leaf node. We compared our method with several classification methods which used image feature to train classifiers. These methods included multi-class SVM (MSVM) [5], hierarchical SVM (HieSVM) [7], tree classifier (TreeSVM) [16] which trained a multi-class SVM at each node, and HESHING [36] which used hash function for each class.

Table 2 shows the experimental results. The results show that our hierarchical hashing method has comparable classification performance as the alternative approaches. With the information of class taxonomy, our method has achieved better results on some subtrees of the dataset, especially for the hierarchy error. By keeping the critical information and the relationship of the hierarchical structure of dataset, our method gets benefit in achieving lower errors. In particular, the effectiveness of the node codes has been well-proven in this experiment.

Table 2. Classification results from various methods on ImageNet.

Subtree		Music	Fur	Tool	Amp	Geo	AWA
Depth		6	4	6	4	5	3
HieSVM	F Error	0.80	0.58	**0.72**	0.64	**0.50**	0.45
	H Error	4.42	2.16	**3.76**	1.86	2.14	1.62
TreeSVM	F Error	0.82	0.64	0.86	0.70	0.52	0.43
	H Error	4.82	3.16	5.54	2.42	2.36	1.78
HESHING	F Error	**0.76**	**0.54**	0.74	0.60	0.50	0.41
	H Error	4.12	2.34	3.96	1.82	2.18	**1.41**
MSVM	F Error	0.80	0.61	0.79	0.62	0.50	0.44
	H Error	4.60	2.75	4.89	1.94	2.24	1.75
Ours	F Error	0.81	0.61	0.73	**0.59**	0.51	**0.40**
	H Error	**3.80**	**1.98**	3.80	**1.81**	**2.05**	1.43

5   Conclusions

In this paper, a hierarchical hashing method for image retrieval has been introduced. This method takes both hierarchical semantic level relationship and feature level relationship of images into consideration and fuses them to a learning objective. The proposed method has been evaluated using specific hierarchical criteria on the hierarchical databases. The experimental results on two hierarchical datasets show that our method has better performance than several approaches on image retrieval and classification tasks.

References

1. Andoni, A., Indyk, P.: Near-optimal hashing algorithms for approximate nearest neighbor in high dimensions. In: IEEE Symposium on Foundations of Computer Science, pp. 459–468 (2006)
2. Arandjelović, R., Zisserman, A.: Three things everyone should know to improve object retrieval. In: IEEE Conference on Computer Vision and Pattern Recognition, pp. 2911–2918 (2012)
3. Berg, A.C., Berg, T.L., Malik, J.: Shape matching and object recognition using low distortion correspondences. In: IEEE Conference on Computer Vision and Pattern Recognition, vol. 1, pp. 26–33 (2005)
4. Charikar, M.S.: Similarity estimation techniques from rounding algorithms. In: ACM Symposium on Theory of Computing, pp. 380–388 (2002)
5. Crammer, K., Singer, Y.: On the algorithmic implementation of multiclass kernel-based vector machines. J. Mach. Learn. Res. **2**, 265–292 (2002)
6. Datar, M., Immorlica, N., Indyk, P., Mirrokni, V.S.: Locality-sensitive hashing scheme based on p-stable distributions. In: Annual Symposium on Computational Geometry, pp. 253–262 (2004)
7. Dekel, O., Keshet, J., Singer, Y.: Large margin hierarchical classification. In: International Conference on Machine Learning, p. 27 (2004)
8. Deng, J., Berg, A.C., Fei-Fei, L.: Hierarchical semantic indexing for large scale image retrieval. In: IEEE Conference on Computer Vision and Pattern Recognition, pp. 785–792. IEEE (2011)
9. Deng, J., Dong, W., Socher, R., Li, L.J., Li, K., Fei-Fei, L.: Imagenet: A large-scale hierarchical image database. In: IEEE Conference on Computer Vision and Pattern Recognition, pp. 248–255 (2009)
10. Erin Liong, V., Lu, J., Wang, G., Moulin, P., Zhou, J.: Deep hashing for compact binary codes learning. In: IEEE Conference on Computer Vision and Pattern Recognition, pp. 2475–2483 (2015)
11. Gong, Y., Lazebnik, S.: Iterative quantization: A procrustean approach to learning binary codes. In: IEEE Conference on Computer Vision and Pattern Recognition, pp. 817–824. IEEE (2011)
12. Gong, Y., Pawlowski, M., Yang, F., Brandy, L., Boundev, L., Fergus, R.: Web scale photo hash clustering on a single machine. In: IEEE Conference on Computer Vision and Pattern Recognition, pp. 19–27 (2015)
13. Hays, J., Efros, A.A.: Scene completion using millions of photographs. Trans. Graph. **26**(3), 4 (2007)
14. Jain, P., Kulis, B., Grauman, K.: Fast image search for learned metrics. In: IEEE Conference on Computer Vision and Pattern Recognition, pp. 1–8 (2008)
15. Jia, Y.: Caffe: An open source convolutional architecture for fast feature embedding (2013)
16. Koller, D., Sahami, M.: Hierarchically classifying documents using very few words (1997)
17. Krizhevsky, A., Sutskever, I., Hinton, G.E.: Imagenet classification with deep convolutional neural networks. In: Proceedings of the Neural Information Processing Systems Conference, pp. 1097–1105 (2012)

18. Kulis, B., Darrell, T.: Learning to hash with binary reconstructive embeddings. In: Proceedings of the Neural Information Processing Systems Conference, pp. 1042–1050 (2009)

19. Kulis, B., Grauman, K.: Kernelized locality-sensitive hashing for scalable image search. In: IEEE Conference on Computer Vision and Pattern Recognition, pp. 2130–2137 (2009)

20. Lampert, C.H., Nickisch, H., Harmeling, S.: Learning to detect unseen object classes by between-class attribute transfer. In: IEEE Conference on Computer Vision and Pattern Recognition, pp. 951–958 (2009)

21. Liu, W., Wang, J., Ji, R., Jiang, Y.G., Chang, S.F.: Supervised hashing with kernels. In: IEEE Conference on Computer Vision and Pattern Recognition, pp. 2074–2081 (2012)

22. Liu, W., Wang, J., Kumar, S., Chang, S.F.: Hashing with graphs. In: International Conference on Machine Learning, pp. 1–8 (2011)

23. Lowe, D.G.: Distinctive image features from scale-invariant keypoints. Int. J. Comput. Vision **60**(2), 91–110 (2004)

24. Nesterov, Y.: Introductory Lectures on Convex Optimization, vol. 87. Springer Science & Business Media, New York (2004)

25. Norouzi, M., Blei, D.M.: Minimal loss hashing for compact binary codes. In: International Conference on Machine Learning, pp. 353–360 (2011)

26. Philbin, J., Chum, O., Isard, M., Sivic, J., Zisserman, A.: Object retrieval with large vocabularies and fast spatial matching. In: IEEE Conference on Computer Vision and Pattern Recognition, pp. 1–8 (2007)

27. Raginsky, M., Lazebnik, S.: Locality-sensitive binary codes from shift-invariant kernels. Adv. Neural Inf. Process. Syst. **22**, 1509–1517 (2009)

28. Salakhutdinov, R., Hinton, G.: Semantic hashing. Int. J. Approximate Reasoning **50**(7), 969–978 (2009)

29. Shen, F., Shen, C., Liu, W., Shen, H.T.: Supervised discrete hashing. In: IEEE Conference on Computer Vision and Pattern Recognition, pp. 37–45 (2015)

30. Strecha, C., Bronstein, A.M., Bronstein, M.M., Fua, P.: LDAHash: Improved matching with smaller descriptors. IEEE Trans. Pattern Anal. Mach. Intell. **34**(1), 66–78 (2012)

31. Torralba, A., Fergus, R., Freeman, W.T.: 80 million tiny images: A large data set for nonparametric object and scene recognition. IEEE Trans. Pattern Anal. Mach. Intell. **30**(11), 1958–1970 (2008)

32. Torralba, A., Fergus, R., Weiss, Y.: Small codes and large image databases for recognition. In: IEEE Conference on Computer Vision and Pattern Recognition, pp. 1–8 (2008)

33. Verma, N., Mahajan, D., Sellamanickam, S., Nair, V.: Learning hierarchical similarity metrics. In: IEEE Conference on Computer Vision and Pattern Recognition, pp. 2280–2287 (2012)

34. Wang, J., Kumar, S., Chang, S.F.: Semi-supervised hashing for large-scale search. IEEE Trans. Pattern Anal. Mach. Intell. **34**(12), 2393–2406 (2012)

35. Weiss, Y., Torralba, A., Fergus, R.: Spectral hashing. In: Advances in Neural Information Processing Systems, pp. 1753–1760 (2009)

36. Zhao, B., Xing, E.P.: Hierarchical feature hashing for fast dimensionality reduction. In: IEEE Conference on Computer Vision and Pattern Recognition, pp. 2051–2058 (2014)

Visual Saliency Fusion Based Multi-feature for Semantic Image Retrieval

Jianan Chen[1], Cong Bai[1(✉)], Ling Huang[1], Zhi Liu[2], and Shengyong Chen[1]

[1] College of Computer ,Science, Zhejiang University of Technology, Hangzhou, China
congbai@zjut.edu.cn

[2] School of Communication and Information Engineering,
Shanghai University, Shanghai, China

Abstract. In this paper, a saliency fusion based content-based image retrieval method is proposed. Different saliency detection methods were conducted firstly and the output saliency maps were fused by double low rank matrix recovery method. Then the images were segmented into foreground and background according to the fusion result. As the foreground and background had the different impacts on the semantic understanding of the image, different features represented in the form of histogram were extracted. Finally, a fusion of z-score normalized Chi-Square distance is adopted as the similarity measurement. This proposal has been implemented on three widely used benchmark databases and the results evaluated in terms of mean Average Precision (mAP), precision, recall, and $F1$-measure show that our proposal outperforms the referred state-of-the-art approaches.

Keywords: Saliency fusion · Bag of words
Semantic image retrieval · Chi-square

1 Introduction

With the development of visual technology, more attention have been put on image retrieval both in the industry and the research community. Image retrieval techniques could be widely classified into two categories: text-based and content-based. The text-based approach index images in database by key-words, which came from manually added annotation. However, manual annotation is an imprecise and time consuming job. Content-based image retrieval (CBIR) searches images by their own visual contents, which has been presented in the early 1990s [15]. However, how to extract meaningful features from the large collections of image data is still a challenging problem due to the deviation of semantic understanding between human and computer [9]. Semantic gap exists between low-level handcrafted features and high-level human perception [8]. The reason is that a highly evolved human brain could transform the visual signals into concrete subject, while the computer couldn't do that with high accuracy up to now.

J. Yang et al. (Eds.): CCCV 2017, Part II, CCIS 772, pp. 126–136, 2017.
https://doi.org/10.1007/978-981-10-7302-1_11

Many researches have been dedicated into semantic image retrieval, comprehensive reviews could be found in [12,15]. Bag-of-words (BOW) framework, which is initially proposed in [17], is the most famous one among them. The key idea of BoW is to quantify each local feature into one or more so-called visual words, and each image is represented as a set of unordered visual words. Our work also used BOW partially. Furthermore, we noted that the semantic understanding of the image could be divided into two parts: the foreground object and the background regions in general. That means that retrieval results possibly met the high level perception could be got if foreground and background were represented individually by different features. In our previous work [1], image segementation based on RC-saliency [3] was used to segment the foreground and background in the image, and got a better performance in semantic image retrieval. Although RC-saliency has a good effect in the direction of color division, it has some limitations in texture and shape. That means it could only perform well in special kinds of images. So it is far from the broadness of human vision. Thus segmentation based on one visual saliency model could have the limitations on its universality, which could decreases the performance of the retrieval. Obviously, better retrieval results could be got if the performance of segmentation improved. So a saliency fusion based multi-feature model is take into consideration in our proposal.

In this paper, a novel semantic image retrieval method named saliency fusion based multi-feature (SFMF) is proposed. Firstly, we figure out seven saliency maps generated by different methods, and fused them by double low rank matrix recovery method (DLRMR) [7]. Secondly, SaliencyCut [3] based on the fused saliency map is used to segment images into foreground objects and background regions. Finally, local features and global features extracted from both foreground objects and background regions with different weights in the similarity fusion are used for retrieval.

The remaining of this paper is organized as follows. Section 2 describes SFMF in details. Experiment results and analysis are given in Sect. 3, followed by conclusion in Sect. 4.

2   The Proposed SFMF

There are three phrases in the SFMF: fusion stage, offline processing and online retrieval. In the fusion stage, segmentation based on DLRMR saliency fusion is performed on each database image, and thus each image is divided into two parts: foreground object and background region. After the segmentation by fusion saliency maps, images are represented by a multi-feature representation. Furthermore, between the foreground objects and background regions, different features and different weights in the similarity are considered in the retrieval.

2.1   Fusion Stage

The framework of the segmentation based on saliency fusion is illustrated in Fig. 1. Different from the onefold saliency map, saliency fusion intended to

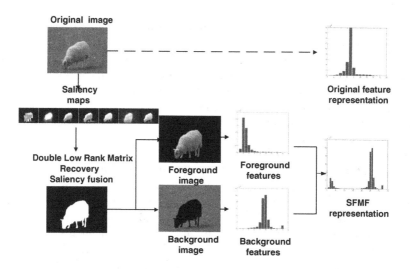

Fig. 1. Segmentation based on saliency fusion. Segmentation is executed by SaliencyCut using saliency map generated by saliency fusion by double low rank matrix recovery. After segmentation, background region and foreground object are obtained and represented by different features.

combine various saliency detection methods makes the fusion results better than each individual saliency detection methods. It highlights the advantages of several algorithms and avoids the weakness of a few algorithms so that the final saliency map obtained as a result of fusion outperforms each of them. For the reasons mentioned above, double low rank matrix recovery(DLRMR)[7] is used to cast object and background decomposition problem. Furthermore, different features are used to represent foreground objects and background regions, which makes the SFMF representation to be more characteristic than the solo feature representation of the whole image.

First of all, seven saliency detection methods: AMC [6], BL [18], BSCA [16], HC [3], MR [20], MS [19], ST [13] have been chosen to obtain saliency maps $\{S_k | 1 \leq k \leq 7\}$.

Then mean shift algorithm [4] segmented the image into regions $\{P_i\}_{i=1,...,n}$, where n is the number of super-pixels. The saliency map S_k could be represented using an n-dimensional vector $\mathbf{X}_k = [x_{1k}, x_{2k}, ..., x_{nk}]^T$, the ith element of the vector corresponds to the mean of the saliency values of pixels in the super-pixel P_i. By arranging $\mathbf{X}_k$ into a matrix, we get the combined matrix representation of individual saliency maps as $\mathbf{X} = [\mathbf{X}_1, \mathbf{X}_2, ..., \mathbf{X}_7].\mathbf{X} \in R^{n \times 7}$. With super-pixel instead of pixel as the smallest unit to calculate more in line with content semantics visual.

Generally, a natural image $\mathbf{I}$ could be decomposed as:

$$\Phi(\mathbf{I}) = \mathbf{A} + \mathbf{E}, \tag{1}$$

where Φ indicates a certain transformation, $\mathbf{A}$ and $\mathbf{E}$ denote matrices corresponding to background and foreground. Then treat matrix $\mathbf{X}$ as a feature representation of the image $\mathbf{I}$ in the saliency feature space, with each row representing a super-pixel feature vector [7]. Equation (1) could be rewritten as:

$$\mathbf{X} = \mathbf{A} + \mathbf{E}, \tag{2}$$

Therefore, saliency fusion could be cast as a low rank affinity pursuit. Given matrix $\mathbf{X} = [\mathbf{X}_1, \mathbf{X}_2, ..., \mathbf{X}_7]$. $\mathbf{X} \in R^{n \times 7}$, the low rank matrix recovery problem could then be formulated as:

$$\min_{\mathbf{A},\mathbf{E}} rank(\mathbf{A}) + \lambda(rank(\mathbf{E})) \quad s.t. \ \mathbf{X} = \mathbf{A} + \mathbf{E}, \tag{3}$$

where parameter $\lambda > 0$ balances the effects between two ranks [7].

Through the alternating direction method of multipliers(ADMM) [2], the final low rank $\mathbf{E}$ measures the contribution of each saliency method and learns an adaptive combination of the maps. It counts the value of E in every region on each saliency map and separate the object area from the background with a suitable threshold, which is recommended 1.4 times around average valuer.

Some comparison examples of the segmentation using the saliency maps fused by DLRMR and generated by one single visual saliency model, such as RC-saliency are shown in Fig. 2. From these examples, we could observe that saliency map fusion could get better segmentation performance than single visual saliency model.

Fig. 2. Given the input image and the ground truth, the segmentation using the saliency maps generated by DLRMR are better than by one saliency model such as RC-saliency, the method used in our previous work.

2.2 Offline Processing

After the images have been segmented into foreground objects and background regions, different feature will be extracted considering their different characters, which is distinct from traditional image retrieval methods.

For the background regions, features are extracted in HSV color space. As they had large areas of similar colors and textures in general, local binary patterns (LBP) in V channel and color histograms in H and S channel were extracted as color and texture features. We choose these two features not only they are simple but also efficient.

For the foreground objects, beside the texture and color features as extracted from the background, local features should also be considered. So the Scale-invariant feature transform (SIFT) [14] is chosen as an instinct choice for its successful achievements in object retrieval task. SIFT feature is packed in the BOW framework for retrieval. That means SIFT features extracted from the images will be compared with a visual words vocabulary clustered by K-means and the frequency of the visual words appearance in the image will be used as the representation of the image.

The features extracted from the images are defined as formula (4),

$$F \begin{cases} F_f = (H_h, H_s, LBP_v, SIFT_g), \\ F_b = (H_h, H_s, LBP_v), \end{cases} \tag{4}$$

where F_f is features of foreground image, and F_b is features of background image, H_h and H_s are histogram features in hue (H) and saturation (S) channel of HSV color space, LBP_v is histogram of local binary patterns statistics in value (V) channel in HSV color space, where $SIFT_g$ is the histogram of visual words in gray level space. We set the parameters in advance, the weight of F_b is less than F_f.

2.3 Online Retrieval

The input query image is also needed to segment based on saliency fusion and represented by different features in the way mentioned in Section 2.2. A fusion of the z-score normalized chi-square distances is proposed to measure the similarity between the query and the images in the database.

The chi-square distance between the histogram of query image H_Q and the histogram of image from the database H_I is defined as:

$$DS(Q, I) = \sum_{j=1}^{K} \frac{(H_Q(j) - H_I(j))^2}{H_Q(j) + H_I(j)}, \tag{5}$$

where K is the number of bins in the histogram.

Since different histograms are constructed for an image, the similarity between the images is measured by fusing the distances of histograms with different weights. However, normalization before fusion is necessary because each histogram is composed of different feature vectors. Additionally, in order to avoid errors introduced by outliers, distances are normalized through the following ways: given a query image, by calculating distances on one type of histogram between this query and all images from the database, one set of distances

$\{DS_A(Q, I_i)\}$ obtained, where $i = \{1, 2, ..., P\}$ and P is the number of images in the database. Thus the normalized distance is defined as:

$$DS_A^N(Q, I_i) = \frac{DS_A(Q, I_i) - \mu_{A_Q}}{\sigma_{A_Q}}, \qquad (6)$$

where μ_{A_Q} and σ_{A_Q} are the mean value and the standard deviation of the distances set $\{DS_A(Q, I_i)\}$ respectively.

Finally, all kinds of distances are fused as one distance to decide the similarly between the images and the query. In the stage of distances fusion, not all of the distances have the same weight, more weights on the LBPs distances of background and foreground are given. In particularly, 3 times on foreground and 2 times on background is the better choice as an experimental choice.

The SFMF algorithm framework is summarized in Algorithm 1.

Algorithm 1. SFMF framework

Input: A query image
1: Conduct individual saliency detection methods and obtain S_k.
2: Compute saliency fusion map by DLRMR and segment the image into foreground and background.
3: Extract visual features F_f from foreground object.
4: Extract visual features F_b from background regions.
5: Combine F_f and F_b to obtain F.
6: Measure the distance between query image F and images in dataset by z-score normalized chi-square distance.
Output:
Image retrieval results ordered by the distance.

---

3   Experiments

3.1   Image Dataset

The experiments were executed on three publicly available and widely used benchmark database: Corel 5k [10], VOC 2006 [5] and Corel 10k [11].

Corel 5k contains 50 themes with 100 images for each of size $192 * 128$ or $128 * 192$ in JPEG format. 8 themes were selected as the compared methods did, which have an obvious object in the image. The selected themes including "bear", "pyramid", "building", "plane", "snowberg", "horse", "tiger" and "train". Among those images, 200 images are used to train the vision dictionary, and 800 images for testing.

In Pascal VOC 2006, ten themes were selected: "sheep", "motorbike", "cow", "horse", "dog", "bus", "car", "person" and "bicycle" and 1000 images are selected randomly covered all these ten themes. Same with the Corel 5k, 200 images are used to train the visual words dictionary, and the rest of them are used for testing.

To prove the versatility of the algorithm, a large scale image database, Corel 10 is chosen. Corel 10k has expanded the number of pictures in the B set to reach 10000 images in total, and contains 100 categories from diverse contents such as sunset, beach, flower, building, car, horses, mountains, fish, food, door, etc.

Some examples of queries from these databases and its retrieval results are demonstrated in Fig. 3. From these examples, we could see that the proposed algorithm could get very good retrieval results obviously for the images containing salient objects with clear background as the salient objects were firstly segmented from the background regions, such as 'Poker' and 'Gun' in the figure. Furthermore, the proposed algorithm could also get good results for the images containing salient objects with complicated background, as 'Hippo' and 'Terraces' in the figure.

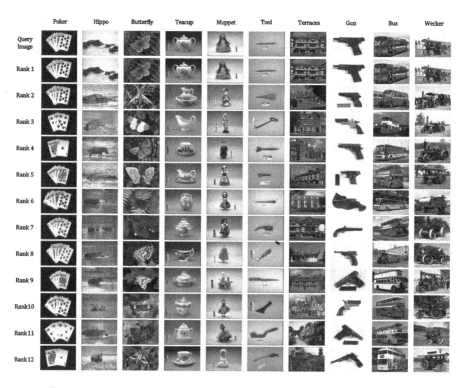

Fig. 3. Examples of query and its top 12 retrieved images

3.2   Experimental Measurement

We compute the bounded mean Average Precision (mAP) to count how many positive images at top K relevant results. The mAP at bound K is defined as follows:

$$mAP_K = \frac{\sum_{i=1}^{N} AP_K}{N}, \qquad 1 \leqslant K \leqslant N, \tag{7}$$

where N is the number of query topics, K is the top K retrieval results considered. And AP at bound K is defined as follows:

$$AP_K = \frac{\sum_{i=1}^{K} \rho_i P_i}{N_{GT}}, \qquad (8)$$

where K is computing depth of AP, ρ_i is a boolean function defined as follows:

$$\rho_i = \begin{cases} 1, & the\ i-th\ query\ result\ is\ correct, \\ 0, & the\ i-th\ query\ result\ is\ incorrect, \end{cases} \qquad (9)$$

and P_i is precision of top i results, N_{GT} is the number of positive samples in top K query results from Ground Truth [21].

Other two metrics are precision and recall. These two metrics are often combined as the weighted harmonic mean, namely F-measure, and it is an overall performance measure [11]. It could be defined as follows:

$$\begin{cases} F = \frac{(1+\beta^2) \times P \times R}{(\beta^2 \times P) + R}, \\ P = \frac{I_N}{N}, \\ R = \frac{I_N}{M}, \end{cases} \qquad (10)$$

In the experiments of image retrieval, $precision(P)$ is the ratio of the number of retrieved similar images to the number of retrieved images, while $recall(R)$ is the ratio of the number of retrieved similar images to the total number of similar images. Where I_N is the number of retrieved similar images, N is the total number of images retrieved and M is the total number of similar images in the database. The coefficient β allows one to weight either precision or recall more heavily, and they are balanced when $\beta = 1$. If there is no particular reason to favor precision or recall, $\beta = 1$ is commonly used to image retrieval or information retrieval. Parameters are set as the same as [11], $N = 12$; $M = 100$ and $\beta = 1$ on Corel-10k dataset, Thus, F-measure is so called F_1-measure.

3.3 Comparison

BOW could be seen as the baseline of content-based image retrieval, especially for the semantic image retrieval. RoI-BOW [21] is a recently reported method that improved the BOW model with the region segmentation. With these segmentation, images could be seen as the combination of the regions of interest (RoI) and the regions of Non-RoI. SSH and MSD are also recently reported by Liu [10, 11]. These algorithms used saliency model also and got a good results in image retrieval field. SBMF is our earlier work [1], thus we selected the BOW, RoI-BOW, SSH, MSD and SBMF for comparisons.

Tables 1 and 2 show the mAP at top ten and top twenty retrieved results for each theme in Corel 5K and VOC 2006 database respectively. From the table, we could conduct the conclusion that the performance of the proposed SFMF outperforms in most of the themes selected and archives the best performance overall.

Table 1. Comparison of the mAP on Corel 5K

Themes	Top 10 (%)				Top 20 (%)			
	SFMF	SBMF	ROI	BOW	SFMF	SBMF	ROI	BOW
Pyramid	**98.57**	97.78	94.08	81.13	**96.91**	96.12	90.33	70.76
Bear	**90.79**	82.44	83.78	73.57	**83.55**	69.71	66.71	60.67
Building	94.21	**94.25**	87.11	76.76	89.55	**89.76**	78.50	63.83
Horse	**99.69**	97.60	98.61	74.54	**98.94**	94.72	95.69	60.44
Plane	**93.14**	90.14	89.26	86.47	**87.96**	84.00	79.79	78.93
Snowberg	**90.54**	89.22	88.17	75.94	**84.79**	82.82	79.68	59.30
Tiger	**95.95**	90.93	94.21	77.22	**92.00**	83.92	86.34	62.99
Train	**90.36**	89.20	90.86	73.73	**85.13**	82.30	86.34	56.69
Average	**94.16**	91.45	90.76	77.42	**89.85**	83.14	82.92	64.20

Table 2. Comparison of the mAP on VOC 2006

Themes	Top10 (%)				Top20 (%)			
	SFMF	SBMF	ROI	BOW	SFMF	SBMF	ROI	BOW
Bicycle	**97.85**	91.80	86.70	85.44	**95.92**	85.07	74.78	73.62
Bus	**92.40**	85.12	83.47	81.90	**86.48**	76.15	70.34	63.65
Car	**99.64**	97.39	84.47	79.42	**99.02**	95.24	69.92	62.17
Cat	**79.56**	77.30	71.82	75.97	**66.13**	61.87	55.21	58.31
Cow	**90.00**	89.62	81.82	81.84	79.59	**81.40**	71.72	61.94
Dog	82.80	**82.98**	76.51	74.99	**66.13**	65.15	60.64	59.90
Horse	**82.19**	76.27	73.19	77.06	**69.20**	61.06	56.43	58.00
Motorbike	**82.06**	78.68	77.72	77.57	**70.41**	66.21	59.33	60.62
Person	80.56	**80.71**	81.43	79.15	**66.20**	65.28	62.48	58.18
Sheep	**96.62**	93.82	84.48	77.06	**93.46**	89.70	68.39	57.82
Average	**88.38**	85.37	80.16	79.04	**79.25**	74.71	64.92	61.42

To further verify the improvement, experiments are conducted on Corel 10k database also, the results are shown in Table 3. From this table, we could see that the proposed SFMF algorithm achieves the better performances than the BOW baseline, MSD, SSH and SBMF.

3.4   Running Efficiency

Although the proposed SFMF uses multiple features in retrieval, this proposal scheme has high efficiency in terms of vector space and runtime. The length of the feature vector of the foreground objects is 366: 256 of LBP, 30 of SIFT, 80 of color histogram; and the length of the feature vector of the background regions is

Table 3. The precision, recall, F-measure of five methods on Corel 10k

	precision(%)	recall(%)	F-measure(%)
BOW	30.36	3.64	6.51
MSD	45.62	5.48	9.78
SSH	54.88	6.58	11.76
SBMF	59.98	7.20	12.85
SFMF	**74.59**	**8.95**	**15.98**

336: 256 of LBP, 80 of color histogram. So the total length of the feature vector for the retrieval is 702.

The running time evaluation experiments are implemented in Matlab2013b on the laptop with a Core I7-4720 processor with 8 GB memory. The running time of segmenting and training visual words reaches about 2.6 s per image and 1.4 s per image respectively, and the running time of retrieval is about 0.5 s per query. It should be noted that the segmentation and training could be done off-line prior to retrieval.

4   Conclusion and Perspectives

A novel semantic image retrieval method named SFMF was proposed, which integrated saliency fusion method and traditional image feature representation. Images were segmented into foreground objects and background regions by saliency maps generated by visual saliency fusion, then different features were extracted in consideration of different characteristics of foreground and background. Experiments implemented on three widely used databases have proved the amelioration of the segmentation based on visual saliency fusion could lead the better performance.

However, it also should be noted that the calculation load of the segmentation and visual saliency fusion is still heavy. Furthermore, more efficient features vector needs to be discovered in the future as the size of LBP constitutes the large proportion of the feature vector. These two issues could be discovered perspectively.

Acknowledgments. This work is supported by Natural Science Foundation of China under Grant No. 61502424, 61471230, U1509207 and 61325019, Zhejiang Provincial Natural Science Foundation of China under Grant No. LY15F020028. The author would like to thanks Dr. Junxia Li and Prof. Jian Yang for providing the source code of DLRMR [7].

References

1. Bai, C., Chen, J.N., Huang, L., Kplama, K., Chen, S.: Saliency based multi-feature modeling for semantic image retrieval. J. Vis. Commun. Image Representation (Accepted)
2. Boyd, S., Parikh, N., Chu, E., Peleato, B., Eckstein, J.: Distributed optimization and statistical learning via the alternating direction method of multipliers. Found. Trends® Mach. Learn. **3**(1), 1–122 (2011)
3. Cheng, M.M., Mitra, N.J., Huang, X., Torr, P.H.S., Hu, S.M.: Global contrast based salient region detection. IEEE Tran. Pattern Anal. Mach. Intel. **37**(3), 569–582 (2015)
4. Comaniciu, D., Meer, P.: Mean shift: a robust approach toward feature space analysis. IEEE Trans. Pattern Anal. Mach. Intel. **24**(5), 603–619 (2002)
5. Everingham, M., Zisserman, A., Williams, C.K.I., Van Gool, L.: The PASCAL Visual Object Classes Challenge 2006 (VOC 2006). http://www.pascal-network. org/challenges/VOC/voc2006/results.pdf
6. Jiang, B., Zhang, L., Lu, H., Yang, C., Yang, M.H.: Saliency detection via absorbing markov chain. In: ICCV, pp. 1665–1672 (2013)
7. Li, J., Lei, L., Zhang, F., Jian, Y., Rajan, D.: Double low rank matrix recovery for saliency fusion. IEEE Trans. Image Process. **25**(9), 4421–4432 (2016)
8. Li, Z., Tang, J.: Weakly supervised deep matrix factorization for social image understanding. IEEE Trans. Image Process. **26**(1), 276–288 (2017)
9. Li, Z., Tang, J., He, X.: Robust structured nonnegative matrix factorization for image representation. IEEE Trans. Neural Networks Learn. Syst. **PP**(99), 1–14 (2017)
10. Liu, G.H., Li, Z.Y., Zhang, L., Xu, Y.: Image retrieval based on micro-structure descriptor. Pattern Recogn. **44**(9), 2123–2133 (2011)
11. Liu, G.H., Yang, J.Y., Li, Z.Y.: Content-based image retrieval using computational visual attention model. Pattern Recogn. **48**(8), 2554–2566 (2015)
12. Liu, Y., Zhang, D., Lu, G., Ma, W.: A survey of content-based image retrieval with high-level semantics. Pattern Recogn. **40**(1), 262–282 (2007)
13. Liu, Z., Zou, W., Le, M.O.: Saliency tree: a novel saliency detection framework. IEEE Trans. Image Process. **23**(5), 1937–1952 (2014)
14. Lowe, D.G., Lowe, D.G.: Distinctive image features from scale-invariant keypoints. Int. J. Comput. Vis. **60**(2), 91–110 (2004)
15. Mei, T., Rui, Y., Li, S., Tian, Q.: Multimedia search reranking. ACM Comput. Surv. **46**(3), 1–38 (2014)
16. Qin, Y., Lu, H., Xu, Y., Wang, H.: Saliency detection via cellular automata. In: Computer Vision and Pattern Recognition, pp. 110–119 (2015)
17. Sivic, J., Zisserman, A.: Video google: A text retrieval approach to object matching in videos. In: IEEE International Conference on Computer Vision, p. 1470 (2003)
18. Tong, N., Lu, H., Xiang, R., Yang, M.H.: Salient object detection via bootstrap learning. In: Computer Vision and Pattern Recognition, pp. 1884–1892 (2015)
19. Tong, N., Lu, H., Zhang, L., Xiang, R.: Saliency detection with multi-scale super-pixels. IEEE Sig. Process. Lett. **21**(9), 1035–1039 (2014)
20. Yang, C., Zhang, L., Lu, H., Ruan, X., Yang, M.H.: Saliency detection via graph-based manifold ranking. In: Computer Vision and Pattern Recognition, pp. 3166–3173 (2013)
21. Zhang, J., Li, D., Zhao, Y., Chen, Z., Yuan, Y.: Representation of image content based on roi-bow. J. Vis. Commun. Image Representation **26**, 37–49 (2015)

Unsupervised Multi-view Subspace Learning via Maximizing Dependence

Meixiang Xu[1,2], Zhenfeng Zhu[1,2(✉)], and Yao Zhao[1,2]

[1] Institute of Information Science, Beijing Jiaotong University, Beijing 100044, China
{15112068,zhfzhu,yzhao}@bjtu.edu.cn
[2] Beijing Key Laboratory of Advanced Information Science and Network Technology,
Beijing 100044, China

Abstract. The recent years have witnessed the great significance of learning from multi-view data in real-world tasks, such as clustering, classification and retrieval. In this paper, we propose an unsupervised dependence (correlation) maximization model, referred to as UDM, for multi-view subspace learning. Our proposed model is based on Hilbert-Schmidt Independence Criterion (HSIC), a kernel-based technique for measuring dependence between two random variables statistically. In the proposed model, sparse constraint on the projection matrix for each view is imposed as regularizations, playing the role of feature selection, which enables to capture more discriminative subspace representations. To efficiently solve the formulated optimization problem, an iterative optimizing algorithm is designed. Experimental results on cross-modal retrieval have shown the superiority of UDM over the compared approaches and the rapid convergence speed of the optimizing algorithm.

Keywords: Subspace learning · Multi-view learning · Dependence

1 Introduction

In recent years, there has been rapid growth of multi-view data and much efforts have witnessed the great significance of learning from multi-view data in many real-world applications. Often, multi-view data, presented in diverse forms or derived from different domains, show heterogeneous characteristics, which is a big challenge for practical tasks such as cross-modal retrieval, machine translation, biometric verification, matching, transfer learning, etc. To address this challenge, two common strategies are mainly adopted. One is to learn distance metrics, the other is to learn a common space. In this paper we focus on the latter, that is, multi-view subspace learning.

Intrinsically, multiple views represent the same underlying semantic data object, therefore, they are inherently correlated to each other. Based on this fact, statical techniques such as canonical correlation analysis (CCA) [11], Kullback-Leibler (KL) divergence [1], mutual information [12] and Hilbert-Schmidt Independence Criterion (HSIC) [6] and so on, to measure correlation (dependence) of

© Springer Nature Singapore Pte Ltd. 2017
J. Yang et al. (Eds.): CCCV 2017, Part II, CCIS 772, pp. 137–148, 2017.
https://doi.org/10.1007/978-981-10-7302-1_12

two random variables, have been investigated and used for multi-view learning. Especially, CCA is the most popular one among the aforementioned measures.

From the point of multi-view learning, CCA can be regarded as finding the projection matrix for each view of the data object, by which the data can be projected into a common subspace where the low dimensional embeddings are maximally correlated. Due to the encouraging success of CCA, CCA-based approaches have attracted much attention during the past decades. Substantial variants of CCA have been developed for multi-view subspace representations, including unsupervised ones [16,18,20], supervised ones [15,19], sparsity-based ones [4,10], DNN-based ones [2,13,22], etc. Like CCA, considerable attention has been gradually paid to the use of HSIC for the dependence-based tasks of multi-view classification [7], clustering [3], dictionary learning [8,9]. Concerning these methods, they are supervised ones, which conformably expect the dependence between multi-view data and the corresponding labels to be maximized. However, labels are unknown beforehand in most cases of multi-view learning tasks.

In this paper, we propose an unsupervised dependence maximization model for multi-view subspace learning, referred to as UDM. The proposed UDM is designed specifically for the case of two views, which can be extended to multiple views. Unlike the supervised HSIC-based approaches, UDM aims at maximizing the dependence between two views under the unsupervised setting. Simultaneously, it incorporates the imposed $\ell_{2,1}$-norm constraint on the projection matrix for each view as regulations, playing the role of feature selection, which enables more discriminative representations. To solve the optimization problem formulated by UDM, an efficient iterative optimizing algorithm is designed. Experimental results on two real-world cross-modal datasets demonstrate the effectiveness and efficiency of UDM, and show the superiority of UDM over the compared approaches. Convergence curves of the objective function demonstrate the rapid convergence speed of the optimization algorithm.

2   Notations and HSIC

2.1   Notations

To begin with, we introduce some notations used in this paper. For any matrix $\mathbf{A} \in \mathbb{R}^{n \times m}$, $\mathbf{A}^{\cdot i}$ and $\mathbf{A}^{\cdot j}$ are used to represent its i-th row and j-th column, respectively. $\|\mathbf{A}\|_{2,1}$ is the $\ell_{2,1}$-norm of $\mathbf{A}$, defined as $\|\mathbf{A}\|_{2,1} = \sum_{i=1}^{n} \|\mathbf{A}^{\cdot i}\|_2$. $\|\mathbf{A}\|_{HS}$ is the Hilbert-Schmidt norm of $\mathbf{A}$, defined as $\|\mathbf{A}\|_{HS} = \sqrt{\sum_{i,j} a_{ij}^2}$. Besides, $tr(\cdot)$ represents the trace operator, $\otimes$ the tensor product and $\mathbf{I}$ an identity matrix with an appropriate size. Throughout the paper, matrices and vectors are represented in bold uppercase and lowercase letters respectively. Variables are represented by conventional letters.

2.2   Hilbert-Schmidt Independence Criteria

Let C_{xy} be the cross-covariance function between x and y, $\varphi(x)$ and $\phi(y)$ two mapping functions with $\varphi(x) : x \in \mathcal{X} \to \mathbb{R}$ and $\phi(y) : y \in \mathcal{Y} \to \mathbb{R}$, $\mathcal{G}$ and $\mathcal{H}$ two Reproducing Kernel Hilbert Spaces (RKHSs) in $\mathcal{X}$ and $\mathcal{Y}$. The associated positive definite kernels k_x and k_y is defined as $k_x(x, x^T) = <\Phi(x), \Phi(x)>_{\mathcal{G}}$ and $k_y(y, y^T) = <\Phi(y), \Phi(y)>_{\mathcal{H}}$. Then cross-covariance C_{xy} is defined as:

$$C_{xy} = E_{xy}\left[(\varphi(x) - u_x) \otimes (\phi(y) - u_y)\right]. \tag{1}$$

where u_x and u_y is the expectation of $\varphi(x)$ and $\phi(y)$ respectively, i.e. $u_x = E(\varphi(x))$ and $u_y = E(\phi(y))$.

Given two independent RKHSs $\mathcal{G}$, $\mathcal{H}$ and the joint distribution p_{xy}, HSIC is the Hilbert-Schmidt norm of C_{xy}, defined as:

$$HSIC(p_{xy}, \mathcal{G}, \mathcal{H}) := \|C_{xy}\|_{HS}^2. \tag{2}$$

In practical applications, the empirical estimate of HSIC is commonly used. Given n finite number of data samples $Z := \{(x_1, y_1), \cdots, (x_N, y_N)\}$, the empirical expression of HSIC is formulated as:

$$HSIC(Z, F, G) = (n-1)^{-2} tr(\mathbf{K}_1 \mathbf{H} \mathbf{K}_2 \mathbf{H}). \tag{3}$$

where $\mathbf{K}_1$ and $\mathbf{K}_2$ are two Gram matrices with $k_{1,ij} = k_1(x_i, x_j)$ and $k_{2,ij} = k_2(y_i, y_j)$ $(i, j = 1, \cdots, N)$. $\mathbf{H} = \mathbf{I} - \frac{1}{n}\mathbf{1}_n\mathbf{1}_n^T$, is a centering matrix, and $\mathbf{1}_n \in \mathbb{R}^n$ is a full-one column vector.

More details about HSIC can be found in literatures [6].

3   Multi-view Subspace Learning Model via Kernel Dependence Maximization

3.1   The Proposed Subspace Learning Model

Multi-view subspace learning approaches aim to project different high-dimensional heterogeneous views into a coherent low-dimensional common subspace in linear or nonlinear ways, where samples with the same or similar semantics have the coherent representation, as illustrated in Fig. 1.

In the following, the case of two views is mainly considered. Suppose that there are n pairs of observation samples $\{\mathbf{x}_1^i, \mathbf{x}_2^i\} \in \mathbb{R}^{1 \times d_1} \times \mathbb{R}^{1 \times d_2}$, where $\{\mathbf{x}_1^i\}_{i=1}^n$ and $\{\mathbf{x}_2^i\}_{i=1}^n$ are from view $\mathbf{X}_1 = [\mathbf{x}_1^1, \cdots, \mathbf{x}_1^n]^T \in \mathbb{R}^{n \times d_1}$ and view $\mathbf{X}_2 = [\mathbf{x}_2^1, \cdots, \mathbf{x}_2^n]^T \in \mathbb{R}^{n \times d_2}$ respectively. $\{\mathbf{x}_1^i, \mathbf{x}_2^i\}$ denotes the i-th pair samples in the sample set $\{\mathbf{x}_1^i, \mathbf{x}_2^i\}_{i=1}^n$. The goal of this paper is to learn the projection matrix $\mathbf{P}_v (v = 1, 2)$ for views $\mathbf{X}_v (v = 1, 2)$ simultaneously. Through $\mathbf{P}_v$, heterogeneous views $\mathbf{X}_v$ are projected into a common subspace $\mathbf{S}$, where samples $\mathbf{x}_1^i$ and $\mathbf{x}_2^j (i, j = 1, \cdots, n)$ with the same and similar semantics have the coherent representation. Correspondingly, the new representation for $\mathbf{X}_1$ and

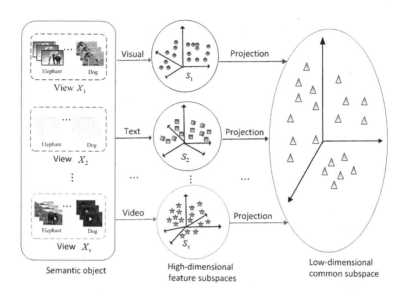

Fig. 1. Sketch map of subspace learning for multi-view data

$\mathbf{X}_2$ in the shared subspace is $\mathbf{X}_1^S = \mathbf{X}_1\mathbf{P}_1$ and $\mathbf{X}_2^S = \mathbf{X}_2\mathbf{P}_2$. Adopting linear kernel as the kernel measure, kernel matrices $\mathbf{K}_{\mathbf{X}_1}$ and $\mathbf{K}_{\mathbf{X}_2}$ can be denoted as $\mathbf{K}_{\mathbf{X}_1} = \langle \mathbf{X}_1^S, \mathbf{X}_1^S \rangle = \mathbf{X}_1\mathbf{P}_1\mathbf{P}_1^T\mathbf{X}_1^T$ and $\mathbf{K}_{\mathbf{X}_2} = \langle \mathbf{X}_2^S, \mathbf{X}_2^S \rangle = \mathbf{X}_2\mathbf{P}_2\mathbf{P}_2^T\mathbf{X}_2^T$. Since multi-view data describe the same semantic object from different levels, they are inherently correlated to each other. Based on HSIC, the proposed unsupervised subspace learning model is formulated as:

$$\max_{\mathbf{P}_1,\mathbf{P}_2} tr(\mathbf{H}\mathbf{X}_1\mathbf{P}_1\mathbf{P}_1^T\mathbf{X}_1^T\mathbf{H}\mathbf{X}_2\mathbf{P}_2\mathbf{P}_2^T\mathbf{X}_2^T)$$
$$s.t.\ \mathbf{P}_1^T\mathbf{P}_1 = \mathbf{I}_1; \mathbf{P}_2^T\mathbf{P}_2 = \mathbf{I}_2, \tag{4}$$

where the orthogonal constraints imposed on $\mathbf{P}_v(v = 1, 2)$ is to avert the trivial solution of all zeros.

As demonstrated in literatures such as [14,21], $\ell_{2,1}$-norm based learning models have capabilities of sparsity, feature selection and robustness to noise. Inspired by this, by imposing the $\ell_{2,1}$-norm constraint on the projection matrix $\mathbf{P}_v(v = 1, 2)$ as regularization terms to learn more discriminative representations for multi-view data, accordingly we have the following formulation:

$$\max_{\mathbf{P}_1,\ \mathbf{P}_2} tr\left(\mathbf{H}\mathbf{X}_1\mathbf{P}_1\mathbf{P}_1^T\mathbf{X}_1^T\mathbf{H}\mathbf{X}_2\mathbf{P}_2\mathbf{P}_2^T\mathbf{X}_2^T\right) - \lambda_1\|\mathbf{P}_1\|_{2,1} - \lambda_2\|\mathbf{P}_2\|_{2,1}$$
$$s.t.\ \mathbf{P}_1^T\mathbf{P}_1 = \mathbf{I}_1;\ \mathbf{P}_2^T\mathbf{P}_2 = \mathbf{I}_2, \tag{5}$$

where λ_1 and λ_2 are the regularization parameters.

3.2 Optimization

Since the optimization objective function involved the $\ell_{2,1}$-norm, which is an intractable problem to handle. Consequently, here we employ the alternative optimization strategy to solve the optimization problem. With $\|\mathbf{A}\|_{2,1} = tr\left(\mathbf{A}^T \mathbf{D} \mathbf{A}\right)$ where $\mathbf{D} = diag\left(\frac{1}{\|A^{\cdot i}\|_2}\right)$, first let us re-express the formulation in Eq. (5) as:

$$\max_{\mathbf{P}_1, \mathbf{P}_2} tr\left(\mathbf{HX}_1\mathbf{P}_1\mathbf{P}_1^T\mathbf{X}_1^T\mathbf{HX}_2\mathbf{P}_2\mathbf{P}_2^T\mathbf{X}_2^T\right) - \lambda_1 tr\left(\mathbf{P}_1^T\mathbf{D}_1\mathbf{P}_1\right) - \lambda_2 tr\left(\mathbf{P}_2^T\mathbf{D}_2\mathbf{P}_2\right)$$
$$s.t.\ \mathbf{P}_1^T\mathbf{P}_1 = \mathbf{I}_1;\ \mathbf{P}_2^T\mathbf{P}_2 = \mathbf{I}_2. \tag{6}$$

Specifically, according to the alternative optimization rules, the optimization problem formulated in Eq. (6) (i.e. Eq. (5)) can be decomposed into the following two sub-maximization ones:

(1) **Solve $\mathbf{P}_1$, fixing $\mathbf{P}_2$:**

$$\max_{\mathbf{P}_1} tr(\mathbf{HX}_1\mathbf{P}_1\mathbf{P}_1^T\mathbf{X}_1^T\mathbf{HX}_2\mathbf{P}_2\mathbf{P}_2^T\mathbf{X}_2^T) - \lambda_1 tr\left(\mathbf{P}_1^T\mathbf{D}_1\mathbf{P}_1\right)$$
$$\Leftrightarrow \max_{\mathbf{P}_1} tr\left(\mathbf{P}_1^T(\mathbf{X}_1^T\mathbf{HX}_2\mathbf{P}_2\mathbf{P}_2^T\mathbf{X}_2^T\mathbf{HX}_1 - \lambda_1\mathbf{D}_1)\mathbf{P}_1\right) \tag{7}$$
$$s.t.\ \mathbf{P}_1^T\mathbf{P}_1 = \mathbf{I}_1.$$

Let $\mathbf{B}_1 = \mathbf{X}_1^T\mathbf{HX}_2\mathbf{P}_2\mathbf{P}_2^T\mathbf{X}_2^T\mathbf{HX}_1 - \lambda_1\mathbf{D}_1$, we can obtain $\mathbf{P}_1$ by solving the eigenvalue problem of $\mathbf{B}_1$, here $\mathbf{P}_1$ consists of the first d eigenvectors corresponding to the d largest eigenvalues of $\mathbf{B}_1$.

(2) **Solve $\mathbf{P}_2$, fixing $\mathbf{P}_1$:**

$$\max_{\mathbf{P}_2} tr(\mathbf{HX}_2\mathbf{P}_2\mathbf{P}_2^T\mathbf{X}_2^T\mathbf{HX}_1\mathbf{P}_1\mathbf{P}_1^T\mathbf{X}_1^T) - \lambda_2 tr\left(\mathbf{P}_2^T\mathbf{D}_2\mathbf{P}_2\right)$$
$$\Leftrightarrow \max_{\mathbf{P}_2} tr\left(\mathbf{P}_2^T(\mathbf{X}_2^T\mathbf{HX}_1\mathbf{P}_1\mathbf{P}_1^T\mathbf{X}_1^T\mathbf{HX}_2 - \lambda_2\mathbf{D}_2)\mathbf{P}_2\right) \tag{8}$$
$$s.t.\ \mathbf{P}_2^T\mathbf{P}_2 = \mathbf{I}_2.$$

Likewise, let $\mathbf{B}_2 = \mathbf{X}_2^T\mathbf{HX}_1\mathbf{P}_1\mathbf{P}_1^T\mathbf{X}_1^T\mathbf{HX}_2 - \lambda_2\mathbf{D}_2$, we can obtain $\mathbf{P}_2$ by solving the eigenvalue problem of $\mathbf{B}_2$, here $\mathbf{P}_2$ consists of the first d eigenvectors corresponding to the d largest eigenvalues of $\mathbf{B}_2$.

To better understand the procedure for solving the proposed method, we summarize in detail the solver for solving the optimization problem in Eq. (5) as Algorithm 1.

3.3 Convergence Analysis

The convergence of the proposed UDM under the iterative optimization algorithm in Algorithm 1 can be summarized by the following Theorem 1.

Algorithm 1. Multi-view Subspace Learning via Dependence Maximizing

---

Input: Multi-view data $\mathbf{X}_v \in \mathbb{R}^{n \times d_v}$, $v = 1, 2$; the regularization parameters λ_1 and λ_2.

Output: The projection matrices $\mathbf{P}_v$, $v = 1, 2$.

1 **Initializing**: Initialize $\mathbf{P}_1$ and $\mathbf{P}_2$ randomly, let $t = 0$;

2 **while** *not converge* **do**

3 $\quad$ Update $\mathbf{D}_1$ and $\mathbf{D}_2$: $\mathbf{D}_1^{(t)} = diag\left(\frac{1}{2\left\|\mathbf{P}_1^{\cdot i(t)}\right\|_2}\right)$, $\mathbf{D}_2^{(t)} = diag\left(\frac{1}{2\left\|\mathbf{P}_2^{\cdot i(t)}\right\|_2}\right)$;

4 $\quad$ Update $\mathbf{P}_1$: obtain $\mathbf{P}_1^{(t+1)}$ by performing eigen-decomposition on

$\quad$ $\mathbf{B}_1 = \mathbf{X}_1^T \mathbf{H} \mathbf{X}_2 \mathbf{P}_2^{(t)} \left(\mathbf{P}_2^{(t)}\right)^T \mathbf{X}_2^T \mathbf{H} \mathbf{X}_1 - \lambda_1 \mathbf{D}_1^{(t)}$. The d eigenvectors

$\quad$ corresponding to the first largest d eigenvalues of $\mathbf{B}_1$ compose $\mathbf{P}_1$;

5 $\quad$ Update $\mathbf{P}_2$: obtain $\mathbf{P}_2^{(t+1)}$ by performing eigen-decomposition on

$\quad$ $\mathbf{B}_2 = \mathbf{X}_2^T \mathbf{H} \mathbf{X}_1 \mathbf{P}_1^{(t)} \left(\mathbf{P}_1^{(t)}\right)^T \mathbf{X}_1^T \mathbf{H} \mathbf{X}_2 - \lambda_2 \mathbf{D}_2^{(t)}$. The d eigenvectors

$\quad$ corresponding to the first largest d eigenvalues of $\mathbf{B}_2$ compose $\mathbf{P}_2$;

6 $\quad$ $t = t + 1$;

7 **return** $\mathbf{P}_1, \mathbf{P}_2$.

Theorem 1. *Under the iterative optimizing rules in Algorithm 1, the objective function defined by Eq. (5) is increasing monotonically, and it can converge to its global maximum.*

Due to space limitation, here we omit the detailed proof of Theorem 1. The convergence curves in Sect. 4.5 can also demonstrate the good convergence behavior of the optimizing algorithm.

4 Experiments

To test the performance of UDM, we conducted experiments on cross-modal retrieval between image and text, i.e. using image to query text (I2T) and using text to query image (T2I), adopting Mean Average Precision (MAP) as the evaluation metric and the normalized correlation (NC) as the distance measure [15].

4.1 Datasets

The follow-ups are brief descriptions on the used two datasets i.e. Wikipedia [23] and NUS-WIDE [5].

- **Wikipedia:** This dataset consists of 2866 image-text pairs labeled with 10 semantic classes in total. For each image-text pair, we extract 4096-dimensional visual features by convolutional neural network to represent the image view, and 100-dimensional LDA textual features to represent the text view. In the experiment, the dataset is partitioned into two parts, one for training (2173 pairs) and the other for testing (693 pairs).

- **NUS-WIDE:** This dataset is a subset from [5], including 190420 image examples totally, each with 21 possible labels. For each image-text pair, we extract 500-dimensional SIFT BoVW features for image and 1000-dimensional text annotations for text. To reduce the computational complexity, further we sample a subset with 8687 pairs of image-text. Likewise, the dataset is divided into two parts, one for training (5212 pairs) and the other for testing (3475 pairs).

4.2   Benchmark Approaches and Experimental Setup

The proposed UDM is unsupervised, kernel-based, correlation-based and sparsity-based. Accordingly, the compared approaches include CCA, KPCA [17], KCCA [16], SCCA [10]. The parameters involved in the compared approaches are kept default following the literatures. For details, please refer to the corresponding literatures. Next, we will present the specific settings for the parameters involved in UDM. First, by fixing λ_1 and λ_2 we determine the optimal d. Specifically, we tune d from the range of $\{5, 10, 20, 40, 60, 80\}$ and $\{50, 100, 150, 200, 250, 300, 350\}$ on Wikipedia and NUS-WIDE respectively, as shown in Fig. 2. It can be seen from Fig. 2, with $d = 40$ on Wikipedia and $d = 50$ on NUS-WIDE, UDM obtains the best performance. Therefore, in the following experiments we set $d = 40$ and $d = 50$ for Wikipedia and NUS-WIDE. Then, with d fixed we decide the optimal λ_1 and λ_2 by tuning them from $\{10^{-5}, 10^{-4}, 10^{-3}, 10^{-2}, 10^{-1}, 1, 10, 10^2, 10^3, 10^4, 10^5\}$ with d fixed. Empirically, we determine $\lambda_1 = 10^{-3}$ and $\lambda_2 = 10^3$ on Wikipedia, as well as $\lambda_1 = \lambda_2 = 10^{-5}$ on NUS-WIDE. In the subsequent section, we will give the parameter sensitivity analysis on λ_1 and λ_2.

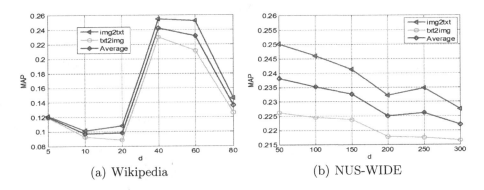

(a) Wikipedia   (b) NUS-WIDE

Fig. 2. MAP vs. varying d on Wikipedia and NUS-WIDE with λ_1 and λ_2 fixed

Table 1. MAP comparison on Wikipedia

Approaches	Image as query	Text as query	Average
KPCA	0.1983	0.1826	0.1905
CCA	0.1222	0.1189	0.1206
KCCA	0.3337	0.3031	0.3184
SCCA	0.2270	0.1961	0.2116
UDM	**0.4204**	**0.4394**	**0.4299**

Table 2. MAP comparison on NUS-WIDE

Approaches	Image as query	Text as query	Average
KPCA	0.2326	0.2215	0.2171
CCA	0.2441	0.2356	0.2399
KCCA	0.2554	0.2451	0.2503
SCCA	0.2415	0.2145	0.2145
UDM	**0.2904**	**0.2498**	**0.2702**

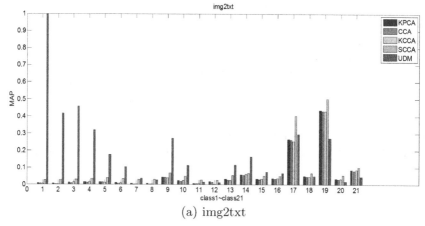

(a) img2txt

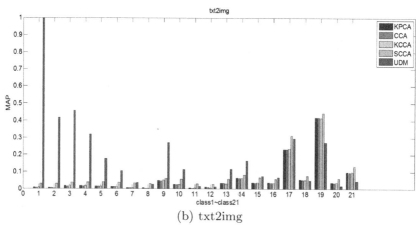

(b) txt2img

Fig. 3. Per class MAP on NUS-WIDE

4.3 Results

Tables 1 and 2 displays the comparison results on two datasets, respectively. As can be seen from Tables 1 and 2, the proposed UDM performs best, followed by KCCA, on both Wikipedia and NUS-WIDE. Besides, Fig. 3 shows the per-class MAP scores of all the compared approaches on NUS-WIDE. From Fig. 3, we can observe that UDM achieves better results on most categories, but it is not always the best on each category. More specifically, it achieves the best result on the first sixteen categories while it is the worst among the five approaches on

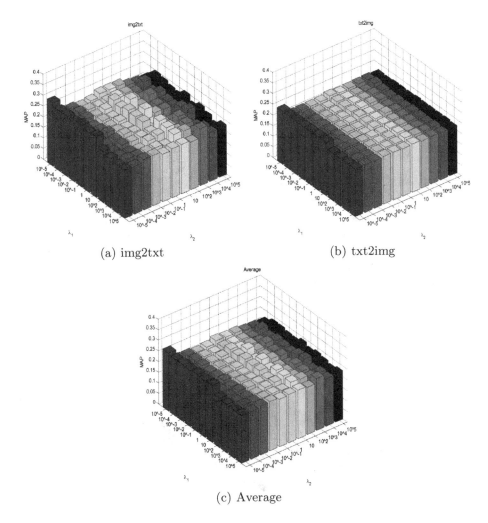

(a) img2txt

(b) txt2img

(c) Average

Fig. 4. MAP vs. varying λ_1 and λ_2 on NUS-WIDE

category 20 and category 21. Therefore, incorporating label supervision information will be considered to improve UDM for each category.

4.4   Parameter Sensitivity Analysis

To show the impacts of λ_1 and λ_2 on UDM, we have carried out experiments on Wikipedia and NUS-WIDE respectively, by tuning them from the same range set as Subsect. 4.2. Figures 4 and 5 shows the retrieval MAP scores versus different

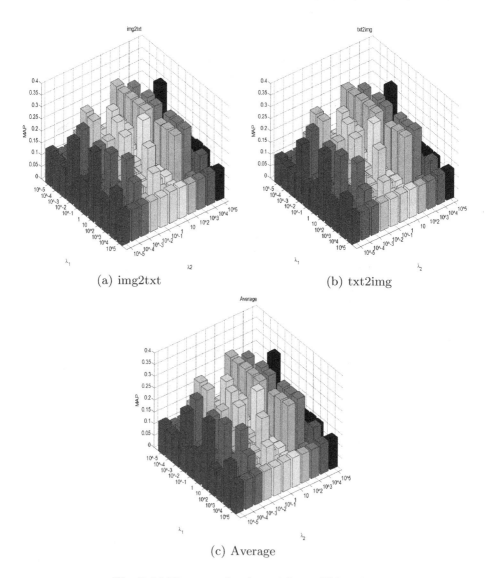

(a) img2txt

(b) txt2img

(c) Average

Fig. 5. MAP vs. varying λ_1 and λ_2 on Wikipedia

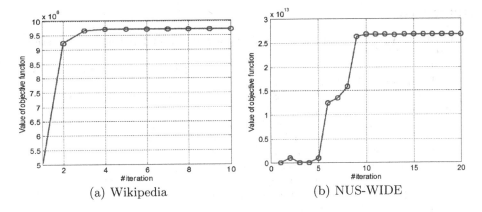

Fig. 6. The objective function vs. the number of iteration

values of λ_1 and λ_2 on NUS-WIDE and Wikipedia respectively. From Figs. 4 and 5, we can see that the performance of UDM varies as λ_1 and λ_2 changes. By contrast, the proposed UDM on Wikipedia is much more sensitive to two parameters than on NUS-WIDE.

4.5 Convergence Study

Figure 6 displays the relationship between the objective function and the number of iteration on Wikipedia and NUS-WIDE, respectively. As can be observed from Fig. 6, for each dataset, the objective function defined in Eq. (5) can rapidly converge to its maximum within about ten iterations, which demonstrates the efficiency of the designed iterative optimization algorithm.

5   Conclusions and Future Work

In this paper, we have proposed a HSIC-based unsupervised learning approach for discovering common subspace representations shared by multi-view data, which is a kernel-based, correlation-based and sparsity-based projection method. To solve the optimization problem, we develop an efficient iterative optimizing algorithm. Cross-modal retrieval results on two benchmark datasets have shown the superiority of the proposed UDM over the compared approaches. Inspired by CCA-like methods, nonlinear extensions of UDM will be considered by incorporating nonlinear kernel and neutral network to expect a better common representation for multi-view data in future work.

Acknowledgments. This work was jointly supported by the National Natural Science Foundation of China (NO. 61572068, NO. 61532005), the National Key Research and Development of China (NO. 2016YFB0800404) and the Fundamental Scientific Research Project (NO. KKJB16004536).

References

1. Cichocki, A., Yang, H.H.: A new learning algorithm for blind signal separation. NIPS **3**, 757–763 (1996)
2. Andrew, G., Arora, R., Blimes, J., Livescu, K.: Deep canonical correlation analysis. In: ICML, pp. 1247–1255 (2013)
3. Cao, X., Zhang, C., Fu, H., Liu, S., Zhang, H.: Diversity-induced multi-view subspace clustering. In: CVPR, pp. 586–594 (2015)
4. Chu, D., Liao, L., Ng, M.K., Zhang, X.: Sparse canonical correlation analysis:new formulation and algorithm. TPAMI **35**(12), 3050–3065 (2013)
5. Chua, T.-S., Tang, J., Hong, R., Li, H., Luo, Z., Zhang, Y.: Nus-wide: a real-world web image database from national university of Singapore. In: CIVR (2009)
6. Principe, J.C.: Information theory, machine learning, and reproducing kernel Hilbert spaces. In: Information Theoretic Learning. Information Science and Statistics, pp. 1–45. Springer, New York (2010)
7. Fang, Z., Zhang, Z.: Simultaneously combining multi-view multi-label learning with maximum margin classification. In: ICDM, pp. 864–869 (2012)
8. Gangeh, M.J., Fewzee, P., Ghodsi, A., Kamel, M.S., Karray, F.: Kernelized supervised dictionary learning. TSP **61**(19), 4753–4767 (2013)
9. Gangeh, M.J., Fewzee, P., Ghodsi, A., Kamel, M.S., Karray, F.: Multi-view supervised dictionary learning in speech emotion recognition. ACM Trans. Audio Speech Lang. Process. **22**(6), 1056–1068 (2014)
10. John, S.-T., Hardoon, D.R.: Sparse canonical correlation analysis. Mach. Learn. **83**(3), 331–353 (2011)
11. Hotelling, H.: Relations between two sets of variates. Biometrika **28**(3/4), 321–377 (1936)
12. Torkkola, K.: Feature extraction by non-parametric mutual information maximization. J. Mach. Learn. Res. **3**(3), 1415–1438 (2003)
13. Ngiam, J., Khosla, A., Kim, M., Nam, J., Lee, H., Ng, A.: Multimodal deep learning. In: ICML, pp. 689–696 (2011)
14. Nie, F., Huang, H., Cai, X., Ding, C.: Efficient and robust feature selection via joint l2,1-norms minimization. In: NIPS, pp. 1813–1821 (2010)
15. Rasiwasia, N., Pereira, J.C., Coviello, E., Doyle, G., Lanckriet, G.R.G., Levy, R., Vasconcelos, N.: A new approach to cross-modal multimedia retrieval. In ICMM, pp. 251–260 (2010)
16. Akaho, S.: A kernel method for canonical correlation analysis. In: IMPS 2001 (2007)
17. Schölkopf, B., Mika, S., Smola, A., Rätsch, G., Müller, K.R.: Kernel PCA pattern reconstruction via approximate pre-images. In: Niklasson, L., Bodén, M., Ziemke, T. (eds.) ICANN 1998. Perspectives in Neural Computing, pp. 147–152. Springer, London (1998). https://doi.org/10.1007/978-1-4471-1599-1_18
18. Sharma, A., Jacobs, D.W.: Bypassing synthesis: PLS for face recognition with pose, low-resolution and sketch. In: CVPR, pp. 593–600 (2011)
19. Tae-Kyun, K., Kittler, J., Cipolla, R.: Discriminative learning and recognition of image set classess using canonical correlation. TPAMI **29**(6), 1005–1018 (2007)
20. Tenenbaum, J.B., Freeman, W.T.: Separating style and content with bilinear models. Neural Comput. **12**(6), 1247–1283 (2000)
21. Wang, K., He, R., Wang, L., Wang, W., Tan, T.: Joint feature selection and subspace learning for cross-modal retrieval. TPAMI **38**(10), 2010–2023 (2016)
22. Wang, W., Arora, R., Livescu, K., Bilmes, J.: On deep multi-view representation learning. In: ICML (2015)
23. Wei, Y., Zhao, Y., Lu, C., Wei, S., Liu, L., Zhu, Z., Yan, S.: Cross-modal retrieval with cnn visual features: A new baseline. TCB **47**(2), 449–460 (2017)

Structured Multi-view Supervised Feature Selection Algorithm Research

Caijuan Shi[1][(✉)], Li-li Zhao[1], Liping Liu[1], Jian Liu[1], and Qi Tian[2]

[1] Information Engineering College, North China University of Science
and Technology, Tangshan 063210, China
shicaijuan2011@gmail.com
[2] The Department of Computer Science, The University of Texas
at San Antonio (UTSA), San Antonio 78249-1604, USA

Abstract. Face more and more multi-view data, how to enhance the feature selection performance has become one of the research issues. However, the most existing multi-view feature selection methods only consider the importance of each view features, but ignore the importance of individual feature in each view in the feature selection progress. In this paper we propose a novel supervised feature selection method based on structured multi-view sparse regularization, namely Structured Multi-view Supervised Feature Selection (SMSFS). SMSFS can realize feature selection by both considering the importance of each view features and the importance of individual feature in each view to boost the feature selection performance. Extensive experiments are performed on two image datasets and the results show the effectiveness of the proposed method SMSFS.

1 Introduction

Recently, with the rapid development of the information technology and computer vision technology, image and video data are often expressed by a lot of different types of visual features, such as the shape, the color, the texture, etc. Each type of features characterizes these data in one specific feature space and has particular physical meaning and statistic property. Conventionally, each type can be regarded as a view and the data represented by different types of features is named as multi-view data [1]. However, these abundant and various types of features not only result in high computational cost, but also often comprise irrelevant and/or redundant features. Therefore, feature selection, as a process of selecting relevant features and reducing dimensionality, has become a research issue. However, confronting with multi-view data, the conditional single-view feature selection methods havent d good feature selection performance. So some multi-view feature selection methods have been widely researched and proposed in recent years. One of the methods is to directly concatenate the multi-view features into a long vector, and then single-view methods are adopted to realize the feature selection [2,3]. This concatenation strategy is easy to realized, but it cannot efficiently explore the complementary of different view features.

© Springer Nature Singapore Pte Ltd. 2017
J. Yang et al. (Eds.): CCCV 2017, Part II, CCIS 772, pp. 149–157, 2017.
https://doi.org/10.1007/978-981-10-7302-1_13

Recently, multi-view learning has been widely applied into the feature selection methods to enhance the feature selection performance by exploring the correlated and complementary information between different views [1,4]. However, these methods consider one view features as a whole and all features in the same view have equally importance, ignoring the importance of individual feature in each view. If we can not only consider the importance of each view features, but also consider the importance of individual feature in each view in the feature selection progress, the feature selection performance can be enhanced. In [5], Wang et al. have proposed group ℓ1-norm (G1-norm), which can discriminate different importance of the features of a specific view. In [6], Wang et al. have proposed a sparse multimodal learning approach to integrate heterogeneous features by using the joint structured sparsity regularizations.

In this paper, we propose a new structured multi-view supervised feature selection framework, namely Structured Multi-view Supervised Feature Selection (SMSFS). SMSFS can enhance the feature selection performance by considering the importance of each view features without ignoring the importance of individual feature in each view based on structured multi-view sparse regularization. SMSFS is applied into image annotation task on two image datasets, NUS-WIDE [7] and MSRA MM 2.0 [8], and the experimental results demonstrate that effectiveness of the proposed algorithm.

2   Related Work

In this section, we discuss two related works on multi-view learning and sparse regularization.

2.1   Multi-view Learning

Recently, multi-view learning has obtained extensive research interest and different types of multi-view learning algorithms have been proposed. These algorithms can be roughly classified four kinds: co-training [9], subspace learning-based algorithm, multiple kernel learning (MKL) and graph ensemble-based multi-view learning.

Co-training [9] trains alternately to maximize the mutual agreement on two distinct different views of data and it can improve the performance when the two views are conditionally independent of each other. Subspace learning-based algorithm aims to obtain a latent subspace shared by multiple views by assuming that the input views are generated from this latent subspace. The representative algorithms include canonical correlation analysis (CCA) [10] and kernel canonical correlation analysis (KCCA) [11]. Multiple kernel learning learns a kernel machine from multiple Gram kernel matrices [12], which naturally correspond to different views of features and are combined either linearly or non-linearly to improve learning performance. Graph ensemble-based algorithms integrate multiple graphs, each of which encodes the local geometry of a particular view, to explore complementary properties of different views [1].

2.2   Sparse Regularization

In order to select the most discriminative features, a variety of sparse regularization has been widely applied into feature selection, including l_1-norm (LASSO), l_p-norm ($0 < p \leq 1$), $l_{2,1}$-norm and $l_{2,p}$-matrix norm ($0 < p \leq 1$). Though l_1-norm (LASSO) [13] is the most well-known sparse regularization, it has not good sparsity. In order to obtain better sparsity, much works [14,15] have extended the l_1-norm to the l_p-norm ($0 < p < 1$) model. In [16], Xu et al. have concluded that when p is 1/2, the l_p-norm, i.e. $l_{1/2}$-norm has the best sparsity. In [17], Nie et al. have introduced a joint $l_{2,1}$-norm minimization on both loss function and regularization for feature selection. In [18], Wang et al. have extended $l_{2,1}$-norm to $l_{2,p}$-matrix norm ($0 < p \leq 1$) to select joint and more sparse features. When p is equal to 1/2, the $l_{2,1/2}$-norm has the best performance.

3   Structured Multi-view Supervised Feature Selection (SMSFS)

In this section, we propose a novel structured multi-view supervised feature selection framework SMSFS. We introduce the SMSFS formulation, and then conduct an effective algorithm for optimizing the objective function.

3.1   SMSFS Formulation

3.1.1   Structured Multi-view Sparse Regularization

Let $W \in \mathbb{R}^{d \times c}$ be the projection matrix, and then W can be expressed as:

$$W = \begin{bmatrix} w_1^1 & \cdots & w_c^1 \\ \cdots & \cdots & \cdots \\ w_1^m & \cdots & w_c^m \end{bmatrix} \in \mathbb{R}^{d \times c} \tag{1}$$

where $W_p^q \in \mathbb{R}^{d_q}$ indicates the weights of all features in the q-th view with respect to the p-th class.

The $l_{2,1/2}$-matrix norm of the projection matrix $W \in \mathbb{R}^{d \times c}$ is defined as [18]:

$$||W||_{2,1/2} = (\sum_{i=1}^{d} ||w_i||_2^{1/2})^2 \tag{2}$$

The group l_1-norm (G_1-norm) is defined as [5]:

$$||W||_{G1} = \sum_{i=1}^{c} \sum_{j=1}^{m} ||w_i^j||_2 \tag{3}$$

Therefore, the structured multi-view sparse regularization is constructed with the group l_1-norm (G_1-norm) and $l_{2,1/2}$-matrix norm in our proposed algorithm SMSFS.

$$||W||_{G1} + \mu ||W||_{2,1/2}^{1/2} \tag{4}$$

This structured multi-view sparse regularization can guarantee the proposed algorithm SMSFS realize feature selection by considering both the importance of each view features and the importance of individual feature in each view. Then the feature selection performance can be boosted.

3.1.2   SMSFS Formulation

The multi-view training data are denoted as $X = [x_1, x_2 \cdots, x_n]^T$ and the ith multi-view datum with m views is denoted as $x_i = [x_i^1, x_i^2 \cdots, x_i^m]^T \in \mathbb{R}^{(\sum_{v=1}^m d_v) \times 1}$. Thus, the feature data matrix of vth view and the feature matrix of all views can be denoted as $X^v = [x_1^v, x_2^v \cdots, x_n^v] \in \mathbb{R}^{d_v \times n}$ and $X = [X^1, X^2, \cdots, X^m]^T \in \mathbb{R}^{d \times n}$ respectively, where $d = \sum_{v=1}^m d_v$. $Y = [y_1, y_2 \cdots, y_n]^T \in {0, 1}^{n \times c}$ is the label of training dataset, where c is the number of classes and $y_i \in \mathbb{R}^{l \times c} (1 \le i \le n)$ is the ith label vector.

A generally sparse feature selection framework to obtain W is to minimize the following regularized empirical error

$$min_W loss(W^T X, Y) + \lambda R(W) \tag{5}$$

where $loss(\cdot)$ is the loss function and $\lambda R(W)$ is the regularization with λ as its regularization parameter.

Here we select the minimizing the prediction error as the loss function and the structured multi-view sparse regularization as the regularization, then the proposed SMSFS can be presented as follows:

$$argmin_W ||X^T W - Y||_F^2 + \lambda ||W||_{G1} + \mu ||W||_{2,1/2}^{1/2} \tag{6}$$

where $\lambda ||W||_{G1} + \mu ||W||_{2,1/2}^{1/2}$ is the structured multi-view sparse regularization-which guarantees SMSFS consider both the importance of each view features and the importance of individual feature in each view, and then achieve good feature selection performance. λ and μ are regularization parameters. $||X^T W - Y||_F^2$ is the loss function.

3.2   Optimization

Because the $l_{2,1/2}$-matrix norm is non-convex and G_1-norm is non-smooth, we propose an efficient algorithm to solve the objective function (6) in this section.

Given $W = [w^1, \cdots, w^d]^T$ and define a diagonal matrix $\widetilde{D}$ with diagonal elements $\widetilde{D}_{ii} = 1/4||w^i||^{3/2}$ then we can get $||W||_{2,1/2}^{1/2} = 4Tr(W^T \widetilde{D} W)$ and $||W||_{G1} = \sum_{i=1}^c \sum_{j=1}^k ||w_i^j||_2$.

So the objective function in (6) can be written as:

$$argmin_W Tr((X^T W - Y)^T (X^T W - Y))$$
$$+ \lambda \sum_{i=1}^c Tr(w_i^T D^i w_i) + \mu Tr(W^T \widetilde{D} W) \tag{7}$$

By setting the derivative of (7) w.r.t to zero, we have

$$X(X^T w_i - y_i) + \lambda D^i w_i + 4\mu \widetilde{D} w_i = 0 \tag{8}$$

where $D^i (1 \leq i \leq c)$ is a block diagonal matrix with the j-th diagonal block as $\frac{1}{2||w_i^j||_2} I_j$ I_j is an identity matrix with size of d_j.

Therefore, we can obtain

$$X(X^T w_i - y_i) + \lambda D^i w_i + 4\mu \widetilde{D} w_i = 0$$
$$w_i = (XX^T + \lambda D^i + 4\mu \widetilde{D})^{-1} X y_i \tag{9}$$

An iterative algorithm is proposed to solve the objective function in Algorithm 1.

Algorithm 1. The SMSFS algorithm.

Input: The vth view feature matrix $X^v \in \mathbb{R}^{d_v \times n}$ and the feature matrix $X \in \mathbb{R}^{d \times n}$; The labels matrix $Y \in \mathbb{R}^{n \times c}$; Regularization parameters λ, μ.

1: Initialize projected matrix $W_0 \in \mathbb{R}^{d \times c}$ randomly;

2: repeat

Compute the diagonal matrix $\widetilde{D}_t = \begin{bmatrix} \frac{1}{4||w_t^1||_2^{3/2}} & & \\ & \cdots & \\ & & \frac{1}{4||w_t^d||_2^{3/2}} \end{bmatrix}$;

Compute the block diagonal matrix $D_t^i (1 \leq i \leq c)$ where the j-th diagonal block $\frac{1}{2||w_i^j||_2} I_j$;

For each $w_i (1 \leq i \leq c), (w_t)_i = (XX^+ \lambda D_t^i + 4\mu \widetilde{D}_t)^{-1} X y_i$

$t = t + 1$;

until convergence;

Output: Optimized projected matrix $W \in \mathbb{R}^{d \times c}$.

4   Experiments

In our paper, we apply the proposed algorithm SMSFS into image annotation task on two image datasets NUS-WIDE dataset [7] and MSRA-MM2.0 dataset [8].

4.1   Datasets and Visual Features

NUS-WIDE dataset includes 269648 real-world images belonging to 81 concepts and MSRA-MM2.0 dataset consists of 50000 images belonging to 100 concepts. In our experiments, we use three types of visual features, including 144-dimension color correlogram, 128-dimension wavelet texture and 73-dimension edge direction histogram for NUS-WIDE dataset or 75-dimension edge direction histogram for MSRA-MM 2.0 dataset.

4.2 Experiment Setup

In our experiments, we randomly sample 3000 images as training data in each dataset. The experiments are independently repeated five times with the average results. The regularization parameters μ and λ in objective function (6) are tuned from 0.00001, 0.001, 0.1, 1, 10, 1000, 100000 and the best results are reported.

We compare our proposed method SMSFS with two supervised feature selection methods, including sub-feature uncovering with sparsity (SFUS) [3] and sparse multimodal learning method by utilizing mixed structured sparsity norms (SMML) [6]. To evaluate the performance, three evaluation metrics, i.e., Mean Average Precision (MAP), MicroAUC and MacroAUC are used in our experiments.

4.3 Performance Evaluation

We compare the proposed method SMSFS with SFUS and SMML on two datasets, and the compared results are listed in Table 1. The best results are shown in bold.

Table 1. Performance comparison

Datasets	Metrics	SFUS	SMML	SMSFS
NUS	MAP	0.096 ± 0.001	0.113 ± 0.002	$\mathbf{0.116 \pm 0.001}$
	MicroAUC	0.781 ± 0.003	0.830 ± 0.002	$\mathbf{0.836 \pm 0.001}$
	MacroAUC	0.677 ± 0.004	0.794 ± 0.003	$\mathbf{0.800 \pm 0.002}$
MSRA	MAP	0.048 ± 0.001	0.072 ± 0.001	$\mathbf{0.076 \pm 0.001}$
	MicroAUC	0.809 ± 0.008	0.880 ± 0.004	$\mathbf{0.885 \pm 0.003}$
	MacroAUC	0.554 ± 0.006	0.664 ± 0.001	$\mathbf{0.669 \pm 0.002}$

From Table 1, we can see that SMSFS has better performance than SFUS and SMML in term of MAP, MacroAUC and MicroAUC on two datasets. This indicates that SMSFS can utilize the structured multi-view sparse regularization to select the most discriminative features, and then to boost the image annotation performance.

4.4 Influence of Selected Features

Here we conduct an experiment to study the performance variation with different selected features number. At the same time, we compare the proposed method SMSFS with SMML and SFUS. The number of selected features is set to 100, 150, 200, 250, 300, and all for NUS-WIDE dataset and MSRA-MM2.0 dataset respectively. MAP is used as the metric and the results of this experiment are shown in Fig. 1.

Figure 1 illustrates that the performance of SMSFS, SFUS and SMML varies when the number of selected features changes. From Fig. 1 we can see: (1) When the number of selected features is too small, MAP is lower than that with all features. This could be attributed to the loss of some useful information. (2) When all the features are selected, MAP is not the best because some noise is included in the initial visual features. (3) Three methods all have the largest MAP with 250 selected features on two datasets, but MAP of SMSFS is higher than those of SFUS and SMML. These results indicate that the proposed method SMSFS can select the more sparse and discriminative features to achieve the good performance based on the structured multi-view sparse regularization.

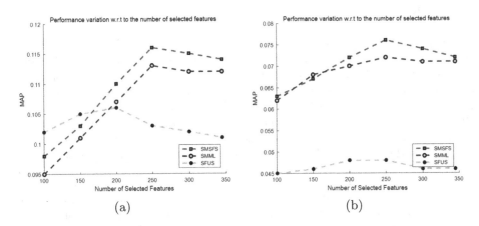

Fig. 1. The performance variation according to the number of selected features of methods SMSFS, compared with SMML and SFUS. (a) NUS-WIDE dataset. (b) MSRA-MM dataset.

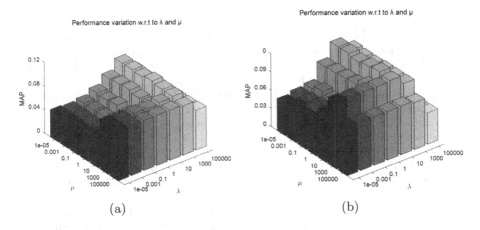

Fig. 2. MAP variation according to μ and λ on different datasets. (a) NUS-WIDE dataset. (b) MSRA dataset.

4.5   Regularization Parameters Analysis

There are two regularization parameters μ and λ in SMSFS objective function
(6). In this section, we use "grid-search" strategy from 0.00001, 0.001, 0.1, 1, 10,
1000, 100000 to learn the parameter sensitivity. Here MAP is used as the metric
and Fig. 2 demonstrates the MAP variation with μ and λ on two datasets.

From Fig. 2 we obtain that the performance of SMSFS is sensitive to regularization parameters μ and λ. SMSFS can obtain the largest MAP by setting
μ to 10 and λ to 1000 on NUS-WIDE dataset, and obtain the largest MAP by
setting μ to 10 and λ to 10 on MSRA dataset respectively.

5   Conclusion

In this paper we propose a novel structured multi-view supervised feature selection framework SMSFS, which can enhance the performance of feature selection
by considering both the importance of features of each view and the importance
of each feature of one view based on the structured multi-view sparse regularization. Because the objective function of SMSFS is non-convex, we introduce an
effective algorithm for optimizing the objective function. Some experiments are
conducted on two datasets for image annotation task and the results demonstrate
that proposed algorithm SMSFS can achieve good feature selection performance.

Acknowledgement. This work was supported partly by the National Natural Science Foundation of China (61502143), Natural Science Foundation of Hebei Province
(F2016209165), Doctoral Research Foundation of North China University of Science
and Technology (201510) and Cultivation Fundation of North China University of Science and Technology (SP201509).

References

1. Feng, Y., Xiao, J., Zhuang, Y., Liu, X.: Adaptive unsupervised multi-view feature
   selection for visual concept recognition. In: Lee, K.M., Matsushita, Y., Rehg, J.M.,
   Hu, Z. (eds.) ACCV 2012. LNCS, vol. 7724, pp. 343–357. Springer, Heidelberg
   (2013). https://doi.org/10.1007/978-3-642-37331-2_26
2. Shi, C.J., Ruan, Q.Q., An, G.Y.: Sparse feature selection based on graph Laplacian
   for web image annotation. Image Vis. Comput. **32**(3), 189–201 (2014)
3. Ma, Z.G., Nie, F.P., Yang, Y., Uijlings, J.R.R., Sebe, N.: Web image annotation via
   subspace-sparsity collaborated feature selection. IEEE Trans. Multimedia **14**(4),
   1021–1030 (2012)
4. Shi, C.J., Ruan, Q.Q., An, G.Y.: Semi-supervised sparse feature selection based
   on multi-view Laplacian regularization. Image Vis. Comput. **41**(9), 1–10 (2015)
5. Wang, H., Nie, F., Huang, H., Risacher, S.L., Saykin, A.J., Shen, L., et al.: Identifying disease sensitive and quantitative trait-relevant biomarkers from multidimensional heterogeneous imaging genetics data via sparse multimodal multitask
   learning. Bioinformatics **28**(12), i127–i136 (2012)
6. Wang, H., Nie, F., Huang, H., Ding, C.: Heterogeneous visual features fusion via
   sparse multimodal machine. In: Proceedings of CVPR, pp. 3097–3102 (2013)

7. Chun, T., Tang, J., Hong, R., et al.: NUS-WIDE: a real-world web image dataset from National University of Singapore. In: Proceedings of CIVR, pp. 1–9 (2009)

8. Li, H., Wang, M., Hua, X.: MSRA-MM2.0: a large-scale web multimedia dataset. In: Proceedings of ICDMW, pp. 164–169 (2009)

9. Blum, A., Mitchell, T.: Combining labeled and unlabeled data with co-training. In: Proceedings of the Workshop on Computational Learning Theory, pp. 92–100 (1998)

10. Hotelling, H.: Relations between two sets of variates. Biometrika **28**(3/4), 321–377 (1936)

11. Akaho, S.: A kernel method for canonical correlation analysis, Arxiv preprint cs/0609071 (2006)

12. Nilufar, S., Ray, N., Zhang, H.: Object detection with DOG scale space: a multiple kernel learning approach. IEEE Trans. Image Process. **21**(8), 3744–3756 (2012)

13. Cawley, G., Talbot, N., Girolami, M.: Sparse multinomial logistic regression via bayesian L_1 regularisation. In: Proceedings of NIPS, pp. 209–216 (2006)

14. Chartrand, R.: Exact reconstruction of sparse signals via nonconvex minimizaion. IEEE Sig. Process. Lett. **14**(10), 707–710 (2007)

15. Chartrand, R.: Fast algorithms for nonconvex compressive sensing: MRI reconstruction from very few data. In: Proceedings of IEEE International Symposium on Biomedical Imaging, pp. 262–265 (2009)

16. Xu, Z.B., Zhang, H., Wang, Y., Chang, X.Y., Liang, Y.: $L_{1/2}$ regularizer. Sci. China. **53**(6), 1159–1169 (2010)

17. Nie, F.P., Xu, D., Hung, T., Zhang, C.: Flexible manifold embedding: a framework for semi-supervised and unsupervised dimension reduction. IEEE Trans. Image Process. **19**(7), 1921–1932 (2010)

18. Wang, L.P., Chen, S.C.: $L_{2,p}$-Matrix Norm and Its Application in Feature Selection. http://arxiv.org/abs/1303.3987 (2013)

An Automatic Shoeprint Retrieval Method Using Neural Codes for Commercial Shoeprint Scanners

Junjian Cui[✉], Xiaorui Zhao, and Daixi Li

Dalian Everspry Sci. & Tech. Co. Ltd., No. 31 Xixian Street,
High-Tech Industrial Zone, Dalian, China
{cuijunjian,zhaoxiaorui,lidaixi}@everspry.com
http://www.footprintmatcher.com

Abstract. In this paper, an automatic shoeprint retrieval method used in forensic science is proposed. The proposed method extracts shoeprint features using recently reported descriptor called neural code. The first step of feature extraction is rotation compensation. Then, shoeprint image is divided into top region and bottom region, and two neural codes for both regions are obtained. Afterwards, a matching score between test image and reference image is calculated. The matching score is a weighted sum of cosine similarities of both regions' neural codes. Experimental results show that our method outperforms other methods on a large-scale database captured by commercial shoeprint scanners. By using PCA, the performance can be improved while the feature dimension is reduced dramatically. To our knowledge, this is the first study using the database collected by commercial shoeprint scanners, and our method obtained a cumulative match score of 88.7% at top 10.

Keywords: Shoeprint retrieval · Convolutional neural network Neural code

1 Introduction

Shoe impressions play very important roles in forensic science since the frequency of shoe impression occurrence at the place of crime is much higher than other evidences (e.g. DNA, fingerprints, and hair) [7]. According to [11], 35% of crime scenes have useful shoeprints for further investigation, and shoe prints can be used in linking cases, knowing the brand and model of the shoes. Thus, an automated shoeprint retrieval or recognition algorithm is necessary for fast and accurate forensic investigations. Shoeprint image retrieval composed of two categories: real crime scene shoeprint retrieval and suspect's shoeprint retrieval, and this paper is focused on the latter category.

Nowadays, several commercial shoeprint scanners have been developed [1,2]. These scanners can scan suspect's shoeprint and acquire a highly detailed shoeprint image. The detailed image may then be uploaded into a database for

© Springer Nature Singapore Pte Ltd. 2017
J. Yang et al. (Eds.): CCCV 2017, Part II, CCIS 772, pp. 158–169, 2017.
https://doi.org/10.1007/978-981-10-7302-1_14

comparison against recovered crime scene shoeprint impressions. For an application of commercial shoeprint scanners in suspect's shoeprint retrieval, please refer to [4].

Recently, several automatic shoeprint retrieval or recognition methods have been reported. Bouridane et al. [8] extracted shoeprint features using fractal transformation and compared features using Mean Square Noise Error (MSNE) method. On a database of 145 images, the recognition accuracy was 88%. A method using Fourier-Mellin transform and power spectral density (PSD) was proposed by Chazal et al. [10]. On a database of 476 images, the method achieved cumulative match scores (CMSs) of 65% and 87% at top 1 and top 5, respectively. Pavlou and Allinson [14] employed Maximally Stable Extremal Region (MSER) to detect keypoints and used SIFT descriptor to describe the keypoints. On a database containing 374 images, the method achieved 92% CMS at top 8. Zhang and Allinson [17] first calculated edge orientation histogram; and then, they used 1D DFT to obtain a rotation invariant feature. On a database of 512 images with 10% Gaussian noise, they achieved a CMS of 97.7% at top 20; however, their algorithm was not robust against rotation and noise. Patil and Kulkarni [13] utilized Radon transform to compensate image rotation and applied Gabor filter to extract multi-resolution features. The method reached a CMS of 91% at top 1 when evaluating on a database containing 200 images. Wang et al. [16] first divided shoeprint image into top region and bottom region and extracted features using Wavelet-Fourier transform. A confidence value of each region was computed, and final matching score was computed based on the confidence value and feature similarities. The method was evaluated on a very large database containing 210,000 real crime scene shoeprint images and achieved a CMS of 90.87% at top 2.

Most of methods mentioned above are suspect's shoeprint retrieval methods except for [16]. They performed pretty well only on their own databases; however, none of them used commercial shoeprint scanners to construct the database. Consequently, the applications of these methods are limited in their own databases.

Differing from these methods, this paper focuses its attention on suspect's shoeprint images collected by commercial shoeprint scanner called EverOS^{TM} [1]. Both test images and reference images in the database are binary images which are captured by EverOS^{TM}. According to [16], most well-known local features (e.g. SIFT, SURF, and MSER) do not perform well on binary images. Therefore, in this paper, we use a deep Convolutional Neural Network (CNN) to extract high-level feature which is called neural code. Experimental results show that the deep CNN-based neural codes outperform other shoeprint features (e.g. [13,16,17]).

The rest of this paper is organized as follows: Sect. 2 briefly reviews CNN and neural code; Sect. 3 explains the proposed method in detail; some experimental results are given in Sect. 4, and Sect. 5 concludes this paper.

2   Related Works

Deep Convolutional Neural Networks (CNNs) have demonstrated the state-of-the-art performance in computer vision, speech, and other areas. Especially in ImageNet Large-Scale Visual Recognition Challenge (ILSVRC), CNNs (e.g. vggNet [15], and ResNet [12]) have exhibited high classification accuracies, and ResNet [12] has outperformed human-level performance.

Our work was inspired by Babenko et al. [6]. In [6], they argued that outputs of the upper layers in a CNN could be served as good features (neural codes) for image retrieval, although the CNN was originally trained to classify images. They extracted neural codes from the last max-pooling layer (L5) and two fully connected layers (L6, and L7). Experimental results showed that the neural code extracted from L5 performed much better than the neural codes extracted from L6 and L7. They also showed that using PCA dimension reduction, the dimension of the neural code (L5) could be reduced from 9216 to 128, with no loss of retrieval accuracy.

In this paper, we employ vgg-verydeep-16 [15] to extract neural code. Vgg-verydeep-16 model is pre-trained using ImageNet and MatConvNet [3], which can be downloaded from MatConvNet. The model comprises five convolutional layers and two fully connected layers followed by a soft-max layer. Each convolutional layer is followed by a max-pooling layer. As the dimensions of neural codes extracted from other layers are too high ($\geq$100,000), in this paper, we only consider five neural codes extracted from the last max-pooling layer and two fully connected layers (each fully connected layer has a pre-activation layer and a ReLu layer).

3   The Proposed Method

The pipeline of the proposed method is illustrated in Fig. 1. The proposed method is divided into two phases: offline database feature extraction phase, and online shoeprint image retrieval phase. In Fig. 1, the dashed box indicates offline database feature extraction phase, and the other parts indicate online image retrieval phase.

In offline database feature extraction phase, we first estimate main axis of each shoeprint image in the database, and each image is rotated so that the main axis is in the vertical position. Afterwards, each image is divided into two semantic regions (top region and bottom region); then, neural codes (features) for both regions are extracted. Extracted neural codes are used in the next phase: online shoeprint retrieval.

In online shoeprint retrieval phase, neural codes for test image are extracted as described above; then, top region similarity and bottom region similarity between test image and an image in the database are calculated respectively. The matching score between two images is a weighted sum of top region similarity and bottom region similarity. After calculating all matching scores between test image and all images in the database, matching scores are sorted so that images most similar to the test image are presented at the top of ranked list. Finally, this ranked list is returned as the retrieval result.

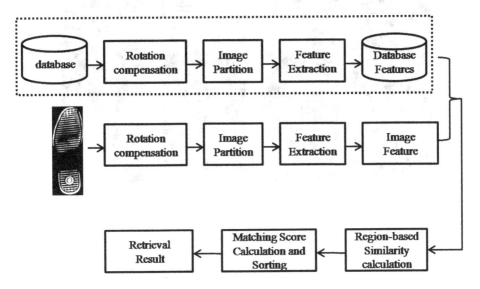

Fig. 1. Pipeline of the proposed method

3.1 Rotation Compensation

Figure 2 illustrates some shoeprint examples provided by EverOS™. EverOS™ provides binary images which are pre-processed interactively. The pre-processing includes three steps: rotation compensation, noise removal, and binary image generation. Among them, rotation compensation has much more effect on the retrieval result than the other two steps. Below, we will explain how to mitigate the effect caused by rotation.

In order to compensate rotation, main axis estimation is necessary. We adopt a computationally inexpensive and effective method called shape orientation algorithm [5] to find the main axis. The shape orientation algorithm is composed of three steps:

Step 1: Generate an edge map of a shoeprint image using Canny edge detector [9].
Step 2: Obtain the mass center (x_c, y_c) of the edge map.
Step 3: Find a straight line which passes through the mass center and summation of the distances from points on the edge map to this line is minimum.

We refer to the straight line described in Step 3 as main axis and will describe how to find it. The equation of a straight line passes mass center (x_c, y_c) with angle θ is given as

$$ax + by = c, \tag{1}$$

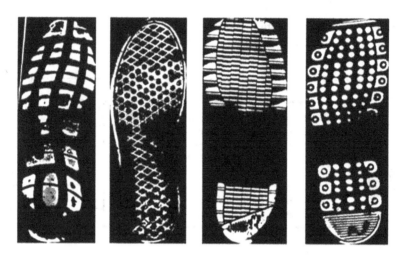

Fig. 2. Shoeprint examples provided by EverOS™

where

$$
\begin{cases}
a = \tan\theta \\
b = -1 \qquad\qquad \text{if } \theta \neq 90°, \\
c = y_c - x_c \times \tan\theta
\end{cases}
\tag{2}
$$

and

$$
\begin{cases}
a = 1 \\
b = 0 \qquad \text{if } \theta = 90°. \\
c = -x_c
\end{cases}
\tag{3}
$$

The distance (d_i) from a point (x_i, y_i) on the edge map to the straight line is given as

$$
d_i = \frac{|ax_i + by_i + c|}{\sqrt{a^2 + b^2}}.
\tag{4}
$$

Thus, the summation of the distances (D) from points on the edge map to the straight line is expressed as

$$
D = \sum_i d_i.
\tag{5}
$$

For each angle θ_j in the range of $0 \leq \theta_j \leq \pi$ with a suitable step size (in 180 steps and 1 degree step size), we can obtain corresponding summation of distances D_j for each angle. Among these D_js, we can find the minimum, and the corresponding angle is the angle of the main axis.

After finding the main axis, we can compensate rotation by rotating the shoeprint image so that the main axis is in the vertical position.

3.2   Image Partition

A shoeprint can be divided into several parts: toe part, sole part, arch part, heel part, and back of heel part. When comparing two shoeprints, these parts can be ranked based on their importance, and the rank is as follows: sole (rank 1), heel, toe, back of heel, and arch (rank 5) [16]. We adopt the method proposed in [16] to divide a shoeprint image into two semantic regions: top region and bottom region. The top region contains toe, sole, and a part of arch, and the bottom region contains heel, back of heel, and the rest part of arch. The ratio between top region and bottom region is set to 6:4. Details are illustrated in Fig. 3.

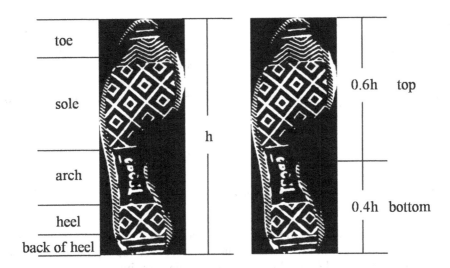

Fig. 3. Shoeprint image partition

3.3   Feature Extraction

The CNN model used in this paper is vgg-verydeep-16 net. The model has five convolutional layers (layer 1–layer 5), and each convolutional layer is followed by a max-pooling layer. At the top of the model, there are two fully connected layers (layer 6, layer 7) and a soft-max layer. Each fully connected layer includes a pre-activation layer (layer 6p for layer 6, and layer 7p for layer 7) and a ReLu transform layer (layer 6r for layer 6, and layer 7r for layer 7). The output of a certain layer can be flattened into a vector, and we call this vector a neural code. Since the neural codes obtained from first four max-pooling layers are of too high dimensions (e.g. a neural code extracted from the forth max-pooling layer is $14 \times 14 \times 512 = 100352$ dimension); therefore, we only consider five neural codes: a neural code extracted from the last max-pooling layer (layer 5m), and four neural codes extracted from fully connected layers (layer 6p, layer 6r, layer 7p, and layer 7r). These five neural codes are denoted by NC5m, NC6p, NC6r, NC7p, and NC7r. The dimension of NC5m is $7 \times 7 \times 512 = 25088$, and the dimensions of the rest neural codes are 4096.

The CNN model is applicable to 224×224 images; thus, top region and bottom region of a shoeprint image are resized to 224×224. Then, for each shoeprint image, we obtain two neural codes for top region and bottom region respectively. We consider these two neural codes as shoeprint image features.

Since the dimension of NC5m is much higher than the other neural codes used in this paper, we use PCA to reduce NC5m's dimension. In Sect. 4, we will show PCA dimension reduction can improve retrieval performance compared to original NC5m.

3.4 Region-Based Similarity Calculation

After feature extraction, feature similarity calculation is needed. While comparing two shoeprints, two similarities (top region similarity and bottom region similarity) are calculated.

In this paper, we use cosine similarity to compare two neural codes. For two neural codes n_1 and n_2, cosine similarity (sim) between n_1 and n_2 is defined as

$$sim = \frac{n_1 \bullet n_2}{\|n_1\|_2 \|n_2\|_2}. \tag{6}$$

3.5 Matching Score Calculation and Sorting

After calculating top region similarity and bottom region similarity, a matching score between two shoeprint images can be calculated. The matching score is a weighted sum of two similarities. Since the ratio between top region and bottom region was set to 6:4, the matching score is calculated using

$$score = 0.6 \times sim_t + 0.4 \times sim_b, \tag{7}$$

where, $score$, sim_t and sim_b are matching score, top region similarity and bottom region similarity, respectively.

Real shoeprints have left prints, right prints, and upside down prints. While comparing two shoe prints, they are compared four times. For a test image, its original version, mirror version, up-down flipped version, and up-down flipped + mirror version are compared to an image in the database and four matching scores are calculated. Among the four matching scores, we pick the maximum as the final matching score between the test image and the image in the database. Figure 4 demonstrates the diagram of final matching score calculation.

Once all matching scores between the test image and all images in the database are calculated, these matching scores are sorted so that images most similar to the test image are presented at the top of the ranked list. Finally, the ranked list is returned as the final retrieval result.

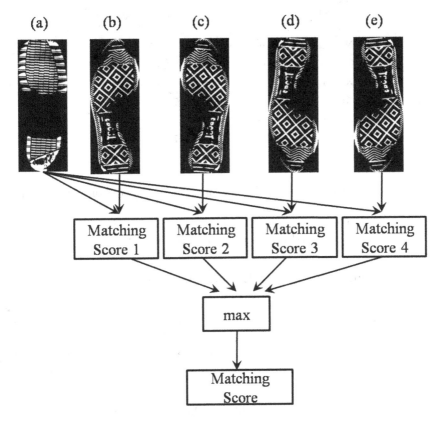

Fig. 4. Diagram of final matching score calculation. (a) An image in the database, (b) test image (original version), (c) test image (mirror version), (d) test image (up-down flipped version), (e) test image (up-down flipped + mirror version)

4   Experimental Results

4.1   Dataset and Evaluation Metrics

In order to evaluate the proposed method, we conducted several experiments. The dataset used in the experiments is composed of the test set and the database. The test set includes 1,000 images and the database includes 37,886 images belonging to 37,886 classes. For each image in the test set, there exists only one matching image in the database. Performance of the shoeprint image retrieval accuracy was evaluated in terms of cumulative match score (CMS).

4.2   Performance of Different Neural Codes

We first evaluated the performance of five neural codes described in Sect. 3. CMSs for top 1, top 5, and top 10 are shown in Table 1.

Table 1. CMS (%) of different neural codes

Neural code	Dimension	Top 1	Top 5	Top 10
NC5m	25088	64.4	83.5	87.9
NC6p	4096	59.0	77.6	81.7
NC6r	4096	59.0	77.4	81.4
NC7p	4096	57.3	76.9	81.3
NC7r	4096	56.7	76.1	80.2

As can be seen from Table 1, the neural code extracted from layer 5m performs much better than the neural codes extracted from fully connected layers. This result is consistent with [6], in which they showed that neural codes extracted from convolutional layers are more discriminative.

4.3   Performance of Different Dimensions of NC5m

In previous subsection, we have shown that neural code extracted from convolutional layer performed much better than neural codes extracted from fully connected layers. From Table 1, we can see that dimension of NC5m is much higher than the other neural codes. Therefore, we use PCA to reduce the dimension of NC5m, and CMSs for top 1, top 5, and top 10 are shown in Table 2.

Table 2. CMS (%) of different dimensions of NC5m

Dimension	top 1	top 5	top 10
Original	64.4	83.5	87.9
95%	64.9	84.2	88.7
4096	64.8	84.2	88.7
2048	64.5	83.9	88.4
1024	64.0	83.0	87.7
512	62.5	82.5	86.8
256	60.6	80.6	85.1
128	57.5	77.8	82.0
64	53.1	73.7	78.5
32	44.4	66.4	72.4
16	33.2	51.5	57.8

In Table 2, "95%" in the third line means that we use 95% of the total principle component covariance. As can be seen from Table 2, with PCA dimension reduction, performance has been improved (e.g. 95%, 4096, and 2048) compared

Table 3. CMS (%) of different methods

Method	top 1	top 5	top 10
[17]	12.8	20.9	24.5
[13]	23.5	32.2	36.6
[16]	54.8	69.8	74.5
NC5m	64.4	83.5	87.9
NC5m + PCA (95%)	64.9	84.2	88.7

to original NC5m. We also can see that 256 dimensional neural code still out-performs neural codes extracted from fully connected layers; however the performance degrades considerably when dimension is reduced to 32.

4.4 Comparison with Other Methods

We also compared our method with three state-of-the-art methods [13,16,17], and the results are illustrated in Fig. 5 and Table 3. From Fig. 5 and Table 3, we can see that the proposed methods (NC5m and NC5m + PCA) reached much higher CMSs than other methods. Among [13,16,17], only [16] has been

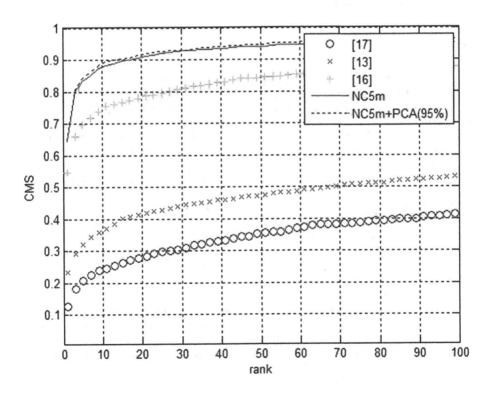

Fig. 5. CMS (%) with respect to rank for different methods

evaluated on a large-scale dataset, and none of them has been evaluated on images captured by commercial shoeprint scanners. Therefore, this experiment also showed that the proposed method is applicable to the databases collected by commercial shoeprint scanners.

5   Conclusion

In this paper, we have proposed a suspect's shoeprint retrieval method using neural codes. The proposed method used neural codes to extract shoeprint features. By using rotation compensation, shoeprint image partition, region-based similarity calculation, and weighted sum of similarities, the proposed method is simple but performs well on shoeprint images captured by commercial shoeprint scanners. As far as we know, this is the first study using the shoeprint database collected commercial scanners. Performance of the proposed method has been evaluated on a database containing 37,886 shoeprint images, and the proposed method performed much better than other state-of-the-art methods. We also showed that by using PCA dimension reduction method, performance can be improved while using a short neural code. Since we used pre-trained CNN model to extract neural codes, training a CNN model using shoeprint images and expanding its application to real crime scene shoeprint images will be the future scopes of this work.

References

1. Dalian Everspry Sci. & Tech. Co. Ltd.: http://www.footprintmatcher.com
2. Hangzhou Chancel Electronic Technology Co. Ltd.: http://www.hzchancel.cn
3. Matconvnet: http://www.vlfeat.org/matconvnet/pretrained
4. Treadfinder Homepage: https://www.treadfinder.uk
5. Abdel-Kader, R.F., Ramadan, R.M., Zaki, F.W., El-Sayed, E.: Rotation-invariant pattern recognition approach using extracted descriptive symmetrical patterns. Int. J. Adv. Comput. Sci. Appl. **3**(5), 151–158 (2012)
6. Babenko, A., Slesarev, A., Chigorin, A., Lempitsky, V.: Neural codes for image retrieval. In: Proceedings of ECCV, pp. 584–599 (2014)
7. Bodziak, W.J.: Footwear Impression Evidence: Detection. Recovery and Examination. CRC Press, Boca Raton (2000)
8. Bouridane, A., Alexander, A., Nibouche, M., Crookes, D.: Application of fractals to the detection and classification of shoeprints. In: Proceedings of IEEE International Conference on Image Processing, pp. 474–477 (2000)
9. Canny, J.: A computational approach to edge detection. IEEE Trans. Pattern Anal. Mach. Intell. **8**(6), 679–698 (1986)
10. Chazal, P.D., Flynn, J., Reilly, R.: Automated processing of shoeprint images based on the Fourier transform for use in forensic science. IEEE Trans. Pattern Anal. Mach. Intell. **27**(3), 341–350 (2005)
11. Girod, A.: Computer classification of the shoeprint of burglar soles. Forensic Sci. Int. **82**, 59–65 (1996)

12. He, K., Zhang, X., Ren, S., Sun, J.: Deep residual learning for image recognition. In: Proceedings of IEEE Conference on Computer Vision and Pattern Recognition, pp. 770–778 (2016)
13. Patil, P., Kulkarni, J.: Rotation and intensity invariant shoeprint matching using gabor transform with application to forensic science. Pattern Recogn. **42**(7), 1308–1317 (2009)
14. Pavlou, M., Allinson, N.: Automated encoding of footwear patterns for fast indexing. Image Vis. Comput. **27**, 402–409 (2009)
15. Simonyan, K., Zisserman, A.: Very deep convolutional networks for large-scale image recognition. In: Proceedings of International Conference on Learning Representations (2015)
16. Wang, X., Sun, H., Yu, Q., Zhang, C.: Automatic shoeprint retrieval algorithm for real crime scenes. In: Proceedings of ACCV, pp. 399–413 (2014)
17. Zhang, L., Allinson, N.: Automatic shoeprint retrieval system for use in forensic investigations. In: UK Workshop on Computational Intelligence (2005)

Uncovering the Effect of Visual Saliency on Image Retrieval

Qinjie Zheng[1,2,3], Shikui Wei[1,2,3]([✉]), Jia Li[1,2,3], Fei Yang[1,2,3], and Yao Zhao[1,2,3]

[1] Institute of Information Science, Beijing Jiaotong University, Beijing 100044, China
shkwei@bjtu.edu.cn
[2] Beijing Key Laboratory of Advanced Information Science and Network Technology, Beijing Jiaotong University, Beijing 100044, China
[3] State Key Laboratory of Virtual Reality Technology and Systems, School of Computer Science and Engineering, Beihang University, Beijing 100191, China

Abstract. Visual saliency modeling has achieved impressive performance for boosting vision-related systems. Intuitively, it should be beneficial to content-based image retrieval task, since the users' query attention is heavily related to the region of interests (ROI) in query image. Although some approaches have been proposed to combine image retrieval systems with visual saliency models, no a comprehensive and systematic study is made to discover the effect of different saliency models on image retrieval in a qualitative and quantitative manner. In this paper, we attempt to concretely investigate the diversity of visual saliency models on image retrieval by making extensive experiments based on nine popular saliency models. To cooperatively mining the complementary information from different models, we also propose a novel approach to effectively involve visual saliency into image retrieval systems by a learning process. Extensive experiments on a generally used image benchmark demonstrate that the new image retrieval system remarkably outperforms the original one and other traditional ones.

Keywords: Visual saliency · Image retrieval · Evaluation

1 Introduction

Visual saliency estimation is a mechanism simulating human vision system to detect the conspicuous content in an image, and lots of excellent saliency models have been proposed [14,15,18,22,28]. Due to good properties of visual saliency, it has been widely used in many fields, such as abstract extraction, classification, compression, monitoring [4]. In recent years, many researchers devote to introduce visual saliency into image retrieval to improve the searching accuracy. Formally, Content-Based Image Retrieval (CBIR) aims at effectively and efficiently finding out the needed images from a large-scale image database, which has achieved great progress in past two decades. Generally speaking, most of

© Springer Nature Singapore Pte Ltd. 2017
J. Yang et al. (Eds.): CCCV 2017, Part II, CCIS 772, pp. 170–179, 2017.
https://doi.org/10.1007/978-981-10-7302-1_15

existing image retrieval methods attempt to improve image retrieval performance from the following three aspects: (1) constructing discriminative image features [6,12,26]; (2) designing good similarity estimation schemes [19,30]; (3) handling large-scale issues [2,5,13,21,31]. Since the user's query attention is heavily related to the specific regions in query image, the visual saliency is considered to be good information for boosting image retrieval accuracy. For example, Acharya et al. [1] directly employed Itti saliency model [11] to generate saliency map and then extracted feature vectors from the saliency map for image retrieval. Similarly, some researchers [3] exploited an image segmentation and color histogram based saliency model to extract saliency map and then extracted image features from the map. Papushoy et al. [23] employed GBVS visual attention model [10] to extract saliency map and introduced the salient information into region-level based image retrieval system. Similar approaches are also reported in [9,27]. In their work, the salient information is involved by weighting different regions according to their perceived saliency. In [20,25], a histogram of saliency map is extracted as separated image feature, and it is integrated into original similarity measure of image retrieval system. Although these approaches have been proposed to combine image retrieval systems with visual saliency models, no a comprehensive and systematic study is made to discover the effect of different saliency models on image retrieval in a qualitative and quantitative manner.

To concretely investigate the diversity of visual saliency models on image retrieval, we conduct extensive experiments based on nine popular saliency models. To cooperatively employing the complementary information from different models, we also propose a novel approach to effectively involve visual saliency into image retrieval systems by a learning process.

Our main contributions can be summarized as follows:

1. **Extensive Experimental Studies:** Through the experimental studies based on nine classic saliency models, we explicitly evaluate the effect of visual saliency on image retrieval and discover some effective manners of involving the salient information, which will be beneficial to many computer vision problems.
2. **A Novel Learning-based Saliency Involving Approach:** We propose a novel learning-based approach to optimally involve visual saliency into image retrieval systems. From the leaning perspective, we provide a new framework for optimally involving salient information.

2   Extensive Experimental Studies

In this section, we conduct extensive experiments based on nine popular saliency models to discover the relationship between visual saliency and image retrieval. More specifically, a popular image retrieval framework based on local image features is employed to implement baseline image retrieval systems. Under this framework, two reasonable schemes are designed to involve salient information into image retrieval process and the optimal combination of saliency model and involving schemes is identified experimentally.

2.1   The Image Retrieval Framework

We employ the popular BoW-based image retrieval framework. As a kind of local based framework, it can not only better balance the effectiveness and efficiency but also provide more flexibility for involving salient information. In Fig. 1, we sketch key steps of the framework. In particular, a visual codebook is first constructed by employing some clustering methods [12,26]. Based on the visual codebook, an orderless collection can be built for one image by replacing the image's local features with their nearest visual words. After inserting each visual word in image database and its corresponding image ID into the inverted table, we can perform image retrieval process after given a query image. We call this scheme BoW model. If Hamming embedding code, which encodes the quantization error between local feature and its visual word, is inserted into the inverted table with image ID, we call it BoW+HE scheme. In our experimental studies, both schemes are exploited. Based on the image retrieval framework, two saliency involving schemes are employed to introduce salient information into image retrieval process.

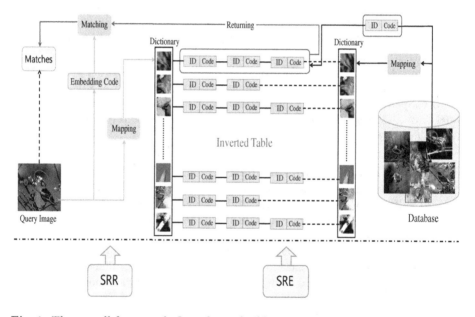

Fig. 1. The overall framework. In order to build an inverted indexing structure, each key point in database images is first mapped to the nearest visual word, and then its image ID without (BoW) or with (BoW+Embedding) embedding code is inserted into the list corresponding to the visual word. In the online query stage, each key point in the query image is also mapped to the nearest visual word, and the items in the corresponding list are returned as matches. If embedding codes are employed, the returned list will be further refined and only top n items whose are most similar to query key point in distances among their codes are returned as matches.

2.2 Dataset and Saliency Involving Schemes

INRIA Holidays Dataset [12]: It is a commonly used image benchmark in image retrieval area, and it contains 500 image groups with different scenes or objects. For each group, there are several images, and the total number of images in all groups is 1491. To evaluate image retrieval task, the first image in each group is treated as query, which results in a query set with 500 images. The other 991 images are treated as database images.

As indicated in Fig. 1, there are two manners to introduce visual salient information into the local-based image retrieval framework. They are listed as follows:

- **Saliency Region Representation Embedding (SRE):** A 4-dimensional vector is extracted for each image based on the whole saliency map. In particular, the first component is the average value of H channel values of all salient points, and the rest three components correspond to S channel, I channel, and sum of three channels, respectively. All the key points in an image are associated with the same vector. Given any key point in the query image, all its similar items returned from the inverted table is sorted in ascending order by the distances between the 4D vector of the query key point and the 4D vectors of database key points. Only top n items are selected and assigned to a big weight.
- **Saliency Region Representation Re-ranking (SRR):** Given a query image, the returned database images are re-ranked by the distances between the 4D vector of the query image and the 4D vectors of database images. Different from SRE, SRR is a kind of global saliency involving scheme.

2.3 A Quantitative Study on Saliency Involving Schemes

As discussed above, different saliency models will result in quite different saliency maps. To evaluate their effectiveness on improving the image retrieval quality, we conduct extensive experiments by individually employing 9 state-of-the-art saliency models (*i.e.*, AWS [8], BMS [29], GBVS [10], HFT [17], ITTI [11], LDS [7], RARE [24], SP [16], SSD [15]). The parameters for all the experiments are optimal. The experimental results are illustrated in Fig. 2. The first row is the results from BoW model with different combinations of saliency models and involving methods, and the second row is from the BoW+Embedding method. In order to clearly show the effect of different saliency models, the cases without any salient information (BL) also illustrated as baselines.

For BOW model, whatever saliency involving methods we adopt, almost all the saliency models outperform the baseline system. This means that the saliency maps extracted from existing models indeed reflect the true human salient information more or less. For these saliency models, LDS achieves the best performance, compared all the other models with any saliency involving scheme. The highest MAP is achieved at the combination of LDS and SRE schemes, which is up to 0.540. It is higher than Baseline (0.456) by 8.4% points.

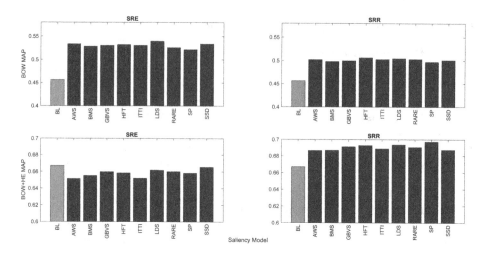

Fig. 2. Evaluation on various combinations of saliency models and saliency involving schemes in two image retrieval methods. The green lines in the first row denote the performance of BoW baseline system, whose MAP is 0.456. The green lines in the second row denote the performance of BoW+HE baseline system, whose MAP is 0.667. The yellow lines mean that the saliency involving schemes are combined with salient information labeled manually. (Color figure online)

For BOW+HE model, SRE involving scheme cannot provide positive effect on the baseline system. That is, BOW+HE model heavily depends on the saliency involving schemes. When we introduce Hamming embedding codes into image retrieval system, the searching accuracy has been improved significantly, compared BoW baseline (0.456) with BoW+HE baseline (0.667). In this situation, if saliency maps are not accurate enough, it will remarkably degrade the positive effect of Hamming embedding codes. According to our experiments, only the SRR scheme can play a positive role in image retrieval boosting. For saliency models, SP model performs the best. The highest MAP value is up to 0.697 and is higher than the baseline (0.667) by 3% points. In fact, LDS model still works well, which achieves nearly the same performance with SP.

From the 4 sub figures in Fig. 2, SRR saliency involving approach provides stable and consistent improvement for all saliency models in both image retrieval methods. That is, saliency involving approach plays an important role when introducing salient information into the image retrieval framework.

3   Learning-Based Saliency Involving Approach

In real-world image retrieval scenario, it is impossible to manually obtain saliency maps for a large-scale image database. Therefore, most of existing methods directly employed one of saliency models to extract saliency map to approximate the human vision system. However, different saliency models will result in quite different saliency maps, which can be treated as different approximations of true saliency map. In addition, the performance of different saliency

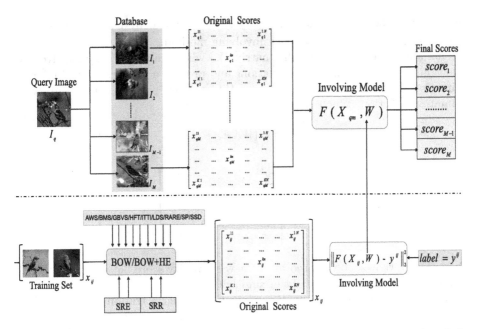

Fig. 3. Illustration of the proposed learning-based saliency involving approaches. It is divided into two parts, *i.e.* online involving and offline learning. In the offline learning stage, the similarity scores of image pairs in training set are calculated by employing all possible combinations of saliency model and involving schemes. Then, they are utilized with pair labels to train an optimal involving model. In the online stage, all scores between query image and any database image are first estimated and then are employed to obtain the a final score by involving model.

involving schemes also varies with different image retrieval schemes. If we can find a saliency involving approach that obtains an optimal complement of various saliency models and involving schemes, we can better improve the image retrieval performance of original search engines. Toward this end, we propose a novel learning-based saliency involving approach. Figure 3 illustrates the key idea of the proposed approach.

Formally, we suppose that there are N different saliency models $M = \{M_1, M_2, ...M_N\}$ and K saliency involving methods $\Phi = \{\Phi_1, \Phi_2, ...\Phi_K\}$. Given an image pair (x_i, x_j), we can get their similarity score matrix $\mathcal{X}_{ij}$ by employing different combinations of saliency models and saliency involving methods, which can be formulated as follows:

$$f_{KN}(x_i, x_j) = \mathcal{X}_{ij} = \begin{bmatrix} x_{ij}^{11} & \cdots & \cdots & \cdots & x_{ij}^{1N} \\ \cdots & \cdots & \cdots & \cdots & \cdots \\ \cdots & \cdots & x_{ij}^{kn} & \cdots & \cdots \\ \cdots & \cdots & \cdots & \cdots & \cdots \\ x_{ij}^{K1} & \cdots & \cdots & \cdots & x_{ij}^{KN} \end{bmatrix} \qquad (1)$$

where x_{ij}^{kn} is the two images' similarity score obtained by combining the nth saliency model and the kth saliency involving method.

Our aim is to learn a weight matrix W, which can optimally involve the salient information from all combinations and provide a more reliable similarity estimation between two images. Toward this end, we propose a new similarity estimation function, which is formulated as follows:

$$\mathcal{F}(\mathcal{X}_{ij}, W) = tr(W^T \mathcal{X}_{ij})$$

$$W = \begin{bmatrix} w^{11} & ... & ... & ... & w^{1N} \\ ... & ... & ... & ... & ... \\ ... & ... & w^{kn} & ... & ... \\ ... & ... & ... & ... & ... \\ w^{K1} & ... & ... & ... & w^{KN} \end{bmatrix} \tag{2}$$

where $\mathcal{F}(\mathcal{X}_{ij}, W)$ can be treated as the new similarity score between two images x_i and x_j, and w^{kn} is the weight of similarity score x_{ij}^{kn}.

To learn the weight matrix W, we must construct a training set of triplets $\{(x_i, x_j; y^{ij})\}_{i,j=0}^{L}$, where y^{ij} is one if images x_i and x_j are truly relevant, otherwise zero. Our approach attempts to approximate the score $\mathcal{F}(\mathcal{X}_{ij}, W)$ to the relevant label y^{ij}, which can be formulated by minimizing the following approximation error:

$$\mathcal{L}(x_i, x_j; y^{ij}) = \|\mathcal{F}(\mathcal{X}_{ij}, W) - y^{ij}\|_2^2 \tag{3}$$

The overall objective function is defined as follows:

$$\min_{\mathbf{W}} \sum_{i=1}^{L} \sum_{j=1}^{L} \mathcal{L}(x_i, x_j; y^{ij}) + \lambda \|W\|_F^2 \tag{4}$$

where L is the number of training images and λ is the parameter controlling the sparse term.

4   Experiments

4.1   Experimental Setup

The INRIA Holidays dataset is employed for evaluation. For each image, 9 state-of-the-art saliency models are employed individually to extract saliency maps. In addition, 2 saliency involving methods above-mentioned are exploited to evaluate the performance of different combinations. To further show the scalability of the proposed learning-based saliency involving approach, a large-scale distracted image dataset, i.e., Flickr1M dataset, is employed in our large-scale image retrieval experiments.

4.2 Evaluation on Learning-Based Saliency Involving Scheme

To obtain an optimal complement of various saliency models and involving approaches, we propose a learning-based scheme. In this section, we conduct some experiments to evaluate its effectiveness. To facilitate the experiments, the saliency involving approach is fixed to re-ranking scheme (*i.e.*, SRR) due to its stable performance. In addition, we only employ the BoW+HE framework, since boosting its performance is more challenging.

Table 1. Performance of 9 saliency models and the learning-based scheme

Method	BL	AWS	BMS	GBVS	HFT	ITTI	LDS	RARE	SP	SSD	Proposed
mAP	0.667	0.687	0.687	0.691	0.693	0.689	0.694	0.691	0.697	0.687	**0.701**

The experimental results are shown in Table 1. Clearly, the proposed learning-based scheme outperforms the best performance of all traditional methods. The possible reason lies in that the optimal process generates a more complete saliency map than GND saliency map. This means that the learning-based scheme can even complement the labeling error from human labelers.

4.3 Large-Scale Image Retrieval Experiments

To evaluate the scalability of the proposed learning-based scheme, we conduct some experiments on large-scale image database. In our experiments, we only employ BoW+HE framework, and all the parameters are set to be optimal. The experimental results are demonstrated in Table 2. As expected, after introducing the salient information by using the proposed scheme, the final retrieval accuracy is remarkably improved comparing the baseline system. This means that the proposed scheme can work well on large-scale image retrieval tasks.

Table 2. Performance on large-scale image retrieval scenario

Method	BOW+HE	**BOW+HE+Learning**
mAP	0.257	**0.296**

5 Conclusion

In this paper, we make comprehensive and systematic study to discover the essential relation between image retrieval and visual saliency. Specially, we explicitly discover the effect of visual saliency on image retrieval in a quantitative manner. The key finding is that salient information indeed has positive effect on image retrieval and the manner of introducing salient information play an important role on performance boosting. According to the finding, we propose a novel approach to effectively involve visual saliency into image retrieval systems by a learning process. Extensive experiments on a generally used image benchmark demonstrate that the new image retrieval system remarkably outperforms the original one and the learning-based visual saliency involving approach is also better than the traditional ones. In addition, large-scale experiments show good scalability of the proposed approach.

Acknowledgments. This work was supported in part by National Natural Science Foundation of China (No. 61572065, No. 61532005, No. 61370113), National Key Research and Development of China (No. 2016YFB0800404), Joint Fund of Ministry of Education of China and China Mobile (No. MCM20160102).

References

1. Acharya, S., Devi, M.R.V.: Image retrieval based on visual attention model. Procedia Eng. **30**, 542–545 (2012)
2. Babenko, A., Lempitsky, V.: Efficient indexing of billion-scale datasets of deep descriptors. In: Proceedings of the IEEE Conference on Computer Vision and Pattern Recognition, pp. 2055–2063 (2016)
3. Boato, G., Dang-Nguyen, D.T., Muratov, O., Alajlan, N., Natale, F.G.B.D.: Exploiting visual saliency for increasing diversity of image retrieval results. Multimedia Tools Appl. **75**(10), 1–22 (2015)
4. Chiappino, S., Mazzu, A., Marcenaro, L., Regazzoni, C.S.: A bio-inspired logical process for saliency detections in cognitive crowd monitoring. In: IEEE International Conference on Acoustics, Speech and Signal Processing, pp. 2110–2114 (2015)
5. Dai, Q., Li, J., Wang, J., Jiang, Y.G.: Binary optimized hashing. In: Proceedings of the 2016 ACM on Multimedia Conference, pp. 1247–1256. ACM (2016)
6. Duan, L., Ma, W., Miao, J., Zhang, X.: Visual saliency based bag of phrases for image retrival. In: The ACM Siggraph International Conference, pp. 243–246 (2014)
7. Fang, S., Li, J., Tian, Y., Huang, T., Chen, X.: Learning discriminative subspaces on random contrasts for image saliency analysis. IEEE Trans. Neural Netw. Learn. Syst. **PP**(99), 1–14 (2016)
8. Garciadiaz, A., Leborn, V., Fdezvidal, X.R., Pardo, X.M.: On the relationship between optical variability, visual saliency, and eye fixations: a computational approach. J. Vis. **12**(6), 17 (2012)
9. Giouvanakis, E., Kotropoulos, C.: Saliency map driven image retrieval combining the bag-of-words model and PLSA. In: 19th International Conference on Digital Signal Processing, pp. 280–285 (2014)
10. Harel, J., Koch, C., Perona, P.: Graph-based visual saliency. In: Advances in Neural Information Processing Systems, pp. 545–552 (2007)
11. Itti, L., Koch, C., Niebur, E.: A model of saliency-based visual attention for rapid scene analysis. IEEE Trans. Patt. Anal. Mach. Intell. **20**(11), 1254–1259 (1998)
12. Jegou, H., Douze, M., Schmid, C.: Hamming embedding and weak geometric consistency for large scale image search. In: Forsyth, D., Torr, P., Zisserman, A. (eds.) ECCV 2008. LNCS, vol. 5302, pp. 304–317. Springer, Heidelberg (2008). https://doi.org/10.1007/978-3-540-88682-2_24

13. Johnson, J., Douze, M., Jégou, H.: Billion-scale similarity search with GPUs. ArXiv Preprint ArXiv:1702.08734 (2017)
14. Lee, G., Tai, Y.W., Kim, J.: Deep saliency with encoded low level distance map and high level features. In: Proceedings of the IEEE Conference on Computer Vision and Pattern Recognition, pp. 660–668 (2016)
15. Li, J., Duan, L.Y., Chen, X., Huang, T., Tian, Y.: Finding the secret of image saliency in the frequency domain. IEEE Trans. Patt. Anal. Mach. Intell. **37**(12), 1 (2015)
16. Li, J., Tian, Y., Huang, T.: Visual saliency with statistical priors. Int. J. Comput. Vis. **107**(3), 239–253 (2014)
17. Li, J., Levine, M.D., An, X., Xu, X., He, H.: Visual saliency based on scale-space analysis in the frequency domain. IEEE Trans. Patt. Anal. Mach. Intell. **35**(4), 996–1010 (2013)
18. Liang, X., Xu, C., Shen, X., Yang, J., Tang, J., Lin, L., Yan, S.: Human parsing with contextualized convolutional neural network. IEEE Trans. Patt. Anal. Mach. Intell. **39**(1), 115–127 (2017)
19. Liao, L., Wei, S., Zhao, Y., Gu, G.: Improving the similarity estimation via score distribution. In: IEEE International Conference on Multimedia and Expo, pp. 1–6 (2016)
20. Liu, G.H., Yang, J.Y., Li, Z.Y.: Content-based image retrieval using computational visual attention model. Patt. Recogn. **48**(8), 2554–2566 (2015)
21. Liu, R., Zhao, Y., Wei, S., Zhu, Z., Liao, L., Qiu, S.: Indexing of CNN features for large scale image search. ArXiv Preprint ArXiv:1508.00217 (2015)
22. Liu, Y.J., Yu, C.C., Yu, M.J., He, Y.: Manifold SLIC: a fast method to compute content-sensitive superpixels. In: Proceedings of the IEEE Conference on Computer Vision and Pattern Recognition, pp. 651–659 (2016)
23. Papushoy, A., Bors, A.G.: Image retrieval based on query by saliency content. Dig. Sig. Process. **36**, 156–173 (2015)
24. Riche, N., Mancas, M., Gosselin, B., Dutoit, T.: Rare: a new bottom-up saliency model. In: 19th IEEE International Conference on Image Processing, pp. 641–644. IEEE (2012)
25. Wan, S., Jin, P., Yue, L.: An approach for image retrieval based on visual saliency. In: International Conference on Image Analysis and Signal Processing, pp. 172–175 (2009)
26. Wei, S., Xu, D., Li, X., Zhao, Y.: Joint optimization toward effective and efficient image search. IEEE Trans. Cybern. **43**(6), 2216–2227 (2013)
27. Wen, Z., Gao, J., Luo, R., Wu, H.: Image Retrieval Based on Saliency Attention. Springer, Heidelberg (2014)
28. Zhang, D., Han, J., Li, C., Wang, J., Li, X.: Detection of co-salient objects by looking deep and wide. Int. J. Comput. Vis. **120**(2), 215–232 (2016)
29. Zhang, J., Sclaroff, S.: Exploiting surroundedness for saliency detection: a boolean map approach. IEEE Trans. Patt. Anal. Mach. Intell. **38**(5), 889–902 (2016)
30. Zheng, L., Wang, S., Wang, J., Tian, Q.: Accurate image search with multi-scale contextual evidences. Int. J. Comput. Vis. **120**(1), 1–13 (2016)
31. Zhou, W., Li, H., Sun, J., Tian, Q.: Collaborative index embedding for image retrieval. IEEE Trans. Patt. Anal. Mach. Intell. **PP**, 1 (2017)

Image Color and Texture

Shape-Color Differential Moment Invariants Under Affine Transforms

Hanlin Mo[1,2(✉)], Shirui Li[1,2], You Hao[1,2], and Hua Li[1,2]

[1] Key Laboratory of Intelligent Information Processing,
Institute of Computing Technology, Chinese Academy of Sciences, Beijing, China
{mohanlin,lishirui,haoyou,lihua}@ict.ac.cn
[2] University of Chinese Academy of Sciences, Beijing, China

Abstract. We propose a general structural formula of shape-color primitive by using partial derivatives of each color channel in this paper. By using shape-color primitive, shape-color differential moment invariants (SCDMIs) can be constructed very easily, which are invariant to shape affine and color affine transforms. And 50 instances of SCDMIs are obtained. In experiments, several commonly used image descriptors and SCDMIs are used in image classification and retrieval for color image databases, respectively. By comparing the results, we find that SCDMIs get better results.

Keywords: Shape-color primitive · Affine transform
Partial derivatives · Shape-color differential moment invariants

1 Introduction

Image classification and retrieval for color images are two hotspots in pattern recognition. How to extract effective features, which are robust to color variations caused by the changes in the outdoor environment and geometric deformations caused by viewpoint changes, is the key issue. A classical approach is to construct invariant features for color images. Moment invariant is one of them.

Moment invariant was first proposed by Hu in [1]. In 1962, He defined geometric moment and constructed 7 geometric moment invariants which were invariant to the similarity transform (rotation, scaling and translation). Researchers applied them to many fields in pattern recognition and achieved good results [2,3]. Nearly 30 years later, Flusser et al. constructed affine moment invariants (AMIs) in [4] which are invariant to affine transform. The geometric deformation of an object, which is caused by the viewpoint change, can be represented by projective transform. However, general projective transform is a kind of complex nonlinear transform. So, it's difficult to construct projective moment invariants. When the distance between the camera and the object is much larger than the size of the object itself, the geometric deformation can be approximated by the affine transform. AMIs have been used in many practical

---

Student is the first author.

© Springer Nature Singapore Pte Ltd. 2017
J. Yang et al. (Eds.): CCCV 2017, Part II, CCIS 772, pp. 183–196, 2017.
https://doi.org/10.1007/978-981-10-7302-1_16

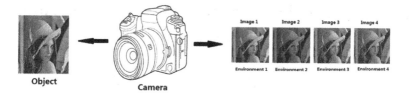

Fig. 1. The changes in color resulting from changes in the outdoor environment. (Color figure online)

applications [5,6]. In order to get more AMIs, researchers designed various methods. Suk et al. presented the graph method in [7], which can be used to construct AMIs with arbitrary orders and degrees. Xu et al. proposed the concept of geometric primitives in [8], including distance, area and volume. AMIs can be constructed by using various geometric primitives. This method made the construction of moment invariants has geometric meaning.

The above-mentioned moment invariants are all designed for gray images. With the popularity of color images, invariants for color images began to appear gradually. Researchers tried to construct features which have invariance for geometric deformations and the changes in color space (Fig. 1).

Geusebroek et al. [9] proved that the affine transform model was the best linear model to simulate changes in color resulting from changes in the outdoor environment. Mindru et al. constructed the moment invariants in [10], which were invariant to shape affine and color diagonal-offset transforms. These invariants were obtained by using the related concepts of Lie group. Some complex partial differential equations had to be solved. Thus, the number of them was limited and difficult to be generalized. Suk et al. [11] put forward affine moment invariants for color images by combining all color channels. But this approach was not intuitive and did not work well for the changes in color space. To solve these problems, Gong et al. [12–14] constructed the color primitive by using the concept of geometric primitive proposed in [8]. Combining the color primitive with some shape primitives, moment invariants that were invariant to shape affine and color affine transforms can be obtained easily, which were named shape-color affine moment invariants (SCAMIs). In [14], they obtained 25 SCAMIs which satisfied the independency of functions. However, we find that a large number of SCAMIs with simple structures and good properties are missed in [14].

In this paper, we propose a general structural formula of shape-color primitives by using partial derivatives of each color channel. Then, two special cases of shape-color primitives are used to construct shape-color differential moment invariants (SCDMIs), which are invariant to shape affine and color affine transforms. We find that SCAMIs proposed in [14] is a special case of our method. Finally, commonly used image descriptors and SCDMIs are used to image classification and retrieval for color images, respectively. By comparing the results, we find that SCDMIs get better results.

2   Related Work

In order to construct image features which are robust to color variations and geometric deformations, researchers have made various attempts. Among them, SCAMIs proposed in [14] are worthy of special attention, which have invariance for shape affine and color affine transforms. Two affine transforms are defined by

$$
\begin{pmatrix} x' \\ y' \end{pmatrix} = SA \cdot \begin{pmatrix} x \\ y \end{pmatrix} + ST = \begin{pmatrix} \alpha_1 \ \alpha_2 \\ \beta_1 \ \beta_2 \end{pmatrix} \cdot \begin{pmatrix} x \\ y \end{pmatrix} + \begin{pmatrix} O_x \\ O_y \end{pmatrix} \tag{1}
$$

$$
\begin{pmatrix} R'(x,y) \\ G'(x,y) \\ B'(x,y) \end{pmatrix} = CA \cdot \begin{pmatrix} R(x,y) \\ G(x,y) \\ B(x,y) \end{pmatrix} + CT = \begin{pmatrix} a_1 \ a_2 \ a_2 \\ b_1 \ b_2 \ b_2 \\ c_1 \ c_2 \ c_2 \end{pmatrix} \cdot \begin{pmatrix} R(x,y) \\ G(x,y) \\ B(x,y) \end{pmatrix} + \begin{pmatrix} O_R \\ O_G \\ O_B \end{pmatrix} \tag{2}
$$

where SA and CA are nonsingular matrices.

For the color image $I(R(x,y), G(x,y), B(x,y))$, let $(x_p, y_p), (x_q, y_q), (x_r, y_r)$ be three arbitrary points in the domain of I. The shape primitive and the color primitive are defined by

$$
S(p,q) = \begin{vmatrix} (x_p - \bar{x}) \ (x_q - \bar{x}) \\ (y_p - \bar{y}) \ (y_q - \bar{y}) \end{vmatrix} \tag{3}
$$

$$
C(p,q,r) = \begin{vmatrix} (R(x_p,y_p) - \bar{R}) \ (R(x_q,y_q) - \bar{R}) \ (R(x_r,y_r) - \bar{R}) \\ (G(x_p,y_p) - \bar{G}) \ (G(x_q,y_q) - \bar{G}) \ (G(x_r,y_r) - \bar{G}) \\ (B(x_p,y_p) - \bar{B}) \ (B(x_q,y_q) - \bar{B}) \ (B(x_r,y_r) - \bar{B}) \end{vmatrix} \tag{4}
$$

where $\bar{X}$ represents the mean value of X, $X \in \{x, y, R, G, B\}$. Then, using (3), the shape core (sCore) can be defined by

$$
sCore(n, m; d_1, d_2, ..., d_n) = \underbrace{S(1,2)S(k,l)...S(r,n)}_{m} \tag{5}
$$

where n and m represent that the sCore is the product of m shape primitives which are constructed by n points $(x_1, y_1), (x_2, y_2), ..., (x_n, y_n)$. $k < l,\ r < n$, $k, l, r \in \{1, 2, ...n\}$. d_i represents the number of times that the point (x_i, y_i) occurs in all shape primitives, $i = 1, 2, ..., n$.

Similarly, the color core (cCore) can be defined by

$$
cCore(N, M; D_1, D_2, ..., D_N) = \underbrace{C(1,2,3)C(G,K,L)...C(P,Q,N)}_{M} \tag{6}
$$

Where N and M represent that the cCore is the product of M color primitives which are constructed by N points $(x_1, y_1), (x_2, y_2), ..., (x_N, y_N)$. $G < K < L$, $P < Q < N$, $G, K, L, P, Q \in \{1, 2, ...N\}$. D_i represents the number of times that the point (x_i, y_i) occurs in all color primitives, $i = 1, 2, ..., N$.

Suppose the color image $I(R(x,y), G(x,y), B(x,y))$ is transformed into the image $I'(R'(x',y'), G'(x',y'), B'(x',y'))$ by two transforms defined by (1) and (2). $(x'_p, y'_p), (x'_q, y'_q)$ and (x'_r, y'_r) in I' are the corresponding points of $(x_p, y_p), (x_q, y_q)$ and (x_r, y_r) in I. Gong et al. [14] have proved

$$S'(p,q) = |SA| \cdot S(p,q) \tag{7}$$

$$C'(p,q,r) = |CA| \cdot C(p,q,r) \tag{8}$$

Further results can be concluded that

$$sCore'(n,m;d_1,d_2,...,d_n) = |SA|^m \cdot sCore(n,m;d_1,d_2,...,d_n) \tag{9}$$

$$cCore'(N,M;D_1,D_2,...,D_N) = |CA|^M \cdot cCore(N,M;D_1,D_2,...,D_N) \tag{10}$$

Therefore, the SCAMIs can be constructed by

$$
\begin{aligned}
&SCAMIs(n,m,N,M;d_1,...,d_n;D_1,...,D_N) \\
&= \frac{In(sCore(n,m,d_1,...,d_n) \cdot cCore(N,M,D_1,...,D_N))}{In(sCore(1,0))^{max(n+N)+m-\frac{3M}{2}} \cdot In(cCore(3,2;2,2,2))^{\frac{M}{2}}}
\end{aligned}
\tag{11}
$$

where $In(X)$ means multiple integrals. Then there is a relation

$$
\begin{aligned}
&SCAMIs'(n,m,N,M;d_1,...,d_n;D_1,...,D_N) \\
&= SCAMIs(n,m,N,M;d_1,...,d_n;D_1,...,D_N)
\end{aligned}
\tag{12}
$$

It must be said that $\max\{n,N\}$, $\max_i\{d_i\}$ and $\max_i\{D_i\}$ are named the degree, the shape order and the color order of SCAMIs, respectively. In fact, (11) can be expressed as the polynomial of shape-color moments. This kind of moments was first proposed in [16] and defined by

$$SCM_{pq\alpha\beta\gamma} = \iint (x-\bar{x})^p (y-\bar{y})^q (R(x,y)-\bar{R})^\alpha (G(x,y)-\bar{G})^\beta (B(x,y)-\bar{B})^\gamma dxdy \tag{13}$$

Gong et al. [14] pointed out that they constructed all SCAMIs of which degrees $\leqslant 4$, shape orders $\leqslant 4$ and color orders $\leqslant 2$. They obtained 24 SCAMIs which are functional independencies using the method proposed by Brown [17]. However, we will point out in Sect. 3 that they omitted many simple and well-behaved SCAMIs.

3   The Structural Framework of SCDMIs

In this section, we introduce the general definitions of shape-color differential moment and shape-color primitive, firstly. Then, using the shape-color primitive, the shape-color core can be constructed. Finally, according to (11) and the shape-color core, we obtain the general structural formula of SCDMIs. Also, 50 instances of SCDMIs are given for experiments in Sect. 4.

3.1   The Definition of General Shape-Color Moment

Definition 1. Suppose the color image $I(R(x,y), G(x,y), B(x,y))$ has the k-order partial derivatives $(k = 0, 1, 2, ...)$. The general shape-color differential moment is defined by

$$SCM^k_{pq\alpha\beta\gamma} = \iint (x - \bar{x})^p (y - \bar{y})^q (R^{(k)}(x,y) - \bar{R}^{\delta(k)})^\alpha (G^{(k)}(x,y) - \bar{G}^{\delta(k)})^\beta$$
$$(B^{(k)}(x,y) - \bar{B}^{\delta(k)})^\gamma dxdy$$

$$(14)$$

where $(R^{(k)}(x,y), G^{(k)}(x,y), B^{(k)}(x,y))$ represent the k-order partial derivatives of $(R(x,y), G(x,y), B(x,y))$. $\bar{R}, \bar{G}, \bar{B}$ represent the mean values of R, G, B. $\delta(k)$ is the impact function.

$$\delta(k) = \begin{cases} 1 & (k = 0) \\ 0 & (k \neq 0) \end{cases} \tag{15}$$

We can find that (13) and (14) are identical, when $k = 0$. Therefore, the shape-color moment is a special case of the general shape-color differential moment.

3.2   The Construction of General Shape-Color Primitive

Definition 2. Suppose the color image $I(R(x,y), G(x,y), B(x,y))$ has the k-order partial derivatives $(k = 0, 1, 2, ...)$. $(x_p, y_p), (x_q, y_q), (x_r, y_r)$ are three arbitrary points in the domain of I. The general shape-color primitive is defined by

$$SCP_k(p,q,r) = \begin{vmatrix} F^k_R(x_p, y_p) & F^k_R(x_q, y_q) & F^k_R(x_r, y_r) \\ F^k_G(x_p, y_p) & F^k_G(x_q, y_q) & F^k_G(x_r, y_r) \\ F^k_B(x_p, y_p) & F^k_B(x_q, y_q) & F^k_B(x_r, y_r) \end{vmatrix} \tag{16}$$

where

$$F^k_C(x,y) = \sum_{i=0}^k \binom{k}{i} (x - \bar{x})^i (y - \bar{y})^{k-i} \frac{\partial^k C(x,y)}{\partial x^i \partial y^{k-i}}, \quad C \in \{R, G, B\}. \tag{17}$$

We can find that $C(p,q,r)$ defined by (4) is a special case of $SCP_k(p,q,r)$, when $k = 0$.

3.3   The Construction of General Shape-Color Core

Definition 3. Using Definition 2, the general shape-color core (scCore) is defined by

$$scCore_k(N, M; D_1, D_2, ..., D_N) = \underbrace{SCP_k(1,2,3) SCP_k(G, K, L) ... SCP_k(P, Q, N)}_{M}$$

$$(18)$$

Where $k = 1, 2, ...,$ N and M represent that the $scCore_k$ is the product of M shape-color primitives constructed by N points $(x_1, y_1), (x_2, y_2), ..., (x_N, y_N)$. $G < K < L, P < Q < N, G, K, L, P, Q \in \{1, 2, ...N\}$. D_i represents the number of times that the point (x_i, y_i) occurs in all shape-color primitives, $i = 1, 2, ..., N$.

Obviously, $cCore(N, M; D_1, D_2, ..., D_N)$ defined by (6) is a special case of $scCore_k(N, M; D_1, D_2, ..., D_N)$, when $k = 0$.

3.4   The Construction of SCDMIs

Theorem 1. Let the color image $I(R(x, y), G(x, y), B(x, y))$ be transformed into the image $I'(R'(x', y'), G'(x', y'), B'(x', y'))$ by (1) and (2), $(x'_p, y'_p), (x'_q, y'_q)$ and (x'_r, y'_r) in I' are the corresponding points of $(x_p, y_p), (x_q, y_q)$ and (x_r, y_r) in I, respectively. Suppose that $R(x, y), G(x, y), B(x, y), R'(x', y'), G'(x', y'),$ $B'(x', y')$ have the k-order partial derivatives $(k = 0, 1, 2, ...)$. Then there is a relation

$$SCP'_k(p, q, r) = |CA| \cdot SCP_k(p, q, r) \tag{19}$$

where

$$SCP'_k(p, q, r) = \begin{vmatrix} F^k_{R'}(x'_p, y'_p) & F^k_{R'}(x'_q, y'_q) & F^k_{R'}(x'_r, y'_r) \\ F^k_{G'}(x'_p, y'_p) & F^k_{G'}(x'_q, y'_q) & F^k_{G'}(x'_r, y'_r) \\ F^k_{B'}(x'_p, y'_p) & F^k_{B'}(x'_q, y'_q) & F^k_{B'}(x'_r, y'_r) \end{vmatrix} \tag{20}$$

Further, the following relation can be obtained

$$scCore'_k(N, M; D_1, D_2, ..., D_N) = |CA|^M \cdot scCore_k(N, M; D_1, D_2, ..., D_N) \tag{21}$$

where

$$scCore'_k(N, M; D_1, D_2, ..., D_N) = \underbrace{SCP'_k(1, 2, 3)SCP'_k(G, K, L)...SCP'_k(P, Q, N)}_{M}$$

$$\tag{22}$$

By using Maple2015, the proof of Theorem 1 is obvious. We can find that (10) is a special case of (21), when $k = 0$. So, when we replace $cCore(N, M; D_1, D_2, ..., D_N)$ in (11) with $scCore_k(N, M; D_1, D_2, ..., D_N)$, (12) is still tenable. Now, we can define SCDMIs.

Theorem 2.

$$SCDMIs_k(n, m, N, M; d_1, ..., d_n; D_1, ..., D_N)$$
$$= \frac{In(sCore(n, m, d_1, ..., d_n) \cdot scCore_k(N, M, D_1, ..., D_N))}{In(sCore(1, 0))^{max(n+N)+m-\frac{3M}{2}} \cdot In(scCore_k(3, 2; 2, 2, 2))^{\frac{M}{2}}} \tag{23}$$

Then there is a relation

$$SCDMIs'_k(n, m, N, M; d_1, ..., d_n; D_1, ..., D_N)$$
$$= SCDMIs_k(n, m, N, M; d_1, ..., d_n; D_1, ..., D_N) \tag{24}$$

where

$$SCDMIs'_k(n, m, N, M; d_1, ..., d_n; D_1, ..., D_N)$$
$$= \frac{In(sCore'(n, m, d_1, ..., d_n) \cdot scCore'_k(N, M, D_1, ..., D_N))}{In(sCore'(1,0))^{max(n+N)+m-\frac{3M}{2}} \cdot In(scCore'_k(3,2;2,2,2))^{\frac{M}{2}}} \quad (25)$$

(23) can be expressed as the polynomial of $SCM^k_{pq\alpha\beta\gamma}$. (11) is a special case of (23) when $k = 0$. The proof of (24) is exactly the same as that of (12) proposed in [14].

3.5   The Instances of SCDMIs

We can use (23) to construct instances of $SCDMIs$ by setting different k values. However, color images are discrete functions. The partial derivatives can't be accurately calculated. As the order of the partial derivative increases, the calculation error is also increasing, which will greatly affect the stability of SCDMIs. So, we only set $k = 0, 1$ in this paper.

When $k = 0$, $SCDMIs_0$ are equivalent to SCAMIs. We construct $SCDMIs_0$ of which degrees $\leqslant 4$, shape orders $\leqslant 4$ and color orders $\leqslant 1$. Gong et al. [14] pointed out that in order to obtain $SCDMIs_0$ of which degrees $\leqslant 4$, shape orders $\leqslant 4$ and color orders $\leqslant 1$, $scCore_0(3, 1; 1, 1, 1)$ must be $C(1, 2, 3)$. This judgment is wrong. In fact, $scCore_0(3, 1; 1, 1, 1)$ can be $C(1, 2, 3)$, $C(1, 2, 4)$, $C(1, 3, 4)$ and $C(2, 3, 4)$. Thus, lots of $SCDMIs_0$ were missed in [14]. By correcting this shortcoming, we get 25 $SCDMIs_0$ that satisfy the independency of functions by using the method proposed by [17].

At the same time, when $k = 1$, $SCP_1(p, q, r)$ is defined by

$$SCP_1(p, q, r) = \begin{vmatrix} F^1_R(x_p, y_p) & F^1_R(x_q, y_q) & F^1_R(x_r, y_r) \\ F^1_G(x_p, y_p) & F^1_G(x_q, y_q) & F^1_G(x_r, y_r) \\ F^1_B(x_p, y_p) & F^1_B(x_q, y_q) & F^1_B(x_r, y_r) \end{vmatrix} \quad (26)$$

where

$$F^1_C(x, y) = (x - \bar{x})\frac{\partial C}{\partial x} + (y - \bar{y})\frac{\partial C}{\partial y} \quad C \in \{R, G, B\}. \quad (27)$$

By replacing $C(1, 2, 3)$, $C(1, 2, 4)$, $C(1, 3, 4)$ and $C(2, 3, 4)$ with $SCP_1(1, 2, 3)$, $SCP_1(1, 2, 4)$, $SCP_1(1, 3, 4)$ and $SCP_1(2, 3, 4)$, 25 $SCDMIs_1$ can be obtained. Therefore, we can construct the feature vector SCDMI50, which is defined by

$$SCDMI50 = [SCDMIs^1_0, ..., SCDMIs^{25}_0, SCDMIs^1_1, ..., SCDMIs^{25}_1] \quad (28)$$

The construction methods of 50 instances are shown in Table 1. In order to more clearly explain that SCDMIs can be expanded into the polynomial of

Table 1. The construction methods of $SCDMI50$

Name	scCore	sCore
$SCDMIs_0^1/SCDMIs_1^1$	$C(1,2,3)/SCP_1(1,2,3)$	$(x_1y_2 - x_2y_1)(x_1y_3 - x_3y_1)^2$
$SCDMIs_0^2/SCDMIs_1^2$	$C(1,2,3)/SCP_1(1,2,3)$	$(x_1y_2 - x_2y_1)(x_1y_3 - x_3y_1)^3$
$SCDMIs_0^3/SCDMIs_1^3$	$C(1,2,3)/SCP_1(1,2,3)$	$(x_1y_2 - x_2y_1)(x_1y_3 - x_3y_1)(x_2y_3 - x_3y_2)$
$SCDMIs_0^4/SCDMIs_1^4$	$C(1,2,3)/SCP_1(1,2,3)$	$(x_1y_2 - x_2y_1)(x_1y_3 - x_3y_1)(x_2y_3 - x_3y_2)^3$
$SCDMIs_0^5/SCDMIs_1^5$	$C(1,2,3)/SCP_1(1,2,3)$	$(x_1y_2 - x_2y_1)^2(x_1y_3 - x_3y_1)^2(x_2y_3 - x_3y_2)$
$SCDMIs_0^6/SCDMIs_1^6$	$C(1,2,4)/SCP_1(1,2,4)$	$(x_1y_2 - x_2y_1)(x_2y_3 - x_3y_2)(x_3y_4 - x_4y_3)$
$SCDMIs_0^7/SCDMIs_1^7$	$C(1,2,4)/SCP_1(1,2,4)$	$(x_1y_2 - x_2y_1)(x_2y_3 - x_3y_2)(x_3y_4 - x_4y_3)^3$
$SCDMIs_0^8/SCDMIs_1^8$	$C(1,3,4)/SCP_1(1,3,4)$	$(x_1y_2 - x_2y_1)(x_2y_3 - x_3y_2)(x_3y_4 - x_4y_3)^3$
$SCDMIs_0^9/SCDMIs_1^9$	$C(1,2,3)/SCP_1(1,2,3)$	$(x_1y_2 - x_2y_1)^2(x_2y_3 - x_3y_2)(x_3y_4 - x_4y_2)$
$SCDMIs_0^{10}/SCDMIs_1^{10}$	$C(1,2,4)/SCP_1(1,2,4)$	$(x_1y_2 - x_2y_1)^2(x_2y_3 - x_3y_2)(x_3y_4 - x_4y_3)^3$
$SCDMIs_0^{11}/SCDMIs_1^{11}$	$C(2,3,4)/SCP_1(2,3,4)$	$(x_1y_2 - x_2y_1)^2(x_2y_3 - x_3y_2)(x_3y_4 - x_4y_3)^3$
$SCDMIs_0^{12}/SCDMIs_1^{12}$	$C(1,2,4)/SCP_1(1,2,4)$	$(x_1y_2 - x_2y_1)^3(x_2y_3 - x_3y_2)(x_3y_4 - x_4y_3)^3$
$SCDMIs_0^{13}/SCDMIs_1^{13}$	$C(1,2,3)/SCP_1(1,2,3)$	$(x_1y_2 - x_2y_1)(x_2y_3 - x_3y_2)^2(x_3y_4 - x_4y_3)^2$
$SCDMIs_0^{14}/SCDMIs_1^{14}$	$C(1,2,4)/SCP_1(1,2,4)$	$(x_1y_2 - x_2y_1)(x_2y_3 - x_3y_2)^2(x_3y_4 - x_4y_3)^2$
$SCDMIs_0^{15}/SCDMIs_1^{15}$	$C(1,2,4)/SCP_1(1,2,4)$	$(x_1y_2 - x_2y_1)(x_2y_3 - x_3y_2)^3(x_3y_4 - x_4y_3)$
$SCDMIs_0^{16}/SCDMIs_1^{16}$	$C(1,3,4)/SCP_1(1,3,4)$	$(x_1y_2 - x_2y_1)(x_2y_3 - x_3y_2)(x_3y_4 - x_4y_3)^2 \cdot (x_4y_1 - x_1y_4)$
$SCDMIs_0^{17}/SCDMIs_1^{17}$	$C(1,2,3)/SCP_1(1,2,3)$	$(x_1y_2 - x_2y_1)^2(x_2y_3 - x_3y_2)(x_3y_4 - x_4y_3)^3 \cdot (x_4y_1 - x_1y_4)$
$SCDMIs_0^{18}/SCDMIs_1^{18}$	$C(1,2,4)/SCP_1(1,2,4)$	$(x_1y_2 - x_2y_1)(x_1y_3 - x_3y_1)(x_1y_4 - x_4y_1)$
$SCDMIs_0^{19}/SCDMIs_1^{19}$	$C(1,2,4)/SCP_1(1,2,4)$	$(x_1y_2 - x_2y_1)(x_1y_3 - x_3y_1)(x_1y_4 - x_4y_1) \cdot (x_3y_4 - x_4y_3)^3$
$SCDMIs_0^{20}/SCDMIs_1^{20}$	$C(2,3,4)/SCP_1(2,3,4)$	$(x_1y_2 - x_2y_1)(x_1y_3 - x_3y_1)^2(x_1y_4 - x_4y_1) \cdot (x_3y_4 - x_4y_3)^2$
$SCDMIs_0^{21}/SCDMIs_1^{21}$	$C(1,2,3)/SCP_1(1,2,3)$	$(x_1y_2 - x_2y_1)^2(x_1y_3 - x_3y_1)(x_1y_4 - x_4y_1) \cdot (x_3y_4 - x_4y_3)$
$SCDMIs_0^{22}/SCDMIs_1^{22}$	$C(1,2,3)/SCP_1(1,2,3)$	$(x_1y_2 - x_2y_1)^2(x_1y_3 - x_3y_1)(x_1y_4 - x_4y_1) \cdot (x_3y_4 - x_4y_3)^3$
$SCDMIs_0^{23}/SCDMIs_1^{23}$	$C(1,2,4)/SCP_1(1,2,4)$	$(x_1y_2 - x_2y_1)^2(x_1y_3 - x_3y_1)(x_1y_4 - x_4y_1) \cdot (x_3y_4 - x_4y_3)^3$
$SCDMIs_0^{24}/SCDMIs_1^{24}$	$C(1,2,4)/SCP_1(1,2,4)$	$(x_1y_2 - x_2y_1)(x_2y_3 - x_3y_2)(x_3y_4 - x_4y_3)^2 \cdot (x_4y_1 - x_1y_4)(x_2y_4 - x_4y_2)$
$SCDMIs_0^{25}/SCDMIs_1^{25}$	$C(1,2,4)/SCP_1(1,2,4)$	$(x_1y_2 - x_2y_1)^2(x_2y_3 - x_3y_2)(x_3y_4 - x_4y_3)^2 \cdot (x_4y_1 - x_1y_4)^2(x_2y_4 - x_4y_2)$

$SCM_{pq\alpha\beta\gamma}^k$, we give the shape-color moment polynomial of $SCMIs_0^3$.

$$
\begin{aligned}
SCDMIs_0^3 =& \{6SCM_{02001}^0 SCM_{11010}^0 SCM_{20100}^0 - 6SCM_{02001}^0 SCM_{11100}^0 SCM_{20010}^0 \\
& - 6SCM_{02010}^0 SCM_{11001}^0 SCM_{20100}^0 + 6SCM_{02010}^0 SCM_{11100}^0 SCM_{20001}^0 \\
& + 6SCM_{02100}^0 SCM_{11001}^0 SCM_{20010}^0 - 6SCM_{02100}^0 SCM_{11010}^0 SCM_{20001}^0 \} \\
& / \{6SCM_{00002}^0 SCM_{00020}^0 SCM_{00200}^0 - 6SCM_{00002}^0 (SCM_{00110}^0)^2 \\
& - 6(SCM_{00011}^0)^2 SCM_{00200}^0 + 12SCM_{00011}^0 SCM_{00101}^0 SCM_{00110}^0 \\
& - 6SCM_{00020}^0 (SCM_{00101}^0)^2 \}
\end{aligned}
\tag{29}
$$

4   Experimental Results

In this section, some experiments are designed to evaluate the performance of SCDMI50. Firstly, we verify the stability and discriminability of SCDMI50 by using synthetic images. Then, some retrieval experiments based on real image databases are performed. Also, we chose some commonly used image descriptors for comparison.

It is worth noting that we have to choose a appropriate method to calculate the first order partial derivatives of $R(x,y)$, $G(x,y)$ and $B(x,y)$. In [21], the 5 points difference formulas were used for approximating the first partial derivatives of discrete grayscale images and achieved good results. For color images, they can be defined by

$$
\begin{aligned}
\frac{\partial C(x,y)}{\partial x} &= C(x-2,y) - 8C(x-1,y) + 8C(x+1,y) - C(x+2,y) \\
\frac{\partial C(x,y)}{\partial y} &= C(x,y-2) - 8C(x,y-1) + 8C(x,y+1) - C(x,y+2)
\end{aligned}
\tag{30}
$$

where $C \in \{R, G, B\}$. We choose this method because it guarantees the computational accuracy of the first partial differential to a certain extent and also maintains a relatively fast calculation speed.

4.1   The Stability and Discriminability of SCDMI50

We select 50 different kinds of butterfly images, which are shown in Fig. 2(a). Then, 5 shape affine transforms and 4 color affine transforms are applied to each image. One image can get 20 transformed versions which are shown in Fig. 2(b). Thus, we obtain the database containing 1000 images. 10% images selected randomly are used as the training data and the rest make up the testing data.

For comparison with $SCDMI50$, we chose some commonly used global features of gray images or color images.

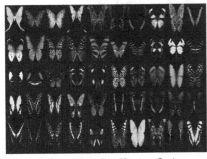
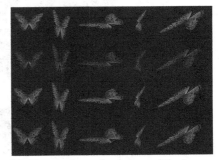

(a) 50 different kinds of butterfly images   (b) 20 transformed versions of the image

Fig. 2. Sample images from the butterfly image database (Color figure online)

- **Hu moments**, which were composed of 7 invariants under the shape similarity transform, proposed in [1]. Hu moments were designed for gray images.
- **AMIs** which were composed of 17 invariants under the shape affine transform proposed in [18]. AMIs were designed for gray images.
- **RGhistogram**, which consisted of 60-dimensional features and was scale-invariant in color space, proposed in [19]. The pixel range of each color channel was divided into 20 intervals for statistics.
- **Transformed color distribution**, which consisted of 60-dimensional features, was proposed in [19]. Transformed color distribution was scale-invariant and sift-invariant in color space. The pixel range of each color channel was divided into 20 intervals for statistics.
- **Color moments** consisted of the first, second and third geometric moments of each channel of the color image, which were proposed in [15]. Color moments are sift-invariant in color space.
- **GPSOs** were invariant under the shape affine transforms and the color diagonal-offset transform, which were proposed in [10]. GPSOs consisted of 21 moment invariants.

Subsequently, image classification is performed on the butterfly database by using different kinds of features. We use the Nearest Neighbor classifier based on the Chi-Square distance to estimate the species of test images. Finally, we list the classification accuracies obtained by using different features in Table 2.

On the one hand, we can find that the classification result obtained by using $SCDMI50$ is better than those obtained by using other features. And, color information is very important for the classification. On the other hand, in order to observe the property of $SCDMI50$ more clearly, we use Chi-Square distance

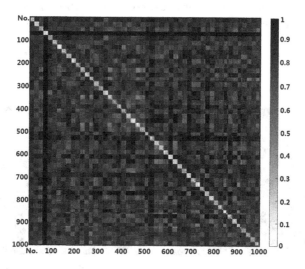

Fig. 3. The visualization of the distance matrix. As the distance increases, the color changes from white to black.

Table 2. The classification accuracies obtained by using different features

Descriptor	Accuracy	Descriptor	Accuracy
SCDMI50	**98.67%**	GPSOs	78.56%
AMIs	50.11%	Hu moments	25.00%
Color moments	74.77%	RGhistogram	80.56%
Transformed color distribution	95.78%		

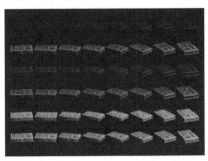

(a) 20 different classes of objects in COIL-100

(b) Each of categories contains 48 images

Fig. 4. Sample images from COIL-100 (Color figure online)

to calculate the feature distance between any two images. So, we can get a 1000×1000 distance matrix which is shown in Fig. 3. Obviously, the color of the area near the diagonal is lighter than those of other regions, indicating that $SCDMI50$ of similar images are similar in values, and vice versa. Also, these results demonstrate that the construction formula of SCDMIs designed in Sect. 3 is correct.

4.2 Image Retrieval on Real Image Database

In order to further test the performance of $SCDMI50$, the database COIL-100 proposed in [20] is chosen for our experiment. COIL-100 contains 7202 images of 100 categories, each of which has 72 images taken from different angles. We choose 20 classes, each class contains 8 images $(0°, 10°, 20°, 30°, 40°, 50°, 60°, 70°)$. For each image, 6 color affine transforms are applied. Thus, 960 images of 20 categories are obtained, each of categories contains 48 images which are shown in Fig. 4.

Then, retrieval experiment is performed on this database. Similar to Sect. 4.1, We choose Chi-Square distance to measure the similarity of two images. The Precision-Recall curves obtained by using different features are shown in Fig. 5(a).

Obviously, the result obtained by using SCDMI50 is far superior to those obtained by using other features. This is because that traditional image features

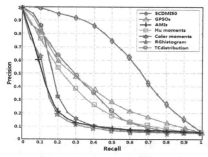
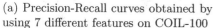
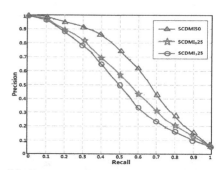

(a) Precision-Recall curves obtained by using 7 different features on COIL-100

(b) Precision-Recall curves obtained by using $SCDMI50$, $SCDMI_025$ and $SCDMI_125$ on COIL-100

Fig. 5. The Precision-Recall curves on COIL-100 (Color figure online)

are constructed by using only color information or shape information. However, we use two kinds of information while constructing SCDMI50. In addition, traditional image descriptors are only stable for simple changes in color space, such as the diagonal-offset transforms. When the color space changes drastically, they are less robust.

Finally, we compare the retrieval results of $SCDMI50$, $SCDMI_025$ and $SCDMI_125$ which are shown in Fig. 5(b). $SCDMI_025$ and $SCDMI_125$ are defined by

$$SCDMI_025 = [SCDMIs_0^1, SCDMIs_0^2, ..., SCDMIs_0^{25}] \tag{31}$$

$$SCDMI_125 = [SCDMIs_1^1, SCDMIs_1^2, ..., SCDMIs_1^{25}] \tag{32}$$

Because of the calculation error of partial derivatives, the result obtained by using $SCDMI_125$ slightly worse than that obtained by using $SCDMI_025$, but far better than those obtained by using traditional features. Meanwhile, $SCDMI_125$ increase the number of invariants which have simple structures and good properties. We combine $SCDMI_125$ and $SCDMI_025$ to get $SCDMI50$ which achieve the best retrieval result in our experiment.

5 Conclusion

In this paper, we propose a kind of shape-color differential moment invariants (SCDMIs) for color images, which are invariant to the shape affine and color affine transforms, by using partial derivatives of each color channel. It is obvious that all SCAMIs proposed in [14] are the special cases of SCDMIs, when $k = 0$. Then, we correct the mistake in [14] and obtain 50 instances of SCDMIs, which have simple structures and good properties. Finally, several commonly used image descriptors and SCDMIs are used in color image classification and retrieval, respectively. By comparing the experimental results, we find that SCD-MIs get better results.

Acknowledgments. This work has been funded by National Natural Science Foundation of China (Grant No. 60873164, 61227802 and 61379082).

References

1. Hu, M.K.: Visual pattern recognition by moment invariants. IRE Trans. Inf. Theory **8**(2), 179–187 (1962)
2. Zhang, Y.D., Wang, S.H., Sun, P., Phillips, P.: Pathological brain detection based on wavelet entropy and Hu moment invariants. Bio-Med. Mater. Eng. **26**(s1), S1283–S1290 (2015)
3. Dudani, S.A., Breeding, K.J., McGhee, R.B.: Aircraft identification by moment invariants. IEEE Trans. Comput. **26**(1), 39–46 (1977)
4. Flusser, J., Suk, T.: Pattern recognition by affine moment invariants. Pattern Recogn. **26**(1), 167–174 (1993)
5. Flusser, J., Suk, T.: Affine moment invariants: a new tool for character recognition. Pattern Recogn. Lett. **15**(4), 433–436 (1994)
6. Renuka, L., Vrushsen, P.: Facial expression recognition based on affine moment invariants. Int. J. Comput. Sci. Issues **9**(6), 388–392 (2012)
7. Suk, T., Flusser, J.: Graph method for generating affine moment invariants. In: Proceedings of the International Conference on Pattern Recognition, pp. 192–195 (2004)
8. Xu, D., Li, H.: Geometric moment invariants. Pattern Recogn. **41**(1), 240–249 (2008)
9. Geusebroek, J.M., Van den Boomgaard, R., Smeulders, A.W.M., Geerts, H.: Color invariance. IEEE Trans. Pattern Anal. Mach. Intell. **23**(12), 1338–1350 (2001)
10. Mindru, F., Tuytelaars, T., Van Gool, L., Moons, T.: Moment invariants for recognition under changing viewpoint and illumination. Comput. Vis. Image Underst. **94**(1), 3–27 (2004)
11. Suk, T., Flusser, J.: Affine moment invariants of color images. In: International Conference on Computer Analysis of Images and Patterns, pp. 334–341 (2009)
12. Gong, M., Hu, P., Cao, W.G., Li, H.: A kind of shape-color moment invariants. In: 12th International Conference on Computer-Aided Design and Computer Graphics, pp. 425–432 (2011)
13. Gong, M., Li, H., Cao, W.G.: Moment invariants to affine transformation of colours. Pattern Recogn. Lett. **34**(11), 1240–1251 (2013)
14. Gong, M., Hao, Y., Mo, H.L., Li, H.: Naturally Combined Shape-Color Moment Invariants under Affine Transformations (2017). http://arxiv.org/abs/1705.10928
15. Stricker, M.A., Orengo, M.: Similarity of color images. In: IS&T/SPIE's Symposium on Electronic Imaging: Science & Technology, pp. 381–392. International Society for Optics and Photonics (1995)
16. Mindru, F., Van Gool, L., Moons, T.: Model estimation for photometric changes of outdoor planar color surfaces caused by changes in illumination and viewpoint. In: Proceedings of the International Conference on Pattern Recognition, pp. 620–623 (2002)
17. Brown, A.B.: Functional dependence. Trans. Am. Math. Soc. **38**(2), 379–394 (1935)
18. Suk, T., Flusser, J.: Affine moment invariants generated by graph method. Pattern Recogn. **44**(9), 2047–2059 (2011)

19. Sande, K.V.D., Gevers, T., Snoek, C.: Evaluating color descriptors for object and scene recognition. IEEE Trans. Pattern Anal. Mach. Intell. **32**(9), 1582–1596 (2010)
20. Nene, S.A., Nayer, S.K., Murase, H.: Columbia object image library ($COIL - 100$). Technical Report CUCS-006-96, CUCS (1996)
21. Wang, Y.B., Wang, X.W., Zhang, B., Wang, Y.: Projective invariants of D-moments of 2D grayscale images. J. Math. Imaging Vis. **51**(2), 248–259 (2015)

Image Composition

A Novel Layer Based Image Fusion Approach via Transfer Learning and Coupled Dictionary

Kai Hu, Bin Sun$^{(\boxtimes)}$, Qiao Deng, and Qi Yang

School of Aeronautics and Astronautics,
University of Electronic Science and Technology of China,
2006 Xiyuan Ave, West Hi-Tech Zone, Chengdu 611731, China
sunbinhust@uestc.edu.cn

Abstract. A novel layer based image fusion method is proposed in this paper. It exploits and utilizes the implicated patterns among source images with two parts: (i) proposed a more precise model roots in transfer learning and coupled dictionary for layering source images; (ii) designed appropriate fusion scheme which bases on multi-scale transformation for recombining layers into final fused image efficiently. Rigorous experimental comparison in subjective and objective demonstrates that proposed image fusion method achieves better result in visual perception and computer process.

Keywords: Image fusion · Layer division · Transfer learning
Coupled dictionary

1 Introduction

Due to limitation of single sensor, multi-sensor images are often used for providing comprehensive description of target. Image fusion combines multi-sensor images into a single more comprehensive fused image for visual perception and computer process. Image fusion has been widely used in many applications, such as visible and infrared photography, medical detection, remote imagery and so on.

In early studies, researchers usually directly define significance and match measure of image content in original or transform domain. A series of approaches work in different domains that multi-scale transformation [1–4] and spare representation [5,6] are the most typical types. And many approaches make effort to define better indexes and fusion rule [7]. More detailed introduction about these algorithm could be found in the excellent review of Li [8].

Since multi-sensor images provide descriptions of same target with different sensor types and environment, some relationships could be expected to exist among them [9]. The same target leads to redundancy while variety sensor types and environment generate complementary in information. Some novel approaches are designed working with redundancy and complementarity of source images. With sophisticated manual definition of this relationship. For

© Springer Nature Singapore Pte Ltd. 2017
J. Yang et al. (Eds.): CCCV 2017, Part II, CCIS 772, pp. 199–209, 2017.
https://doi.org/10.1007/978-981-10-7302-1_17

example, satellites multi-spectrum and panchromatic images fusion of SIRF [10], infrared-visible image fusion [11] and MRSR in multi-focus image fusion [12] are proved superior to previous work in result. Some approaches design algorithm automatically reveal implicated relationship among source images [13,14].

In this paper, inspired by [13,15], a novel algorithm is proposed for exploiting redundancy and complementarity via transfer learning, coupled dictionary and some specific prior knowledge. After separated information of source images, a fusion scheme was designed for obtaining final fused image.

The rest of this paper is organized as follows. Introduction and discussion about related works are exhibited in Sect. 2. Section 3 describes our work in detail and Sect. 4 compares proposed method with state-of-the-art algorithm.

2   Related Work

2.1   Layer Division Based Image Fusion

The relationship of redundancy and complementarity could be mathematically modeled in the following form with probabilistic view. In order to simplify discussion, we only consider the situation with only two source images.

$$P(x_1) = P(x_1\overline{x_2}) + P(x_1x_2) \tag{1}$$

$$P(x_2) = P(x_2\overline{x_1}) + P(x_1x_2) \tag{2}$$

This formulation divides original image content into two parts. Left part of both equations represent marginal distribution of source images x_1 and x_2. $P(x_1\overline{x_2})$ and $P(x_2\overline{x_1})$ are individual components while $P(x_1x_2)$ is correlated component. Essentially, the core task of exploiting redundancy and complementarity is to estimate individual and correlated components across two images.

Yu [13] actually extracts features of source images through K-SVD dictionary and labels them by proposed joint spare representation (JSR). Finally, correlated layer and individual layer are reconstructed with labeled features respectively. It is an approximation of correlated components and individual components with K-SVD dictionary atoms respectively.

$$P(x) = P(f(\theta_C)) + P(f(\theta_I)), \theta_C + \theta_I = \theta \tag{3}$$

θ is all atoms in K-SVD dictionary while θ_C and θ_I are correlated and individual features. $f(\cdot)$ represents spare reconstruction procedure for estimating ideal layer division.

The more precise dictionary, the better divided layers [15]. Unfortunately, existing dictionary training algorithm cannot guarantee its feature is orthogonal, exactly precise and no reconstructive error. As illustrated in Fig. 1, these defects make its model not very precise.

In consideration of difficulties in estimating individual parts and correlated parts directly, we propose a novel model to estimate posterior probability since marginal probability is known. We reformulate (1) and (2) to the following form.

$$P(x_1) = P(x_1\overline{x_2}) + P(x_1|x_2) \cdot P(x_2) \tag{4}$$

$$P(x_2) = P(x_2\overline{x_1}) + P(x_1|x_2) \cdot P(x_1) \tag{5}$$

In this form, more precise estimation of union probability is obtained with fusion of posterior probability. This task has been conducted for many years in transfer learning technique and coupled dictionary which are introduced in Sect. 2.2.

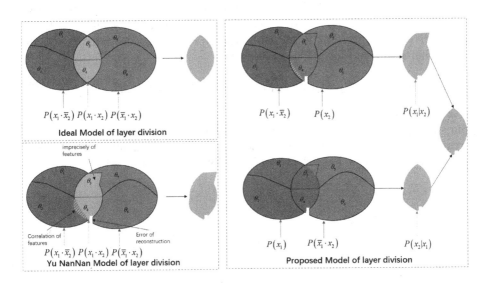

Fig. 1. Layer division model

2.2 Introduction of Transfer Learning and Coupled Dictionary

Transfer learning is a technique which aims to help improving performance in target task through transfer knowledge from a source task [16]. The core of transfer learning is the mechanism to extract and transfer knowledge of source to target. Some work [17,18] focus on transferring knowledge across two unlabeled domains. It is reported that transfer learning hasn't been used in image fusion tasks though it highly matches image fusion.

Coupled dictionary training is often used to observe feature spaces for associating cross-domain image data and jointly improve presentative ability of each dictionary [19]. Rui Gao proposed a multi-focus image fusion approach with coupled dictionary [20].

3 Proposed Method

3.1 Transfer Learning and Coupled Dictionary Based Layer Division

In Sect. 2.1, the problem is converted to estimation of $P(x_1|x_2)$ and $P(x_2|x_1)$ in (4) and (5). In this model, both images will be divided into two layers due to asymmetric of $P(x_1|x_2)$ and $P(x_2|x_1)$.

Let D_1 is the feature dictionary which is trained from image x_1, it is reasonable to treat this dictionary D_1 as source knowledge in transfer learning [18]. Reconstructing x_2 with D_1, correlated component could be reconstructed well while individual component presumably lose. With same process in x_1 and x_2, the optimal goal could formulate as following:

$$min\Big\{\|X_1 - D_2 \cdot A_1\|_2 + \|X_2 - D_1 \cdot A_2\|_2 + l(A_1, A_2, D_1, D_2)\Big\} \qquad (6)$$

X_1 is the sample matrix of x_1 which consists of vectorized patches obtained from sliding windows. $A1$ and $A2$ are the coefficient matrixes. $l(A_1, A_2, D_1, D_2)$ is regularization condition presents some expected properties for dictionary and reconstructive process. The scheme of layer division is illustrated in Fig. 2.

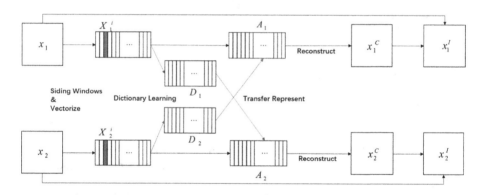

Fig. 2. Flow diagram of layer division based on transfer learning and coupled dictionary

3.2 Feature Extraction and Exchanged Representation

Learning dictionary D is the process for obtaining a series features which determine the accuracy in layer division. Unfortunately, based on discussion in Sect. 2.1, reconstructive error still influences the result of layer division in proposed model. Instead of designing a brand new sophisticated algorithm for seeking better dictionary, inspired by [12], an alternative method is taken in this paper.

Undoubtedly, original image patches could transfer knowledge between images and do not lose any information. However, in procedure of knowledge transfer that original image patches are not good at distinguishing features. In order to ensure that each component could be presented correctly in any layer images, it is necessary to ensure that elements in all layer images is positive for consistency with original patches.

$$min\left\{\|X_1 - D_2 \cdot A_1\|_2 + \|X_2 - D_1 \cdot A_2\|_2 + l(A_1, A_2, D_1, D_2)\right\},$$
$$(X_l^h)_{ij} \geq 0; l = 1, 2; h = C, I \tag{7}$$

Our optimal formulation is changed to the following form. Logical notation means the logical relation between corresponding elements in both matrixes:

$$min\left\{\|X_1 - D_2 \cdot A_1\|_2 + \|X_2 - D_1 \cdot A_2\|_2 + l(A_1, A_2, D_1, D_2)\right\},$$
$$(X_l^h)_{ij} \geq 0; l = 1, 2; h = C, I \tag{8}$$

A general condition in image fusion field is that source images has been pre-registered before further processing, and so does our work. This condition guarantees a property in image fusion that correlated knowledge which depicts the same phenomenon existing in same region of both source images. So, the j-th patch of X_1 only share correlated knowledge with j-th patch of X_2. We could rewrite our optimal formulation (7) with this property as the following form.

$$min\left\{\|X_1 - D_2 \cdot A_1\|_2 + \|X_2 - D_1 \cdot A_2\|_2 + \lambda_1 \sum_{j=1}^{j=m} (\|\alpha_{1_{j,j}}\|_1 + \|\alpha_{2_{j,j}}\|_1 - 2)\right.$$
$$\left. + \lambda_2 \sum_{j=1}^{j=m} (\|\alpha_{1_{i,j}}\|_1 + \|\alpha_{2_{i,j}}\|_1)\right\}, i \neq j \tag{9}$$

In this formulation, α_{1_j} is the j-th column of A_1. The first regularization with λ_1 ensures a patch accept knowledge from corresponding patch in another image. λ_2 controls tolerance that patch accept knowledge from other patch. Third and fourth regularization determine penalty on negative element of layer images. When these parameter are infinite, the optimal solve actually occur when any element is equal between each column of X_1 and X_2.

Definition of coupled dictionary problem is similar to (8), except for exchanged position of D_1 and D_2. In this paper, we proposed a novel algorithm to solve this optimization reference to property mentioned in [19].

3.3   Fusion Scheme Base on Proposed Layer Division Method

Indeed, correlated layer and individual layer store redundant and complementary information respectively. The ideal situation is that fused image inherits all of redundant and complementary information directly. Due to possible incompatibility between complementary components, the better fusion rule in reality is to chose more informative one. Furthermore, complementary information is no longer influenced by redundant information when its information is measured. Hence the complementary component in fused image looks enhanced compare with source images.

Instead of spare representation, which cannot handle high frequency information efficiently [7], multi-scale transformation is much better to corresponding layers. Our fusion scheme is designed with DTCWT [2] and NSCT [4] as Fig. 3.

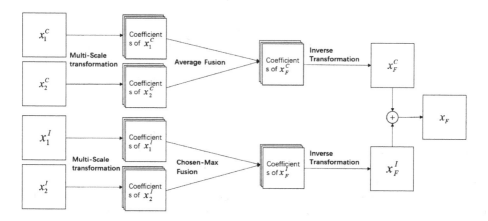

Fig. 3. Flow diagram of fusing divided correlated layer images and individual layer images

4   Experiment

Infrared-visible (IR-VI) image pair and multi-focus image pair are chosen in experiment due to their distinctive property in distribution of individual and correlated components. A comparison of divided layers between work in [13] and proposed method is held in the first part. Then subjective and objective comparisons of final fused image are held.

4.1   Layer Division Results and Discussion

Approach of [13] works with sliding windows in size 8×8 and K-SVD dictionary in size 64×500. In proposed method, on account of low similarity in IR-VI images but high similarity in multi-focus images, sliding windows with size 3×3 and 16×16 are applied in them respectively.

In Figs. 4 and 5, divided layers of proposed method and [13] are illustrated. Obviously, more clear pattern is emerged in proposed methods divided layers and no clutter information exists in them. Individual layers of proposed method show some apparent dissimilarity compare to JSR approach.

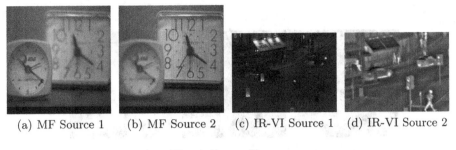

(a) MF Source 1   (b) MF Source 2   (c) IR-VI Source 1   (d) IR-VI Source 2

Fig. 4. Source Images

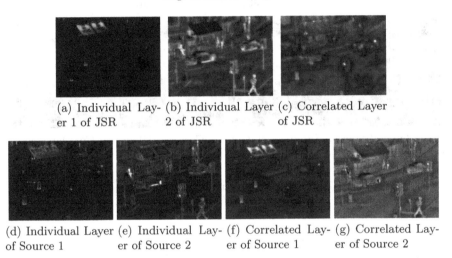

(a) Individual Lay- (b) Individual Layer (c) Correlated Layer
er 1 of JSR   2 of JSR   of JSR

(d) Individual Layer (e) Individual Lay- (f) Correlated Lay- (g) Correlated Lay-
of Source 1   er of Source 2   er of Source 1   er of Source 2

Fig. 5. Division layer images of IR-VI

4.2 Fusion Results and Comparison

Divided layers finally integrated into fused image with multi-scale transformation fusion scheme, so discrete wavelet transform (DWT), non-sampled contourlet transform (NSCT) are chosen for a comparative study. Level of multiscale decomposition is 4 in both DWT and NSCT. Besides, a comparison among some approaches is held with optimal parameters as descriptions in their papers [7,13].

In accordance with the point that multi-sensor images are used for comprehensively describe target, complementary information may be more important than redundant information. Because proposed approaches decrease interference between complementary information and redundant information in some degree, there are some details enhanced of results in Figs. 6 and 7. Dissimilar with some image enhanced algorithm, this emphasis of individual component does not generate artificial component.

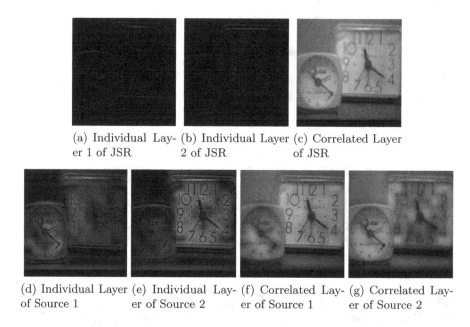

(a) Individual Lay- (b) Individual Layer (c) Correlated Layer
er 1 of JSR   2 of JSR   of JSR

(d) Individual Layer (e) Individual Lay- (f) Correlated Lay- (g) Correlated Lay-
of Source 1   er of Source 2   er of Source 1   er of Source 2

Fig. 6. Division layer images of multi-focus

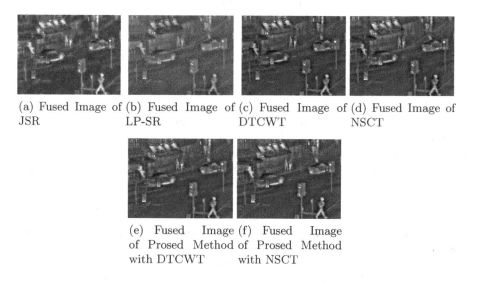

(a) Fused Image of (b) Fused Image of (c) Fused Image of (d) Fused Image of
JSR   LP-SR   DTCWT   NSCT

(e) Fused   Image (f) Fused   Image
of Prosed Method   of Prosed Method
with DTCWT   with NSCT

Fig. 7. IR-VI Fused Result Comparison (Color figure online)

In Fig. 6, proposed method inherit the elliptical structure of fence in the gallery, bracket of left traffic light and line on the road when other approaches lost them due to only reserving information of single one source. We mark these regions with red box. Besides, fused image of proposed method clearly depicts the details of stools in front of bar, windows lattice and backpack of pedestrian who walk pass the bar.

In the fused scenario consisted of multi-focus images, proposed method obtain sharper and clearer results than other methods. By carefully comparison border of focus regions, no shadow or oversharp edge exists in results. We also mark some important regions with red box. Proposed model makes gradient information more significant while gradient is individual component among source images.

According to the work of Liu [21], four objective metrics which emphasis on different views are chosen for evaluating and comparing proposed method. Q_{VIFF} [22] simulates human vision when Q_e comes from information theory. Q_{SF} [23] presents the richness of gradient information and Q_{SSIM} [22] measure the similarity of structure between fused image and source images.

According to Table 1, high scores in Q_{VIFF}, Q_e and Q_{SF} are obtained by results of proposed method in both IR-VI and multi-focus scenarios. Due to the emphasis of individual component, proposed method does not keep a consistent structure with single source image decreases its score in Q_{SSIM} (Fig. 8).

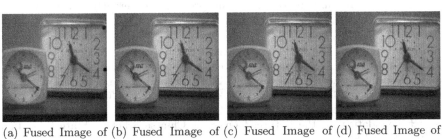

(a) Fused Image of (b) Fused Image of (c) Fused Image of (d) Fused Image of
JSR   NSCT-SR   DTCWT   NSCT

(e) Fused   Image (f) Fused   Image
of Prosed Method of Prosed Method
with DTCWT   with NSCT

Fig. 8. MF Fused Result Comparison (Color figure online)

Table 1. Objective metric scores of IR-VI fusion and multi-focus fusion

	IR-VI fusion				Muti-focus fusion			
	Q_{SSIM}	Q_e	Q_{SF}	Q_{VIFF}	Q_{SSIM}	Q_e	Q_{SF}	Q_{VIFF}
JSR	0.5831	6.578	12.32	0.489	**0.9153**	7.281	12.70	0.951
LP-SR/NSCT-SR	0.5556	6.841	16.88	**0.657**	0.8926	7.2978	15.8621	0.9622
DTCWT	0.6026	6.408	16.25	0.538	0.8935	7.336	16.32	0.945
NSCT	**0.6111**	6.452	16.48	0.541	0.8994	7.332	16.22	0.963
Ours(DTCWT)	0.5952	6.836	**22.47**	0.568	0.8431	7.644	26.51	1.207
Ours(NSCT)	0.5910	**6.841**	20.48	0.585	0.8403	**7.674**	**27.47**	**1.232**

5   Conclusion

In this paper, we proposed a novel approach for dividing image into correlated layer and individual layer. Compared with previous work, our layer is better in revealing implicit pattern among source images. For layer division task, we make use of existing morphology transformation to fuse divided layer respectively and combine them into final fused image. The experimental results show that proposed method is competitive to state-of-the-art approaches. Since misalignment, noise and moving objection are presented differently in each images, robust approach is the focus in our future work.

Acknowledgments. The authors thank Glenn Easley, Yu Liu, Zheng Liu for sharing the code of their works. The work is supported by the Fundamental Research Funds for the Central Universities of China (No. ZYGX2015J122 and No. ZYGX2015KYQD032), New Characteristic Teaching Material Construction (No. Y03094023701019427), National Natural Science Foundation of China (No.61701078), and the Scientific Research Foundation for the Returned Overseas Chinese Scholars, State Education Ministry [2015] 1098.

References

1. Zhang, Z., Blum, R.S.: A categorization of multiscale-decomposition-based image fusion schemes with a performance study for a digital camera application. In: Proceedings of the IEEE, vol. 87, p. 1315 (1999)
2. Lewis, J.J., O'Callaghan, R.J., Nikolov, S.G., Bull, D.R., Canagarajah, N.: Pixel- and region-based image fusion with complex wavelets. Inf. Fusion **8**, 119 (2007)
3. Nencini, F., Garzelli, A., Baronti, S., Alparone, L.: Remote sensing image fusion using the curvelet transform. Inf. Fusion **8**, 143 (2007)
4. Li, T., Wang, Y.: Biological image fusion using a NSCT based variable-weight method. Inf. Fusion **12**, 85 (2011)
5. Yang, B., Li, S.: Multifocus image fusion and restoration with sparse representation. IEEE Trans. Instrum. Meas. **59**, 884 (2010)
6. Kim, M., Han, D.K., Ko, H.: Joint patch clustering-based dictionary learning for multimodal image fusion. Inf. Fusion **27**, 198 (2016)

7. Liu, Y., Liu, S., Wang, Z.: A general framework for image fusion based on multi-scale transform and sparse representation. Inf. Fusion **24**, 147 (2015)
8. Li, S., Kang, X., Fang, L., Hu, J., Yin, H.: Pixel-level image fusion: a survey of the state of the art. Inf. Fusion **33**, 100 (2017)
9. Kong, W.W., Lei, Y., Ren, M.M.: Fusion technique for infrared and visible images based on improved quantum theory model. In: Zha, H., Chen, X., Wang, L., Miao, Q. (eds.) CCCV 2015. CCIS, vol. 546, pp. 1–11. Springer, Heidelberg (2015). https://doi.org/10.1007/978-3-662-48558-3_1
10. Chen, C., Li, Y., Liu, W., Huang, J.: SIRF: simultaneous satellite image registration and fusion in a unified framework. IEEE Trans. Image Process. **24**, 4213 (2015)
11. Ma, J., Chen, C., Li, C., Huang, J.: Infrared and visible image fusion via gradient transfer and total variation minimization. Inf. Fusion **31**, 100 (2016)
12. Zhang, Q., Levine, M.D.: Robust multi-focus image fusion using multi-task sparse representation and spatial context. IEEE Trans. Image Process. **25**, 2045 (2016)
13. Yu, N., Qiu, T., Bi, F., Wang, A.: Image features extraction and fusion based on joint sparse representation. IEEE J. Sel. Top. Signal Process. **5**, 1074 (2011)
14. Son, C., Zhang, X.: Layer-based approach for image pair fusion. IEEE Trans. Image Process. **25**, 2866 (2016)
15. Panagakis, Y., Nicolaou, M.A., Zafeiriou, S., Pantic, M.: Robust correlated and individual component analysis. IEEE Trans. Pattern Anal. Mach. Intell. **38**, 1665 (2016)
16. Pan, S.J., Yang, Q.: A survey on transfer learning. IEEE Trans. Knowl. Data Eng. **22**, 1345 (2010)
17. Ando, R.K., Zhang, T.: A framework for learning predictive structures from multiple tasks and unlabeled data. J. Mach. Learn. Res. **6**, 1817 (2005)
18. Dai, W., Yang, Q., Xue, G., Yu, Y.: Self-taught clustering. In: ICML 2008, 8 p. Helsinki, USA (2008)
19. Huang, D., Wang, Y.F.: Coupled dictionary and feature space learning with applications to cross-domain image synthesis and recognition. In: IEEE International Conference on Computer Vision, 2496 p. (2013)
20. Gao, R., Vorobyov, S.A., Zhao, H.: Multi-focus image fusion via coupled dictionary training. In: International Conference on Acoustics Speech and Signal Processing, 1666 p. (2016)
21. Liu, Z., Blasch, E., Xue, Z., Zhao, J., Laganiere, R., Wu, W.: Objective assessment of multiresolution image fusion algorithms for context enhancement in night vision: a comparative study. IEEE Trans. Pattern Anal. Mach. Intell. **34**, 94 (2012)
22. Han, Y., Cai, Y., Cao, Y., Xu, X.: A new image fusion performance metric based on visual information fidelity. Inf. Fusion **14**, 127 (2013)
23. Zheng, Y., Essock, E.A., Hansen, B.C., Haun, A.M.: A new metric based on extended spatial frequency and its application to DWT based fusion algorithms. Inf. Fusion **8**, 177 (2007)

Local Saliency Extraction for Fusion of Visible and Infrared Images

Weiping Hua, Jufeng Zhao$^{(\boxtimes)}$, Guangmang Cui, Xiaoli Gong, and Liyao Zhu

School of Electronics and Information, Hangzhou Dianzi University,
Hangzhou 310018, China
dabaozjf@hdu.edu.cn

Abstract. In this paper, a local saliency extraction-based dual-band image fusion algorithm is proposed. Combing the variable computational windows, the local gray distance is designed for saliency analysis. And saliency map is further obtained by considering spatial weight. For dual-band image fusion, firstly, we design several local windows named different levels, and get the corresponding saliency maps. Secondly, achieve weighted fusion under different levels with saliency maps. Finally, all fused images are compounded into one fused result. According to experimental results, the proposed method could produce a fused image with good visual effect, preserving even enhancing the details effectively. Comparing with other seven methods, both subjective evaluation and objective metric indicate that the proposed algorithm performs best.

Keywords: Image fusion · Visible and infrared · Local saliency

1 Introduction

Image fusion aims to combine salient information from source images. There exist abundant object details in Visible (VI) image, while infrared (IR) one has particular target characteristic. Fusion for VI and IR images expect to maintain the both advantages of VI and IR images, which is useful for target detection, monitoring, etc.

Lots of methods are designed for image fusion. The fusion based wavelet [1] and curvelet transform are famous multiple resolution-based algorithms. Meanwhile, those pyramid-based approaches also play an important role in image fusion, such as Laplace pyramid [2], ratio pyramid [3], morphological pyramid [4]. Those methods would smooth details because of their down sampling and up sampling, which is time consuming. To achieve image fusion, multiscale directional nonlocal means(MDNLM) filter is used [5]. MDNLM is a multiscale, multidirectional, and shift-invariant image decomposition method. There is an algorithm based on compressive sensing [6], and the sparse coefficients of the source images are obtained by discrete wavelet transform. Bai propose an outstanding algorithm for IR and VI image fusion [7], which utilizes region detection through multi scale center-surround top-hat transform. But some parameters are difficult

© Springer Nature Singapore Pte Ltd. 2017
J. Yang et al. (Eds.): CCCV 2017, Part II, CCIS 772, pp. 210–221, 2017.
https://doi.org/10.1007/978-981-10-7302-1_18

to select for new users. To keep the thermal radiation and appearance information simultaneously, people design a fusion algorithm, named Gradient Transfer Fusion(GTF) [8], based on gradient transfer and total variation (TV) minimization. Since human visual system(HVS) is the best system for judging quality of image, saliency preserving is also popular in image fusion [9]. And we also have developed dual-band image fusion using saliency analysis [10, 11]. With saliency characteristic highlighting, the fused results have good visual effect, preserving details well. But how to well extract salient features seems a little difficult for VI and IR image fusion [12]. The saliency extraction method should be designed.

In our paper, we propose a local saliency extraction-based dual-band image fusion method. The local-window-based gray distance idea is used for saliency analysis. And spatial weight is also imposed on the design. The dual-band images are weighted fused based on saliency maps for each band. Furthermore, the fusion is considered under different levels, which are determined by the size of local window to extract features with different sizes. We could finally obtain details enhanced results, inherit-ing the important information from source images.

2 Local Window-Based Visual Saliency Extraction

HVS could rapidly finish saliency extraction from the scene, as eyes would focus on those areas they concern. Different regions attract different attention, which means HVS would give different weight to the regions. We desire to design algorithms to simulate this ability of HVS to improve the effect of image processing.

Since HVS is sensitive to contrast in visual signal, the histogram-based contrast (HC) method could extract salient object well using color statistics of the input image [13, 14]. In that paper, they operated algorithm in L*a*b* space. This kind of saliency map V is useful for our fusion. V means the weight distribution that HVS pays attention to original image f. And $V \in [0, 1]$, the larger value in V the more attention HVS pays to.

In VI and IR image fusion, we need to highlight characteristic of different sizes, especially small size. Current saliency extraction method with global idea would be out of action. Here, we intend to design a local-based approach to solve this problem. Inspired by the work of Chen [13, 14], we expect to obtain saliency value within a local window with size of $W \times W$. When the value of W changes, the saliency value slightly varies.

At arbitrary pixel (i, j), the corresponding saliency value $V(i, j)$ for image f is defined as:

$$V(i, j) = \sum_{(i,j)=\Omega_C, \forall(x,y)\in\Omega} [\Gamma(f_{ij}, f_{xy}), D_g(f_{ij}, f_{xy})]. \tag{1}$$

where Ω denotes a local window with size of $W \times W$, whose center is Ω_C located at pixel pixel (i, j). And (x, y) represents arbitrary pixel in Ω. f_{xy} and f_{ij} are gray value at (x, y) and (i, j), respectively. In Eq. (1), $D_g(\bullet)$ is gray distance function, which measures the gray distance between the two pixel:

$$D_g(f_{ij}, f_{xy}) = |f_{ij} - f_{xy}|. \tag{2}$$

And $\Gamma(\bullet)$ denotes spatial weight. Here we design this weight because we expect to show the visual difference when the spatial distance changes. $\Gamma(\bullet)$ is defined by the following formula:

$$\Gamma(ij, xy) = e^{\frac{-D_s(ij, xy)}{\sigma^2}}. \tag{3}$$

where $D_s(\bullet)$ means the spatial distance between two pixel using numbers of pixels. And σ^2 is dilatation factor. A large σ^2 makes the distant pixels (x, y) impose large influence on current pixel (i, j). From the equation, we could find that the closer between (x, y) and (i, j), the larger influence f_{xy} affect f_{ij}.

The saliency map V is calculated using Eqs.(1), (2) and (3). With this design, we could get salient areas with changing of W. With different levels of $\{W_k\}$, we could obtain corresponding saliency map $\{V_k\}$, and $k = 1, 2, ...N$. This could help extracting different features into fused images.

The local computation is operated within a $W \times W$ window as shown in Fig. 1. The saliency value at (i, j) is $V(i, j)$, which is calculated as follows:

(1) Extract a $W \times W$ image patch P_{ij}, whose center pixel is (i, j). P_{ij} is segmented from original image f, forming neighborhood Ω.
(2) Saliency value of center pixel is obtained using P_{ij}. And this saliency value is treated as $V(i, j)$.

Therefore, $V(i, j)$ is computed by following the above two steps. Then, moving this local window pixel-by-pixel, the whole saliency map V could be achieved. Proper size W would emphasize important object in image. And W is usually odd, which could easily determine center pixel. Finally, the function for V is rewritten as $g(\bullet)$:

$$V = g\left(W, f, \sigma^2\right). \tag{4}$$

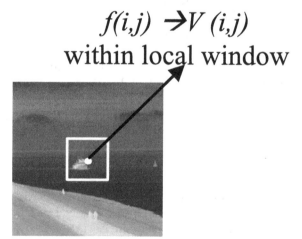

Fig. 1. A local $W \times W$ window for pixel (i, j) to calculate $V(i, j)$

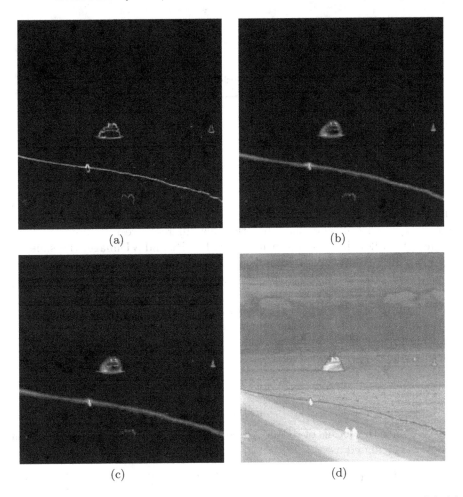

Fig. 2. Saliency maps with local idea and global method: (a) is original image, (b)–(c) are saliency maps with local window ($W = 5$, 19, $\sigma^2 = 3.5$), (d) saliency map using global method.

Figure 2 shows an example of saliency maps, which are generated by our method and global idea [13]. (a) is original infrared image. (b) and (c) are maps created using Eq. (4) with $W = 5$ and 19, respectively. (d) is the result of global method. We could conclude that, we would extract those characteristic with different sizes, which is very useful in next image fusion.

3 Fusion for Dual-Band Images

According to local-based saliency analysis, the saliency map V could be obtained. Those areas and pixels with large saliency values are expected to be highlighted

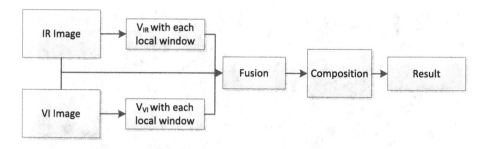

Fig. 3. Flowchart for our fusion

in fused result. We try to utilize multi-window to generate multi-saliency maps, to enhance the details and characteristic have different sizes.

The whole flowchart is shown in Fig. 3. For IR and VI image, the saliency map V_{IR} and V_{VI} are computed under different size levels. And they were fused at each level. The last fused result is achieved by combining those fused image of different levels with proper rules.

With different levels of $\{W_k\}$, we could obtain corresponding saliency map $\{V_k\}$, and $k = 1, 2,...N$. N is the number of size we selected. And $W_{k-1} < W_k$. Following Eq. (4), the kth multiple saliency maps and are created for IR and VI images f_{IR} and f_{VI}, respectively:

$$V_k^{IR} = g\left(W_k, f^{IR}, \sigma^2\right). \tag{5}$$

$$V_k^{VI} = g\left(W_k, f^{VI}, \sigma^2\right). \tag{6}$$

To enhance the details of IR and VI, respectively, our fusion rule for kth local window is calculated as:

$$F_k = \frac{1}{2}\left\{\left[f^{IR}V_k^{IR} + f^{VI}\left(1 - V_k^{IR}\right)\right] + \left[f^{IR}\left(1 - V_k^{VI}\right) + f^{VI}V_k^{VI}\right]\right\}. \tag{7}$$

Through above rules, the details are enhanced. Meanwhile, the energy of original images could be preserved.

Finally, we expect to combine those fused images using Eq. (7). In our paper, the fused results need high contrast. Thus, we should emphasize characteristic as more as possible. The final image composition is operated as follows:

$$F = \max_{k=1,2...N}\left\{F_k\right\}. \tag{8}$$

With Eq. (8), the largest value in each pixel of fused images under different levels would be preserved to create a high contrast result. This rule could help produce a details enhanced fused result with good visual effect.

4 Experiment and Discussion

We adopt two image pairs for experiment. The two are downloaded from the weblink: http://www.google.com. The images are shown in Fig. 4, which named

Fig. 4. Source images named 'boat' (320 × 320) and 'road' (320 × 250): (a) and (b) are IR and VI images named 'boat', (c) and (d) are IR and VI images named 'road'.

'boat' and 'road'. (a) and (b) are IR and VI images named 'boat'. While the other two are corresponding images named 'road'.

To prove the validness of the proposed method, seven fusion approaches are introduced for comparison. Those algorithms includes direct average algorithm (Direct), wavelet-based algorithm (Wavelet) [1], Laplacian pyramid method (LapP) [2], ratio pyramid approach (RatioP) [3], morphological pyramid (MorP) [4], multi scale center-surround top-hat transform based algorithm (MSCT) [7], saliency preserving method (SalPr) [9].

In order to do further analysis, more data is considered. The other four famous image databases that used in Ref. [10] are adopted. These four datasets are named "UNcamp", "Dune", "Trees" and "Octec", respectively. This four databases are downloaded from the websites www.imagefusion.org in 2012.

All the methods are coded using MATLAB. And the experiments are run on a personal computer (i5-2310 with 2.9 GHz CPU, 4 GB memory).

4.1 Experimental Setting

The parameters W_k and σ^2 in Eqs. (5) and (6) should be determined. The size W_k is expected to cover different salient regions. A large N would make the algorithm run slow, but highlight more objects. According to experimental experience, the number of windows $N = 3$, and $\{W_k\}$ $(k = 1, 2, 3) = 5$, 9, 15. This three window sizes are enough. σ^2 determines the influence of surrounding pixels imposing on center one. We think ten pixels distant would decrease to 0.1 of the neighbor of center pixel. Then $e^{-10/\sigma^2} = 0.1$, we get $\sigma^2 \simeq 4.34$.

4.2 Objective Evaluation for Fusion Image

Besides subjective assessment, we adopt Entropy and Joint Entropy as the objective evaluation for image fusion. X, Y are treated as the two source images. And F is fused result. The fused image quality assessment (FQA) is described as follows.

Since the fused image has combined details of two source images, the fused result should own more information. Thus, Entropy is usually used as FQA [7,12],

$$En = -\sum_{i=0}^{L-1} p_F(i)\log_2(P_F(i)).\tag{9}$$

In Eq. (9), L denote gray level. Usually $L = 256$ for 8 bit-depth image. This formula shows the probability for gray value i in F. If the En is larger, the fused result seems better.

Joint Entropy is another metric, which show the information of fused image inherited from source ones. The joint entropy is defined by [7,12],

$$JE_{FXY} = -\sum_{i=0}^{L-1}\sum_{j=0}^{L-1}\sum_{k=0}^{L-1} p_{FXY}(i,j,k)\log_2(p_{FXY}(i,j,k)).\tag{10}$$

where $P_{FXY}(i,j,k)$ is the joint probability. This show the probability that gray values in F is i, in image X is j and gray value in image Y is k, respectively. This metric is similar as Entropy. When JE becomes larger, the fused image is believed to be better.

4.3 Experimental Results and Comparison

We have shown the results of image 'boat' in Fig. 5. In this figure, the unique information are two boats, several people, and cloud, etc. (a)–(h) are results of our method and other seven algorithms. The Direct method only takes the average of source images, this leads to the decrease of contrast. The wavelet algorithm makes some artifacts, and the object seems blurred. The result of MorP create a terrible beach, which seems an error. MSCT produce a good result, but the two people is not highlighted enough. Result of LapP, RatioP and

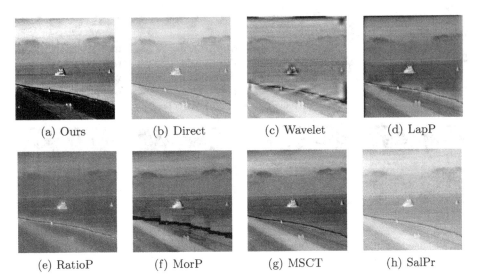

(a) Ours   (b) Direct   (c) Wavelet   (d) LapP

(e) RatioP   (f) MorP   (g) MSCT   (h) SalPr

Fig. 5. Fused results of 'boat': (a)–(h) are results of proposed method and other seven algorithms, respectively.

Table 1. Quantitative comparison using En, JE for Fig. 5.

FQA/method	Ours	Direct	Wavelet	LapP	RatioP	MorP	MSCT	SalPr
En	7.74	5.67	6.62	6.28	5.72	6.40	6.41	6.31
JE	6.79	6.10	6.42	6.30	6.12	6.34	6.35	6.31

SalPr has low contrast, some characteristic nearly disappeared. The two boats has much higher contrast in our result than other fused ones. In (a), the people, the cloud looks striking. And the beach between sea and land is enhanced. The whole image has better visual effect than other ones.

The result of FQA is listed in Table 1. Our result has the largest value of En and JE, which indicates that our algorithm performs best, inheriting abundant information from the two source images.

The results of 'road' are shown in Fig. 6. These two source images own more information than the boat image, such as those people, street lamps, the cars, advertisement words, etc. Observing these results, only our method creates a fused result with high contrast, emphasizing all these features. Table 2 lists the corresponding FQA results. According to judgement of En and JE, the proposed algorithm is outstanding. The largest values of both two metrics demonstrate that our method performs best in dual-band fusion.

4.4   Computational Speed

One of the key metric to evaluate an algorithm is computational speed. In above sections, we have pointed out that the size of boat and road are 320×320 and

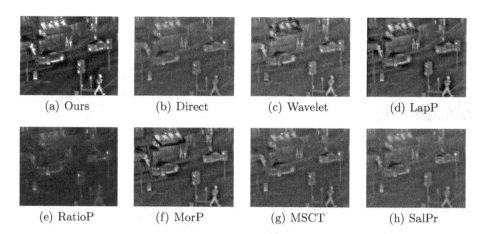

(a) Ours   (b) Direct   (c) Wavelet   (d) LapP

(e) RatioP   (f) MorP   (g) MSCT   (h) SalPr

Fig. 6. Fused results of 'road': (a)–(h) are results of proposed method and other seven algorithms, respectively.

Table 2. Quantitative comparison using En, JE for Fig. 6.

FQA/method	Ours	Direct	Wavelet	LapP	RatioP	MorP	MSCT	SalPr
En	6.76	5.91	6.40	6.41	4.95	6.30	5.72	5.86
JE	6.05	5.76	5.93	5.93	5.44	5.90	5.71	5.75

Table 3. Comparison of processing time.

Image/methods	Size	Ours	Direct	Wavelet	LapP	RatioP	MorP	MSCT	SalPr
Fig. 5	320 × 320	1.81	0.02	0.29	0.66	0.03	0.38	3.31	1.62
Fig. 6	320 × 250	1.62	0.02	0.22	0.52	0.02	0.31	3.09	1.38

320×250, respectively. All eight algorithms are tested in this section, giving the processing time in Table 3. We could find the fastest method is Direct algorithm. Our method runs slowly because $N = 3$, and the code also need improved and accelerated. If only take one proper local window, the time would be shortened. For example, only use $W = 9$, it only takes 0.4 s. When we only use $W = 9$, the fused result is still better than other seven methods. But if we need a much better result, we take $N = 3$ for enhanced. Whatever, in our future work, the algorithm should be further accelerated.

4.5   Visual Performance Discussion

Since HVS is the best criteria, we design subjective evaluation experiment for further visual performance assessment. The detail subjective metric includes "Description of IR image (DIR)", "Description of VI Image (DVI)",

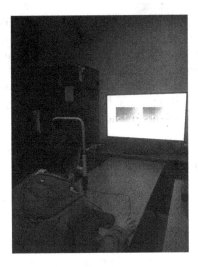

Fig. 7. Subjective evaluation device

Table 4. Evaluation results based on eight methods for Fig. 5, using subjective method.

Methods	DIR	DVI	Con	EF
Ours	A	A	A	A
Direct	B	C	C	C
Wavelet	B	C	B	D
LapP	B	C	C	C
RatioP	C	C	C	B
MorP	C	C	B	D
MSCT	B	B	B	A
SalPr	B	B	C	C

Table 5. Evaluation results based on eight methods for Fig. 6, using subjective method.

Methods	DIR	DVI	Con	EF
Ours	A	A	A	A
Direct	B	C	C	C
Wavelet	B	B	C	B
LapP	B	B	B	C
RatioP	C	D	C	C
MorP	B	B	B	A
MSCT	B	B	C	B
SalPr	B	C	C	C

"Contrast(Con)" and "Edge Feature(EF)". DIR and DVI metrics means the information inherits from the dual-band source images. Con is short for contrast, which is an important metric for fused image. To evaluate details preserving and enhancement, EF is adopted to evaluate sharpness of edges and contours. Four levels are set to describe the quality of images, from worst to the best corresponding to A to D. "D" means very bad, while "C" represents not good, and "B" corresponding to acceptable but not the best, "A" denotes the best level of visual performance.

The subjective results are given by the 20 researchers in our laboratory, the Subjective evaluation device is shown in Fig. 7. And these results are listed in Tables 4 and 5.

Learning from Tables 4 and 5, the proposed method obtains the highest score. Compared with other approaches, the fused result of our approach not only has best global and local contrast, but also sharpest edges. Inspired by subjective evaluation, the proposed method has achieved the best performance in visual effect and details inheritance even enhancement.

5   Conclusion

In this paper, the authors have designed a local saliency extraction-based fusion algorithm for IR and VI images. Utilizing multi-window saliency extraction, the saliency map-based image fusion could create fused image with salient object and information highlighted. Those characteristic with different sizes have been all enhanced. The experiments is proved that this local saliency-based method could well extract feature information of source images, which is effective for dual-band image fusion. The important information will be inherited and enhanced. Both objective assessment and subjective evaluation indicate that the outstanding performance of the proposed method.

It is easy to use our algorithm to achieve dual-band image fusion, enhancing those areas and pixels HVS interests. And the algorithm is suitable for further target detection, scene surveillance and other relative fields, which are our next try.

In future, the local window selection will be automatic on basis of image content. This automation will greatly improve the applicability of algorithm. Meanwhile, how to accelerate the method will be also focused.

Acknowledgments. We thank the reviewers for helping us to improve this paper. Many thanks to Alexander Toet and the TNO Human Factors Research Institute. This work is supported by National Natural Science Foundation of China (Grant No. 61405052). The work is also partly supported by the key technologies R&D Program of Guangzhou city (No. 201704020182). This work is partly supported by State Key Laboratory of Pulp and Paper Engineering (No. 201537), science and technology plan of Zhejiang province (No. 2017C01033) and Key Laboratory Open Fund for RF Circuits and Systems (Hangzhou Dianzi University), Ministry of Education.

References

1. Pajares, G., De La Cruz, J.M.: A wavelet-based image fusion tutorial. Pattern Recognit. **37**(9), 1855–1872 (2004)
2. Burt, P.J., Adelson, E.H.: The Laplacian pyramid as a compact image code. IEEE Trans. Commun. **31**(4), 532–540 (2003)
3. Toet, A.: Image fusion by a ratio of low-pass pyramid. Pattern Recognit. Lett. **9**(4), 245–253 (1989)
4. Matsopoulos, G.K., Marshall, S.: Application of morphological pyramids: fusion of MR and CT phantoms. J. Vis. Commun. Image Represent. **6**(2), 196–207 (1995)
5. Yan, X., Qin, H., Li, J., Zhou, H., Zong, J., Zeng, Q.: Infrared and visible image fusion using multiscale directional nonlocal means filter. Appl. Opt. **54**(13), 4299 (2015)
6. Liu, Z., Yin, H., Fang, B., Chai, Y.: A novel fusion scheme for visible and infrared images based on compressive sensing. Opt. Commun. **335**, 168–177 (2015)
7. Bai, X., Zhou, F., Xue, B.: Fusion of infrared and visual images through region extraction by using multi scale center-surround top-hat transform. Opt. Express **19**(9), 8444 (2011)
8. Ma, J., Chen, C., Li, C., Huang, J.: Infrared and visible image fusion via gradient transfer and total variation minimization. Inf. Fusion **31**(C), 100–109 (2016)
9. Hong, R., Wang, C., Wang, M., Sun, F.: Salience preserving multifocus image fusion with dynamic range compression. Int. J. Innov. Comput. Inf. Control IJICIC **5**(8), 2369–2380 (2009)
10. Zhao, J., Gao, X., Chen, Y., Feng, H., Wang, D.: Multi-window visual saliency extraction for fusion of visible and infrared images. Infrared Phys. Technol. **76**, 295–302 (2016)
11. Shi, Z., Jiangtao, X., Zhang, Y., Zhao, J., Xin, Q.: Fusion for visible and infrared images using visual weight analysis and bilateral filter-based multi scale decomposition. Infrared Phys. Technol. **71**, 363–369 (2015)
12. Guihong, Q., Zhang, D., Yan, P.: Information measure for performance of image fusion. Electron. Lett. **38**(7), 313–315 (2002)
13. Cheng, M.-M., Mitra, N.J., Huang, X., Torr, P.H.S., Hu, S.-M.: Global contrast based salient region detection. IEEE TPAMI **37**(3), 569–582 (2015)
14. Cheng, M.M., Zhang, G.X., Mitra, N.J., Huang, X., Hu, S.M.: Global contrast based salient region detection. In: Computer Vision and Pattern Recognition, pp. 409–416 (2011)

Distributed Compressive Sensing for Light Field Reconstruction Using Structured Random Matrix

Ningkai Yang[1], Guojun Dai[1], Wenhui Zhou[1], Hua Zhang[1,2(✉)],
and Renbin Yang[1]

[1] School of Computer Science and Technology, Hangzhou Dianzi University,
Hangzhou 310018, China
`zhangh@hdu.edu.cn`
[2] Key Laboratory of Network Multimedia Technology of Zhejiang Province,
Zhejiang University, Hangzhou 310027, China

Abstract. Taking advantage of the strong correlation between the sequences of light field images, a distributed compressive sensing for light field reconstruction method using structured random matrix is proposed. Firstly, the sequence of light field images is superimposed to form the 3D image matrix. Since the angle difference of the optical field camera array is fixed, the 2D slice images of each 3D image matrix will show the characteristics of the stepped stripes, and the angles of the stripes are same. Secondly, slices are rearranged according to the inclination angle to obtain a 2D vertical stripe. Thirdly, structured random matrix (SRM) is used as the measurement matrix to reconstruct these images. SRM-DCS algorithm are proposed for reconstruction of light field images. The experimental results represented that the proposed algorithm performs better in subjective visualization and objective evaluation compared with other compressive sensing algorithms.

Keywords: Light field · Distributed compressive sensing
Structured random matrix

1 Introduction

In recent years, with the development of 3D display technology, great numbers of researchers paid more and more attention to light field technology, which is an effective mean achieving real-time rendering of 3D scene. The main research contents of the light field technology include the theory and technology of light filed rendering, the construction and application of light field array and the acquisition and reconstruction of signal [1]. And the acquisition and reconstruction of light field signal is one of the core problems need to be solved in the light field. As the light field consists of large-scale camera arrays, the system will have hundreds of images per second to store and transmit in order to meet the real-time needs of 3D scene interaction.

© Springer Nature Singapore Pte Ltd. 2017
J. Yang et al. (Eds.): CCCV 2017, Part II, CCIS 772, pp. 222–233, 2017.
https://doi.org/10.1007/978-981-10-7302-1_19

At present, light field acquisition and reconstruction technology can be divided into compressed and non-compressed light field technology. The former captures light field signals through dense camera array [1,2], generate high precision light field images without compression and reconstruction. However, the efficiency of this light field acquisition system reconstruction is not high due to the large transmission and storage of data. And the latter implements the compression coding of the light field by installing a microlens array or optical mask and other devices in the camera [3]. The light field images are restored by a linear or nonlinear algorithm during the reconstruction phase. Sparse light field arrays can improve transmission and reconstruction efficiency in this way, but with high cost of such a light field acquisition system and complex structure.

Now, the method of light field acquisition and reconstruction mainly face the contradiction between reconstruction precision and efficiency. In this paper, a distributed compressive sensing field reconstruction method using structured measurement matrix are proposed to improve the accuracy and efficiency of reconstructing the images using distributed compression sensing. In the signal acquisition stage, the K-SVK algorithm is used to train the adaptive dictionary of the light field, which makes light field signal sparse coding. In the reconstruction phase of the image, since the slices of the 3D image matrix formed by the light field images sequence have the characteristics of the stepped stripes, the slices are rearranged to obtain vertical stripe and then reconstruct the light field image sequences by combining structured measurement matrix. The experimental results represented that the reconstructed light field images is more effective in subjective visual and objective evaluation compared with other compressive sensing algorithms.

2   Related Work

Light field theory and application is the core research content of computational photography, which is studied for nearly a hundred years. At the beginning, Lippmann put forward the concept of integral photography, and then Gershun proposed the use of light to describe the radiation properties of light and formed a prototype of the light field. Adelson et al. [4] of MIT proposed a seven-dimensional plenoptic function $P(x, y, z, \theta, \varphi, \lambda, t)$ by using a formal description of the light in space, where (x, y, z) is the position of the receiving light side in the three-dimensional space, (θ, φ) is the azimuth and inclination of the light, λ represent the wavelength of the light and t indicates the moment. On this basis, Levoy et al. [5] of Stanford proposed a biplane parametric representation of the light field, that is, the light field information is described by the coordinates (s, t) and (u, v) of the intersection of a ray with the spatial plane and the angle plane, the light field information briefly and intuitively through four-dimensional light field function $L(s, t, u, v)$. And Debevec et al. [5] of USC collected the dynamic field of the periodic motion successfully. These series of work set off a wave of light field research.

These days in the field of image processing, the demand for fast and efficient algorithms and low hardware requirements is becoming more and more prominent, and the use of compressive sensing to capture and reconstruct light fields is attracting more and more researchers. Compressive sensing (CS) was proposed by Donoho and Candes et al. [6,7], which using a measurement matrix to project a sparsible or compressible high-dimensional signal onto a low-dimensional space, and the measured values obtained after projection can be reconstructed by a certain linear or non-linear decoding model [7,8]. The compressive sensing has loog been applied to the image acquisition. Derin Babacan et al. [8] have proposed a new camera sampling model and reconstructed the light field image using the Bayesian method combined with compressive sensing. Shu and Ahuja [9] put forward an idea of three-dimensional scene and used it in reducing the sampling density of the light field camera array. Gan et al. [10] proposed the method utilizing block compressive sensing to collect and reconstruct image signals, which is use the independent measurement matrix on the same size of the image block for projection and reconstruction, which improves the efficiency of image reconstruction to a certain extent.

3   Proposed Approach

The image sequences of light field camera array are not much different, resulting in a large number of common parts, thus making full use of the correlation between the light field images can effectively improve the efficiency of light field reconstruction. In this paper, a distributed compressive sensing field reconstruction method using structured measurement matrix is proposed. Firstly, all images are superimposed as 3D image matric. Since the slices of the 3D image matrix formed by the sequence of the light field images have the characteristics of the stepped stripes, the slices are rearranged to obtain 2D vertical stripe image, and then the measurement matrix is modified into a Structured Random Matrix (SRM) by using a compressive sensing reconstruction algorithm. Finally, the light field images sequence is recovered from the 3D image matrix. The sketch is shown in Fig. 1.

3.1   Problem Formulation

Compressive sensing is a new type of signal processing method for reconstructing the original signal from sparse signals. The sparse representation of the signal is the primary problem of compressive sensing. Since the compressive sense is based on the original signal that can be sparse or compressed, it is necessary to sparse and transform the original signal on some transform domains, where the original signal $x \in \mathbb{R}^n$ can be represented by the sparse matrix $\Psi \in \mathbb{R}^{n \times m}$ as $x = \Psi\alpha$, Ψ is an orthogonal matrix, α is the coefficient vector of the original signal. If there are only s non-zero elements in α, then α is the sparse representation of the s-sparse of the signal x [11].

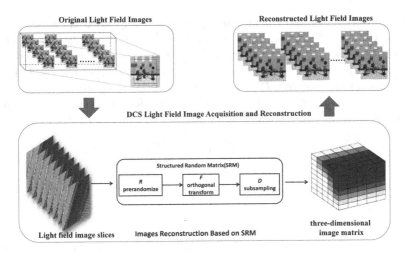

Fig. 1. Proposed algorithm for acquisition and reconstruction of light field images

Distributed compressive sensing aims at the correlation between distributed signals to explore the correlation between signals, and achieve multi-signal distributed compression sampling and joint reconstruction. Distributed compressive sensing divides the signals collected by distributed sources into common component and innovation [12], where the common components are the same and the innovation components are different. When the signal is sparse and reconstructed, the common only need to be processed once. For the distributed source system of the light field array, the distributed compressive sensing technique can reduce the sampling rate by accurately reconstructing the image signal and improving light field image reconstruction efficiency (Fig. 2).

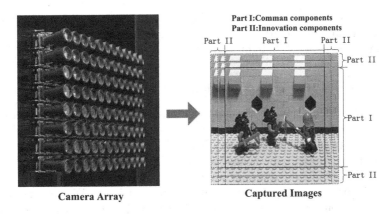

Fig. 2. Light field image capture used DCS method sketch

Assuming that the light field camera array is equipped with k camera, the measured value of the image signal capture by the camera $p(p = 1, 2, ..., k)$ can be expressed as $y_p = \Phi_p x_p$, where $y_p \in \mathbb{R}^m$ is the vector of the measured value of the camera p, $\Phi_p \in \mathbb{R}^{m \times n}(m \ll n)$ is the sensing matrix of the camera, m is called the sampling volume, $R = m/n$ represents the sampling rate, $x_p \in \mathbb{R}^n$ is a vectorized representation of the captured image for camera p, so the camera array signal and the set of measured values can be expressed as:

$$X = \begin{bmatrix} x_1 \\ x_2 \\ \vdots \\ x_k \end{bmatrix}, Y = \begin{bmatrix} y_1 \\ y_2 \\ \vdots \\ y_k \end{bmatrix}, \Phi = \begin{bmatrix} \Phi_1 & 0 & ... & 0 \\ 0 & \Phi_2 & ... & 0 \\ \vdots & \vdots & \ddots & \vdots \\ 0 & 0 & ... & \Phi_k \end{bmatrix} \tag{1}$$

Where $X \in \mathbb{R}^{kn}$, $Y \in \mathbb{R}^{km}$, $\Phi \in \mathbb{R}^{kn \times km}$, then the measured values of camera arrays can be expressed as $Y = \Phi X$. In order to reconstruct the s-sparse coefficient signal α from the measured value y_p, where the measurement matrix Φ and sparse matrix Ψ need to satisfy the RIP criterion [9], which is used to limit the column vectors of these two matrices to irrelevant. For any s-sparse signal α and fitting error threshold $\delta \subset (0, 1)$ is satisfied as:

$$(1 - \delta)\|\alpha_p\|_2^2 \le \|\Phi_p x_p\|_2^2 \le (1 + \delta)\|\alpha_p\|_2^2 \tag{2}$$

After completing the design of the measurement matrix, it is necessary to design the reconstruction algorithm to solve the problem of reconstructing the original signal x from the measured value y. The essence of the problem is to solve an under-determined equation, which is theoretically infinite solution, but because of the RIP standard, the problem can be transformed into a convex optimization problem:

$$arg\ min\|\alpha\|_1\ s.t.\ \|Y - \Phi\Psi\alpha\|_2 \le \varepsilon \tag{3}$$

Where Φ is the distributed compression sensing matrix, ε is the noise value of the measured value, which determines the accuracy of the reconstructed image, and the image sequence X of light field camera array is reconstructed by $X = \Psi\alpha$.

3.2   Light Field Images Acquisition and Sparse

The sparse representation of the light field signal is the basis and premise of the compressive sensing application in the reconstruction of the light field. The sparse representation of the signal is a linear combination of the primitive signals as a few radians through the sparse matrix, which transforms the signal over a sparse domain to reduce redundant information in the original signal, it can also save storage space and improve signal processing efficiency. Dictionary learning solute the problem of the atomic library can find the best linear combination of m atoms to represent the signal. The core issue is to optimize the following questions:

$$min\ \|x - D\alpha\|_2^2\ s.t.\ \|\|_0 \le s \tag{4}$$

The light field image sequence has a particularity, that is, a collection object is generally fixed, so the gap between the collected images is small. In the sparse representation of the light field signal, the adaptive dictionary is obtained by training the light field signal. The proposed method uses the K-SVK algorithm [13,14,16] to train the adaptive dictionary of the light field image sequence named Lego Knights provided by the Stanford University Computer Graphics Laboratory signal acquisition phase.

The K-SVK algorithm is a commonly used algorithm for dictionary learning, and the resulting dictionary has a lower average representation error and a faster convergence rate. The K-SVK dictionary learning algorithm produces a dictionary of segmented smooth texture atoms, the visualization results of the adaptive dictionary obtained by training the light field are shown in Figure 3, which is capable of effectively sparse representation of other light field images.

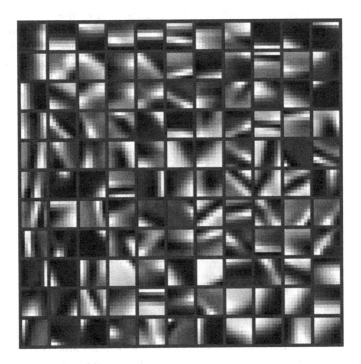

Fig. 3. Light field dictionary learning visualization results

3.3   Light Field Images Reconstruction

The light field images is a sequence of images obtained by photographing the same scene from different views through the camera array. There is great correlation between images. Compared with traditional compressive sensing, the distributed compressive sensing makes full use of the correlation between the light field images to reconstruct the image. According to the characteristics of camera parallax fixation in the light field array, all the images are superimposed into 3D image matrix and the image matrix is cut along each line to get the same stepped stripes with the same inclination angle. And the slice image pixels are rearranged along the inclination angle θ of the stepped stripes, then the rearranged image is maked to form into a vertical stripe image slice. Finally, these images are reconstructed by compressive sensing [15]. Figure 4 illustrates the light field image sequence rearrangement: (a) The 3D image matrix consists of a sequence of light field images and is cut along each row to obtain slices of stepped fringe features. (b) According to the characteristics of the slice image, the inclination angle θ of the reorder line L_θ is solved. (c) The 2D vertical stripe image is rearranged along the L_θ.

Another key problem to reconstruct light field images using DCS is to select a measurement matrix Φ. At present non-structured measurement matrix (NSRM) such as Gaussian matrix, Bernoulli matrix and Hadamard matrix were generally selected to reconstruct image when use CS. This kind of measurement matrix satisfies the RIP criterion and is simple to construct, but it has the drawbacks of its unfixed inherent elements and large required storage space, which is difficult to meet in a large scale of data transmission and real-time reconstruction.

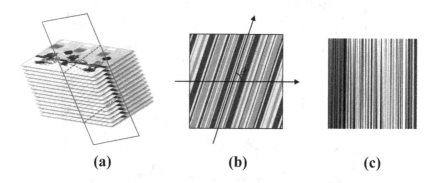

(a)   **(b)**   **(c)**

Fig. 4. Light field images sequence reconstruction using DCS

In this paper, a structured random matrix [16] (SRM) with high computational speed and storage efficiency is selected for the measurement matrix according to the mass data feature of light field, which has following advantages: (1) SRM is almost uncorrelated to all orthogonal matrices and multiple sparse signals and satisfy RIP criteria; (2) SRM accurately reconstructs the original signal with fewer number of measurements than NSRM; (3) SRM can be decomposed into the product of many structured sub-matrices of block diagonalized matrices, facilitating block processing and linear filtering, with low complexity and fast computational properties. SRM is defined as:

$$\Phi_B = \sqrt{N/M} \cdot D \cdot F \cdot R \qquad (5)$$

Where $R \in \mathbb{R}^{n \times n}$ is a random permutation matrix or diagonal random matrix, which randomly scrambles the order of the target signal, the diagonal element R_{ii} is Bernoulli random variables in the same distribution. $F \in \mathbb{R}^{n \times n}$ is a standard orthogonal matrix, the role of which is to disperse the sampling signal information to all measurement points, usually with Fast Fourier Transform (FFT), Discrete Cosine Transform (DCT), Walsh Hadamar Transform (WHT). $D \in \mathbb{R}^{n \times n}$ is called the subsampling matrix, which randomly selects the M-row subset of the matrix FR. The coefficient $\sqrt{N/M}$ is in order to make the energy of the measurement vector consistent with the original signal, also known as the compression sampling rate. Using the SRM as the measurement matrix to carry out the signal reconstruction consists of three steps. First, the target signal is pre-randomized by R; Then, the target orthogonal matrix F is applied to the random signal of the previous step; Finally, randomly select the N column from FR as the measurement matrix.

In this paper, taking advantage of SRM can be used for large-size images directly observe the characteristics of sampling, the SRM-DCS algorithm, which is suitable for light field signal acquisition, is proposed by using SRM as the measurement matrix (Table 1).

Table 1. Proposed SRM-DCS algorithm

Input: Original images L, Termination condition ε, Image sequence number k;
Output: Reconstructed images I;
step1. k images with a size of $n \times n$ are superimposed into a three-dimensional image matrix, and the image matrix is cut along the $i(i = 1, 2, ..., m)$ line to obtain a new image x_i;
step2. Image x_i is rearranged along the inclination angle θ of the stripes to obtain a vectorized vertical stripe image $\widehat{x}_i$;
step3. Constructs the SRM as the measurement matrix Φ_i of the image $\widehat{x}_i$, where $R \in \mathbb{R}^{n \times n}$ is the random permutation matrix, $F \in \mathbb{R}^{n \times n}$ is the transformation matrix, $D \in \mathbb{F}^{M \times N}$ is the M row subsampling of the random selection matrix FR;
step4. Judging whether $\widehat{I}_{(i)}$ is satisfies the termination condition: $\| D^{(i)} - D^{(i-1)} \| \leq \varepsilon$, if the termination condition is satisfied, output image $I_{(i)}^*$, otherwise return to step 3 to continue execution, where $D^{(i)} = \| \widehat{I}_{(i)} - \widehat{I}_{(0)} \|_2$;
step5. The image sequence $\widehat{I}_{(i)}$ is synthesized into a three-dimensional image matrix I, k images with a size of $n \times n$ is extracted to complete the reconstruction of the light field image.

4   Experiments

In order to compare the performance of SRM-DCS algorithm with other compressive sensing image reconstruction algorithms, the light field image (Fig. 5) Lego Knights provided by Stanford University Computer Graphics Laboratory [17] was selected to experiment. The experimental hardware equipment is a computer that has Intel i5-4590 CPU (3.30 GHz) and memory for 8 GB. The experiments were conducted using MATLAB 2014b on Windows.

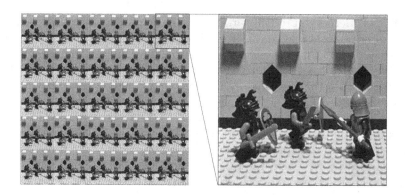

Fig. 5. 5×5 Lego Knights light field images

NSRM-DCS (Gaussian) in the experiment represents that the image is reconstructed with a random Gaussian measurement matrix, NSRM-DCS (Bernoulli) represents the use of random Bernoulli measurement matrix for image reconstruction, and NSRM-DCS (PartHadamard) represents the use of part of the Hadamard measurement matrix for image reconstruction. These three methods are commonly used algorithm implementation to do the measurement matrix compression sensing (NSRM-DCS) with non-structured measurement matrix (NSRM). SRM-DCS (FFT) represents a fast Fourier transform (FFT) as an orthogonal matrix F in a structured random matrix, SRM-DCS (DCT) denotes a discrete cosine transform (DCT) as an orthogonal matrix F in a structured random matrix, SRM-DCS (WHT) represents the use of the Walsh Hadamard transform (WHT) as the orthogonal matrix F in the structured random matrix. And these three methods are commonly used algorithm implementation to do the measurement matrix compression sensing (NSRM-DCS) with structured measurement matrix (SRM).

Equation 6 shows with the increase in sample volume, the relationship between the sampling volume and the reconstruction success rate using NSRM-DCS and SRM-DCS. The number of experimental iterations is 200 times, the image quality is excel-lent if the general image SNR is higher than 60 dB, the reconstructed image whose SNR greater than 60 dB is called reconstruction successful image in the experiment. X in this experiment is the original image, X_r the reconstructed image, and the SNR can be expressed as:

$$SNR = 20 \lg(\|X\|_2 / \|X - X_r\|_2) \tag{6}$$

Figure 6 represented that the success rate using SRM is better than using NSRM acted as a measurement matrix, and the former algorithm has earlier convergence. The success rate of SRR-DCS is lower than that of SRM-DCS, which has a close to the success rate using the SRM acted as the measurement matrix only when using the Part Hadamard acted as measurement matrix and the sampling volume is less than 90. In the sampling volume of 130, the use of SRM has higher success rate of about 20% compared with NSRM. The convergence time of SRM-DCS is earlier than NSRM-DCS, and Fig. 6 shows the measurement matrix algorithm using SRM start-ed convergence when the sampling volume reaches 110, but the measurement matrix algorithm using NSRM don't begin to converge when the sampling volume is 130.

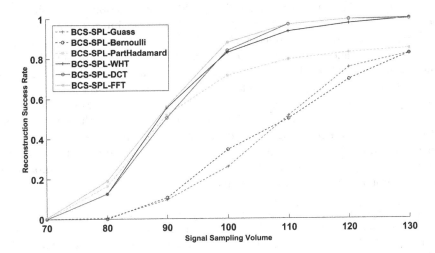

Fig. 6. The Relationship between the Sampling Volume and the reconstruction success rate using NSRM-DCS and SRM-DCS

Table 2 shows the PSNR values of the Lego Knights field images sequence, when use NSRM-DCS and SRM-DCS to reconstruct the light field image. Among them, the sampling rate of R increases from 0.1 to 0.9, and the result represented that the quality of image reconstruction using SRM-DCS is better than that of NSRM-DCS. SRM acted as the measurement matrix reconstructed more efficiently than the NSRM measurement matrix, with the average quality increase by 2.01 dB, which can further show that the reconstruction accuracy of SRM-DCS algorithm is better than the NSRM-DCS algorithm.

Table 2. PSNR value (dB) of reconstructed images using NSRM-DCS and SRM-DCS

Sampling rate	0.1	0.2	0.3	0.4	0.5	0.6	0.7	0.8	0.9
NSRM-DCS (Gaussian)	23.13	24.32	26.63	29.52	30.95	34.26	36.30	39.54	44.59
NSRM-DCS (Bernoulli)	22.96	24.19	26.30	29.50	30.95	33.44	36.44	39.72	44.58
NSRM-DCS (PartHadamard)	23.43	24.12	26.70	29.34	31.07	34.56	36.46	40.40	44.53
SRM-DCS (DCT)	24.46	27.68	29.76	31.89	33.96	36.02	37.86	41.08	45.22
SRM-DCS (FFT)	24.31	27.88	29.84	31.81	33.56	35.47	38.33	41.05	45.23
SRM-DCS (WHT)	24.34	27.69	29.90	31.49	33.26	35.56	38.33	41.10	45.20

5 Conclusion

In this paper, a distributed compressive sensing field reconstruction method (SRM-DCS) using structured measurement matrix was proposed. In the light field image acquisition stage, the K-SVD algorithm was selected to train the adaptive light field dictionary. In the light field reconstruction phase, since the slices of the 3D image matrix formed by the light field images sequence have the characteristics of the stepped stripes, slices were rearranged according to the inclination angle to obtain a 2D vertical stripe image, and the light field images sequence is reconstructed in combination with the structured measurement matrix (SRM), the experimental results indicated that the algorithm has faster reconstruction precision and reconstruction efficiency. However, the reconstructed light field images using DCS did not make full use of the a prior information of the light field images sequence, so it needs to be optimized in the reconstruction method. In the future, we can spend more attention to the reconstruction of the light field by using the prior information of the compressive sensing combined with the images sequence.

Acknowledgments. This work is supported by the National Natural Science Foundation of China (Grant No. 61471150), the Funds for International Cooperation and Exchange of the National Natural Science Foundation of China (Grant No. 2014DFA12040), the National High-tech R&D Program of China (863 Program, 2015AA015901).

References

1. Liu, Y., Dai, Q., Xu, W.: A real time interactive dynamic light field transmission system. In: IEEE International Conference on Multimedia and Expo, pp. 2173–2176 (2006)
2. Zhou, W., Pan, J., Li, P., Wei, X., Liu, Z.: A distributed stream computing architecture for dynamic light-field acquisition and rendering system. In: Pan, Z., Cheok, A.D., Müller, W., Zhang, M. (eds.) Transactions on Edutainment XIII. LNCS, vol. 10092, pp. 123–132. Springer, Heidelberg (2017). https://doi.org/10.1007/978-3-662-54395-5_11
3. Derin Babacan, S., Ansorge, R., Luessi, M., Molina, R., Katsaggelos, A.K.: Compressive sensing of light fields. In: IEEE International Conference on Image Processing, pp. 2337–2340 (2010)
4. Adelson, E.H., Wang, J.Y.A.: Single lens stereo with a plenoptic camera. IEEE Trans. Pattern Anal. Mach. Intell. **14**(2), 99–106 (1992)
5. Levoy, M., Hanrahan, P.: Light field rendering. In: Conference on Computer Graphics and Interactive Techniques, pp. 31–42 (1996)
6. Wenger, A., Gardner, A., Tchou, C., Unger, J., Hawkins, T., Debevec, P.: Performance relighting and reflectance transformation with time-multiplexed illumination. ACM Trans. Graph. **24**(3), 756–764 (2005)
7. Cands, E.J.: Compressive sampling. Marta Sanz Sol **17**(2), 1433–1452 (2007)
8. Foucart, S., Rauhut, H.: A mathematical introduction to compressive sensing. Appl. Numer. Harmon. Anal. **44** (2013)

9. Derin Babacan, S., Ansorge, R., Luessi, M., Mataran, P.R., Molina, R., Katsaggelos, A.K.: Compressive light field sensing. IEEE Trans. Image Process. 21(12), 4746–4757 (2012). A Publication of the IEEE Signal Processing Society
10. Do, T.T., Tran, T.D., Gan, L.: Fast compressive sampling with structurally random matrices. In: IEEE International Conference on Acoustics, Speech and Signal Processing, pp. 3369–3372 (2008)
11. Hong, T., Bai, H., Li, S., Zhu, Z.: An efficient algorithm for designing projection matrix in compressive sensing based on alternating optimization. Signal Process 125(C), 9–20 (2016)
12. Baron, D., Duarte, M.F., Wakin, M.B., Sarvotham, S., Baraniuk, R.G.: Distributed compressive sensing. (3), 2886–2889 (2009)
13. Aharon, M., Elad, M., Bruckstein, A.: K-SVD: an algorithm for designing overcomplete dictionaries for sparse representation. IEEE Trans. Signal Process. 54(11), 4311–4322 (2006). IEEE Press
14. Elad, M.: Sparse and Redundant Representations: From Theory to Applications in Signal and Image Processing. Springer, New York (2010). https://doi.org/10.1007/978-1-4419-7011-4. Incorporated
15. Kamal, M.H., Golbabaee, M., Vandergheynst, P.: Light field compressive sensing in camera arrays. In: IEEE International Conference on Acoustics, Speech and Signal Processing, pp. 5413–5416 (2012)
16. Duarte, M.F., Eldar, Y.C.: Structured compressed sensing: from theory to applications. IEEE Trans. Signal Process. 59(9), 4053–4085 (2011)
17. Stanford University Computer Graphics Laboratory: The (new) Stanford light field archive. http://lightfield.stanford.edu/lfs.html

Image Quality Assessment and Analysis

Stereoscopic Image Quality Assessment Based on Binocular Adding and Subtracting

Jiachen Yang[1], Bin Jiang[1(⊠)], Chunqi Ji[1], Yinghao Zhu[1], and Wen Lu[2]

[1] School of Electrical and Information Engineering,
Tianjin University, Tianjin 300072, China
{yangjiachen,jiangbin,jcq_,zhuyinghao}@tju.edu.cn
[2] School of Electronic Engineering, Xidian University, Xi'an 710071, China
luwen@xidian.edu.cn

Abstract. There has been a great concern on blind image quality assessment in the field of 2D images, however, stereoscopic image quality assessment (SIQA) is still a challenging task. In this paper, we propose an efficient blind image quality assessment model for stereoscopic images according to binocular adding and subtracting channels. Different from other SIQA methods which focus on complex binocular visual properties, we simply use the visual information from adding and subtracting to describe binocularity (also known as ocular dominance) which is closely related to distortion types. To better evaluate the contribution of each channel in SIQA, a dynamic weighting is introduced according to local energy. Meanwhile, distortion-aware features based on wavelet transform are utilized to describe visual degradation. Experimental results on 3D image databases demonstrate the potential of the proposed framework in predicting stereoscopic image quality.

Keywords: Ocular dominance · Binocularity
Blind image quality assessment · Adding and subtracting

1 Introduction

In recent years, there has been great progress in Image Quality Assessment (IQA). The emergence of various IQA databases and the proposal of IQA theory greatly enrich the way to evaluate image quality [1]. At the beginning, 2D-IQA metrics mainly focus on the difference between reference and distorted images (known as Full-Reference (FR) methods). For example, the well-known Structural Similarity Index Measurement (SSIM) measures image quality from the perspective of image formation, and Visual Information Fidelity (VIF) explores the consistence between image information and distortion. The rising of blind (No-Reference, NR) IQA methods has a huge influence on IQA, such as Distortion Identification-based Image Verity and INtegrity Evaluation (DIIVINE) [2], Blind/Referenceless Image Spatial Quality Evaluator (BRISQUE) [3], Blind Image Integrity Notator using DCT Statistics-II (BLIINDS-II) [4]. They utilize

© Springer Nature Singapore Pte Ltd. 2017
J. Yang et al. (Eds.): CCCV 2017, Part II, CCIS 772, pp. 237–247, 2017.
https://doi.org/10.1007/978-981-10-7302-1_20

distortion-aware/sensitive features to evaluate image quality, and therefore the availability of visual features is of great importance. Meanwhile, there are also deep learning based frameworks [5]. These newly proposed IQA methods achieve great success in predicting image quality.

However, Bosc et al. have proved that 2D-IQA metrics are not applicable to SIQA because of the weak correlation with binocular visual properties [6]. To solve this problem, scholars and experts have been long focusing on binocular visual characteristics. Chen et al. designed an intermediate image closely resembles the cyclopean image [8]. Shao et al. classified the stereoscopic images into non-corresponding, binocular fusion and binocular suppression regions [9]. Ryu et al. models the binocular quality perception in context of blurriness and blockiness [10]. Yang et al. proposed a quality index by evaluating binocular subtracting and adding.

Recently, it has been found that the Human Visual System has separately adaptable channels for adding and subtracting the visual signals from two views. Compared with previous research on binocular interactions, we simply focus on the adding and subtracting channels to demonstrate binocular visual properties. In addition, ocular dominance which produces binocularity is considered to characterize the receptive field properties from monocular response to binocular response, and it is closely related to different distortions. Therefore, we try to use the information from subtracting and adding to measure the ocular dominance. With this method, we also greatly reduce the amount of computation in modeling complicated binocular visual properties. However, to what extent each channel contributes to binocular visual perception is still undiscovered. To solve this problem, we take the local energy response as a weighting index to balance their performance in stereo perception, since energy maps provide useful local binocular rivalry information, which may be combined with the qualities of single-view images to predict 3D image quality [12].

Another key point would be how to extract distortion-aware features from the adding and subtracting channels to describe visual degradation. There have been a great number of effective ways to extract Natural Scene Statistics (NSS) features. For example, General Gaussian Distribution (GGD) model has been used to fit the non-Gaussian distribution from frequency domain, and statistical properties of MSCN coefficients are also explored in spatial domain. However, sometimes these properties would change with different visual content. Taking this shortcoming of these methods into account, He et al. studied the exponential attenuation characteristics of the magnitude, variance and entropy in different wavelet subbands. In this paper, those features from wavelet transform are utilized to represent the properties of natural scenes since they are less sensitive to the image content. Meanwhile, the feature extraction procedure is rather computationally efficient.

Our framework is based on the adding and subtracting theory which describes ocular dominance. As a result, NSS features are extracted from the adding and subtracting channels. Compared with the time-consuming methods, the proposed framework achieves a good balance between efficiency and prediction accuracy.

The rest of the paper is organized as follows. Section 2 presents the framework of the proposed metric. The experimental results and analysis are given in Sect. 3, and finally conclusions are drawn in Sect. 4.

2   Background and Motivation

Before the proposed method, we refers some basic theory about stereo image quality, which will give us some motivation. As for stereo image quality, it refers to the machine method to keep consistent with subjective feelings of humans. In the processing of acquisition and display, the stereo images will be affected by the existing noise interference, which will make the images have a certain difference with the original images. In this way, it will give a bad visual perception.

For human eyes, it can make use of emitted light to see the objects, and the final images will be sent to the visual center through internal light system. In this way, it can produce the visual perception. Based on the previous researches, there are three important visual neural pathways for people brains. Specially, they are ventral pathway, dorsal pathway and visual pathway. For ventral pathway, it contains the primary visual cortex named $V1$, secondary visual cortex named $V2$ and $V3$. And it also contains the ventral extrastriate cortex named $V4$.

When people make use of consciousness and perception, all of the above parts will combine to determine the movement and location of the object. Judging from above, human visual system(HVS) is very complex and it contains physiological structure. In addition, some high level visual cortex interaction should be used for the visual pathway in order to finish the circulation mechanism and the feedback types.

Through the production of random point of view, the absence of any cue will happen in the human eyes, which will affect the perception of the depth. In stereo images perception, the human eye's binocular disparity and convergence can be made use of. And then the binocular disparity can be regarded as the most important part for the three-dimensional technology.

In general, the horizontal distance of the two human eyes is about 6 cm, which can help people get the certain difference in scene. In detail, the projected object in left and right eyes are different respectively. The difference can be named binocular disparity. As for disparity, it can be divided into two parts: vertical and horizontal parallax. specially, the horizontal parallax is the most important for the depth perception. And vertical parallax can only determine the perception comfort. In other word, if the vertical parallax is bad, the human will feel uncomfortable perception.

In this connection, a lot of researchers to study the visual characteristics of the human eye binocular fusion, and establish a corresponding stereo vision model, the corresponding mathematical model is as follows:

Model of Eye-Weighting (EW): Engel *et.al* proposed a binocular weighted (EW) model. The model also contains a weighting factor for the binocular fusion process, but unlike the simple model, the model coefficients can be obtained by integrating the square root of each eye signal autocorrelation function. And it can be described as

$$C = ((W_L \cdot E_L)^2 + (W_R \cdot E_R)^2)^{\frac{1}{2}} \tag{1}$$

Model of Vector Summation (VS): Therefore, Curtis *et al.* Found that the binocular fusion graph is the sum of the two normalized orthogonal vectors, and the vector sum (VS) model is proposed.

$$C = \sqrt{I_L^2 + I_R^2 + I_L \cdot I_R} \tag{2}$$

Among them, IL and IR represent the left and right view information.

Model of Gain Control (GC): Ding et al. on the basis of previous research proposed the corresponding gain control model, they pointed out that people each eye visual gain information not only from in its input signal, but also from in the other eye gain control signal energy.

$$C = \frac{I_L}{1 + I_R} + \frac{I_R}{1 + I_L} + 0.1 \cdot I_L \cdot I_R \tag{3}$$

Model of Neural Network (NN): A neural network (NN) model is proposed by Cogan et al.

$$C = \frac{E_L}{1 + E_L + E_R} \cdot I_L + \frac{E_R}{1 + E_L + E_R} \cdot I_R \tag{4}$$

He proposed that the $log - Gabor$ model can well reflect the simple visual cell function of the human eye. Using $log - Gabor$ filter to volume response calculation of gain control system model of weighted value, according to the US in the literature, we define $[\eta_{s,o}, \zeta_{s,o}]$ as a different direction and size of the filter response, this chapter of the $log - Gabor$ filter $G_{s,o}(\omega, \theta)$ is defined as shown.

$$G_{s,o}(\omega, \theta) = exp[-\frac{(log(\omega/\omega_s))^2}{2\sigma_s^2}] \cdot exp[-\frac{\theta - \theta_0}{2\sigma_o^2}] \tag{5}$$

Among them, s and o is respectively spatial scale and orientation information, θ said direction angle information, ω_s and ω_o is used to determine the filter energy, ω and ω_s represent the normalized radial filter frequency and the center frequency. According to this, we can calculate the signal X position in the size of s, o direction of the local energy information for the:

$$E_o(X) = \sqrt{F_o(X)^2 + H_o(X)^2} \tag{6}$$

Among them,

$$F_o(X) = \sum_s \eta_{s,o}(X) \tag{7}$$

$$H_o(X) = \sum_s \zeta_{s,o}(X) \tag{8}$$

And the energy of the filter is:

$$E(X) = \sum E_o(X) \tag{9}$$

At last ,we can see the energy value is the weight value of the left eye and right eye.

Motivated by the above model, we find that binocular adding and subtracting is another obvious model which can affect the stereo perception. However, there is few research results that pay attention on this issue.

3   The Proposed Framework

The proposed framework contains two parts: binocular adding and subtracting and NSS feature extraction. At last, SVM is used to connect the NSS features with image quality. Detailed information about the framework is shown in Fig. 1.

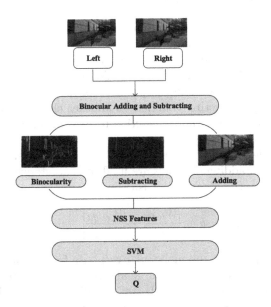

Fig. 1. Framework of the proposed image quality assessment metric

3.1   Binocular Adding and Subtracting

HVS has separately adaptable channels for adding and subtracting the neural signals from the two views. Encoding the adding channel A and subtracting channel S between two stereo-halves can be used for stereopsis.

$$\begin{aligned} A &= L + R \\ S &= |L - R| \end{aligned} \tag{10}$$

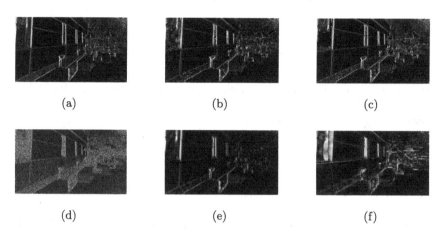

Fig. 2. Examples of binocularity. (a) Original image pair. (b) JP2K compressed image pair.(c) JPEG compressed image pair. (d) White noised image pair. (e) Fast faded image pair. (f) Gaussian blur image pair

Examples of binocular adding and subtracting are shown in Fig. 2. It can be seen that subtracting and adding show totally different visual information from each other, the subtracting channel contains the difference between two views because of the viewing angle, which could be considered as an alternative of disparity. Considering that the computation of disparity is extremely time-consuming, we take subtracting channels as another choice to achieve the balance between better efficiency and prediction results. In comparison, the adding channel is more similar to a information-enriched map.

In order to model neural mechanism for binocular processing, the responses from subtracting and adding channels are used to characterize the ocular dominance. To be specific, the binocularity is defined as :

$$b = \frac{|W_{left}| - |W_{right}|}{|W_{left}| + |W_{right}|} \tag{11}$$

where W_{left} and W_{right} are the monocular response from each view, and b represents the degree of binocularity. A large absolute value of b represents a weak binocular response, and vice versa. To better visualize ocular dominance in terms of distortion, Fig. 2 shows that ocular dominance changes with regard to different distortions, and therefore it could be an index which characterize binocular visual properties.

However, to the best of our knowledge, which channel contributes more to visual perception is still unsolved. Therefore we proposed a weighting scheme to combine the two channels together. Based on the fact that distortion in either view may affect the consistency between the two views and lead to binocular rivalry, a reasonable way to characterize this property is the local energy map since higher-energy regions are more likely to attract visual attention. Therefore, a local energy based weighting scheme is adopted to balance the adding and subtracting channels. Here, the energy for each channel is obtained by summing the local variances using an 11×11 circular-symmetric Gaussian weighting function $w = \{\omega i | i = 1, 2, ..., N\}$, with standard deviation 1.5 samples, normalized to unit sum $(\sum_{i=1}^{N} \omega i = 1)$. And then local energy is calculated by

$$e_{in} = \left(\sum_{i=1}^{N} \omega_i (x_i - \mu_{in})^2 \right)^{\frac{1}{2}} \tag{12}$$

where $\mu_{in} = \sum_{i=1}^{N} \omega_i x_i$ is the local mean value. Finally the energy map is computed from 4scales by:

$$e_l = \frac{1}{n_s \times M} \sum_{i=1}^{n_s} \sum_{j=1}^{M} e_{in, j-l} \tag{13}$$

3.2 NSS Feature Extraction

He *et al.* have demonstrated that the tertiary properties of the wavelet coefficients reflect the self-similar property of scenes. In particular, the exponential attenuation characteristics of the magnitude, variance and entropy in different wavelet subbands are less specific in their representation of an image, and thus are utilized to represent the generalized behaviors of natural scenes.

Specifically, the magnitude m_k is used to encode the generalized spectral behavior, the variance v_k to describe the fluctuations of energy, and the entropy e_k to represent the generalized information, following Eqs. 14–16.

$$m_k = \frac{1}{N_k \times M_k} \sum_{j=1} \sum_{i=1} \log 2 \, |C_k(i,j)| \tag{14}$$

$$v_k = \frac{1}{N_k \times M_k} \sum_{j=1} \sum_{i=1} \log 2 \, |C_k(i,j) - m_k| \tag{15}$$

$$e_k = \sum_{j=1}^{N_k} \sum_{i=1}^{M_k} p \, [Ck(i,j)] \ln p \, [C_k(i,j)] \tag{16}$$

where $C_k(i,j)$ represents the (i,j) coefficient of the k-th subband, M_k and N_k are the length and width of the k-th subband, respectively; $p[\cdot]$ is the probability density function of the subband.

In our framework, an image is decomposed into 4 scales and 8 wavelet subbands without distinguishing the the low-high and high-low subbands because of their similarity in statistics. The vertical and horizontal subbands with an identical mark in the same scale are combined through averaging after the above process. Finally, there are 24 features extracted from each channel in total

$$f = [m_1, m_2, ..., m_8, v_1, v_2, ..., v_8, e_1, e_2, ..., e_8]^T \tag{17}$$

4   Experimental Results and Analyses

4.1   Stereo Database

To verify the performance of the proposed method, the LIVE 3D Image Quality Databases (Phase I and Phase II) of the University of Texas at Austin are used [13], which are shown in Fig. 3. Database Phase I database contains 365 stereopairs with symmetric distortion and Database Phase II contains 360 stereopairs with both asymmetric and symmetric distortion, including JPEG compression, JP2K compression, white noise (WN), Gaussian blur (Blur) and fast fading (FF).

Fig. 3. LIVE 3D image quality databases

4.2   Performance Measure

For performance evaluation, four commonly used indicators are adopted: Pearson Linear Correlation Coefficient (PLCC), Spearman Rank Order Correlation Coefficient (SROCC), Kendall Rank-order Correlation Coefficient (KROCC), and Root Mean Squared Error (RMSE) between subjective scores and objective scores after nonlinear regression. For nonlinear regression, we use a 4-parameter logistic mapping function:

$$DMOS_P = \frac{\beta_1 - \beta_2}{e^{(Q - \beta_3)/|\beta_4|} + 1} + \beta_2 \tag{18}$$

where β_1, β_2, β_3 and β_4 are the parameters to be fitted. A better match is expected to have higher PLCC, SROCC, KROCC and lower RMSE.

In the prediction phase, the stereopairs in each database were randomly divided into two parts, with 80% for training and 20% for test. To ensure that the proposed framework is robust, 1000 iterations of training are performed by varying the splitting of data over the training and test sets, and the median value of all iterations is chosen as the final prediction model. All the parameters of our SVM model are the same for different databases.

In order to demonstrate its efficiency, the proposed method is compared with several existing state-of-art IQA metrics, including five 2D metrics (DIIVINE [2], BLIINDS-II [4] and BRISQUE [3]), and four 3D metrics (Lin's scheme [7], Chen's scheme [8], Shao's scheme-A [9] and Shao's scheme-B [11]). For those 2D-extended BIQA metrics, feature vectors extracted from the left and right views are averaged, and SVM is used to train a regression function. For Lin's scheme, the FI-PSNR metric is adopted into the comparison. For Chen's scheme, the adopted 2D metric is MS-SSIM which performs the best. For Shao's schemes, experimental results in the reference are directly adopted.

4.3 Overall Performance on LIVE 3D Image Database

As shown in Table 1, the values of PLCC, SROCC, KROCC, and RMSE of each metric are reported, where the indicator that gives the best performance is highlighted in bold. It can be observed that the overall performance of the proposed framework on Database Phase-I significantly outperforms other IQA metrics. Shao's scheme also achieves rather competitive performances. The outstanding performance partially demonstrates the efficiency of the proposed framework.

Table 1. Overall performances on LIVE 3D image database

Criteria	LIVE 3D image database phase I				LIVE 3D image database phase II			
	PLCC	SROCC	KROCC	RMSE	PLCC	SROCC	KROCC	RMSE
DIIVINE [2]	0.9252	0.9233	0.7474	6.2245	0.7758	0.7707	0.6023	7.1221
BLIINDS-II [4]	0.9117	0.9087	0.7220	6.7368	0.7865	0.7167	0.5427	6.9708
BRISQUE [3]	0.9119	0.9083	0.7197	6.7314	0.7513	0.7242	0.5481	7.4496
Lin [7]	0.8645	0.8559	0.6559	8.2424	0.6584	0.6375	0.4701	8.4956
Chen [8]	0.9161	0.9153	0.7360	6.5740	0.9067	0.9068	0.7314	4.7587
Shao-A [9]	0.9350	0.9251	-	5.8155	0.8628	0.8494	-	5.7058
Shao-B [11]	0.9565	0.9449	-	4.7552	0.9265	**0.9106**	-	4.3381
Proposed	**0.9594**	**0.9447**	**0.8004**	**4.5086**	**0.9350**	0.9097	**0.7347**	**4.2739**

Results on asymmetric distorted stereopairs are also listed. According to the experimental results on Database Phase II which contains both symmetric and asymmetric distortion, the proposed framework overtakes the other metrics by a large extent. The significant difference further confirms the previous conclusion that our framework can effectively predict the quality of stereoscopically viewed images. However, predicting the quality of asymmetric distorted stereopairs is still a challenge, since all the metrics show less consistency with subjective test.

Considering that samples of training and test are selected from the same dataset, experiments on individual databases are not sufficient to explain the

Table 2. Cross-database performances on LIVE 3D image database (training/test)

Criteria	LIVE I /LIVE II		LIVE II /LIVE I	
	PLCC	SROCC	PLCC	SROCC
DIIVINE [2]	0.6161	0.5177	0.5375	0.4954
BLIINDS-II [4]	0.6986	0.6716	0.7652	0.7398
BRISQUE [3]	0.5949	0.4582	0.5718	0.5564
Shao-B [11]	0.7791	0.7514	**0.8936**	**0.8917**
Proposed	**0.8127**	**0.7809**	0.8464	0.8280

generality and stability of the evaluation model, and therefore cross-database experiments are also carried out in this part. Table 2 also gives the detailed information of cross-database test. Note that two FR metrics (Lin, Chen and Shao's schemes) are not reported. Obviously, the performance of the metrics significantly declines compared with individual dataset, because the source images and distortion types are not consistent. Another probable reason would be that Database Phase-I only contains both symmetrically distorted stereopairs. However, our metric still has a relatively good predictive ability, when metrics are trained on LIVE phase I and tested on LIVE phase II.

Table 3. Running time

Criteria	DIIVINE [2]	BLIINDS-II [4]	BRISQUE [3]	Lin [7]	Chen [8]	Ours
Time(s)	6800	21923	**37**	82	5143	145

In addition to the excellent performance on predicting image quality, the proposed method is also computationally efficient, as shown in Table 3. The total running time on LIVE 3D Image Database-Phase I (365 stereopairs with a resolution of 640× 360 for each view) is only 145 s on a computer with a Core i7 CPU. It means that predicting a stereopair costs less than 0.4 s. Although it is not the most efficient one, it achieves better balance between accuracy and time. Compared with those metrics which spend a lot of time in modeling complex binocular visual properties (e.g. Chen's scheme) and computing NSS features (BLIINDS-II), the proposed method is much less time-consuming. However, designing an quality assessment index applicable to real-time image (video) processing is still challenging.

5   Conclusions

In this paper, an ocular dominance based quality index for stereoscopic images is proposed, in which the NSS features are used to describe the visual degradation on stereopairs. Based on the fact that binocular adding and subtracting are closely related to stereo perception, we use the NSS features from three channels (namely, binocularity, adding, and subtracting) to predict visual quality. Experiments further confirm that the proposed framework is highly consistent with subjective test, and the computation is relatively efficient.

In the future, we will pay much attention to the research on binocular visual properties, and explore more effective ways to describe image degradation.

Acknowledgments. The heading should be treated as a This research is partially supported by National Natural Science Foundation of China (No. 61471260), and Natural Science Foundation of Tianjin (No. 16JCYBJC16000).

References

1. Lin, W., Kuo, C.C.J.: Perceptual visual quality metrics: a survey. J. Vis. Commun. Image Represent. **22**(4), 297–312 (2011)
2. Moorthy, A.K., Bovik, A.C.: Blind image quality assessment: from natural scene statistics to perceptual quality. IEEE Trans. Image Process. Publ. IEEE Sig. Process. Soc. **20**(12), 3350–3364 (2011)
3. Mittal, A., Moorthy, A.K., Bovik, A.C.: No-reference image quality assessment in the spatial domain. IEEE Trans. Image Process. Publ. IEEE Signal Process. Soc. **21**(12), 4695 (2012)
4. Saad, M.A., Bovik, A.C., Charrier, C.: Blind image quality assessment: a natural scene statistics approach in the DCT domain. IEEE Trans. Image Process. Publ. IEEE Signal Process. Soc. **21**(8), 3339 (2012)
5. Fei, G., Tao, D., Gao, X., et al.: Learning to rank for blind image quality assessment. IEEE Trans. Neural Netw. Learn. Syst. **26**(10), 2275–2290 (2015)
6. Bosc, E., Pepion, R., Callet, P.L., et al.: Towards a new quality metric for 3-D synthesized view assessment. IEEE J. Sel. Top. Signal Process. **5**(7), 1332–1343 (2011)
7. Lin, Y.H., Wu, J.L.: Quality assessment of stereoscopic 3D image compression by binocular integration behaviors. IEEE Trans. Image Process. Publ. IEEE Signal Process. Soc. **23**(4), 1527 (2014)
8. Chen, M.J., Su, C.C., Kwon, D.K., et al.: Full-reference quality assessment of stereopairs accounting for rivalry. Image Commun. **28**(9), 1143–1155 (2013)
9. Shao, F., Li, K., Lin, W., et al.: Full-reference quality assessment of stereoscopic images by learning binocular receptive field properties. IEEE Trans. Image Process. Publ. IEEE Signal Process. Soc. **24**(10), 2971–2983 (2015)
10. Ryu, S., Sohn, K.: No-reference quality assessment for stereoscopic images based on binocular quality perception. IEEE Trans. Circuits Syst. Video Technol. **24**(4), 591–602 (2014)
11. Shao, F., Tian, W., Lin, W., et al.: Toward a blind deep quality evaluator for stereoscopic images based on monocular and binocular interactions. IEEE Trans. Image Process. **25**(5), 2059–2074 (2016)
12. Wang, J., Rehman, A., Zeng, K., et al.: Quality prediction of asymmetrically distorted stereoscopic 3D images. IEEE Trans. Image Process. Publ. IEEE Signal Process. Soc. **24**(11), 3400–3414 (2015)
13. Sheikh, H., Wang, Z., Cormack, L., Bovik, A.: Live image quality assessment database release 2 (2005)

Quality Assessment of Palm Vein Image Using Natural Scene Statistics

Chunyi Wang[1,2], Xiongwei Sun[3], Wengong Dong[3], Zede Zhu[3], Shouguo Zheng[3], and Xinhua Zeng[3(✉)]

[1] Institute of Intelligent Machines, Chinese Academy of Sciences, Hefei 230031, China
[2] University of Science and Technology of China, Hefei 230026, China
chunyi@mail.ustc.edu.cn
[3] Institute of Technology Innovation, Hefei Institutes of Physical Science, Chinese Academy of Sciences, Hefei 230088, China
xhzeng@iim.ac.cn

Abstract. Image quality has a great influence on the performance of non-contact biometric identification system. In order to acquire palm vein image with high-quality, an image quality assessment algorithm for palm vein is presented based on natural scene statistics features. Moreover, the effect of uneven illumination caused by palm tilt are taken into account. A large number of experiments are presented and the results based on proposed algorithm are accordance with human subjective assessment.

Keywords: Quality assessment · Palm vein image
Natural scene statistics · Biometric identification

1 Introduction

In recent years, biometric identification technology has received extensive attention in research and application. Some examples of biometric identifiers are fingerprint, facial, iris, voice and palm print features. Identification systems based on these bio-metric features have been applied to mobile phones, personal computers and access control. As these identifiers are exposed to the environment and likely to be damaged or forged, palm vein pattern as an internal identifier is more reliable than these externals. Palm vein image are taken with an infrared camera by means of contactless way, the image has poor contrast and nonuniform brightness. Since palm vein image quality have a considerable influence on recognition performance, the primary work of image preprocessing is quality assessment of palm vein image [1–3].

Many scholars and institutes have proposed several hand vein image quality evaluation methods. Ming et al. [4,5] used mean gray value and gray variance as quality index of dorsal hand vein image. In [6,7],four parameters were selected as the reference for image quality, including gray variance, information entropy,

© Springer Nature Singapore Pte Ltd. 2017
J. Yang et al. (Eds.): CCCV 2017, Part II, CCIS 772, pp. 248–255, 2017.
https://doi.org/10.1007/978-981-10-7302-1_21

cross point and area of effect. They integrated these parameters with weights to get image quality score. Wang et al. [8] focus on the impact of the palm vein preprocessing and match when palm distance change from far to near. They used TenenGrad to evaluate image clarity and used SSIM to show structure difference of the same palm with the changes of distance increasing. Wang et al. [9] proposed relative contrast and relative definition for dorsal hand vein image quality assessment.

The above studies have not considered the influence of uneven illumination on the palm vein image. In this paper, statistics of Harr-Like features are introduced. We use natural scene statistics (NSS) features presented in [10] to estimate image quality. In addition, the effects of uneven illumination caused by palm tilt are taken into ac-count.

2   Quantitative Evaluation of Palm Vein

In palm vein identification system, palm vein images are taken by contactless way with near-infrared camera. These images usually have poor contrast and strong noise, some of them have uneven illumination. Our work is to quantify the quality of palm vein images to reject low-quality images.

2.1   Illumination Uniformity of ROI Image

Firstly, the vein ROI image is partitioned into several patches and the gray mean of each patch of the image is calculated. Then, we use the difference between the minimum and the maximum gray mean among all patches of ROI image to evaluate the uniformity of image brightness. The larger the difference, the more uneven the brightness of ROI image.

We get ROI image F from palm vein image and partition it into several patches equally. If the matrix of $M \times N$ represents gray ROI image F and $F(i,j)$ denotes the gray value of the pixel point at the ith line and the jth row, we compute mean gray $M[n]$ of each part:

$$M[n] = \frac{1}{k * k} \sum_{i=0}^{k-1} \sum_{j=0}^{k-1} F_n(i,j) \qquad (1)$$

Where k denotes the size of each patch, n is the index of each patch, F_n stands for the patch of the ROI image whose index is n.

The score of brightness uniformity can be defined as:

$$Q_m = Min(\frac{M_{\max} - M_{\min}}{M(F)}, 1) \qquad (2)$$

Where $M_{\max}$ and $M_{\min}$ are the maximum and minimum gray mean $M[n]$ among all patches of the ROI image, $M(F)$ is the gray mean of the ROI image.

Figure 1 shows brightness nonuniformity of 30 palm vein images from the same person that tilt palm from zero to $60°$. The size of ROI image is 128×128

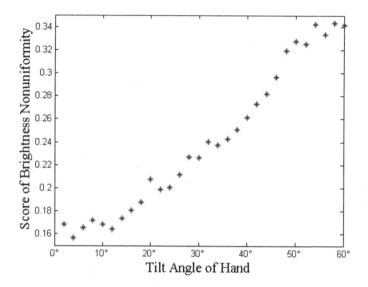

Fig. 1. Trend of brightness nonuniformity with different tilt angles of palm.

pixels, we partition ROI image into 64 patches equally. Thus k equals 16 and n is defined as a variable set from domain $N = \{0, 1, 2 \ldots 63\}$. The vertical axis represents the magnitude of Q_m, and Horizontal axis represents the tilt angle of the palm. Brightness nonuniformity gradually increased with the tilt angle of palm gradually increased, so we can use Q_m as an index to evaluate brightness uniformity.

2.2 Statistics of Pixel Intensities and Their Products

Anish Mittal et al. [10] proposed a natural image quality evaluator (NIQE). They selected 90 natural pictures with high quality and learned the distribution models of pixel intensities and their products of these nature pictures, the parameters of distribution models were regarded as quality aware NSS features. Then, they fitted these NSS features with a multivariate Gaussian (MVG) model and called this model natural MVG model. A MVG model of the image to be quality analyzed is fitted with the same method, the quality of the distorted image is expressed as the distance between the natural MVG model and the MVG model of the distorted image.

NIQE does not need distorted images or human subjective scores on them to train the model and has a better performance on several natural image databases, so it is very suitable for realtime identification system. We put NIQE into our quality assessment algorithm, 120 palm vein images with high quality are chosen from PolyU database [11] to train the natural MVG model.

In our algorithm, the size of patch is 16×16 pixels and Q_n stands for the result of NIQE.

2.3   Statistics of Harr-Like Features

Haar-like features are digital image features used in object recognition [12]. A simple rectangular Haar-like feature can be defined as the difference of the sum of pixels of areas inside the rectangle, which can be at any position and scale within the original image. In gray image, palm vein pattern is composed of dark lines with different orientations, we can use 3-rectangular Harr-Like feature to detect palm vein. Figure 2 illustrates the process flow when computing the quality index based on statistics of Harr-Like features.

The Harr-Like feature windows we have used are showing in Fig. 2. Since the width of palm vein is about 6 pixels, windows with size of 12×12 pixels are chosen to ensure the width of black parts of windows are 6 pixels or so. We use four kinds of windows showing in Fig. 3 to compute Harr-Like features, the formulation is as follows:

$$h = \frac{SumW}{NumW} - \frac{SumB}{NumB} \tag{3}$$

Where h is Harr-Like feature, and stand for the sum of pixel intensities of white and black part of window respectively and denote the number of white and black pixels in the window respectively.

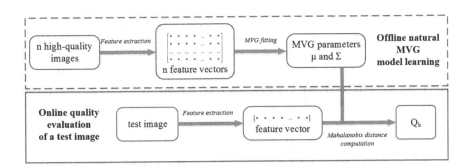

Fig. 2. Flow chat of computation of quality index based on statistics of Harr-Like features.

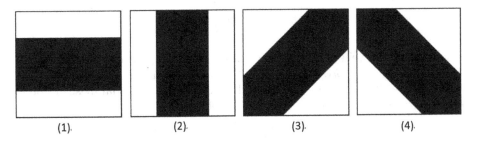

Fig. 3. Four kinds of Harr-Like feature windows we have used in this paper.

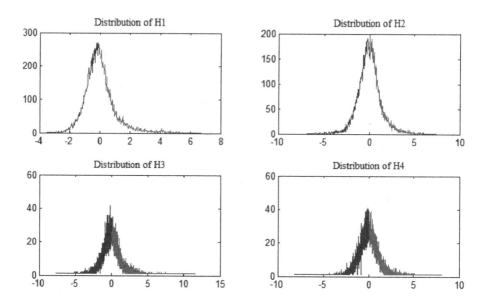

Fig. 4. Four kinds of Harr-Like feature windows we have used in this paper.

Then, we move the window with a step of one pixel over the palm vein image, and for each subsection of the image the Haar-like feature is calculated. We can get four vectors of Harr-Like features with four kinds of windows, and use H1, H2, H3 and H4 to denote this four vectors. The distribution of H1, H2, H3, H4 are showing in Fig. 4. From Fig. 4, we can see that the distribution of Harr-like features can be well-modeled as following a zero mode asymmetric generalized Gaussian distribution (AGGD) [13]:

$$g(x; \gamma, \beta_l, \beta_\gamma) = \begin{cases} \frac{\gamma}{(\beta_l + \beta_\gamma)\Gamma(\frac{1}{\gamma})} \exp(-(\frac{-x}{\beta_l})^\gamma), \forall x \le 0 \\ \frac{\gamma}{(\beta_l + \beta_\gamma)\Gamma(\frac{1}{\gamma})} \exp(-(\frac{x}{\beta_\gamma})^\gamma), \forall x > 0 \end{cases} \qquad (4)$$

The mean of the distribution is defined as follows:

$$\eta = (\beta_r - \beta_l)\frac{\Gamma(\frac{2}{\gamma})}{\Gamma(\frac{1}{\gamma})} \qquad (5)$$

Different palm vein images have different AGGD, we can use the parameters $x(\gamma_1, \beta_{l1}, \beta_{\gamma1}, \eta_1, ...\gamma_4, \beta_{l4}, \beta_{\gamma4}, \eta_4)$ as NSS feature to quantify quality.

We choose 120 palm vein image with high quality from the database of PolyU [11] to learn the Multivariate Gaussian model:

$$f_X(x_1, ..., x_k) = \frac{1}{(2\pi)^{m/2}|\Sigma|^{1/2}} \exp(-\frac{1}{2}(X - \mu)^T \Sigma^{-1}(X - \mu)) \qquad (6)$$

Where $(x_1, ..., x_k)$ are the NSS features computed in (4)-(5), and μ and Σ denote the mean and covariance matrix of the MVG model, which are estimated using a standard maximum likelihood estimation procedure. We use Mahalanobis distance to measure the distortion level of palm vein image, the formulation as:

$$Q_h = \sqrt{(y - \mu)^T \Sigma^{-1}(y - \mu)} \qquad (7)$$

Where y stand for the NSS feature of distorted palm vein image. The greater the distance, the worse the image quality.

2.4 Fusion of Quality Indexes

We combine Q_m, Q_n and Q_h to quantify the quality of palm vein image, the Q can be defined as follows:

$$\begin{cases} Q = Q_m * (Q_n * \alpha + Q_h * \beta) \\ \qquad \alpha + \beta = 1 \end{cases} \qquad (8)$$

Where α and β are weights of Q_n and Q_h respectively, we choose $\alpha = \beta = 0.5$. Since Q_m, Q_n and Q_h are inversely proportional to the image quality, Q is inversely proportional to the image quality.

3   Experimental Analysis

We use a vein image acquisition system to obtain palm vein image, the system is consists of three modules: near infrared light source, the cameral, and the filter. The palm vein image of twenty one colleagues are obtained, 60 palm vein images are taken in different situations for each person. The proposed algorithm has been successfully tested on a variety of test images and only a few of the results are shown in this paper. Please note that the score is inversely proportional to the image quality.

In order to learn the effect of uneven illumination caused by palm tilt, a set of ROI images with different tilt angles from same person are chosen for experiment. We compute every index of quality Q, the results are showing in Table 1.

From Table 1 we can find that the Q and Q_m are decrease with the tilt angle of palm increase, while Q_n and Q_h are fluctuating. The results prove that Q_m can well reflect the effect of uneven illumination on image quality caused by palm tilt.

We choose a series of palm vein images with different qualities from same person, these images have different kinds and level of distortion. We have some experiments with these images to test our algorithm, the results are showing in Table 2.

From Table 2, we can find that the results based on proposed algorithm are accordance with human subjective assessment. Form Q_m of image 2 and image 3, we can see that the illumination uniformity of image 2 and image 3 are similar to

Table 1. Results with different tilt angles of palm.

Vein Image						
Tilt Angle	0°	10°	20°	30°	40°	50°
Q_m	0.183	0.206	0.257	0.290	0.369	0.394
Q_n	7.468	7.801	7.714	8.028	7.997	8.520
Q_h	10.257	11.762	13.359	14.263	12.050	12.750
Q	1.622	2.015	2.708	3.232	3.699	4.190

Table 2. Results with different kinds and level of distortion.

Vein Image						
ID	1	2	3	4	5	6
Q_m	0.257	0.357	0.358	0.552	0.544	0.550
Q_n	7.966	7.875	8.040	8.523	8.301	8.297
Q_h	9.657	10.629	15.872	14.788	19.792	23.009
Q	2.261	3.299	4.284	6.431	7.645	8.605

each other. But image 2 include more useful details than image 3, so the quality of image 1 is better than image 3. Image 3 and image 4 have similar Q_h which means they have similar detail information, but they have a great difference on illumination uniformity, so the quality of image 3 is better than image 4.

From [10] we can see that NIQE has a good performance on nature scene pictures. However, it is not suitable for palm vein images. The method presented in this paper combines NIQE and two other indexes which based on characteristics of palm vein images, thus it has a better performance on palm vein images.

4   Conclusion

In this paper, an image quality assessment algorithm for palm vein is presented based on natural scene statistics features. In addition, we use gray mean difference between different parts of palm vein image to estimate the brightness uniformity. Some experiments are given at the end. From the results of experiments we can see that the image quality assessment score based on proposed algorithm is accordance with human subjective assessment.

References

1. Chen, P., Lan, X., Jin, F., Shi, J.: Research of high quality palm vein image acquisition and roi extraction. Chin. J. Sens. Actuators **28**(7), 7 (2015)
2. Kauba, C., Uhl, A.: Robustness evaluation of hand vein recognition systems. In: 2015 International Conference of the Biometrics Special Interest Group (BIOSIG), pp. 1–5. IEEE (2015)
3. Nayar, G.R., Bhaskar, A., Satheesh, L., Kumar, P.S., Aneesh, R.P.: Personal authentication using partial palmprint and palmvein images with image quality measures. In: 2015 International Conference on Computing and Network Communications (CoCoNet), pp. 191–198. IEEE (2015)
4. Sheng, M.-Y., Zhao, Y., Zhang, D.-W., Zhuang, S.-L.: Quantitative assessment of hand vein image quality with double spatial indicators. In: 2011 International Conference on Multimedia Technology (ICMT), pp. 642–645. IEEE (2011)
5. Zhao, Y., Sheng, M.-Y.: Application and analysis on quantitative evaluation of hand vein image quality. In: 2011 International Conference on Multimedia Technology (ICMT), pp. 5749–5751. IEEE (2011)
6. Cui, J.J., Li, Q., Jia, X.: An image quality assessment algorithm for palm-dorsa vein based on multi-feature fusion. In: Advanced Materials Research, vol. 508, pp. 96–99. Trans Tech Publications (2012)
7. Jia, X., Cui, J., Xue, D., Pan, F.: Near infrared vein image acquisition system based on image quality assessment. In: 2011 International Conference on Electronics, Communications and Control (ICECC), pp. 922–925. IEEE (2011)
8. Wang, J., Yu, M., Qu, H., Li, B.: Analysis of palm vein image quality and recognition with different distance. In: 2013 Fourth International Conference on Digital Manufacturing and Automation (ICDMA), pp. 215–218. IEEE (2013)
9. Wang, Y.-D., Yang, C.-Y., Duan, Q.-Y.: Multiple indexes combination weighting for dorsal hand vein image quality assessment. In: 2015 International Conference on Machine Learning and Cybernetics (ICMLC), vol. 1, pp. 239–245. IEEE (2015)
10. Mittal, A., Soundararajan, R., Bovik, A.C.: Making a completely blind image quality analyzer. IEEE Signal Process. Lett. **20**(3), 209–212 (2013)
11. http://www4.comp.polyu.edu.hk/biometric
12. https://en.wikipedia.org/wiki/haar-like_features
13. Lasmar, N.-E., Stitou, Y., Berthoumieu, Y.: Multiscale skewed heavy tailed model for texture analysis. In: 2009 16th IEEE International Conference on Image Processing (ICIP), pp. 2281–2284. IEEE (2009)

An Error-Activation-Guided Blind Metric for Stitched Panoramic Image Quality Assessment

Luyu Yang[1](✉), Jiang Liu[2], and Chenqiang Gao[3]

[1] Kandao Technology, Shenzhen, China
yly@kandaovr.com
[2] Carnegie Mellon University, Pittsburgh, USA
jiang1@cs.cmu.edu
[3] Chongqing Key Laboratory of Signal and Information Processing,
Chongqing University of Posts and Telecommunications, Chongqing, China
gaocq@cqupt.edu

Abstract. Image stitching is one key enabling component for recent immersive VR technology. The quality of the stitched images greatly affects VR experiences. Evaluation of stitched panoramic images using existing assessment tools is insufficient for two reasons. First, conventional image quality assessment (IQA) metrics are mostly full-referenced, while panorama reference is hard to obtain. Second, existing IQA metrics are not designed to detect and evaluate errors typical in stitched images. In this paper, we design an IQA metric for stitched images, where ghosting and shape inconsistency are the most common visual distortions. Specifically, we first locate the error with a fine-tuned convolutional neural network (CNN), and later refine the locations using an error-activation mapping generated from the network. Each located error is defined by both its size and distortion level. Extensive experiments and comparisons confirm the effectiveness of our metric, and indicate the network's remarkable ability to detect error patterns.

Keywords: Image quality assessment · Multi-view synthesis Virtual reality

1 Introduction

Recent rapid development of virtual reality (VR) technologies has led to a new 360-degree look-around visual experience. By displaying stereoscopic 360 scene in head-mounted rigs like Occulus Rift, users can perceive an immersed sensation of reality. Image stitching is typically used to construct a seamless 360 view from multiple captured viewpoint images, and thus the quality of the stitched scene is crucial in determining the level of immersive experience provided. The widely-adopted stitching process [5,22,23] can be broadly divided into the following steps: (i) register the capturing cameras and project each captured scene accordingly, (ii) merge overlapping spatial regions based on corresponding capturing

© Springer Nature Singapore Pte Ltd. 2017
J. Yang et al. (Eds.): CCCV 2017, Part II, CCIS 772, pp. 256–268, 2017.
https://doi.org/10.1007/978-981-10-7302-1_22

camera parameters, and (iii) smooth/blend over the merged scene. Although errors may be introduced at each step, noticeable distortions are usually introduced at the misaligned overlapping regions around the scene objects, which we call *shape breakage*. The following blending step is then employed to alleviate the breakage by imposing a consistency constraint over the entire scene [7,16]. The misaligned scene objects, after being blended, are exposed as ghosting or object fragment [15,19]. Because of the uniqueness of two most common error types during image stitching—shape breakage (including object fragment) and ghosting, stitched image quality assessment is fundamentally different from conventional IQA for images distorted from compression artifacts or network packet losses [17,18,24]. Specifically, conventional IQA methods focus on various noise type's global influence upon visual comfort or information integrity, while in SIQA aims at local distortions that damage object or scene integrity. Further, the most common distortions in conventional IQA come form compression losses, which is not applied to SIQA tasks.

Hence, the process of assessing stitched image quality can be understood as searching for stitched errors over the composed scene—a process of locating

Fig. 1. Comparison of example dataset used for conventional IQA and SIQA experimentations. It is easily observed that conventional IQA samples are evenly distorted over the image, while the SIQA distortions come in local patches. Patch b and c are patches with ghosting error type, a and d are undistorted patches with high image quality.

and assessing particular error types rather than overall assessment of every local spatial region. Thus, it is necessary to study SIQA as a new problem apart from the conventional IQA. Figure 1 illustrates the comparison between some typical samples used for IQA and SIQA. It is clear that most parts of the stitched image have approximately reference quality as in IQA tasks, and the noisy IQA samples do not have prominent local shape distortions as in SIQA tasks; the two groups of samples hardly share a comparable stand.

In this paper, we propose to assess the stitched image quality based on an error-localization-quantification algorithm. First, we detect the potential error regions by searching through the entire stitched image in local patches of unified size, each patch is to be decided as "intact" or "distorted". The decision is based on an intelligent agent trained via a convolutional neural network (CNN) [21]. Then the detected regions are refined to finer regions according to the extent of error. This process is conducted within each potential region in finer pixel patches, which are later retained or removed from the coarse region according to the contributions made towards the region being tagged as distorted. Finally, after obtaining refined regions that well bound the distortions, a quantized metric is formulated on refined patches assessing both the error range and extent.

Contributions: Our contributions are twofold. First, for the SIQA task we propose a new algorithm. The proposed error-localization-quantification metric is simple, straightforward and requires no reference images. Further, our method outputs the explicit locations of error, this is far more meaningful for stitching algorithm optimization than just an evaluation score. Second, the successful localization of multiple error types in our pipeline demonstrates that the CNN is enabled to have remarkable ability to detect spatial patterns, which is beyond scene object detection. The observation implies the possibility for generic classification, localization and concept discovery.

The paper is organized as follows. Section 2 discusses previous related works in SIQA. Section 3 introduces our proposed method. Experimentation is presented in Sect. 4, and Sect. 5 draws the conclusion.

2   Related Work

This paper has two lines of work related to the proposed the method: previous SIQA methods, and deep features for discriminative localization.

Previous SIQA methods: In contrast with the emergence of panoramic techniques, the works to evaluate the stitched panoramic image quality seem insufficient and slow in development. Here, we introduce the previous SIQA methods. Much previous SIQA metrics pay more attention to photometric error assessment [12,13,20] rather than errors caused by misalignment. In [12] and [20], misaligned error types are omitted and the metrics focus on color correction and intensity consistency, which are low-level representation of overall distortion level. [13] try to quantify the error by computing the structure similarity index (SSIM) of high-frequency information of the stitched and unstitched image difference in the overlapping region. However, since unstitched images used for test are directly cropped from the reference, the effectiveness of the method is not validated. In [10], the work pays more attention to assessing video consistency among subsequent frames and only adopted a luminance-based metric around the seam. In [14], the gradient of the intensity difference between the stitched and reference image is adopted to assess the geometric error, however, the experiments are conducted on mere 6 stitched scenes and references, which is in sufficient for a designed metric. We observe that the design of most previous SIQA metrics require full reference [6], which are difficult to obtain in panorama-related applications. Moreover, there seems hardly any SIQA method directly indicates where the distortion is, thus limit the metric's guidance for stitching algorithms.

Differently in our work, the assessment is handled under the error detection algorithm, which directly indicates the location of error and naturally requires no reference, the method is described in the next section.

Deep feature-based discriminative localization: The implementation of Convolutional Neural Networks (CNNs) has led to impressive performance on a variety of visual recognition tasks [8,9,26]. Much recent work show its remarkable ability to localize objects, and the potential of being transferred to other generic classification, localization and concept discovery [2,25]. Most of the related works are based on the weakly-supervised object localization. In [3], the regions that causes the maximal activations are masked out with a self-taught object localization technique. In [11] a method is proposed for transferring mid-level image representations, and achieve object localization by evaluating the CNNs output on patches with overlap. [25] uses the class activation map to refer to the weighted activation maps generated for each image. In [2], a method for performing hierarchical object detection is proposed under the guidance of a deep reinforcement learning agent.

While global average pooling is not a novel technique that we propose here, the observation that it can be applied for nonphysical spatial patter – error localization and the implementation to solve image quality related problems, to the best of our knowledge, is unique to our work. We believe the effectiveness and simplicity of the proposed method will make it generic for other IQA tasks.

3 Proposed Method

The proposed method is to assess the quality of any stitched image and locate distorted regions. We construct it with three steps: Coarse error localization, error-activation-guided refinement, and error quantification.

3.1 Coarse Error Localization

There are two common error types in a stitched scene – ghosting and shape breakage. We employ a ResNet model, a state-of-the-art architecture, to obtain

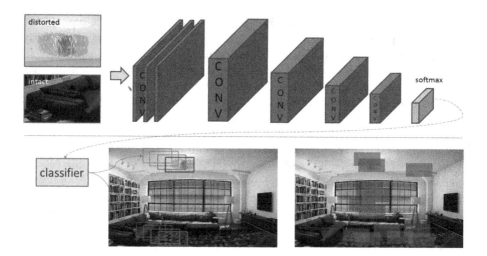

Fig. 2. The coarse error localization pipeline. A ResNet model is truncated and followed by a flatten layer and softmax layer of 2 classes. The trained classifier is applied to a top-down search which categorize local patches from distorted to intact. The detected patches are labeled in red-shadowed bounding boxes.

a two-class classifier between "intact" and "distorted". The fine-tuned model is later utilized for error localization refinement. Even though it is possible for a single patch to hold two types of error at the same time, the detection of each error type is done separately for later assessment. As shown in Fig. 2, we feed the model with labeled bounding boxes containing error as "distorted" examples, and the perfectly aligned areas as "intact". With ResNet, we achieved a remarkable classification accuracy. With the classifier, we coarsely localize the error through the stitched image.

To protect the potential continuous distortion regions, while preserve the fineness of search, we make a trade-off between the window size and sliding step size. In a complex scene constructed by multiple objects, the object volume has prominent effect on visual saliency [1,4]. We assume this also implies to texture patterns like shape breakage or ghosting, thus the integrity of distorted region must be preserved. To this end, we merge the adjacent patches with the same tag, as illustrated in Fig. 2. The merged patches form the coarse error localization.

3.2 Error-Activation Guided Refinement

After obtaining the coarse error, a refined localization that more precisely describe the range of error is required for accurate error descriptions. We find that the class activation mapping considerably discriminative in describing image regions with errors, as a result, we trim the coarse regions with error-activation-guided refinement. The network we fine-tuned for coarse error detection – ResNet architecture largely consists of convolutional layers, similar to [25], we project back the weights of the output layer on to the convolutional feature maps, thus obtaining the importance of each pixel batch that activates a region to be categorized as containing error or no error, the process is called error-activation-mapping.

The error-activation-mapping is obtained by computing the weighted sum of feature maps of the last convolutional layer. For a stitched image with error type T, the error activation mapping E at spatial location (x, y) is computed by Eq. 1:

$$E_T(x,y) = \sum_i \omega_i^T f_i(x,y),\qquad(1)$$

where $f_i(x,y)$ is the activation of unit i in the last convolutional layer at (x,y), and ω_i^T indicates the importance of the global average pooling result for error type T. The score of an image being diagnosed with error type T can be presented in Eq. 2:

$$S_T = \sum_{x,y} \sum_i \omega_i^c f_i(x,y).\qquad(2)$$

Hence the error-activation mapping $E_T(x,y)$ directly represents the importance of the activation at (x, y) leading to the image being diagnosed with error type T. The obtained error-activation mapping will serve as a guidance towards error localization refinement.

For each coarsely localized region, we apply the error-activation mapping as a filter. The threshold is adaptive according to how rigid the filter is, here we adopt the global average. Despite its simplicity, the refinement process integrates the global activation information into the locally categorized patches, which naturally protect the overall integrity of distorted regions. The entire refinement process is as demonstrated in Fig. 3.

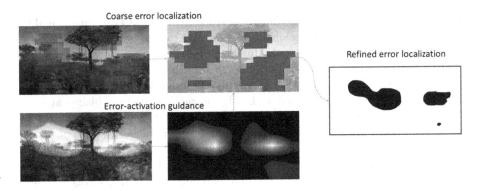

Fig. 3. After the coarse error localization, error-activation mapping further guide the refinement of locally detected error patches. The refined localization better shapes the distorted areas.

3.3   Error Quantification

To quantify the error and form a unified metric, we think it necessary to combine a twofold evaluation, the error range and distortion level. The range is easily represented by the area of the refined location, while the distortion level is represented by the error-activation-mapping weights.

The range index M_r^j of a refined error location j is formulated as follows:

$$M_r^j = A^j/A, \tag{3}$$

where A^j is the area of the refined error location j and A indicates the total area of the image. The distortion level M_d^j of a refined error location j is represented as the sum of error-activation mapping within:

$$M_d^j = \sum_{x,y} E_T(x,y). \tag{4}$$

The quantification of error for location j is represented as:

$$M^j = [M_r^j]_j^\alpha \cdot [M_d^j]_j^\beta, \tag{5}$$

the exponents α_j and β_j are used to adjust the relative importance of range and distortion level. Finally, the quantification of error for an entire stitched image M is formulated as Eq. 6:

$$M = \sum_j M^j = \sum_j [M_r^j]_j^\alpha \cdot [M_d^j]_j^\beta \tag{6}$$

4   Experimentation

Experiment data: All experiments are conducted on our stitched image quality assessment dataset benchmark called SIQA dataset, which is based on synthetic virtual scenes, since we try to evaluate the proposed metric for various stitching algorithms under ideal photometric conditions. The images are obtained by establishing virtual scenes with the powerful 3D model tool—Unreal Engine. As illustrated in Fig. 4. A synthesized 12-head panoramic camera is placed at multiple locations of each scene, covering 360° surrounding view, and each camera has an FOV (field of view) of 90°. Exactly one image is taken for each of the 12 cameras at one location simultaneously. SIQA dataset utilized twelve different 3D scenes varying from wild landscapes to structured scenes, stitched images are obtained using a popular off-the-shelf stitching tool Nuke, altogether 408 stitched scenes, the original images are in high-definition with $3k - by - 2k$ in size.

We label the two error types manually in each scene, a scene might contain multiple regions with a single or both error types, or there might be no distortion at all. For ghosting 297 bounding boxes are labeled, and for shape breakage 220 bounding boxes are labeled.

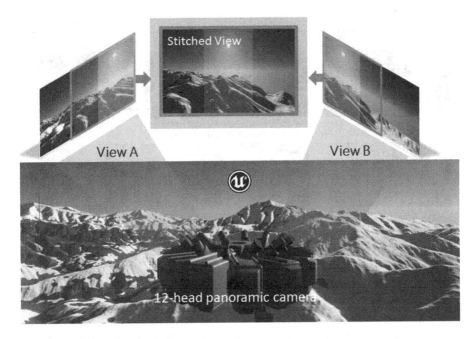

Fig. 4. The 12-head panoramic camera established in a virtual scene using the Unreal Engine, the stitched view is composed of the adjacent camera views with overlap.

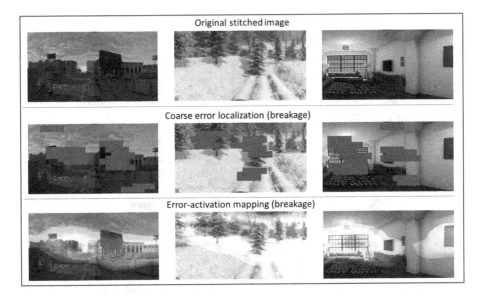

Fig. 5. The coarse error localization result (example error type: breakage), and the corresponding error-activation mapping.

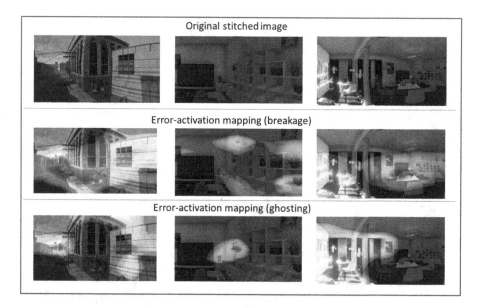

Fig. 6. The error-activation mapping for each error type under the same scene, it is clearly demonstrated that the mapping provide differentiated guidance for each error type.

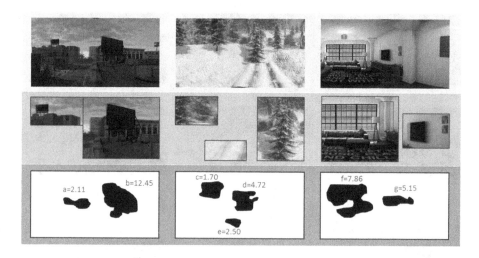

Fig. 7. The refined error localization using the error-activation guidance, and the quantification result M^j of each error location.

Table 1. Test results of classifying each error type using the fine-tuned ResNet architecture.

Error type	Classifier accuracy
Breakage	0.955
Ghosting	0.965

Coarse error localization: Our fine-tuned model to categorize "intact" and "distorted" is the ResNet 50 architecture using Tensor-Flow backed. We truncate the layer after *bn5c branch2c* and follow a flatten layer and a softmax layer of 2 classes. We choose *epoch* = 50 and *batchsize* = 16 as the parameters, the model is fine-tuned separately for the two error types, and the classifier achieves remarkable accuracy of 95.5% for shape breakage and 96.5% for ghosting, as illustrated in Table 1.

The test result of the fine-tuned classifier is illustrated in Table 1. With the remarkable ability to classify distortion and undistorted regions, we impose a top-down search for distorted patches through the entire image. Considering the object size with respect to the image size, we choose 400×400 for ghosting and two window shapes of 200×800 and 800×200 for shape breakage. The differentiated window shapes are chosen according to our analysis. To tell whether a region is ghosted, one must refer to the nearest object to decide where the duplicated artifact comes from, which mostly come in square patches. However, to see if there exists shape breakage, one must refer to the adjacent edge or silhouette to examine whether the shape integrity is damaged, in this case we design both vertical and horizontal window shapes to allow breakage detection. As mentioned earlier, we choose a small sliding step size in order to protect region continuity, here we implement *stepsize* = 100 for both error types. Then we merge the adjacent patches with the same type of error, thus obtaining the coarse localization of error for the entire scene. As Fig. 5 shows, the integrity of continuous distorted region is basically preserved.

Error-activation guided refinement: By projecting back the weights of the output layer on to the convolutional feature maps, we obtain the error-activation mapping for the image, as demonstrated in Fig. 6. We can see that the discriminative regions of the images for each error type is high-lighted. We also observe that the discriminative regions for different error types are different for a given image, this suggest that the error-activation guidance works as expected.

We apply the error-activation guidance as a filter, the input is the coarsely localized regions. The results are quite impressive as Fig. 7 shows. The regions containing errors are prominently refined, which explicitly describe the distorted regions. Based on the properly refined regions, the following quantification is enabled to be reliable defining each error type.

Error quantification: We compute the quantified error for each location of error, and then for the entire image, according to Eqs. 5 and 6 which we introduced in the last section. The relative importance parameters we choose are $\alpha = 1$ and $\beta = 10$. To illustrate the objectiveness of the metric, we make extensive comparisons among the error patches. As demonstrated in Fig. 7, we compare local errors of similar size with differentiated level of distortion, and those with similar score. Location a and g are both from structured scenes with similar size, however, the shape of television in g is much more polluted than a, thus obtaining a relatively higher score of error. Similarly, location f has extensive error range but relatively slight distortion, thus the quantified score is reduced by the distortion level. We also compare the results of various stitched scenes. An interesting observation is that in natural scenes with less structured context, the metric is still capable of locating distortions that are much less noticeable for human vision. The phenomenon reveals the error-localization ability of our method.

5   Conclusion

In this paper we propose an error-activation-guided metric for stitched panoramic image quality assessment, which requires complete no reference. Our method not only provides a proper evaluation of the stitched image quality, but also directly indicates the explicit locations of error. The method is constructed by three main powerful steps: coarse error localization, error-activation-guided refinement and error quantification. Results reveal the error localization ability of the proposed method, and the extensive comparisons also suggest the effectiveness of the our metric and its ability to distinguish minor distortion levels in detail.

Acknowledgements. This work is supported by the National Natural Science Foundation of China (No. 61571071), Wenfeng innovation and start-up project of Chongqing University of Posts and Telecommunications (No. WF201404).

References

1. Achanta, R., Hemami, S., Estrada, F., Susstrunk, S.: Frequency-tuned salient region detection. In: 2009 IEEE Conference on Computer Vision and Pattern Recognition, CVPR 2009, pp. 1597–1604. IEEE (2009)
2. Bellver, M., Giró-i Nieto, X., Marqués, F., Torres, J.: Hierarchical object detection with deep reinforcement learning. arXiv preprint arXiv:1611.03718 (2016)
3. Bergamo, A., Bazzani, L., Anguelov, D., Torresani, L.: Self-taught object localization with deep networks. arXiv preprint arXiv:1409.3964 (2014)
4. Borji, A., Cheng, M.M., Jiang, H., Li, J.: Salient object detection: a benchmark. IEEE Trans. Image Process. **24**(12), 5706–5722 (2015)
5. Brown, M., Lowe, D.G.: Automatic panoramic image stitching using invariant features. Int. J. Comput. Vis. **74**(1), 59–73 (2007)
6. Chen, M.J., Su, C.C., Kwon, D.K., Cormack, L.K., Bovik, A.C.: Full-reference quality assessment of stereopairs accounting for rivalry. Sig. Process. Image Commun. **28**(9), 1143–1155 (2013)

7. Dessein, A., Smith, W.A., Wilson, R.C., Hancock, E.R.: Seamless texture stitching on a 3D mesh by Poisson blending in patches. In: 2014 IEEE International Conference on Image Processing (ICIP), pp. 2031–2035. IEEE (2014)

8. Girshick, R., Donahue, J., Darrell, T., Malik, J.: Rich feature hierarchies for accurate object detection and semantic segmentation. In: Proceedings of the IEEE Conference on Computer Vision and Pattern Recognition, pp. 580–587 (2014)

9. Krizhevsky, A., Sutskever, I., Hinton, G.E.: ImageNet classification with deep convolutional neural networks. In: Advances in Neural Information Processing Systems, pp. 1097–1105 (2012)

10. Leorin, S., Lucchese, L., Cutler, R.G.: Quality assessment of panorama video for videoconferencing applications. In: 2005 IEEE 7th Workshop on Multimedia Signal Processing, pp. 1–4. IEEE (2005)

11. Oquab, M., Bottou, L., Laptev, I., Sivic, J.: Learning and transferring mid-level image representations using convolutional neural networks. In: Proceedings of the IEEE Conference on Computer Vision and Pattern Recognition, pp. 1717–1724 (2014)

12. Paalanen, P., Kämäräinen, J.-K., Kälviäinen, H.: Image based quantitative mosaic evaluation with artificial video. In: Salberg, A.-B., Hardeberg, J.Y., Jenssen, R. (eds.) SCIA 2009. LNCS, vol. 5575, pp. 470–479. Springer, Heidelberg (2009). https://doi.org/10.1007/978-3-642-02230-2_48

13. Qureshi, H., Khan, M., Hafiz, R., Cho, Y., Cha, J.: Quantitative quality assessment of stitched panoramic images. IET Image Process. 6(9), 1348–1358 (2012)

14. Solh, M., AlRegib, G.: MIQM: a novel multi-view images quality measure. In: 2009 International Workshop on Quality of Multimedia Experience, QoMEx 2009, pp. 186–191. IEEE (2009)

15. Szeliski, R.: Image alignment and stitching: a tutorial. Found. Trends® Comput. Graph. Vis. 2(1), 1–104 (2006)

16. Szeliski, R., Uyttendaele, M., Steedly, D.: Fast Poisson blending using multi-splines. In: 2011 IEEE International Conference on Computational Photography (ICCP), pp. 1–8. IEEE (2011)

17. Wang, Z., Bovik, A.C., Sheikh, H.R., Simoncelli, E.P.: Image quality assessment: from error visibility to structural similarity. IEEE Trans. Image Process. 13(4), 600–612 (2004)

18. Wang, Z., Simoncelli, E.P., Bovik, A.C.: Multiscale structural similarity for image quality assessment. In: Conference Record of the Thirty-Seventh Asilomar Conference on Signals, Systems and Computers, 2004, vol. 2, pp. 1398–1402. IEEE (2003)

19. Xiong, Y., Pulli, K.: Fast panorama stitching for high-quality panoramic images on mobile phones. IEEE Trans. Consum. Electron. 56(2), 298–306 (2010)

20. Xu, W., Mulligan, J.: Performance evaluation of color correction approaches for automatic multi-view image and video stitching. In: 2010 IEEE Conference on Computer Vision and Pattern Recognition (CVPR), pp. 263–270. IEEE (2010)

21. Xu, Z., Yang, Y., Hauptmann, A.G.: A discriminative CNN video representation for event detection. In: Proceedings of the IEEE Conference on Computer Vision and Pattern Recognition, pp. 1798–1807 (2015)

22. Zaragoza, J., Chin, T.J., Brown, M.S., Suter, D.: As-projective-as-possible image stitching with moving DLT. In: Proceedings of the IEEE Conference on Computer Vision and Pattern Recognition, pp. 2339–2346 (2013)

23. Zhang, F., Liu, F.: Parallax-tolerant image stitching. In: Proceedings of the IEEE Conference on Computer Vision and Pattern Recognition, pp. 3262–3269 (2014)

24. Zhang, L., Shen, Y., Li, H.: VSI: a visual saliency-induced index for perceptual image quality assessment. IEEE Trans. Image Process. **23**(10), 4270–4281 (2014)
25. Zhou, B., Khosla, A., Lapedriza, A., Oliva, A., Torralba, A.: Learning deep features for discriminative localization. In: Proceedings of the IEEE Conference on Computer Vision and Pattern Recognition, pp. 2921–2929 (2016)
26. Zhou, B., Lapedriza, A., Xiao, J., Torralba, A., Oliva, A.: Learning deep features for scene recognition using places database. In: Advances in Neural Information Processing Systems, pp. 487–495 (2014)

Image Aesthetic Quality Evaluation Using Convolution Neural Network Embedded Fine-Tune

Yuxin Li, Yuanyuan Pu[✉], Dan Xu, Wenhua Qian, and Lipeng Wang

School of Information Science and Engineering, Yunnan University,
Kunming 650504, China
km_pyy@126.com

Abstract. A way of convolution neural network (CNN) embedded fine-tune based on the image contents is proposed to evaluate the image aesthetic quality in this paper. Our approach can not only solve the problem of small-scale data but also quantify the image aesthetic quality. First, we chose Alexnet and VGG_S to compare which is more suitable for image aesthetic quality evaluation task. Second, to further boost the image aesthetic quality classification performance, we employ the image content to train aesthetic quality classification models. But the training samples become smaller and only using once fine-tune can not make full use of the small-scale dataset. Third, to solve the problem in second step, a way of using twice fine-tune continually based on the aesthetic quality label and content label respective, is proposed. At last, the categorization probability of the trained CNN models is used to evaluate the image aesthetic quality. We experiment on the small-scale dataset Photo Quality. The experiment results show that the classification accuracy rates of our approach are higher than the existing image aesthetic quality evaluation approaches.

Keywords: Image aesthetic quality evaluation · Image content
CNN · Embedded fine-tune

1 Introduction

Along with the widespread use of networks and mobile devices such as mobile phone, the number of images increases rapidly. A large number of images are loaded on various social networks every day. To help people exhibit higher aesthetic quality images and explore the aesthetic cognitive ability of computers, image aesthetic quality evaluation is becoming more and more important. Image aesthetic quality evaluation aims to classify the images to high or low aesthetic quality. As shown in Fig. 1, high aesthetic quality images bring more comfortable visual effect to people than the low aesthetic quality images.

In the recent decade, how to make the computers distinguish the image aesthetic quality from the mass images by themselves becomes the main research

© Springer Nature Singapore Pte Ltd. 2017
J. Yang et al. (Eds.): CCCV 2017, Part II, CCIS 772, pp. 269–283, 2017.
https://doi.org/10.1007/978-981-10-7302-1_23

Fig. 1. The first row is high aesthetic quality images, the second row is low aesthetic quality images

direction. The approaches of image aesthetic quality evaluation can be divided into traditional hand-craft [3,4,15,21,22,24,26,32,33] and deep learning CNN [5,8,20,23,30,31,38].

Traditional hand-craft approaches have been aiming at some objective factors of affecting the image aesthetic quality. People have extracted many visual features, including low-level image statistics, such as edge distribution and color histograms, and high-level image graphic rules, such as the rule of thirds and golden ratio. In [3], Datta et al. proposed 56 global features referred to structure, color, light and so on. They used linear regression to quantify the image aesthetics. In [15], Ke et al. proposed a principled approach to design high level global features. They used the perceptual factors that distinguish between professional photos and snapshots to design high level global semantic features and measure the perceptual differences. In [22], Luo et al. began to use local features. They proposed the salient regions from a photo based on professional photography techniques to formulate a number of high-level semantic features based on the quotient of salient and background. The classification rate of their method was 93%. In [21], Luo et al. proposed three classes of local feature and two classes of global feature to automatically classify image aesthetic quality. They used receiver operating characteristic (ROC) curve to prove that these features are excellent for image aesthetic quality evaluation. Shao et al. [26] used Gabor wave transformation, class imbalance and total scene understanding to extract the main part and local features of the different category images. They used hue histogram and color pie to extract global color features at the same time. Dhar et al. [4] mainly analyzed the high-level features that can reflect the image aesthetic quality. They proposed three classes of feature: structure, content and light. They also used this approach to predict the interest feelings of images. In [24], Obrador et al. took focused on the features of structure. They proposed 55 structure features to evaluate the image aesthetics and the classification accuracy was close to the benchmark. In [32,33], Wang et al. proposed 41 features, which referred to structure, color, light, global and local information to evaluate the aesthetic quality of image. Experiment results showed that these 41 features had high classification accuracy of the image aesthetic quality evaluation.

In 2006, Hinton et al. [10] has restarted the research direction of deep learning. In the next few years, CNN has achieved an excellent performance. Motive by the excellent performance in features extracting and autonomous Learning, CNN is wide employed in computer vision and image processing, such as handwritten numerals recognition [17], ImageNet Large Scale Visual Recognition Competition (ILSVRC) [9,16,27,29], object detection and semantic segmentation [7,19,25], face recognition [28,37], emotion recognition [2,18,35], image style recognition and transformation [6,13], sentiment analysis [36]. Researchers are trying to use CNN for the aesthetic calculation and connotation exploration [5,8,20,23,30,31,38]. In [20], Lu et al. proposed a novel double-column CNN and used global view and fine-grained view to train. Their approach produced significantly 10% better results than the results achieved by hand-craft features on the AVA dataset [23]. Guo et al. [8] proposed PDCNN, a Paralleled CNN architecture. The results of their approach were better than the results of [21] on Photo Quality. The PDCNN can also overcome the problem of over-fitting and under-fitting. In [38], Zhou et al. used N-grams to describe the image text feature and used SVM to learn the weights of N-grams. Then they joined the text features extracted by N-grams into the visual features extracted by CNN, to get excellent classification accuracy on AVA dataset [23]. Besides, [5,30,31] use CNN to evaluate image aesthetic quality too.

At present, in many vision tasks, the main problem of using CNN is that the data scale is small, comparing with the large-scale dataset ImageNet which has million images. The problem also lingers in image aesthetic quality evaluation. The best ways of solving the problem are data-augmentation and fine-tune at present. Fine-tune from the pre-trained models on ImageNet is found to yield state-of-the-art performance for many vision tasks such as visual tracking [34], action recognition [14], object recognition [25], human pose estimation [19]. Therefore, this paper mainly use fine-tune to evaluate the images aesthetic quality. The main innovations and contributions in this paper can be summarized as follows:

1. Image contents are employed to further boost the classification performance. We use fine-tune based on the image content to train the aesthetic quality classification models and analyze the influence of the image contents to aesthetic quality classification.
2. Embedded fine-tune is proposed to solve the training data becoming smaller problem due to the limitation of image contents and further boost image classification performance. Embedded fine-tune has twice fine-tune and each fine-tune has different training samples. The results of the experiment show that embedded fine-tune improves the image aesthetic quality classification accuracy rates.
3. The classification accuracy probability of the trained CNN models is used to make specific evaluation to image aesthetic quality, break up the situations that images only have high or low aesthetic label.

2   Relate Work

2.1   Alexnet and VGG_S Architecture

Feature learning is unified with classifier training using RGB images in CNN. As shown in Fig. 2 and Table 1, Alexnet [16] is the champion in ILSVRC2012. It has five convolution layers, two pooling layers and two full connect layers. The input image is divided into R, G, B three channels and the size of each channel is initialized 256 × 256. Rectified Linear Unit (ReLU) as the active function is used in Alexnet. Local Response Normalization (LRN) following ReLU aids generalization.

Data augmentation and Dropout are used to reduce over-fitting in Alexnet. The first data-augmentation way is that 256 × 256 input images are cut to ten 227 × 227 patches from four corners and the center and then mirrored them horizontally. These ten patches are predicted by the network's softmax layer averagely. The second data-augmentation way is that alter the intensities of the R, G, B channels in training images. Specifically, PCA is performed on the set of RGB pixel values. Multiples of the found principal components with a random variable drawn from a Gaussian with mean 0 and standard deviation 0.1 are added to each training image.

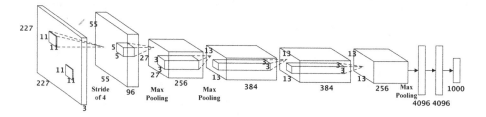

Fig. 2. The parameters and architecture of Alexnet

Table 1. The comparison between Alexnet and VGG_S

Arch	Conv1	Conv2	Conv3	Conv4	Conv5	Fc6	Fc7	Fc8
Alexnet	96 × 11 × 11	256 × 5 × 5	384 × 3 × 3	384 × 3 × 3	384 × 3 × 3	–	–	lr:10
	Stride 4	Stride 1	Stride 1	Stride 1	Stride 1	–	–	base_lr: 20
	padding 0	padding 2	padding 1	padding 1	padding 1	–	–	–
	ReLU	ReLU	ReLU	ReLU	ReLU	ReLU	ReLU	ReLU
	LRN	LRN	–	–	–	Dropout	Dropout	Softmax
	3 × 3 pool	3 × 3 pool	–	–	3 × 3 pool	4096	4096	2
VGG_S	96 × 7 × 7	256 × 5 × 5	512 × 3 × 3	512 × 3 × 3	512 × 3 × 3	–	–	lr:10
	Stride 2	Stride 1	Stride 1	Stride 1	Stride 1	–	–	base_lr: 20
	padding 0	padding 1	padding 1	padding 1	padding 1	–	–	–
	ReLU	ReLU	ReLU	ReLU	ReLU	ReLU	ReLU	ReLU
	LRN	–	–	–	–	Dropout	Dropout	Softmax
	3 × 3 pool	3 × 3 pool	–	–	3 × 3 pool	4096	4096	2

In [11], Hiton et al. proposed dropout. If it's set 0, the output probability of each hidden layer is 0.5. Then the neurons are "dropped out" in this way and they don't contribute to the forward propagation or the back-propagation.

In [1], Chatfield et al. proposed VGG_S which also has five convolution layers, three pooling layers and two full connect layers. The differences are that the size of the extracted patches is changed from 227×227 to 224×224 and the kernel size, stride of the first convolution layer are smaller. But the error rate of VGG_S on ImageNet is lower than that of Alexnet.

2.2 Fine-Tune from ImageNet

As Fig. 3 shows, fine-tune takes a well pre-trained model, adapts the same architecture and retrains on the pre-trained model weights. It reflects a kind of semantic transfer from general to specific. Fine-tuning the last layer of CNN, the parameter is set higher than other layers because this layer is starting from random while the others are already trained and the output number is set to satisfy the target task when we use fine-tune. With the help of the excellent existing models, fine-tune will save much resource in some new researches.

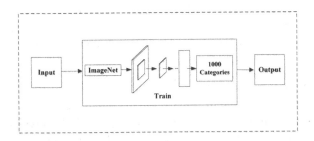

Fig. 3. Fine-tune from the large-scale dataset ImageNet

3   The Proposed Approaches

3.1   Comparing Alexnet with VGG_S

To compare Alexnet and VGG_S which is more suitable for image aesthetic quality evaluation task and boost the image aesthetic quality evaluation classification performance, first, we mix Animal, Architecture, Human, Landscape, Night, Plant and Static images of the dataset together as training samples which are divided into high and low aesthetic quality. Second, we train Alexnet_All and VGG_S_All based on all training samples. Third, to get Alexnet_FT_All and VGG_S_FT_All, we fine-tune from Alexnet_Model and VGG_S_Model which are trained on ImageNet. At last, we use the test samples of each kind of content to test these four models respectively and find out which architecture is more suitable for image aesthetic quality classification task. The way of comparing Alexnet and VGG_S is described in Algorithm 1.

Algorithm 1. Comparing Alexnet with VGG_S

Training:
Input:

$X = \{x_1, x_2, x_3 \cdots x_n\}$ a set of training images of size 256×256;

$Y = \{y_1, y_2, y_3 \cdots y_n\}$ image aesthetic quality labels of X;

1: Get Alexnet_Model and VGG_S_Model trained on the ImageNet;

2: Train Alexnet_All and VGG_S_All with X and Y;

3: Fine-tune Alexnet_Model and VGG_S_Model respectively with X and Y to train Alexnet_FT_All and VGG_S_FT_All;

Testing and Comparing:

Input: $X' = \{x'_1, x'_2, x'_3 \cdots x'_k\}$ a set of animal/ architecture/ human/ landscape/ night/ plant/ static testing images;

$Y' = \{y'_1, y'_2, y'_3 \cdots y'_k\}$ image aesthetic quality label of X';

4: Test Alexnet_FT_All, VGG_S_FT_All, Alexnet_All and VGG_S_All with X' and Y' to achieve the classification accuracy of each category;

5: Compare the classification accuracy rate of Alexnet_FT_All, VGG_S_FT_All, Alexnet_All and VGG_S_All;

3.2 Training Models Based on Image Contents

To further improve the image aesthetic quality evaluation classification performance, as Fig. 4 shows, we use fine-tune to train seven models with the training samples of seven kinds of content based on Alexnet_model or VGG_S_Model. The inside box is training Alexnet_model or VGG_S_Model and the outside box is training the models based on the image content. The CNN architecture used in training is the better one from Alexnet and VGG_S. We use CNN_i to represent the seven models we trained in all, where the i represents Animal, Architecture, Human, Landscape, Night, Plant or Static and CNN represents Alexnet or VGG_S. We also use testing samples of each kind of content to test the models of the same content.

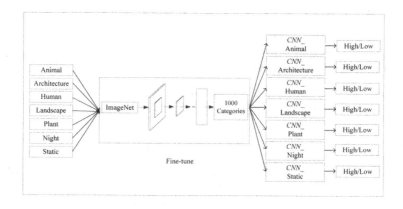

Fig. 4. Train models based on the image content

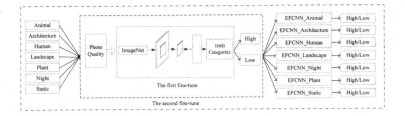

Fig. 5. Use embedded fine-tune to train models

3.3 Using Embedded Fine-Tune to Train Models

When we train the aesthetic quality classification models based on image contents, the training samples become smaller, so we propose embedded fine-tune. As Fig. 5 shows, the inside dashed box shows Alexnet_FT_All or VGG_S_FT_All training process. The outside dashed box shows fine-tune from Alexnet_FT_All or VGG_S_FT_All with training samples of each kind of content. We use EFCNN_i to represent the seven models we trained, where the i also represents Animal, Architecture, Human, Landscape, Night, Plant or Static and EFCNN represents Embedded Fine-tune Convolution Neural Network. The difference between these two fine-tune is that the first fine-tune uses the aesthetic quality label and the second fine-tune uses both aesthetic quality label and content label. The algorithm is as follows:

Algorithm 2. Training EFCNN models

Input:

$X = \{x_1, x_2, x_3 \cdots x_n\}$ a set of training images of size 256×256;

$Y = \{y_1, y_2, y_3 \cdots y_n\}$ image aesthetic quality labels of X;

$X' \subset X$ be a set of Animal /Architecture /Human /Landscape /Night /Plant /Static images;

Y' is their aesthetic quality labels;

1: Get VGG_S_Model trained on the Imagenet;

2: Fine-tune VGG_S_Model with X and Y to create VGG_S_FT_All;

3: Use Embedded Fine-tune to train EFCNN_Animal /Architecture /Human /Landscape /Night /Plant /Static respectively based on VGG_S_FT_All with input X' and Y';

3.4 Image Aesthetic Quality Evaluation

The traditional way of image aesthetic quality evaluation only divides the images into high or low aesthetic quality. We will score images to make the image aesthetic quality evaluation more specific. The score is calculated by the output probability of trained models. We use Softmax with loss to calculate the probability at last, and use the probability to score the image aesthetic quality.

4   Experiment Results and Analysis

4.1   Photo Quality Dataset and Experiment Platform

In this paper, the images used in experiment come from a image dataset named photo quality. Photo Quality built by [21] is a small-scale dataset for photo aesthetic quality evaluation. The images are from a website which photos are taken by professional and amateurs photographers. A total of 17,613 photos are divided into high or low aesthetic quality after eight of the ten observers make the same judgment. According to the photo contents, all photos are divided into 7 categorization, named Animal, Architecture, Human, Landscape, Night, Plant and Static. As Table 2 shows, we filtered the dataset again and chose 15,562 images at last.

Table 2. Number of high quality and low quality images of seven contents in Photo Quality

Number	Contents						
	Animal	Architecture	Human	Landscape	Night	Plant	Static
High quality images	991	533	751	753	416	560	430
Low quality images	2184	1043	1523	1474	1356	1649	1899

We use Caffe, a professional deep learning platform developed by Jia [12], as our experiment platform. Our GPU is GTX 1070 with 8G memory. In order to ensure the generality of the experiment results, we use three cross approach.

4.2   The Comparison Results Between Alexnet and VGG_S

For each content, we randomly extract 300 low aesthetic quality images and 100 high quality aesthetic images from photo quality as our testing samples. The rest of images are training samples. And we use the data-augmentation way of Alexnet to expand the training samples. As Table 3 shows, the classification results of Alexnet_FT_All and VGG_S_FT_All are better than Alexnet_All and VGG_S_All, and the classification accuracy rate of VGG_S_FT_All is an average

Table 3. The comparison results between Alexnet and VGG_S

Models	Contents							
	Animal	Arch	Human	Landscape	Night	Plant	Static	Overall
Alexnet_ALL	0.9307	0.8542	0.929	0.8831	0.89	0.8825	0.8695	0.8913
VGG_S_ALL	0.9454	0.8907	0.944	0.8963	0.901	0.8865	0.8921	0.9093
Alexnet_FT_ALL	0.9543	0.8954	0.95	0.9056	0.9107	0.8987	0.8994	0.9163
VGG_S_FT_ALL	**0.9552**	**0.9121**	**0.9575**	**0.9100**	**0.92**	**0.9115**	**0.911**	**0.925**

of 0.87% higher than Alexnet_FT_All. These comparison results show that we can use fine-tune to improve the classification accuracy rate and VGG_S is more suitable for image aesthetic quality evaluation no matter with or without fine-tune. So we will use VGG_S in next experiments.

4.3 The Results Based on the Image Contents

We use the same training, testing samples and data-augmentation as the experiment of comparing Alexnet with VGG_S. Table 4 shows the comparison results between using image contents and not using image contents. VGG_S_FT_i is trained by VGG_S architecture where the i represents Animal, Architecture, Human, Landscape, Night or Static. The results show that the classification accuracy rate of each image content is higher than VGG_S_FT_All. It proves that training models based on image content can boost the classification performance. Although the training sample size is reduced, the quality of training samples is higher, the categorization accuracy rate is improved.

We also build a confuse matrix to prove the image content is an important factor in image aesthetic quality evaluation. As Fig. 6 shows, the highest clas-

Table 4. The results of image aesthetic quality based on image contents

Models	Contents						
	Animal	Arch	Human	Landscape	Night	Plant	Static
VGG_S_FT_All	0.9552	0.9121	0.9575	0.9100	0.92	0.9115	0.911
VGG_S_FT_i	**0.9677**	**0.9297**	**0.9596**	**0.943**	**0.9244**	**0.9189**	**0.9219**

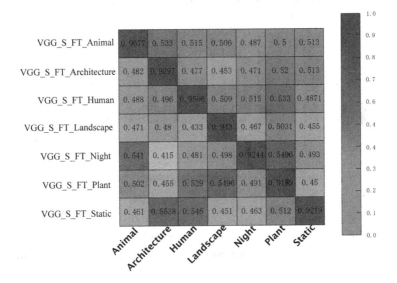

Fig. 6. Confuse matrix of image aesthetic quality evaluation based on image content

sification accuracy rates of the matrix are on its diagonal while the others are much lower. Training models based on image content and using the same content image to testing models can improve the classification accuracy rate. It proves not only that the importance of image content in training process but also the necessity of classification according to image content before we use the model to classify images.

4.4   The Results of Embedded Fine-Tune

Embedded fine-tune is proposed through summarizing conclusions and problems of the first two experiments and its training/testing samples and data-augmentation are same as the first two experiments. We compare the results of embedded fine-tune with the first two experiments and then compare with the existing traditional approaches and CNN approaches.

As Fig. 7 and Table 5 show, the classification accuracy of EFCNN_i is average higher 0.88% than VGG_S_FT_i. The EFCNN_Plant is the highest with the classification accuracy of 1.71% higher than VGG_S_FT_Plant. The classification accuracy of EFCNN_i is average higher 2.08% than VGG_S_FT_All. The EFCNN_Landscape is the highest with the classification accuracy of 4.2% higher than VGG_S_FT_All. Embedded fine-tune can improve the classification performance because the second fine-tune is based on a binary classification problem as the same as the image aesthetic quality classification task. Besides, image contents are jointed to embedded fine-tune and embedded fine-tune makes full use of the small-scale dataset. In Fig. 8, we visualize the first convolution layer to show that embedded fine-tune can improve the classification accuracy rate. The textures of feature extracted by embedded fine-tune are clearer than those textures extracted no using embedded fine-tune.

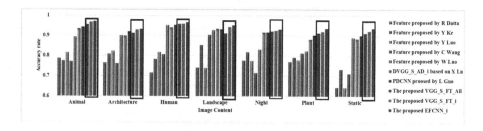

Fig. 7. The histogram of the comparison results

Traditional approaches are more complex and they depend on the features which have been designed according to a particular dataset. The classification accuracies of each content of embedded fine-tune are much better than those results of [3, 15, 21, 22, 32, 33]. The gap between traditional approaches and the CNN approaches is growing ever wider.

Table 5. The comparison results between embedded fine-tune and traditional and CNN approaches

Approaches	Contents						
	Animal	Arch	Human	Landscape	Night	Plant	Static
Features proposed by [3]	0.786	0.7638	0.7147	0.7386	0.7753	0.7694	0.6421
Features proposed by [15]	0.7751	0.8093	0.7829	0.8526	0.817	0.7908	0.7321
Features proposed by [22]	0.8161	0.8238	0.8174	0.7386	0.7753	0.7794	0.6421
Features proposed by [32,33]	0.772	0.761	0.8067	0.903	0.714	0.812	0.711
Features proposed by [21]	0.8937	0.9004	0.9527	0.9273	0.8309	0.8238	0.889
DVGG_S_AD_i	0.9354	0.898	0.941	0.9338	0.9156	0.881	0.8834
PDCNN proposed by [8]	0.9423	0.9188	0.9525	0.93	0.916	0.902	0.8988
Our VGG_S_FT_All	0.9552	0.9121	0.9575	0.91	0.92	0.9115	0.911
Our VGG_S_FT_i	0.9677	0.9297	0.9596	0.943	0.9244	0.9189	0.9219
Our EFCNN_i	**0.9712**	**0.9325**	**0.966**	**0.952**	**0.93**	**0.936**	**0.935**

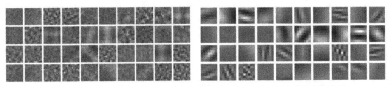

(a) Before using embedded fine-tune   (b) After using embedded fine-tune

Fig. 8. The visualization of the first convolution layer

We also compare embedded fine-tune with the existing CNN approaches. DVGG_S_i (Double VGG_S Adaption) is that we double the VGG_S and use the parallel networks adaption method proposed by Lu [20] where the i represents Animal, Architecture, Human, Landscape, Night or Static and PDCNN (Parallel Deep Convolution Neutral Network) proposed by Guo [8].

The categorization accuracy of EFCNN_i is an average of 2.33% higher than that of PDCNN. The EFCNN_Static is the highest with the categorization accuracy of 3.62% higher than PDCNN. The classification accuracy of EFCNN_i is on average 3.35% higher than that of DVGG_S_AD_i. The EFCNN_Plant is the highest with the classification accuracy of 5.5% higher than DVGG_S_AD_Plant.

PDCNN and DVGGS_AD_i use parallel network architectures to train models. The features extracted by the paralleled CNN architecture from the small-scale dataset are limit too because the data scale is too small to offer more features unless expand the number of the data. Fine-tune from the pre-trained on the large-scale dataset can get more useful features and transfer these features to the target tasks. Embedded fine-tune can not only get the features from the large-scale dataset but also from the target task dataset. So embedded fine-tune can break up the limitation of small-scale data.

Fig. 9. The scores of image aesthetic quality. The two columns on the left are high aesthetic quality images and the two columns on the right are low aesthetic quality images (Color figure online)

4.5 Image Aesthetic Quality Evaluation

We use the classification probability of the trained EFCNN_i to score the image aesthetic quality. The score represents the level of high or low image aesthetic quality. As Fig. 9 shows, blue represents the high aesthetic quality score probability and orange represents the low. If high aesthetic quality score is higher than the low, this image is a high aesthetic quality image. Otherwise it's a low aesthetic quality image.

5 Conclusions

This paper analyzes the effect of image content to image aesthetic quality evaluation. We propose embedded fine-tune to solve the problems that data become smaller and using once fine-tune can not make full use of all data of small-scale dataset when we train the image aesthetic quality classification models. The experiment results show that embedded fine-tune can solve the problem of small-scale and boost the image aesthetic quality evaluation performance. At last, classification probability is used to evaluate the image aesthetic quality. The evaluation makes image aesthetic quality more specific.

Acknowledgments. It is a project supported by Natural Science Foundation of China (No. 61271361, 61163019, 61462093, 61761046), the Research Foundation of Yunnan Province (2014FA021, 2014FB113), and Digital Media Technology Key Laboratory of Universities in Yunnan.

References

1. Chatfield, K., Simonyan, K., Vedaldi, A., Zisserman, A.: Return of the devil in the details: delving deep into convolutional nets. Comput. Sci. (2014)
2. Chu, X., Ouyang, W., Yang, W., Wang, X.: Multi-task recurrent neural network for immediacy prediction. In: IEEE International Conference on Computer Vision, pp. 3352–3360 (2015)
3. Datta, R., Joshi, D., Li, J., Wang, J.Z.: Studying aesthetics in photographic images using a computational approach. In: European Conference on Computer Vision, pp. 288–301 (2006)
4. Dhar, S., Ordonez, V., Berg, T.L.: High level describable attributes for predicting aesthetics and interestingness. IEEE Comput. Soc. **42**(7), 1657–1664 (2011)
5. Dong, Z., Shen, X., Li, H., Tian, X.: Photo quality assessment with DCNN that understands image well. In: He, X., Luo, S., Tao, D., Xu, C., Yang, J., Hasan, M.A. (eds.) MMM 2015. LNCS, vol. 8936, pp. 524–535. Springer, Cham (2015). https://doi.org/10.1007/978-3-319-14442-9_57
6. Gatys, L.A., Ecker, A.S., Bethge, M.: Image style transfer using convolutional neural networks. In: IEEE Conference on Computer Vision and Pattern Recognition, pp. 2414–2423 (2016)
7. Girshick, R.B., Donahue, J., Darrell, T., Malik, J.: Rich feature hierarchies for accurate object detection and semantic segmentation. In: Computer Vision and Pattern Recognition, pp. 580–587 (2013)

8. Guo, L., Li, F.: Image aesthetic evaluation using paralleled deep convolution neural network. Comput. Sci. (2015)
9. He, K., Zhang, X., Ren, S., Sun, J.: Deep residual learning for image recognition. In: IEEE Conference on Computer Vision and Pattern Recognition, pp. 770–778 (2016)
10. Hinton, G.E., Osindero, S., Teh, Y.W.: A fast learning algorithm for deep belief nets. Neural Comput. 18(7), 1527–1554 (2014)
11. Hinton, G.E., Srivastava, N., Krizhevsky, A., Sutskever, I., Salakhutdinov, R.R.: Improving neural networks by preventing co-adaptation of feature detectors. Comput. Sci. 3(4), 212–223 (2012)
12. Jia, Y., Shelhamer, E., Donahue, J., Karayev, S., Long, J., Girshick, R.B., Guadarrama, S., Darrell, T.: Caffe: convolutional architecture for fast feature embedding. CoRR (2014)
13. Karayev, S., Trentacoste, M., Han, H., Agarwala, A., Darrell, T., Hertzmann, A., Winnemoeller, H.: Recognizing image style. Comput. Sci. (2013)
14. Karpathy, A., Toderici, G., Shetty, S., Leung, T., Sukthankar, R., Li, F.F.: Large-scale video classification with convolutional neural networks. In: Computer Vision and Pattern Recognition, pp. 1725–1732 (2014)
15. Ke, Y., Tang, X., Jing, F.: The design of high-level features for photo quality assessment. In: IEEE Computer Society Conference on Computer Vision and Pattern Recognition, pp. 419–426 (2006)
16. Krizhevsky, A., Sutskever, I., Hinton, G.E.: Imagenet classification with deep convolutional neural networks. In: International Conference on Neural Information Processing Systems, pp. 1097–1105 (2012)
17. Lecun, Y., Boser, B., Denker, J.S., Henderson, D., Howard, R.E., Hubbard, W., Jackel, L.D.: Backpropagation applied to handwritten zip code recognition. Neural Comput. 1(4), 541–551 (2014)
18. Levi, G., Hassner, T.: Emotion recognition in the wild via convolutional neural networks and mapped binary patterns. In: ACM on International Conference on Multimodal Interaction, pp. 503–510 (2015)
19. Long, J., Shelhamer, E., Darrell, T.: Fully convolutional networks for semantic segmentation. IEEE Trans. Pattern Anal. Mach. Intell. 39(4), 640 (2014)
20. Lu, X., Lin, Z., Jin, H., Yang, J., Wang, J.Z.: Rapid: rating pictorial aesthetics using deep learning. IEEE Trans. Multimed. 17(11), 2021–2034 (2015)
21. Luo, W., Wang, X., Tang, X.: Content-based photo quality assessment. In: IEEE International Conference on Computer Vision, pp. 2206–2213 (2011)
22. Luo, Y., Tang, X.: Photo and video quality evaluation: focusing on the subject. In: European Conference on Computer Vision, pp. 386–399 (2008)
23. Murray, N., Marchesotti, L., Perronnin, F.: AVA: a large-scale database for aesthetic visual analysis. In: Computer Vision and Pattern Recognition, pp. 2408–2415 (2012)
24. Obrador, P., Schmidt-Hackenberg, L., Oliver, N.: The role of image composition in image aesthetics. In: IEEE International Conference on Image Processing, pp. 3185–3188 (2010)
25. Ouyang, W., Loy, C.C., Tang, X., Wang, X., Zeng, X., Qiu, S., Luo, P., Tian, Y., Li, H., Yang, S.: DeepID-Net: deformable deep convolutional neural networks for object detection. IEEE Trans. Pattern Anal. Mach. Intell. PP(99), 1 (2016)
26. Shao, J., Zhou, Y.: Photo quality assessment in different categories. J. Comput. Inf. Syst. 9(8), 3209–3217 (2013)
27. Simonyan, K., Zisserman, A.: Very deep convolutional networks for large-scale image recognition. Comput. Sci. (2014)

28. Sun, Y., Wang, X., Tang, X.: Deep learning face representation from predicting 10,000 classes. In: IEEE Conference on Computer Vision and Pattern Recognition, pp. 1891–1898 (2014)
29. Szegedy, C., Liu, W., Jia, Y., Sermanet, P., Reed, S., Anguelov, D., Erhan, D., Vanhoucke, V., Rabinovich, A.: Going deeper with convolutions. In: Computer Vision and Pattern Recognition, pp. 1–9 (2015)
30. Tian, X., Dong, Z., Yang, K., Mei, T.: Query-dependent aesthetic model with deep learning for photo quality assessment. IEEE Trans. Multimed. **17**(11), 2035–2048 (2015)
31. Veerina, P.: Learning good taste: classifying aesthetic images. Technical report, Stanford University (2015)
32. Wang, C., Pu, Y., Xu, D., Zhu, J., Tao, Z.: Evaluating aesthetics quality in portrait photos. J. Softw. 20–28 (2015)
33. Wang, C., Pu, Y., Xu, D., Zhu, J., Tao, Z.: Evaluating aesthetics quality in scenery images. In: Proceeding of National Conference on Multimedia Technology, pp. 141–149 (2015)
34. Wang, L., Ouyang, W., Wang, X., Lu, H.: Visual tracking with fully convolutional networks. In: IEEE International Conference on Computer Vision, pp. 3119–3127 (2016)
35. You, Q., Yang, J.: Building a large scale dataset for image emotion recognition: the fine print and the benchmark. In: Thirtieth AAAI Conference on Artificial Intelligence, pp. 308–314 (2016)
36. You, Q., Yang, J., Yang, J., Yang, J.: Robust image sentiment analysis using progressively trained and domain transferred deep networks. In: Twenty-Ninth AAAI Conference on Artificial Intelligence, pp. 381–388 (2015)
37. Zhang, Z., Luo, P., Chen, C.L., Tang, X.: Facial landmark detection by deep multitask learning. In: European Conference on Computer Vision, pp. 94–108 (2014)
38. Zhou, Y., Lu, X., Zhang, J., Wang, J.Z.: Joint image and text representation for aesthetics analysis. In: ACM on Multimedia Conference, pp. 262–266 (2016)

High Capacity Reversible Data Hiding
with Contrast Enhancement

Yonggwon Ri[1(✉)], Jing Dong[1,2(✉)], Wei Wang[1], and Tieniu Tan[1]

[1] National Laboratory of Pattern Recognition, Institute of Automation, Chinese
Academy of Sciences, Beijing 100190, China
yg.ri@cripac.ia.ac.cn, {jdong,wwang,tnt}@nlpr.ia.ac.cn
[2] State Key Laboratory of Information Security, Institute of Information
Engineering, Chinese Academy of Sciences, Beijing 100093, China

Abstract. Reversible data hiding aims at recovering exactly the cover
image from the marked image after extracting the hidden data.
Reversible data hiding with contrast enhancement proposed by *Wu et al.*
achieved a good effect in improving visual quality with considerable
embedding capacity while PSNR of the marked image is relatively low.
In contrast, Prediction error based reversible data hiding does not reveal
obvious change of visual quality while keeping high embedding capacity and PSNR. In this paper, we propose a novel reversible data hiding
method with contrast enhancement based on the combination property
of the above two methods.

Keywords: Reversible data hiding · Contrast enhancement
Prediction error · Watermarking

1 Introduction

With the increase of data exchanges on the internet, data hiding is essentially
important for applications such as copyright protection, secret communication,
authentication and so on. So far, most of works have been focused on digital
images and the priority demand of the embedding system is visual quality of
marked image produced by embedding secret data into original (cover) image.

In the recent years, reversible data hiding (RDH) has been widely investigated
as a special kind of data hiding technique which requires not only extraction
of hidden data but also exact restoration of the cover image from the marked
image [11]. Two most important metrics in evaluating the RDH performance are
embedding capacity (EC) and marked image quality quantified by bit per pixel
(bpp) and Peak Signal to Noise Rate (PSNR) respectively [13,16].

Up to now, many RDH algorithms have been proposed. Nevertheless, in the
technical sense of application, RDH has been supported by two pillars of RDH
with contrast enhancement (RDH-CE) and prediction error expansion based
RDH (RDH-PE) with high capacity and low distortion among the current mechanisms, performing differently than the inherent data hiding in which slight
degradation is tolerable.

© Springer Nature Singapore Pte Ltd. 2017
J. Yang et al. (Eds.): CCCV 2017, Part II, CCIS 772, pp. 284–294, 2017.
https://doi.org/10.1007/978-981-10-7302-1_24

Histogram shifting based RDH, which is considered as one of typical RDH techniques, was first proposed by *Ni et al.* [11]. After that, this technique has been extended to various applications, one of which is the RDH with contrast enhancement (RDH-CE) proposed by *Wu et al.* [16]. In the RDH-CE method, histogram equalization is realized during the embedding procedure due to histogram bin shifting and splitting, which results in the image contrast enhancement. As a result, the *Wu's* method has been recognized as a useful technique in improving visual quality especially for poorly illustrated images including medical and military map images as image contrast is one of the main factors in evaluating the image quality. However, the EC and PSNR of the RDH-CE are relatively low because using more peak pairs to increase the EC may result in over-enhanced marked image and rapid descent of PSNR curve. There exist several works for improving the performance of RDH-CE [1–3,7,8,12]. They archived a considerable improvement in terms of PSNR and EC while avoiding visual distortion.

Nowadays, PE based RDH (RDH-PE) has been extensively exploited in recent years due to the higher performance compared with RDH based on expanding difference values of pixel pairs for embedding secret data [13]. The RDH-PE provides the best embedding performance among the existing RDH techniques in terms of EC and PSNR [4,5]. Many researches have been focused on improving the statistical property of prediction errors [2,6,9,10,15]. Although the PSNR resulted by RDH-PE is kept high, the improvement of visual quality is not observed easily in the sense of human visual perception.

The inherent properties of the above two methods motivate us to seek an effective way to archive satisfactory embedding performance, i.e., the high embedding capacity keeping the good effect of image contrast enhancement. In this paper, we propose a novel RDH-CE method with high capacity by determining two characteristic parameters for the combination of RDH-CE and RDH-PE based on analyzing the change property of its relative contrast error (RCE) [4].

The rest of this paper is organized as follows. The performance properties of RDH-CE and RDH-PE are analyzed in Sect. 2, and the detail of the proposed method is presented in Sect. 3. Finally, experiment and conclusion are given in Sects. 4 and 5 respectively.

2   Related Work

In this section, the RDH-CE algorithm proposed in [16] is first explained briefly, and then the change properties of image contrast caused by RDH-CE and RDH-PE are analyzed with experimental results.

2.1   *Wu's* RDH-CE Method

The RDH-CE is performed by pair-wisely expanding the highest two peak bins in histogram of an image. Two bins with the highest frequency values in histogram G for a given image, denoting the left and right bins by G_a and G_b, are chosen and split outward in each time for embedding secret data. The embedding procedure is performed according to Eq. 1,

$$
i' = \begin{cases}
i - 1 & \text{for } i < G_a \\
G_a - m_k & \text{for } i = G_a \\
i & \text{for } G_a < i < G_b \\
G_b + m_k & \text{for } i = G_b \\
i + 1 & \text{for } i > G_b
\end{cases} \tag{1}
$$

where m_k is the k-th binary bit of the data to be embedded, i and i' are the original and modified pixel values respectively. To avoid overflow or underflow by histogram shifting, a location map is used to record locations of pre-shifted pixels in pre-processing. The procedures of data extraction and image recovery start with extracting the last peak-pair, and continue to extract the embedded data and recover the cover image by applying inversely the embedding procedure. While the embedding procedure usually repeats several times for increasing EC, the range of existing histogram bins is widely extended by histogram shifting, and their heights get similar gradually by histogram bin splitting. Figure 1 shows an example of the over-enhanced image and its histogram. As seen in the figure, the marked image is over-enhanced by the continuous process of histogram equalization, by which the EC in RDH-CE is restrained. The visual quality of marked image is usually evaluated by relative contrast error (RCE) rather than in terms of PSNR.

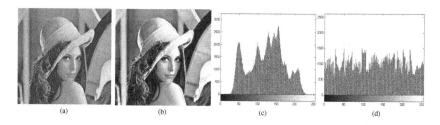

(a)   (b)   (c)   (d)

Fig. 1. Visual perception by RDH-CE for *Lena* image. (a) Original image. (b) Over-enhanced image, PSNR: 19.5 dB, EC: 173632 bits. (c) Histogram of original image. (d) Histogram of marked image.

2.2   Change of the Image Contrast Caused by RDH-PE

A RDH-PE method is based on expanding the prediction error (PE) from the neighborhood of a pixel for embedding data. A prediction error e_i is the difference between each pixel value x_i and its predicted value $\hat{x}_i$, $e_i = x_i - \hat{x}_i$. Each element of the obtained prediction error sequence is modified according to Eq. 2,

$$
\tilde{e}_i =
\begin{cases}
e_i + m_k & \text{for } e_i = 0 \\
e_i - m_k & \text{for } e_i = -1 \\
e_i + 1 & \text{for } e_i > 0 \\
e_i - 1 & \text{for } e_i < -1
\end{cases}
\tag{2}
$$

The marked image is produced by adding the modified error $\tilde{e}_i$ to the predicted value of the cover pixel x_i, $\tilde{x}_i = \hat{x}_i + \tilde{e}_i$. The data extraction and image restoration procedures are performed by inverse order of the embedding procedure.

According to many experiments, the change of image contrast caused by the RDH-PE is extremely small. For example, the RCE of the intermediate-marked image with EC 83321 bits resulted by the RDH-CE is 0.55016 for *Lena* image and the RCE of the final marked image with EC 127282 bits embedded by RDH-PE after applying RDH-CE is 0.55053 for the same image. The difference of two RCEs is only 0.00037 as compared with the increased EC (43961 bits), which reveals a possibility to improve the EC by RDH-CE effectively without the severe visual distortion caused by the histogram equalization. Similarly, the change value of two PSNRs is only 0.0661 at the EC.

3   Proposed Method

The proposed method can be considered as a combined version of RDH-CE and RDH-PE (RDH-CE-PE). The RDH-CE-PE method contains two parts: the first part is the RDH-CE embedding under the restriction of a threshold preselected for preventing over-enhanced image, and then RDH-PE as the second part performs the sequence embedding operator on the image produced by RDH-CE to increase the EC. The data extraction and image recovery are performed by the reverse order of embedding procedure.

3.1   Embedding Algorithm by RDH-CE

Suppose that N peak-pairs are selected for a given 8-bit gray-level image C. The RDH-CE embedding procedure is performed as follows:

Step 1. Histogram bins from 0 to $N-1$ are pre-shifted to right side by N and those from $256 - N$ to 255 to left side by N excluding the first 16 pixels in the bottom row. Locations of pre-shifted pixels are saved within a location map for image recovery.

Step 2. Histogram of the pre-shifted image is calculated excluding the first 16 pixels in the bottom row, and then one highest peak-pair (G_a and G_b) is selected in the histogram.

Step 3. Data embedding is performed by applying Eq. 1 to all pixels in the whole image in sequential order excluding the 16 pixels, and bin values of the peak-pair are kept as the side information (Fig. 2).

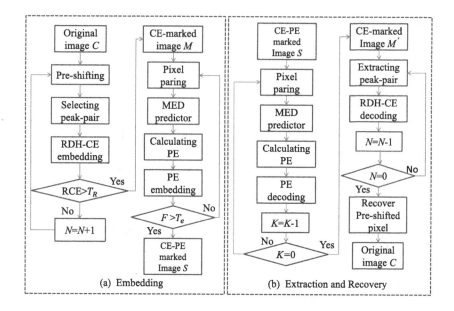

Fig. 2. Flowchart of proposed RDH algorithm

Step 4. Step 2–Step 3 are repeated while $RCE \leq T_R$.

Step 5. The location map (binary bits) is embedded before message bits. The amount of peak-pairs, the length of bitstream in the location map, the LSBs of excluded 16 pixels and the previous peak-pair values are embedded when the last peak-pair is split.

Step 6. Bin values of the last N-th peak-pair are embedded into LSBs of excluded 16 pixels by LSB replacement to generate an intermediate-marked-image M.

3.2   Embedding Algorithm by RDH-PE

The sequence embedding procedure by RDH-PE is performed on the intermediate-marked image M as follows:

Step 1. The prediction value of each x_i in M is calculated by applying Eq. 3 to all pixels in raster scanning order.

$$\hat{x}_i = \begin{cases} min(u, v) & \text{if } w \geq max(u, v) \\ max(u, v) & \text{if } w \leq min(u, v) \\ u + v - w & \text{otherwise} \end{cases} \tag{3}$$

where u, v and w are context elements of x_i in the medium predictor [10].

Step 2. Each prediction error is computed as $e_i = x_i - \hat{x}_i$.

Step 3. Prediction-error histogram (PEH) is generated based on the prediction-error sequence $(e_1, \ldots, e_n)$.

Step 4. The PEH is modified by Eq. (2) to embed data.

Step 5. The modified PE values are added to the prediction values of their corresponding pixels as $\tilde{x}_i = \hat{x}_i + \tilde{e}_i$.

Step 6. Step 1–Step 5 are repeated K times (K levels) while RCE gradient $F \leq T_e$ to produce the final-marked image S.

3.3   Data Extraction and Image Recovery in RDH-CE-PE

The RDH-CE-PE extraction and recovery procedures are performed in the reverse order of embedding. At the beginning of the extraction, the RDH-PE extraction is first performed on the final-marked image S.

Step 1. Calculate the prediction value $\hat{x}_i$ from marked pixel $\tilde{x}_i$ for $i \in \{1, \ldots, N_e\}$ using the medium predictor in the reverse scanning order. Where, N_e is the index of the binary data embedded in cover pixels.

Step 2. Modified prediction error $\tilde{e}_i$ can be determined by subtracting the prediction $\hat{x}_i$ from the marked pixel $\tilde{x}_i$. Then, the embedded data can be extracted as

$$m'_k = \begin{cases} 0 & \text{if } \tilde{e}_i \in \{-1, 0\} \\ 1 & \text{if } \tilde{e}_i \in \{-2, 1\} \end{cases} \tag{4}$$

where m'_k is the k-th binary bit of the extracted data.

Step 3. The prediction error e'_i is recovered by applying Eq. 5 to each modified prediction error $\tilde{e}_i$.

$$e'_i = \begin{cases} \tilde{e}_i & \text{if } \tilde{e}_i \in \{-1, 0\} \\ \tilde{e}_i - 1 & \text{if } \tilde{e}_i > 0 \\ \tilde{e}_i + 1 & \text{if } \tilde{e}_i < -1 \end{cases} \tag{5}$$

Step 4. Finally, the cover pixel is fully restored as $x_i = \hat{x}_i + e'_i$ to restore the intermediate-marked image M' exactly.

Sequentially, the RDH-CE extraction and recovery procedures are performed on M' as follows:

Step 5. The extraction procedure starts with extracting the N-th peak-pair by collecting LSBs of excluded 16 pixels. Then, the N-th peak-pair is used for obtaining the side information including the number of peak-pairs, the length of bitstream in the location map and LSBs of excluded 16 pixels.

Step 6. After generating the histogram of the image M', the embedded data are extracted from pixels with values of $G_a - 1$, G_a, G_b, $G_b + 1$ by Eq. 6.

$$m'_k = \begin{cases} 1 & \text{for } i' = G_a - 1 \\ 0 & \text{for } i' = G_a \\ 0 & \text{for } i' = G_b \\ 1 & \text{for } i' = G_b + 1 \end{cases} \tag{6}$$

Step 7. The image recovery procedure described by Eq. 7 is performed on all pixels excluding the first 16 pixels in the bottom row.

$$
i = \begin{cases} i' + 1 & \text{for } i' < G_a - 1 \\ G_a & \text{for } i' = G_a - 1 \text{ or } i' = G_a \\ G_b & \text{for } i' = G_b \text{ or } i' = G_b + 1 \\ i' - 1 & \text{for } i' > G_b + 1 \end{cases} \tag{7}
$$

Step 8. Step 6–Step 7 are repeated until all data are extracted.

Step 9. Using the location map reconstructed from the side information, pixel values corresponding to pre-shifted histogram bins are retrieved exactly.

Step 10. LSBs of the first 16 pixels in the bottom row are rewritten to original locations, and the cover image is fully recovered without any distortion.

4 Experiments

Our experiment is carried out by embedding/extracting pseudo random binary bits generated by the Matlab function rand() for grey-level images with the size of 512 * 512 [14] and recovering fully original image. The embedding performance of the proposed method is compared with Wu's RDH-CE in terms of EC, PSNR and RCE because it is the combination of RDH-CE and RDH-PE.

4.1 Test Reversibility of the Proposed Method

Figure 3 shows original, marked and recovered images of *Plane* image processed by the proposed method, demonstrating that the proposed method is exactly reversible. Surprisingly, although the amount of data (314599 bits) embedded in Fig. 3(c) is 84740 bits more than 229859 bits in Fig. 3(b), but it is difficult to distinguish between them in visual quality and a little changes of PSNR and RCE are measured. We have observed similar results for many other images.

Fig. 3. Reversibility test of proposed RDH algorithm for *Plane* image. (a) Original image. (b) Intermediate-marked image by RDH-CE, EC: 229859 bits, PSNR: 22.073 dB, RCE: 0.552. (c) Final-marked image by RDH-PE, EC: 314599 bits, PSNR = 21.78 dB, RCE: 0.554. (d) Intermediate-marked image recovered by RDH-PE. (e) Restored original image.

4.2 Embedding Performance and Determination of Reasonable Parameters for Combination of RDH-CE and RDH-PE

The proposed RDH-CE-PE has two important parameters for the combination beside the inherent ones for the RDH-CE or the RDH-PE: the first is a threshold T_R of RCE pre-defined in RDH-CE and the other is a threshold T_e for determining additional EC by RDH-PE without visual distortion. According to experiment results, these two parameters vary from image to image. The parameter T_R ($T_R \in [5.1, 5.9]$) can be first determined arbitrarily for different applications, and the performance curves for *Lena* image are plotted by increasing EC with differently selected T_R in Fig. 4. As one can see from Fig. 4, the sequence embedding by RDH-PE after performing RDH-CE while $RCE \leq T_R$ provides better PSNR and RCE than performing RDH-CE continuously at the same EC. The gradient of curves of PSNR and RCE in RDH-CE-PE is rapidly changed at a specific point of EC considering as the second parameter to approach curves in Wu's RDH-CE degrading the visual quality severely. From the observation, it is noted that the second parameter T_e can be used for restraining the secondary contrast over-enhancing which is described as RCE gradient F quantified by ratio of difference values in RCE and EC according to Eq. 8.

$$(RCE_K - RCE_{K-1})/(EC_N - EC_{K-1}) \leq T_e \tag{8}$$

Here, EC_K and RCE_K indicate the EC (bpp) and RCE at the K-th level in RDH-PE respectively. Through experiments for many images, as T_e is set as 1.2, visual distortion is not perceived in the marked image produced by proposed method. Due to the lack of space, results for only three test images are shown in Fig. 5. Marked images shown in Fig. 5 are resulted as follows.

(a) Original image (Baboon). (b) RDH-CE marked image (Baboon) EC = 84593 bits, RCE = 0.55, PSNR = 25.702933 dB. (c) RDH-CE-PE marked image (Baboon) EC = 115660 bits, RCE = 0.557956, PSNR = 23.312608 dB.

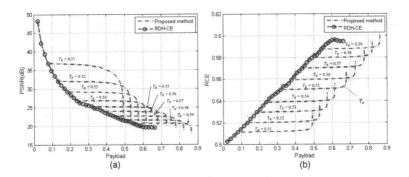

Fig. 4. Performance comparison ((a) PSNR and (b) RCE) of proposed method and RDH-CE for *Lena* image.

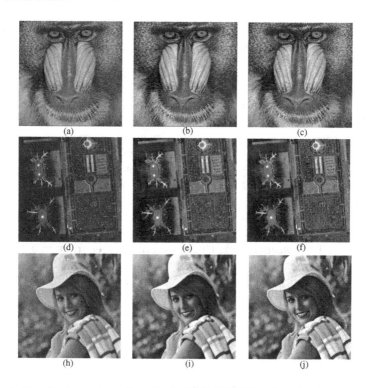

Fig. 5. Results for proposed method (RDH-CE-PE) using three test images.

(d) Original image (Airport). (e) RDH-CE marked image (Airport) EC = 179240 bits, RCE = 0.55, PSNR = 22.82467 dB. (f) RDH-CE-PE marked image (Airport) EC = 231060 bits, RCE = 0.560469, PSNR = 21.259652 dB.

(h) Original image (Elaine). (i) RDH-CE marked image (Elaine) EC = 71280 bits, RCE = 0.55, PSNR = 25.0237 dB. (j) RDH-CE-PE marked image (Elaine) EC = 129244 bits, RCE = 0.557343, PSNR = 23.2630 dB.

Experiment results fully demonstrate superior performance of our proposed method compared with Wu's RDH-CE.

5   Conclusions

In this paper, a high capacity RDH-CE is proposed to improve the performance of Wu's RDH-CE. The proposed method can embed much more data into the contrast enhanced image by the RDH-CE while avoiding visual distortion by contrast over enhancing at the high EC. The proposed method provides the best performance among existing RDH methods with contrast enhancement, which can be supported by suitable combination of RDH-CE and RDH-PE based on restraining the degree of contrast enhancement. The further work will be intended to optimal determination of two specific parameters of RDH-CE-PE.

Acknowledgement. This work is supported by NSFC (No. U1536120, U1636201, 61502496).

References

1. Chen, H., Hong, W., Ni, J., Yuan, J.: A tunable bound of the embedding level for reversible data hiding with contrast enhancement. In: Sun, X., Liu, A., Chao, H.-C., Bertino, E. (eds.) ICCCS 2016. LNCS, vol. 10039, pp. 134–144. Springer, Cham (2016). https://doi.org/10.1007/978-3-319-48671-0_13
2. Chen, H., Ni, J., Hong, W., Chen, T.S.: Reversible data hiding with contrast enhancement using adaptive histogram shifting and pixel value ordering. Signal Process. Image Commun. **46**, 1–16 (2016)
3. Gao, G., Shi, Y.Q.: Reversible data hiding using controlled contrast enhancement and integer wavelet transform. IEEE Signal Process. Lett. **22**(11), 2078–2082 (2015)
4. Gao, M.Z., Wu, Z.G., Wang, L.: Comprehensive evaluation for he based contrast enhancement techniques - volume 2. Advances in Intelligent Systems and Applications. Smart Innovation, Systems and Technologies, vol. 21, pp. 331–338. Springer, Heidelberg (2013). https://doi.org/10.1007/978-3-642-35473-1_33
5. Hong, W., Chen, T.S., Shiu, C.W.: Reversible Data Hiding for High Quality Images Using Modification of Prediction Errors. Elsevier Science Inc., New York (2009)
6. Hwang, H.J., Kim, H.J., Sachnev, V., Sang, H.J.: Reversible watermarking method using optimal histogram pair shifting based on prediction and sorting. KSII Trans. Internet Inf. Syst. **4**(4), 655–670 (2010)
7. Jiang, R., Zhang, W., Xu, J., Yu, N., Hu, X.: Reversible image data hiding with local adaptive contrast enhancement. In: Park, J., Jin, H., Jeong, Y.S., Khan, M. (eds.) Advanced Multimedia and Ubiquitous Engineering. Lecture Notes in Electrical Engineering, vol. 393, pp. 445–452. Springer, Singapore (2016). https://doi.org/10.1007/978-981-10-1536-6_58
8. Li, X., Li, B., Yang, B., Zeng, T.: General framework to histogram-shifting-based reversible data hiding. IEEE Trans. Image Process. **22**(6), 2181–2191 (2013)
9. Li, X., Yang, B., Zeng, T.: Efficient reversible watermarking based on adaptive prediction-error expansion and pixel selection. IEEE Trans. Image Process. **20**(12), 3524–33 (2011). A Publication of the IEEE Signal Processing Society
10. Li, X., Zhang, W., Gui, X., Yang, B.: Efficient reversible data hiding based on multiple histograms modification. IEEE Trans. Inf. Forensics Secur. **10**(9), 2016–2027 (2015)
11. Ni, Z., Shi, Y.Q., Ansari, N., Su, W.: Reversible data hiding. IEEE Trans. Circuits And Syst. Video Technol. **16**(3), 354–362 (2006)
12. Ri, Y., Dong, J., Wang, W., Tan, T.: Adaptive histogram shifting based reversible data hiding. In: Pan, J.-S., Tsai, P.-W., Watada, J., Jain, L.C. (eds.) IIH-MSP 2017. SIST, vol. 82, pp. 42–50. Springer, Cham (2018). https://doi.org/10.1007/978-3-319-63859-1_6
13. Tian, J.: Reversible data embedding using a difference expansion. IEEE Trans. Circuits Syst. Video Technol. **13**(8), 890–896 (2003)

14. Weber, A.G.: The USC-SIPI image database version 5. USC-SIPI Report 315, 1–24 (1997)
15. Wu, H.Z., Wang, H.X., Shi, Y.Q.: Dynamic content selection-and-prediction framework applied to reversible data hiding. In: IEEE International Workshop on Information Forensics and Security (2017)
16. Wu, H.T., Dugelay, J.L., Shi, Y.Q.: Reversible image data hiding with contrast enhancement. IEEE Signal Process. Lett. **22**(1), 81–85 (2014)

Rank Learning for Dehazed Image Quality Assessment

Jingjing Yao, Wen Lu$^{(\boxtimes)}$, Lihuo He, and Xinbo Gao

School of Electronic Engineering, Xidian University, Xi'an 710071, China
yaojingjing@stu.xidian.edu.cn, {luwen,lhhe,xbgao}@mail.xidian.edu.cn

Abstract. Dehazed image quality assessment algorithm is aimed to evaluate the quality of dehazed images. However, the existing dehazed image quality evaluation algorithms are overly dependent on the dehazed images database with accurate subjective quality scores, which are inaccurate, biased and time-consuming, and it is difficult to obtain a large dehazed images database, or extent the existing database. To overcome these problems, it is by using the subjective quality preference that we propose a rank learning algorithm to evaluate the dehazed image quality, here, the subjective quality preference stands for the information such as "the quality of Image I_a is better than that of image I_b", and the rank learning is aimed to learn a function that can predict the corresponding rank sequence for a given set of input stimulus. In our algorithm, we transform the problem of dehazed image quality evaluation into the classification problem of quality preference learning and then use and random forest and pairwise comparison in turn to learn the function that can predict the corresponding quality rank sequence for a given set of dehazed images. The experimental results show that our algorithm is highly consistent with the subjective feeling of human eye and is superior to the traditional dehazed image quality evaluation algorithms. Moreover, our algorithm has a strong expansibility.

Keywords: Dehazed image quality assessment · Preference learning · Rank learning · Random forest

1 Introduction

Dehazed image quality assessment algorithm is aimed to evaluate the quality of dehazed images, which indirectly reflects the performance of the dehaze algorithm. Dehazed image distortion caused by dehazed algorithm tends to appear contrast distortion, noise pollution, and image color cast, such as the more common image super-saturation enhancement phenomenon, which seriously affects the perception of the image for human eye. Therefore, there has been more focus on dehazed image quality assessment.

Recently, there are some dehazed image quality evaluation algorithms while majority of them such as [1–3], trained models by a large number of dehazed images with accurate subjective quality scores. However, the acquisition of the

© Springer Nature Singapore Pte Ltd. 2017
J. Yang et al. (Eds.): CCCV 2017, Part II, CCIS 772, pp. 295–308, 2017.
https://doi.org/10.1007/978-981-10-7302-1_25

dehazed images subjective scores has some problems: First, the subjective quality scores are not accurate. Observers are usually not sure which score can describe the dehazed image quality accurately, so randomly select a score from a small rough range. As a result, the subjective quality scores are difficult to accurately reflect the small difference between images. Second, the subjective quality scores are easily influenced by the observers' preference to the image content, which further reduces the reliability of the subjective scores. Third, it's difficult to build a large-scale dehazed image quality evaluation database which limits the practicality of [4, 5], or extend the existing database because of the inconvenience of the database construction process: (1) the database must contain a variety of distortion types, and there must include a number of images with different distortion degree or different contents for each type. (2) In order to reduce the impact of personal preference, the organizer need to arrange multiple observers to judge its quality for each image, which greatly increases the manpower, material and time consumption.

In summary, the subjective quality scores are inaccurate, biased, time consuming, which limit the reliability and expansibility of these dehazed image quality evaluation algorithms.

In order to overcome these problems, it is by using the preference information that we propose a rank learning algorithm to evaluate the dehazed image quality. Here, the ranking learning is a key issue in application areas such as page ordering, text retrieval and image search, and is aimed to learn a function that can predict its rank sequence for a given set of input stimulus, and the subjective quality preference stands for the information such as "Image I_a quality is better than image I_b". Given a pair of images, we would call it "Preference Image Pair" (PIP) if the relative quality of the two images is known. Meanwhile, the relative quality of the two images is represented by a Preference Label. In our algorithm, we transform the problem of dehazed image quality evaluation into the classification problem of quality preference learning, and then use random forest and pairwise comparison in turn to learn the function that can predict the corresponding quality rank sequence for a given set of dehazed images. The experimental results show that our algorithm is highly consistent with the subjective feeling of human eye and is superior to the traditional dehazed image quality evaluation algorithms. Moreover, our algorithm has a strong expansibility.

2   The Acquisition of Preference Images Pairs

So far, researchers have constructed some databases for dehazed image quality evaluation. Therefore, designing a proper method to obtain reliable PIPs from the existing database is pretty meaningful. From those existing databases, we select dehazed images with large difference in quality scores, then construct preference image pairs and get the preference labels based on their subjective quality scores. Moreover, in order to get more PIPs, we use different dehazing algorithm to get dehazing images with different quality, then build the preference image pairs and get the corresponding preference labels, which also proves the proposed methods expansibility.

If an existing dehazed image quality evaluation database contains n images, we can get the set of preference image pairs, we can get the set of preference image pairs P_1, which size is N_1:

$$P_1 \subseteq \{(I_i, I_j) \,|\, |s_i - s_j| > T, i, j = 1, ..., n\} \tag{1}$$

where, T is the threshold of the difference in subjective quality scores, and s_i is the quality score of the image I_i. $|s_i - s_j|$ is the absolute value of $(s_i - s_j)$. For each preference image pair $p_k = (I_i, I_j) \epsilon P_1$, we can get the preference label l_k, $k = 1, 2, ...N_1$ based on $(s_i - s_j)$:

$$l_k = \begin{cases} sign(s_i - s_j) \\ -sign(s_i - s_j) \end{cases} \tag{2}$$

Moreover, we get more PIPs using following method. First, selected some original hazy images, then use method of Fattal13 [6], He09 [7], Choi [4,8,9] to obtain dehazed images, and get some preference images pairs and the corresponding preference labels though our subjective evaluation. In order to simplify the process of the preference image pairs acquiring, we specify that each sub-images pair conforms to the same preference label of the corresponding preference image pair. For each pair of images, randomly select 50 pieces of $n * n$ non-overlapping sub-image blocks, where n is 64. In this way, we get pairs of preference images.

Finally, we get our preference images pairs as follows:

$$P = \{(I_{k1}, I_{k2}), k = 1, ..., N\} \tag{3}$$

where, $N = N_1 + N_2$, I_{k1} and I_{k2} are the two images of the k-th preference image pair. And for each preference pair, we get its preference label $l_k, k = 1, ..., N$, if the quality of dehazed image I_{k1} is better than I_{k2}, the preference label $l_k = 1$, and $l_k = -1$ if the quality of dehazed image I_{k1} is worse than I_{k2}. Then we can get the preference labels set:

$$L = \{l_1, ..., l_N\} \subset \{-1, +1\}^N \tag{4}$$

In the process of PIPs acquisition, we can put the preference image pairs from different databases together without any data correction. In addition, we can also add PIPs from our subjective experiment to the total set of PIPs. So it can be seen that the acquisition of the preference image pairs is very simple and convenient, and extending the existing PIPs is pretty easy, which can effectively overcome the problem of the dehazed images subjective scores' acquisition and the database building or expanding, and has great significance to the popularization and application of dehazed image quality evaluation algorithms.

3   The Acquisition of Preference Images Pairs

This section details the proposed quality evaluation algorithm. In our algorithm, we transform the problem of dehazed image quality evaluation into the classification problem of quality preference learning, and then, based on our database of preference image pairs, we learn the mapping relationship between the preference image pairs and the corresponding preference labels using the random forest classification model, finally get ranking result through pairwise comparison in turn, which is predicted base on voting strategy (Fig. 1).

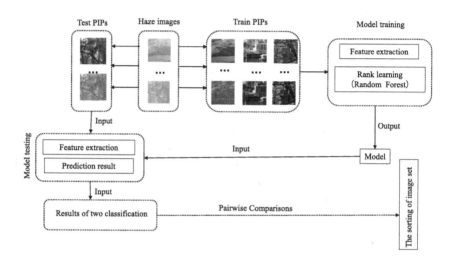

Fig. 1. The framework of dehazed image quality assessment.

3.1   Features Extraction

In this paper, features are extracted from the following two aspects. On the one hand, we extract the features that can represent the degree of haze density. Image dehazing is a process of image clarity. Thus, the big difference between dehazed image quality assessment with conventional IQA is the consideration of the haze removal degree, and we will extract the image haze density features from the image sharpening degree, the texture detail richness and the contrast index. On the other hand, those features that can represent the degree of over-enhanced image distortion are also extracted. For the over-enhanced images, not only there exist the difference of haze density, but also exist the image distortion caused by contrast distortion, noise pollution, image color cast and so on, which seriously affects the visual perception comfort of the human eye. For the natural image dehazing, we should keep the similarity of color tone between images before and after dehazing as much as possible when achieving the purpose of haze removal. Therefore, we extract dehazed images' features using two indicators of perceived comfort and the similarity of color tone between images before and after dehazing. In the following part, we detail these features and demonstrate that these features we extracted can well represent the haze density and over-enhanced image distortion of dehazed images.

(a) Features of haze density

Ruderman et al. [10] found that the operation of brightness normalization simulates the contrast gain mechanism of the human visual cortex, which is called the MSCN coefficient [11] as

$$I_{MSCN}(i,j) = \frac{I_{gray}(i,j) - \mu(i,j)}{\sigma(i,j) + 1} \tag{5}$$

$$\mu(i,j) = \sum_{k=-K}^{K} \sum_{l=-L}^{L} \omega_{k,l} I_{gray}(i+k, j+l) \tag{6}$$

$$\sigma(i,j) = \sqrt{\sum_{k=-K}^{K} \sum_{l=-L}^{L} \omega_{k,l} \left[I_{gray}(i+k, j+l) - \mu(i,j) \right]^2} \tag{7}$$

$$\tilde{J}_{dark} = 1 - (\min_{y \in \Omega(x)} (\min_{c} \frac{I^c(y)}{A})) \tag{8}$$

where $i \in \{1, 2, ...M\}, j \in \{1, 2, ...N\}$, M and N are the image size, and $\omega = \{\omega_{k,k} | k = -K, ...K, l = -L, ...L\}$ is the local Gaussian symmetric convolution window corresponding to pixel (i,j), and K and L denote respectively the length and width of convolution window.

For natural haze images, the variance of the MSCN coefficients decrease as the haze density increase, reflecting the degree of image haze density to a

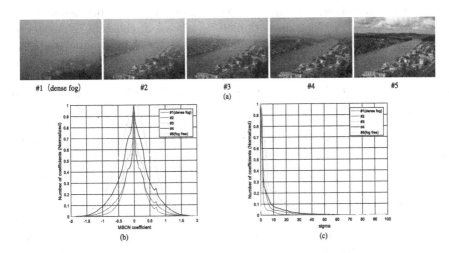

Fig. 2. MSCN coefficient histogram: (a) natural fog images at different fog density in the same scene. The fog density from image #1 to image #5 is sequentially decreased. (b) MSCN coefficient histogram of the images in (a). (c) histogram of the parameters sigma in the MSCN coefficients. (Color figure online)

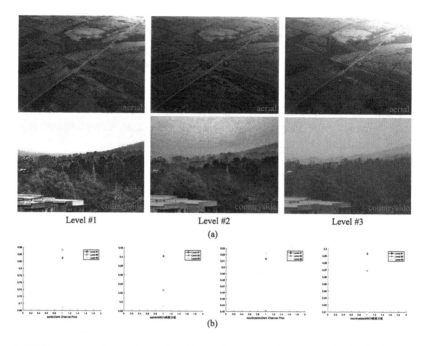

Fig. 3. Differences in image haze density evaluation using respectively the image dark channel features and the MSCN variance coefficient: (a) natural haze images at different haze density. The haze density from image #1 to image #3 is sequentially decreased. (b) haze density rank of the images in (a) by using respectively the dark channel feature and the MSCN variance coefficient. (Color figure online)

certain extent, as shown in Fig. 2(b). For images with low brightness values but with partial dark block distortion, the variance of the MSCN coefficients is more accurate for haze density evaluation than the dark channel statistical features. Figure 3 compares the differences in image haze density estimate using respectively the image dark channel features and the MSCN variance coefficient. The dark channel statistical feature is defined as (8), which is similar to the MSCN characteristic. For both of them, the larger the eigenvalue is, the better the image dehazing result is. When evaluating the haze density of the "aerial" Level #2 and countryside Level #2 in Fig. 3(a), the dark channel feature detects that the number of dark pixels in the image is larger, concluding that the two images have the lowest haze density and the result of image dehazing is best which is not consistent with the actual estimate; while MSCN coefficient variance can make an accurate estimate, as shown in Fig. 3 (b).

The local standard deviation parameter $\sigma(i, j)$ in the MSCN coefficients can accurately measure the sharpness degree of the local structure in the image, which can reflects the image haze density. As shown in Fig. 2(c), the parameter $\sigma(i, j)$ decrease as the fog density increases. Therefore, we use the local standard deviation parameter $\sigma(i, j)$ in the MSCN coefficient as a feature of evaluating image haze density.

The texture information can reflect the spatial distribution and the structure information of images, which is the basis of the visual system for image perceiving. For dehazed images, the richness of the texture information indirectly reflects the image clarity and visibility, so we use it to evaluate the image haze density. The gray covariance matrix can represent texture information well, and the entropy value can accurately reflect the amount of information contained in the image and the complexity of the texture. The greater the entropy value, the richer the image texture. In order to make the feature independent of images content and direction, it is defined as

$$E = (ENT^{0°} + ENT^{45°} + ENT^{90°} + ENT^{135°}) \tag{9}$$

The contrast value reflects the brightness changes in the gray scale of image, and it can well represent the image clarity and detail. High contrast images tend to be sharper and richer, and vice versa. The contrast energy CE, as an approximation of the parameter β in the Weibull function, is a description of the image contrast distribution, which reflects the local contrast changes in the image. CE can convolute the image I by using the Gaussian second derivative filter, and the filter response is normalized to simulate the nonlinear contrast gain control process in human visual cortex. The image contrast is defined in three color channels (grayscale, yellow-blue: yb, red-green: rg):

$$CE(I_c) = \frac{\alpha \cdot Z(I_c)}{Z(I_c) + \alpha \cdot \kappa} - \tau_c \tag{10}$$

$$Z(I_c) = \sqrt{(I_c \otimes h_h)^2 + (I_c \otimes h_v)^2} \tag{11}$$

where $\otimes$ represents the convolution operation, h_h and h_v are Gaussian second derivatives in the horizontal and vertical directions respectively, and $c \in \{gray, yb, rg\}$, gray $= 0.299R + 0.587G + 0.114B$, yb $= 0.5(R + G) - B$, rg $= R - G$. In addition, α is maximum value of $Z(I_c)$, κ is the contrast gain and it is 0.1 in our paper, and τ_c defines the noise threshold of each color channel, the values are 0.2553, 0.2287, 0.0528 respectively.

(b) Over-enhanced distortion feature of image

One of the significant features of over-enhanced image is the contrast distortion caused by high contrast. The paper [12] indicated that the skewness and kurtosis of images can effectively reflect the comfort degree of the human visual perception. For natural images, the skewness and kurtosis distribution of the image conform to the corresponding Gaussian distribution. And the high contrast images show the statistical characteristics of positive skewness, whereas the darker or smoother images show statistical characteristics of negative skewness. Kurtosis is an indicator of measuring the symmetry of variable distribution, and it can also reflect the changes of images contrast. The higher the image's kurtosis is, the stronger the images gloss and the less natural the visual perception is. For those images with contrast distortion, the skew and absolute kurtosis values are higher. Therefore, we take the images skew and kurtosis as features of contrast distortion perceived by human eye, and they are respectively defined in the following formula (12) and (13).

$$skewness(I) = \frac{E[I - E(I)]^3}{\sigma^3(I)} \tag{12}$$

$$kurtosis(I) = \frac{E[I - E(I)]^4}{\sigma^4(I)} - 3 \tag{13}$$

If a image is too dark or too bright, the human eye's feelings to it will be influenced. For these contrast-distorted image of this type. The literature [13] used the Gaussian kernel function to define a first-order statistic, representing the visual comfort of the image, as defined in Eq. (14).

$$Comfort = \exp[-(\frac{E(I) - \mu}{\upsilon})^2] \tag{14}$$

where, μ and υ is the fixed parameter of this model, and the values in our paper is 130 and 300 respectively, $\sigma(I)$ is the variance of image I, and E (I) is the expectation of I.

Naturalness, as a standard of human visual feeling, affects the subjective evaluation result of human eye. In [14], a factor of the naturalness was proposed to enhance the image quality, and it can be used as an image evaluation standard of naturalness degree without reference images. For any image I, we obtain the naturalness value of each channel separately. The closer the value is to 1, the better the images naturalness. The image naturalness degree is defined as follows:

$$N_f = (1 - \theta)\frac{T_1}{T_1^{pr}} + \theta\frac{T_2}{T_2^{pr}} \tag{15}$$

where Θ is weighting factor and it belongs to 0–1, C is the color channel, $c \in \{R, G, B\}$, and T1 and T2 are respectively the gradient distribution model and the Laplace distribution model in [14]. In addition, the values of T_1^{pr} and T_2^{pr} is respectively 0.38 and 0.14 in our paper.

It is not the better as the higher color intensity of the dehazed image. The image dehazing operation is aimed at the image clarity. Therefore, it is necessary to keep the maximum similarity between the two images before and after dehazing. Enhanced images often show a higher degree of color distortion, seriously affecting the image quality and the perception feelings of human eye. We convert the image to the YIQ color space, and define the image hue similarity feature as the fidelity of I and Q color channels. In our paper, we use the eigenvalue f_{IQ} to measure the hue similarity of images before and after dehazing, and it is defined as follows.

$$f_I = \frac{1}{N} \sum_x \frac{2I_r(x) \cdot I_d(x) + c_0}{I_r^2(x) + I_d^2(x) + c_0} \tag{16}$$

$$f_Q = \frac{1}{N} \sum_x \frac{2Q_r(x) \cdot Q_d(x) + c_0}{Q_r^2(x) + Q_d^2(x) + c_0} \tag{17}$$

$$f_{IQ} = \frac{1}{N} \sum_x \left(\frac{2I_r(x) \cdot I_d(x) + c_1}{I_r^2(x) + I_d^2(x) + c_1} \cdot \frac{2Q_r(x) \cdot Q_d(x) + c_2}{Q_r^2(x) + Q_d^2(x) + c_2} \right) \tag{18}$$

where x is the pixel coordinate, I_r, Q_r, I_d, and Q_d are I and Q color channel values of the hazy image and the dehazed image respectively. In addition, c_0, c_1, and c_2 are constants in order to maintain the validity of the feature.

3.2 The Training Data Set

According to the method of Sect. 2, we can get the database of the preference image pairs with preference labels. Then, using the method of Sect. 3.1, we calculate feature vector for each image included in the image pairs, and calculate the feature difference vector for each preference image pair (Table 1).

Table 1. List of the features used in our algorithm.

Features	Features representation	The formula
(1)	Variance of MSCN	(5)
(2)	Sharpening degree	(7)
(3)	Detail of texture	(9)
(4) (5) (6)	Contrast	(10)
(7)	The image skewness	(12)
(8)	The image kurtosis	(13)
(9)	The degree of visual comfort	(14)
(10)	The degree of images naturalness	(15)
(11)	Hue similarity	(16)

- Our database of preference images pairs:

$$P = \{(I_{k1}, I_{k2}), k = 1, ..., N\} \tag{19}$$

Where, I_{k1} and I_{k2} are the two images of the k-th preference image pair.
- The feature vector f_{I_k} for each image I_k.
- For each preference image pair $P_k = (I_{k1}, I_{k2}) \in P$ the feature difference vector x_k and the preference label l_k, $k = 1, ..., N$:

$$x_k = f_{I_{k1}} - f_{I_{k2}} \tag{20}$$

$$l_k \in \{-1, +1\} \tag{21}$$

- Let X denotes the set of feature difference vectors and Y denotes the set of the preference labels

$$X = \{x_1, ..., x_k, ..., x_N\} \tag{22}$$

$$L = \{l_1, ..., l_k, ..., l_N\} \tag{23}$$

Then, we train two-class stochastic forest model by $\{X, L\}$ to study the mapping relation between the feature difference vectors and the preference labels.

3.3　Rank and Preference Learning by Random Forest

The preference labels are $+1$ or -1, so the problem of learning mapping relation between the feature difference vectors and the preference labels is transformed into a two classification problem.

Because Random forest has high classification precision, ability to avoid overfitting, and simpleness to implement, we choose random forest as our two classification model in our algorithm. We train the Random forest model based on the establishing database, and the training process is described as follows. First, randomly generate multiple training sets from the established dehazed images set. Then construct a decision tree $g_i(x, \Theta_i)$ for each image set, and $i = 1, ..., M$, where M is the total number of decision trees, Θ_i is a mechanism used in the training data. And random forest set $G = \{g_1(x), g_2(x), ..., g_M(x)\}$ is constituted by different decision trees. When using the constructed stochastic forest model to classify hazy images, the classification result is determined by voting mechanism, that is, the mode of different classification results evaluated by decision trees determines the final classification result. For the two classification problem, the output is -1 or $+1$, as shown in Eq. (24).

$$C = \arg\max(p(c|x)), c \in \{-1, 1\} \tag{24}$$

$$p(c|x) = \frac{1}{M} \sum_{i=1}^{M} p_i(c|x) \tag{25}$$

For given test data of preference dehazed pairs, we can use the trained Random forest model to get the preference labels based on its feature difference vectors. And if a feature difference vector is 0, we think the two images of this pair have a same quality, and set the corresponding preference label to 0. Finally, we can get the final ranking result through pairwise comparison in turn.

4　Experimental Results

In order to verify the relevance of the established PIPs and the features mentioned, we conduct the following experiment. First, 80% of established PIPs is selected for training the random forest classification model, and the remaining 20% is used for testing this model. Repeat the above steps and use the average result of 50 repetitive training as the final classification result. We compare the final results using SVM and random forest classifiers in Table 2.

Table 2. The results classification using SVM and random forest model.

Method	Accuracy
Our method using SVM	80.97%
Our method using random forest	90.58%

It can be seen that the classification result of random forest model is superior to SVM. This result also proved the validity of our extracted features.

In order to verify the accuracy of our method in evaluating image haze density, we subjectively sort the images with different haze density in the same scene, and sorting results are shown in Fig. 4. Then we compare the subjective sorting result with the evaluation result of our algorithm. The evaluation result of our algorithm is: Level #1 < Level #2 < Level #3 < Level #4, and is consistent with the subjectively sorting result, which fully demonstrates the effectiveness of the algorithm in image haze density evaluation.

In order to verify the effectiveness of our algorithm in evaluating over enhancement distortion, we select several hazy images and respectively use the algorithm [6, 7, 15, 16] to remove haze, and the dehazing result is shown in Fig. 5. Then, the results of our algorithm are compared with the result of standard evaluation algorithm proposed in [1, 4], and the comparison results are respectively shown in Tables 3 and 4.

The haze density standard proposed in [4] can evaluate the change of image haze density to a certain extent. However, it excessively used the image color bright features as the haze density evaluation criteria, so the evaluation result was overly dependent on the image color information and was not accurate, as shown in "y01". At the same time it did not take into account the factor of over-enhanced distortion, so the evaluation results did not match human visual perception. The evaluation results using the blind contrast evaluation algorithm in [1] is shown in Table 4 and it can be seen that the evaluation algorithm cannot detect the over-enhancement phenomenon. In our paper, through a large number of experimental analysis, we select features closely related to the changes in haze density, color distortion features, and features of human visual perception which can reflect over-enhanced distortion. Using the random forest classification

Table 3. The ranking results of images in Fig. 5 using respectively our evaluation algorithm and the image haze density evaluation algorithm in [4], and the quality is more better if it is in the more forward position of the sequence.

	FADE	Our paper	Reference rank sequence
ny17	(c) (a) (b) (d)	(a) (b) (d) (c)	(a) (b) (d) (c)
y01	(c) (b) (a) (d)	(a) (d) (b) (c)	(a) (d) (b) (c)

Table 4. The evaluation results of images in Fig. 5 using contrast evaluation algorithm [1].

	ny17				y1			
	Fattal13	He09	Tan08	Kopf08	Fattal13	He09	Tan08	Kopf08
e	0.04	0.01	−0.06	0.01	0.03	0.08	0.08	0.09
r	1.95	1.65	2.22	1.62	1.88	1.33	2.28	1.62
Σ	0.00	0.00	0.01	0.01	0.00	0.01	0.01	0.00

| Original image | Level #1 | Level #2 | Level #3 | Level #4 |

Fig. 4. The hazy images at different haze density in the same scene: the haze density decrease in turn from Level #1 to Level #4.

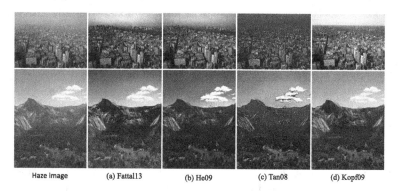

| Haze image | (a) Fattal13 | (b) He09 | (c) Tan08 | (d) Kopf09 |

Fig. 5. The dehazing results using different algorithm.

model, not only can we make an accurate estimate of image haze density, as shown the sorting result of (a) (b) (d) in Fig. 5, but also can make a more accurate estimate of image over-enhanced distortion; and our evaluation results are largely consistent with the human subjective feelings.

5 Conclusion

From the problem of subjective quality scores that are inaccurate, biased and time-consuming, and the difficulty in building a large dehazed images database or extenting the existing database. It is by using the preference information that we propose a rank learning algorithm to evaluate the dehazed image quality. In our algorithm, we transform the problem of dehazed image quality evaluation into the classification problem of quality preference learning, and then use the feature fusion and random forest to solve it, finally get ranking result through pairwise comparison in turn. The experimental results show that our algorithm is highly consistent with the subjective feeling of human eye and is superior to the traditional dehazed image quality evaluation algorithms. Moreover, our algorithm has a strong expansibility. Further research is needed to explore the dehazed image quality-relevant features, and generate preference dehazed image pairs with preference labels.

Acknowledgments. This research was supported partially by the National Natural Science Foundation of China (Nos. 61372130, 61432014, 61501349, 61571343), supported by the Key Industrial Innovation Chain Project in Industrial Domain (No. 2016KTZDGY04-02).

References

1. Hautiere, N., Tarel, J.P., Aubert, D.: Blind contrast enhancement assessment by gradient ratioing at visible edges. Image Anal. Stereol. **27**(2), 87–95 (2011)
2. Li, D., Yu, J.: No-reference quality assessment method for defogged images. J. Image Graph. **16**(9), 1753–1757 (2011)
3. Hou, W., Gao, X., Tao, D., Li, X.: Blind image quality assessment via deep learning. IEEE Trans. Neural Netw. Learn. Syst. **26**(6), 1275–1286 (2017)
4. Yao, B., Huang, L., Liu, C.: Research on an objective method to compare the quality of defogged images. In: Chinese Conference on Pattern Recognition, pp. 1–5 (2009)
5. Choi, L.K., You, J., Bovik, A.C.: Referenceless prediction of perceptual fog density and perceptual image defogging. IEEE Trans. Image Process. **24**(11), 3888–3901 (2015)
6. Dong, N.A.N., Du-Yan, B.I., Shi-Ping, M.A.: A quality assessment method with classified-learning for dehazed images. Acta Automatica Sinica **42**(2), 270–278 (2016)
7. Wu, Q., Li, H., Meng, F., Ngan, K.N., Luo, B.: Blind image quality assessment based on multichannel feature fusion and label transfer. IEEE Trans. Circuits Syst. Video Technol. **26**(3), 425–440 (2016)
8. Lin, J.Y., Hu, S., Wang, H., Wang, P., Katsavounidis, I.: Statistical study on perceived JPEG image quality via MCL-JCI dataset construction and analysis. Electron. Imaging **5**(13), 1–9 (2016)
9. Fattal, R.: Dehazing using color-lines. ACM Trans. Graph. (TOG) **34**(1), 1–14 (2014)
10. He, K., Sun, J., Tang, X.: Single image haze removal using dark channel prior. IEEE Trans. Pattern Anal. Mach. Intell. **33**(12), 2341–2353 (2011)
11. Ren, W., Liu, S., Zhang, H., Pan, J., Cao, X., Yang, M.-H.: Single image dehazing via multi-scale convolutional neural networks. In: Leibe, B., Matas, J., Sebe, N., Welling, M. (eds.) ECCV 2016. LNCS, vol. 9906, pp. 154–169. Springer, Cham (2016). https://doi.org/10.1007/978-3-319-46475-6_10
12. Berman, D., Avidan, S.: Non-local image dehazing. In: IEEE Conference on Computer Vision and Pattern Recognition, pp. 1674–1682 (2016)
13. Ni, W., Gao, X., Wang, Y.: Single satellite image dehazing via linear intensity transformation and local property analysis. Neurocomputing **175**(6), 25–39 (2016)
14. Ruderman, D.L.: The statistics of natural images. Netw. Comput. Neural Syst. **5**(4), 517–548 (2009)
15. Mittal, A., Moorthy, A.K., Bovik, A.C.: No-reference image quality assessment in the spatial domain. IEEE Trans. Image Process. **21**(12), 4695–4708 (2012)
16. Fang, Y., Ma, K., Wang, Z., et al.: No-reference quality assessment of contrast-distorted images based on natural scene statistics. IEEE Signal Process. Lett. **22**(7), 838–842 (2015)
17. Gu, K., Zhai, G., Lin, W., et al.: The analysis of image contrast: from quality assessment to automatic enhancement. IEEE Trans. Cybern. **46**(1), 284 (2016)

18. Gong, Y., Sbalzarini, I.F.: Image enhancement by gradient distribution specification. In: Jawahar, C.V., Shan, S. (eds.) ACCV 2014. LNCS, vol. 9009, pp. 47–62. Springer, Cham (2015). https://doi.org/10.1007/978-3-319-16631-5_4
19. Tan, R.T.: Visibility in bad weather from a single image. In: IEEE Conference on Computer Vision and Pattern Recognition, pp. 1–8 (2008)
20. Kopf, J., Neubert, B., Chen, B.: Deep photo: model-based photograph enhancement and viewing. ACM Trans. Graph. **27**(5), 1–10 (2008)

Image Restoration

A Low-Rank Total-Variation Regularized Tensor Completion Algorithm

Liangchen Song[1], Bo Du[2(✉)], Lefei Zhang[2], and Liangpei Zhang[1]

[1] State Key Laboratory of Information Engineering in Survey,
Mapping, and Remote Sensing, Wuhan University, Wuhan 430079,
People's Republic of China
[2] School of Computer, Wuhan University, Wuhan, Hubei, China
gunspace@163.com

Abstract. In this paper, we first model visual data as a tensor and then impose both low-rank and total-variation constraint to complete the tensor. More specifically, we adopt a novel tensor-tensor production framework (also known as t-product) and its theory of low-rank based completion. By using the concept of t-product, it is the first time that we extend classic Total-Variation (TV) to a t-product and $l_{1,1,2}$ norm based constraint on the gradient of visual data. After proposing our model, we derive a iterative solver based on alternating direction method of multipliers (ADMM). We show the effectiveness of our method and compare our method with state-of-art algorithms in the experimental section.

Keywords: Low-rank tensor completion · Tensor nuclear norm
Total-variation · Visual data completion

1 Introduction

Recent years has witnessed the development of acquisition techniques, and the visual data tends to contain more and more information and thus should be treated as a complex high-dimensional data, i.e. tensors. For example, a hyperspectral image or multispectral image can be represented as a third-order tensor since the spatial information has two dimensions and the spectral information takes one dimension. Using a tensor to model visual data can handle the complex structure better and has become a hot topic in computer vision community recently, for instance, face recognition [15], color image and video in-painting [7], hyperspectral image processing [16,17], gait recognition [12].

Visual data may have missing values in the acquisition process due to mechanical failure or man-induced factors. If we process the multi-way array data as a tensor instead of splitting it into matrices, estimating missing values for visual data is also known as tensor completion problem. Motivated by the successful achievements of low-rank matrix completion methods [8,10], many tensor completion problems could also be solved through imposing low-rank constraint.

© Springer Nature Singapore Pte Ltd. 2017
J. Yang et al. (Eds.): CCCV 2017, Part II, CCIS 772, pp. 311–322, 2017.
https://doi.org/10.1007/978-981-10-7302-1_26

Decomposition and subspace methods has been widely studied [13,14] and a common way to impose low-rank constraint is to decompose tensors and get its factors first, then by limiting the sizes of factors we attain low-rank property. CP decomposition and Tucker decomposition are two classic decomposition methods to compute the factors of a tensor [5]. However, both the two methods mentioned above have crucial parameters that are supposed to be determined by users. In this paper, we adopt a novel tensor-tensor product framework [1], and a singular value decomposition formula for tensors [3,4], which is parameter-free. The rencently proposed singular value decomposition formula for tensors is also referred to as t-SVD, and t-SVD has been proved effective by many influential papers [18].

When trying to restore degraded visual data, it is reasonable to utilize the local smoothness property of visual data as prior knowledge or regularization. A common constraint is Total-Variation (TV) norm, which is computed as the l_1 of the gradient magnitude, and this norm has been qualified to be effective to preserve edges and piecewise structures. If we take the low-rank assumption as a global constraint and TV norm as a local prior, then it is natural to combine this two constraints together to estimate missing values in visual data. So more recently, some related methods that take tensor and TV into consideration are proposed [2,6,11]. [2] aims to design a norm that considers both the inhomogeneity and the multi-directionality of responses to derivative-like filters. [11] proposes an image super-resolution method that integrates both low-rank matrix completion and TV regularization. [6] proposes integrating TV into low-rank tensor completion, but their methods use the same rank definition as [7] or Tucker decomposition, and as mentioned before, the parameters are hard to deal with. Moreover, TV is l_1 norm on matrix but we are processing visual tensors, so we need to extend TV to tensor cases. Also, we employ a new tensor norm namely $l_{1,1,2}$ which stems from multi-task learning [9].

In this paper, we seek to design a parameter-free method for low-rank based completion procedure, which means parameters only exist in TV regularization. To achieve this goal, simply using t-SVD framework is not enough, since t-SVD is based on t-product and a free module so we need to reinvent TV with t-product system. Our main contributions are as follows:

1. Using the concept of t-product, we design difference tensors $\mathcal{A}$ and $\mathcal{B}$ that when multiplied by a visual tensor $\mathcal{X}$ takes the gradient of $\mathcal{X}$, i.e. $\nabla \mathcal{X}$.
2. Motivated by $l_{1,1,2}$ norm which is originally designed to model sparse noise in visual data [18], we use $l_{1,1,2}$ norm to ensure sparsity of the gradient magnitude, then traditional TV is extended to tensor cases.

2   Notations and Preliminaries

In this section, we introduce some notations of tensor algebra and give the basic definitions used in the rest of the paper.

Scalars are denoted by lowercase letters, e.g., a. Vectors are denoted by boldface lowercase letters, e.g., **a**. Matrices are denoted by capital letters, e.g., A.

Higher-order tensors are denoted by boldface Euler script letters, e.g., $\mathcal{A}$. For a third-order tensor $\mathcal{A} \in \mathbb{R}^{r \times s \times t}$, the ith frontal slice is denoted A_i. In terms of MATLAB indexing notation, we have $A_i = \mathcal{A}(:,:,i)$.

The definition of new tensor multiplication strategy [1] begins with converting $\mathcal{A} \in \mathbb{R}^{r \times s \times t}$ into a block circulant matrix. Then

$$
\texttt{bcirc}(\mathcal{A}) = \begin{pmatrix} A_1 & A_n & A_{n-1} & \cdots & A_2 \\ A_2 & A_1 & A_n & \cdots & A_3 \\ A_3 & A_2 & A_1 & \cdots & A_4 \\ \vdots & \vdots & \vdots & \ddots & \vdots \\ A_n & A_{n-1} & A_{n-2} & \cdots & A_1 \end{pmatrix}
$$

is a block circulant matrix of size $rt \times st$. And the $\texttt{unfold}$ command rearrange the frontal slices of $\mathcal{A}$:

$$
\texttt{unfold}(\mathcal{A}) = \begin{pmatrix} A_1 \\ A_2 \\ \vdots \\ A_n \end{pmatrix}, \quad \texttt{fold}(\texttt{unfold}(\mathcal{A})) = \mathcal{A}
$$

Then we have the following new definition of tensor-tensor multiplication.

Definition 1 (t-product). *Let $\mathcal{A} \in \mathbb{R}^{r \times s \times t}$ and $\mathcal{B} \in \mathbb{R}^{s \times p \times t}$. Then the t-product $\mathcal{A} * \mathcal{B}$ is a $r \times p \times t$ tensor*

$$
\mathcal{A} * \mathcal{B} = \texttt{fold}(\texttt{bcirc}(\mathcal{A}) \cdot \texttt{unfold}(\mathcal{B})) \tag{1}
$$

An important property of the block circulant matrix is the observation that a block circulant matrix can be block diagonalized in the Fourier domain [3]. Before moving on to the definition of t-SVD, we need some more definitions from [4].

Definition 2 (Tensor Transpose). *Let $\mathcal{A} \in \mathbb{R}^{r \times s \times t}$, then $\mathcal{A}^T \in \mathbb{R}^{s \times r \times t}$ and*

$$
\mathcal{A}^T = \texttt{fold}([A_1, A_n, A_{n-1}, \cdots, A_2]^T)
$$

Definition 3 (Identity Tensor). *The identity tensor $\mathcal{I} \in \mathbb{R}^{m \times m \times n}$ is the tensor whose first frontal slice is the $m \times m$ identity matrix, and whose other frontal slices are all zeros.*

Definition 4 (Orthogonal Tensor). *A tensor $\mathcal{Q} \in \mathbb{R}^{m \times m \times n}$ is orthogonal if $\mathcal{Q}^T * \mathcal{Q} = \mathcal{Q} * \mathcal{Q}^T = \mathcal{I}$.*

Definition 5 (f-diagonal Tensor). *A tensor is called f-diagonal if each of its frontal slices is a diagonal matrix.*

Definition 6 (Inverse Tensor). *A tensor $\mathcal{A} \in \mathbb{R}^{m \times m \times n}$ has an inverse tensor $\mathcal{B}$ if*

$$
\mathcal{A} * \mathcal{B} = \mathcal{I} \quad and \quad \mathcal{B} * \mathcal{A} = \mathcal{I}
$$

Using these new definitions, we are able to derive a new decomposition method named t-SVD and an approximation theorem on this decomposition.

Theorem 1 (t-SVD). *Let* $\mathcal{A} \in \mathbb{R}^{r \times s \times t}$, *then the t-SVD of* $\mathcal{A}$ *is*

$$\mathcal{A} = \mathcal{U} * \mathcal{S} * \mathcal{V}^T \tag{2}$$

where $\mathcal{U} \in \mathbb{R}^{r \times r \times t}$, $\mathcal{V} \in \mathbb{R}^{s \times s \times t}$ *are orthogonal and* $\mathcal{S} \in \mathbb{R}^{r \times s \times t}$ *is f-diagonal.*

By using the idea of computing in the Fourier domain, we can efficiently compute the t-SVD factorization, and for more details see [3,4]. Then nuclear norm for $\mathcal{A}$ is defined as $\|\mathcal{A}\|_* = \sum_{j=1}^{t} \sum_{i=1}^{\min(r,s)} |\hat{\mathcal{S}}(i,i,k)|$, where $\hat{\mathcal{S}}$ is the result of taking the Fourier transform along the third dimension of $\mathcal{S}$.

3   Proposed t-SVD-TV

In this section, we will first give the details of our model, and then derive a solver with alternating direction method of multipliers (ADMM). Since we combine t-SVD and TV together, we name our method as t-SVD-TV.

3.1   Low-Rank Regularization

Before discussing low-rank model, we need some notations for our problem first. Suppose there is a tensor $\mathcal{M} \in \mathbb{R}^{n_1 \times n_2 \times n_3}$ representing some visual data with missing entries, and we use Ω indicating the set of indices of observations, $\mathcal{X}$ denoting the desired recovery result, then we have

$$\mathcal{P}_\Omega(\mathcal{X}) = \mathcal{P}_\Omega(\mathcal{M})$$

where $\mathcal{P}_\Omega()$ is the projector onto the known indices Ω. So we have the following vanila model [18]:

$$\min_{\mathcal{X}} \quad \|\mathcal{X}\|_* \tag{3}$$
$$\text{s.t.} \quad \mathcal{P}_\Omega(\mathcal{X}) = \mathcal{P}_\Omega(\mathcal{M})$$

Our method takes both low-rank and TV into consideration, which means we also have terms to ensure sparsity on gradient field. We use $\Psi(\cdot)$ as a function promoting sparsity, so our model is

$$\min_{\mathcal{X}} \quad \|\mathcal{X}\|_* + \lambda_1 \Psi\left(\frac{\partial}{\partial_x}\mathcal{X}\right) + \lambda_2 \Psi\left(\frac{\partial}{\partial_y}\mathcal{X}\right) \tag{4}$$
$$\text{s.t.} \quad \mathcal{P}_\Omega(\mathcal{X}) = \mathcal{P}_\Omega(\mathcal{M})$$

where λ_1, λ_2 are tunable parameters. The forms of $\frac{\partial}{\partial_x}\mathcal{X}$, $\frac{\partial}{\partial_y}\mathcal{X}$ and $\Psi(\cdot)$ are given in the following texts.

3.2   TV Regularization

When we need to compute the gradient matrix of a 2D image M, a common way is to design a *difference matrix* A, B. If we assume vertical direction is x and horizontal direction is y, then from AM we get $\frac{\partial}{\partial_x}M$ and from MB we get $\frac{\partial}{\partial_y}M$. And for a 3D data $\mathcal{X}$ we can derive similar means implemented by t-product system. Without loss of generality, we assume $\mathcal{X} \in \mathbb{R}^{n \times n \times k}$. By extending difference matrices to tensors, we define a difference tensor as follows:

$$\mathcal{A} = \texttt{fold}\begin{pmatrix} A \\ 0 \\ 0 \end{pmatrix} \quad \text{and} \quad \mathcal{B} = \texttt{fold}\begin{pmatrix} B \\ 0 \\ 0 \end{pmatrix}$$

where A and B are $n \times n$ difference matrix:

$$A = \frac{1}{2}\begin{pmatrix} -2 & 2 & 0 & \cdots & 0 & 0 \\ -1 & 0 & 1 & \cdots & 0 & 0 \\ 0 & -1 & 0 & \cdots & 0 & 0 \\ \vdots & \vdots & \vdots & \ddots & \vdots & \vdots \\ 0 & 0 & 0 & \cdots & 0 & 1 \\ 0 & 0 & 0 & \cdots & -2 & 2 \end{pmatrix} \quad \text{and} \quad B = \frac{1}{2}\begin{pmatrix} -2 & -1 & 0 & \cdots & 0 & 0 \\ 2 & 0 & -1 & \cdots & 0 & 0 \\ 0 & 1 & 0 & \cdots & 0 & 0 \\ \vdots & \vdots & \vdots & \ddots & \vdots & \vdots \\ 0 & 0 & 0 & \cdots & 0 & -2 \\ 0 & 0 & 0 & \cdots & 1 & 2 \end{pmatrix}$$

For a visual tensor $\mathcal{X}$, we get its gradient tensor by multiplying with $\mathcal{A}, \mathcal{B}$

$$\frac{\partial}{\partial_x}\mathcal{X} = \mathcal{A} * \mathcal{X}, \quad \frac{\partial}{\partial_y}\mathcal{X} = \mathcal{X} * \mathcal{B} \tag{5}$$

Note that $\mathcal{A} * \mathcal{X}, \mathcal{X} * \mathcal{B}$ are third order tensors which have the same size as $\mathcal{X}$. In order to promote sparsity of the gradient, we use the $l_{1,1,2}$ norm for 3D tensors as penalty function $\Psi(\cdot)$. $l_{1,1,2}$ norm is introduced in [18] to model the sparse noise, and for a third order tensor $\mathcal{G}$, $\|\mathcal{G}\|_{1,1,2}$ is defined as $\sum_{i,j}\|\mathcal{G}(i,j,:)\|_F$. Then our optimization problem (4) becomes

$$\min_{\mathcal{X}} \quad \|\mathcal{X}\|_* + \lambda_1\|\mathcal{A} * \mathcal{X}\|_{1,1,2} + \lambda_2\|\mathcal{X} * \mathcal{B}\|_{1,1,2}$$
$$\text{s.t.} \quad \mathcal{P}_\Omega(\mathcal{X}) = \mathcal{P}_\Omega(\mathcal{M}) \tag{6}$$

The terms in (6) are interdependent, so we adopt a widely used splitting scheme which is known as ADMM.

3.3   Optimization by ADMM

The first step of applying ADMM is introducing auxiliary variables. Specifically, let $\mathcal{S}, \mathcal{Y}, \mathcal{Z}_1, \mathcal{Z}_2$ have the same size as $\mathcal{X}$, then our optimization problem (6) becomes

$$\min_{\mathcal{X}} \quad \|\mathcal{S}\|_* + \lambda_1\|\mathcal{Y}_1\|_{1,1,2} + \lambda_2\|\mathcal{Y}_2\|_{1,1,2}$$
$$\text{s.t.} \quad \mathcal{Y}_1 = \mathcal{A} * \mathcal{Z}_1, \mathcal{Y}_2 = \mathcal{Z}_2 * \mathcal{B}$$
$$\mathcal{S} = \mathcal{X}, \mathcal{Z}_1 = \mathcal{X}, \mathcal{Z}_2 = \mathcal{X}, \tag{7}$$
$$\mathcal{P}_\Omega(\mathcal{X}) = \mathcal{P}_\Omega(\mathcal{M})$$

So the augmented Lagrangian is

$$\mathcal{L} = \|\mathcal{S}\|_* + \frac{\rho_1}{2}\left\|\mathcal{S} - \mathcal{X} + \frac{\mathcal{U}}{\rho_1}\right\|_F^2$$

$$+ \lambda_1\|\mathcal{Y}_1\|_{1,1,2} + \frac{\rho_2}{2}\left\|\mathcal{Y}_1 - \mathcal{A}*\mathcal{Z}_1 + \frac{\mathcal{V}_1}{\rho_2}\right\|_F^2$$

$$+ \lambda_2\|\mathcal{Y}_2\|_{1,1,2} + \frac{\rho_3}{2}\left\|\mathcal{Y}_2 - \mathcal{Z}_2*\mathcal{B} + \frac{\mathcal{V}_2}{\rho_3}\right\|_F^2 \qquad (8)$$

$$+ \frac{\rho_4}{2}\left\|\mathcal{Z}_1 - \mathcal{X} + \frac{\mathcal{W}_1}{\rho_4}\right\|_F^2 + \frac{\rho_5}{2}\left\|\mathcal{Z}_2 - \mathcal{X} + \frac{\mathcal{W}_2}{\rho_5}\right\|_F^2$$

$$\text{s.t.} \quad \mathcal{P}_\Omega(\mathcal{X}) = \mathcal{P}_\Omega(\mathcal{M})$$

where tensors $\mathcal{U}, \mathcal{V}_1, \mathcal{V}_2, \mathcal{W}_1, \mathcal{W}_2$ are Lagrange multipliers and $\rho_i(i = 1, ..., 5)$ are positive numbers. We solve (8) by alternatively minimize each variable, so (8) are turned into several subproblems:

Computing $\mathcal{S}$. The subproblem for minimize $\mathcal{S}$ is

$$\mathcal{S}^{k+1} = \arg\min_{\mathcal{S}} \|\mathcal{S}\|_* + \frac{\rho_1}{2}\left\|\mathcal{S} - \mathcal{X}^k + \frac{\mathcal{U}^k}{\rho_1}\right\|_F^2 \qquad (9)$$

Note that (9) can be solved by the singular value thresholding method in [18], so

$$\mathcal{S}^{k+1} := \arg\min_{\mathcal{S}} \|\mathcal{S}\|_* + \frac{\rho_1}{2}\left\|\mathcal{S} - \mathcal{X}^k + \frac{\mathcal{U}^k}{\rho_1}\right\|_F^2 \qquad (10)$$

Computing $\mathcal{Y}_1$ and $\mathcal{Y}_2$. The subproblem for minimize $\mathcal{Y}_1$ is

$$\mathcal{Y}_1^{k+1} = \arg\min_{\mathcal{Y}_1} \lambda_1\|\mathcal{Y}_1\|_{1,1,2} + \frac{\rho_2}{2}\left\|\mathcal{Y}_1 - \mathcal{A}*\mathcal{Z}_1^k + \frac{\mathcal{V}_1^k}{\rho_2}\right\|_F^2 \qquad (11)$$

The closed form solution to (11) is given by

$$\mathcal{Y}_1^{k+1} := \left(1 - \frac{\lambda_1}{\rho_2}\left\|\mathcal{A}*\mathcal{Z}_1^k + \frac{\mathcal{V}_1^k}{\rho_2}\right\|_F^{-1}\right)_+ \left(\mathcal{A}*\mathcal{Z}_1^k + \frac{\mathcal{V}_1^k}{\rho_2}\right) \qquad (12)$$

where $(x)_+ = \max(x, 0)$. Updating strategy for $\mathcal{Y}_2$ is similar to $\mathcal{Y}_1$, so

$$\mathcal{Y}_2^{k+1} := \left(1 - \frac{\lambda_2}{\rho_3}\left\|\mathcal{A}*\mathcal{Z}_1^k + \frac{\mathcal{V}_1^k}{\rho_3}\right\|_F^{-1}\right)_+ \left(\mathcal{Z}_2^k*\mathcal{B} + \frac{\mathcal{V}_2^k}{\rho_3}\right) \qquad (13)$$

Input: an incomplete tensor $\mathcal{M}$, index of known entries Ω, iterarion number N, parameters λ_1, λ_2 and $\rho_i (i = 1, ..., 5)$
Output: a tensor $\mathcal{X}$ after completion

1 Set $\mathcal{P}_\Omega(\mathcal{X}) = \mathcal{P}_\Omega(\mathcal{M}), \mathcal{P}_{\bar{\Omega}}(\mathcal{X}) = 0$, randomly initialize $\mathcal{S}, \mathcal{Y}_1, \mathcal{Y}_2, \mathcal{Z}_1, \mathcal{Z}_2$;
2 **for** $n = 1$ *to* N **do**
3   Update $\mathcal{S}$ by singular value thresholding [18];
4   Update $\mathcal{Y}_1, \mathcal{Y}_2$ by Eq.(12) and Eq.(13) respectively;
5   Update $\mathcal{Z}_1, \mathcal{Z}_2$ by Eq.(15) and Eq.(16) respectively;
6   Update $\mathcal{X}$ by Eq.(18);
7   Update multipliers by

$$\mathcal{U}^{k+1} := \mathcal{U}^k + \rho_1(^k - \mathcal{X}^k)$$

$$\mathcal{V}_1^{k+1} := \mathcal{V}_1^k + \rho_2 \left(\mathcal{Y}^k - \mathcal{A} * \mathcal{Z}_1^k\right)$$

$$\mathcal{V}_2^{k+1} := \mathcal{V}_2^k + \rho_3 \left(\mathcal{Y}^k - \mathcal{Z}_2^k * \mathcal{B}\right)$$

$$\mathcal{W}_1^{k+1} := \mathcal{W}_1^k + \rho_4(\mathcal{Z}_1^k - \mathcal{X}^k)$$

$$\mathcal{W}_2^{k+1} := \mathcal{W}_2^k + \rho_5(\mathcal{Z}_2^k - \mathcal{X}^k)$$

8 **end**
9 Return $\mathcal{X}$.

Algorithm 1. t-SVD-TV

Computing $\mathcal{Z}_1$ and $\mathcal{Z}_2$. The subproblem for minimize $\mathcal{Z}_1$ is

$$\mathcal{Z}_1^{k+1} = \arg\min_{\mathcal{Z}_1} \frac{\rho_2}{2} \left\| \mathcal{Y}_1^k - \mathcal{A} * \mathcal{Z}_1 + \frac{\mathcal{V}_1^k}{\rho_2} \right\|_F^2 + \frac{\rho_4}{2} \left\| \mathcal{Z}_1 - \mathcal{X} + \frac{\mathcal{W}_1^k}{\rho_2} \right\|_F^2 \quad (14)$$

Then Update $\mathcal{Z}_1$ and $\mathcal{Z}_2$ by

$$\mathcal{Z}_1^{k+1} := \left(\rho_4 \mathcal{I} + \rho_2 \mathcal{A}^T * \mathcal{A}\right)^{-1} * \left(\rho_4 \mathcal{I} - \mathcal{W}_1^k + \rho_2 \mathcal{A}^T * \mathcal{Y}_1^k + \mathcal{A}^T * \mathcal{V}_1^k\right) \quad (15)$$

Similarly, we have update rules for $\mathcal{Z}_2$:

$$\mathcal{Z}_2^{k+1} := \left(\rho_5 \mathcal{I} - \mathcal{W}_2 + \rho_3 \mathcal{Y}_2^k * \mathcal{B}^T + \mathcal{V}_2^k * \mathcal{B}^T\right) * \left(\rho_5 \mathcal{I} + \rho_3 \mathcal{B} * \mathcal{B}^T\right)^{-1} \quad (16)$$

Note that the calculation of inverse tensors in (15) and (16) are not hard, since only the first front slices of $\rho_4 \mathcal{I} + \rho_2 \mathcal{A}^T * \mathcal{A}$ and $\rho_5 \mathcal{I} + \rho_3 \mathcal{B} * \mathcal{B}^T$ are non-zero.

Computing $\mathcal{X}$. The subproblem for $\mathcal{X}$ is

$$\mathcal{X}^{k+1} = \arg\min_{\mathcal{X}} \frac{\rho_1}{2} \left\| \mathcal{S}^k - \mathcal{X} + \frac{\mathcal{U}^k}{\rho_1} \right\|_F^2 + \frac{\rho_4}{2} \left\| \mathcal{Z}_1^k - \mathcal{X}^k + \frac{\mathcal{W}_1^k}{\rho_4} \right\|_F^2$$
$$+ \frac{\rho_5}{2} \left\| \mathcal{Z}_2^k - \mathcal{X}^k + \frac{\mathcal{W}_2^k}{\rho_5} \right\|_F^2 \quad (17)$$

So update $\mathcal{X}$ by

$$\mathcal{P}_{\bar{\Omega}}(\mathcal{X}) := \left[\frac{1}{\rho_1 + \rho_4 + \rho_5}(\rho_1{}^k + \mathcal{U}^k + \rho_4 \mathcal{Z}_1^k + \mathcal{W}_1^k + \rho_5 \mathcal{Z}_2^k + \mathcal{W}_2^k) \right]_{\bar{\Omega}} \quad (18)$$

$$\mathcal{P}_\Omega(\mathcal{X}) = \mathcal{P}_\Omega(\mathcal{M})$$

Updating Multipliers

$$\mathcal{U}^{k+1} := \mathcal{U}^k + \rho_1({}^k - \mathcal{X}^k)$$
$$\mathcal{V}_1^{k+1} := \mathcal{V}_1^k + \rho_2\left(\mathcal{Y}^k - \mathcal{A} * \mathcal{Z}_1^k\right)$$
$$\mathcal{V}_2^{k+1} := \mathcal{V}_2^k + \rho_3\left(\mathcal{Y}^k - \mathcal{Z}_2^k * \mathcal{B}\right) \quad (19)$$
$$\mathcal{W}_1^{k+1} := \mathcal{W}_1^k + \rho_4(\mathcal{Z}_1^k - \mathcal{X}^k)$$
$$\mathcal{W}_2^{k+1} := \mathcal{W}_2^k + \rho_5(\mathcal{Z}_2^k - \mathcal{X}^k)$$

With these update formulae, we conclude the solver in Algorithm 1.

4   Experiments

In this section, we evaluate our methods on eight benchmark RGB-color images with different types of missing entries. The eight benchmark images are showed in Fig. 1. Each image is 256×256 and has 3 color channels, so it is a $256 \times 256 \times 3$ tensor. We compare our method (t-SVD-TV), with five state-of-art methods: HaLRTC [7], FBCP [19], t-SVD [18], LRTC-TV-I and LRTC-TC-II [6]. Relative Square Error (RSE) and Peak Signal to Noise Ratio (PSNR) are used to

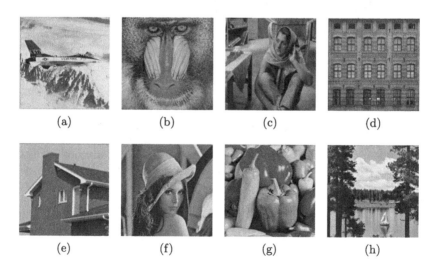

(a)   (b)   (c)   (d)

(e)   (f)   (g)   (h)

Fig. 1. RGB-color images used in our experiment. From left to right they are (a) Airplane (b) Baboon (c) Barbara (d) Facade (e) House (f) Lena (g) Peppers (h) Sailboat

assess the recovery result, and if we denote true data as $\mathcal{T}$, RSE and RSNR are defined as

$$\text{RSE} = \frac{\|\mathcal{X} - \mathcal{T}\|_F}{\|\mathcal{T}\|_F} \tag{20}$$

$$\text{PSNR} = 10 \log_{10} \frac{\mathcal{T}_{\max}^2}{\|\mathcal{X} - \mathcal{T}\|_F^2} \tag{21}$$

where $\mathcal{T}_{\max}$ is the maximum value in $\mathcal{T}$. Better recovery result will have a smaller RSE and larger PSNR.

Parameter Settings. The key parameters in our methods are λ_1 and λ_2. Since they are balancing the weights of vertical and horizontal gradient, and in general vertical and horizontal gradient are of the same importance, so we set $\lambda_1 = \lambda_2$. And by experience, we set $\lambda_1 = \lambda_2 = 0.01$ in our experiments. Other parameters such as $\rho_i (i = 1, ..., 5)$ are concerned with the convergence property of the algorithm, and in our experiments we set $\rho_1 = \rho_2 = 0.001$, $\rho_3 = \rho_4 = \rho_5 = 0.1$.

Color Image Inpainting. We first compare our methods with the five state-of-art methods under different missing rates. We use Baboon in Fig. 1 to illustrate the different inpainting performances with random missing entries. As is shown in Fig. 2, our method performs well for both low and high missing rates while others have their limitations. For example, FBCP only performs well when missing rate is high while results of LRTC-TV-I and LRTC-TV-II are not so good with such a high missing rate.

Then we test our method with the eight images in Fig. 1 with 50% random missing entries. The results is shown in Table 1, from which we can see that our method outperforms others in most pictures and only the result on Peppers is an exception.

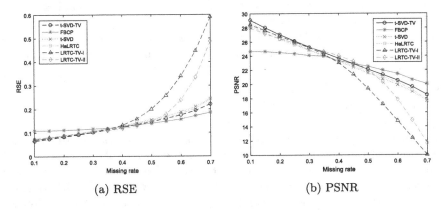

(a) RSE   (b) PSNR

Fig. 2. Result of recovering Baboon with random missing entries

Table 1. Result of recovering different images with 50% missing entries

		Airplane	Baboon	Barbara	Facade	House	Lena	Peppers	Sailboat
PSNR	Ours	**31.1781**	**25.2686**	**31.5562**	**35.0539**	**33.4708**	**31.6148**	29.4887	27.9553
	FBCP	29.5188	24.0503	29.3032	29.7581	31.3138	30.6669	29.1078	27.0701
	t-SVD	30.4749	24.8544	30.8871	34.5542	32.5710	30.8560	28.6451	27.3461
	HaLRTC	30.2067	24.7912	29.7100	33.5470	32.0453	30.0387	29.2286	27.6037
	LRTC-TV-I	29.7039	25.1704	30.3483	30.9438	32.6083	30.9768	30.9704	**27.9658**
	LRTC-TV-II	29.6746	25.2043	29.8204	27.8308	32.4180	31.0287	**31.2462**	27.9656
RSE	Ours	**0.0392**	**0.1016**	**0.0554**	**0.0342**	**0.0363**	**0.0480**	0.0680	**0.0756**
	FBCP	0.0463	0.1166	0.0717	0.0630	0.0466	0.0528	0.0699	0.0819
	t-SVD	0.0423	0.1064	0.0599	0.0363	0.0403	0.0523	0.0748	0.0805
	HaLRTC	0.0440	0.1073	0.0685	0.0407	0.0430	0.0574	0.0703	0.0796
	LRTC-TV-I	0.0463	0.1028	0.0635	0.0550	0.0404	0.0514	0.0573	0.0767
	LRTC-TV-II	0.0463	0.1026	0.0675	0.0787	0.0417	0.0511	**0.0559**	0.0766

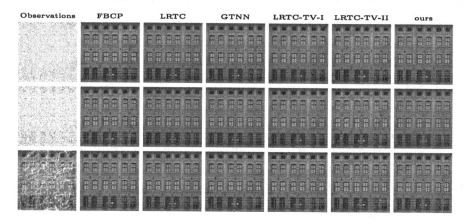

Fig. 3. Comparison of inpainting image Facade. The first line is random missing entries with rate 30%. The second line is random missing pixels with rate 30%. The third line is simulated scratches.

Finally, we present the visual effect of the inpainting algorithms in Fig. 3. The first line is random missing entries with rate 30%. The second line is random missing pixels with rate 30%. The third line is simulated scratches. And we can see that our method performs well in local details while preserving global structures.

5 Conclusions

In this paper, we aim to take both low-rank and total-variation constraint into consideration to complete visual data with missing entries. We propose a novel tensor named gradient tensor by using the t-product framework, then we fuse the novel tensor framework and classic TV together and we verify the effectiveness of our method by experiments. Our future work will focus on design a tensor framework directly that ensure global low-rank property and local smoothness.

Acknowledgements. This work was supported in part by the National Natural Science Foundation of China under Grants U1536204, 60473023, 61471274, and 41431175, China Postdoctoral Science Foundation under Grant No. 2015M580753.

References

1. Braman, K.: Third-order tensors as linear operators on a space of matrices. Linear Algebra Appl. **433**(7), 1241–1253 (2010)
2. Guo, X., Ma, Y.: Generalized tensor total variation minimization for visual data recovery. In: The IEEE Conference on Computer Vision and Pattern Recognition (CVPR), June 2015
3. Kilmer, M.E., Braman, K., Hao, N., Hoover, R.C.: Third-order tensors as operators on matrices: a theoretical and computational framework with applications in imaging. SIAM J. Matrix Anal. Appl. **34**(1), 148–172 (2013)
4. Kilmer, M.E., Martin, C.D.: Factorization strategies for third-order tensors. Linear Algebra Appl. **435**(3), 641–658 (2011)
5. Kolda, T.G., Bader, B.W.: Tensor decompositions and applications. SIAM Rev. **51**(3), 455–500 (2009)
6. Li, X., Ye, Y., Xu, X.: Low-rank tensor completion with total variation for visual data inpainting. In: Thirty-First AAAI Conference on Artificial Intelligence (2017)
7. Liu, J., Musialski, P., Wonka, P., Ye, J.: Tensor completion for estimating missing values in visual data. IEEE Trans. Patt. Anal. Mach. Intell. **35**(1), 208–220 (2013)
8. Liu, T., Tao, D.: On the performance of manhattan nonnegative matrix factorization. IEEE Trans. Neural Netw. Learn. Syst. **27**(9), 1851 (2016)
9. Liu, T., Tao, D., Song, M., Maybank, S.J.: Algorithm-dependent generalization bounds for multi-task learning. IEEE Trans. Patt. Anal. Mach. Intell. **39**(2), 227 (2017)
10. Recht, B.: A simpler approach to matrix completion. J. Mach. Learn. Res. **12**, 3413–3430 (2011)
11. Shi, F., Cheng, J., Wang, L., Yap, P.T., Shen, D.: LRTV: MR image super-resolution with low-rank and total variation regularizations. IEEE Trans. Med. Imaging **34**(12), 2459–2466 (2015)
12. Tao, D., Li, X., Wu, X., Maybank, S.J.: General tensor discriminant analysis and gabor features for gait recognition. IEEE Trans. Patt. Anal. Mach. Intell. **29**(10), 1700–1715 (2007)
13. Tao, D., Li, X., Wu, X., Maybank, S.J.: Geometric mean for subspace selection. IEEE Trans. Patt. Anal. Mach. Intell. **31**(2), 260–274 (2009)
14. Tao, D., Tang, X., Li, X., Wu, X.: Asymmetric bagging and random subspace for support vector machines-based relevance feedback in image retrieval. IEEE Trans. Patt. Anal. Mach. Intell. **28**(7), 1088–1099 (2006)
15. Yan, S., Xu, D., Yang, Q., Zhang, L., Tang, X., Zhang, H.J.: Multilinear discriminant analysis for face recognition. IEEE Trans. Image Process. **16**(1), 212–220 (2007)
16. Zhang, L., Zhang, L., Tao, D., Huang, X.: A multifeature tensor for remote-sensing target recognition. IEEE Geosci. Remote Sens. Lett. **8**(2), 374–378 (2011)
17. Zhang, L., Zhang, L., Tao, D., Huang, X.: Tensor discriminative locality alignment for hyperspectral image spectral-spatial feature extraction. IEEE Trans. Geosci. Remote Sens. **51**(1), 242–256 (2013)

18. Zhang, Z., Ely, G., Aeron, S., Hao, N., Kilmer, M.: Novel methods for multilinear data completion and de-noising based on Tensor-SVD. In: 2014 IEEE Conference on Computer Vision and Pattern Recognition (CVPR), pp. 3842–3849. IEEE (2014)
19. Zhao, Q., Zhang, L., Cichocki, A.: Bayesian CP factorization of incomplete tensors with automatic rank determination. IEEE Trans. Patt. Anal. Mach. Intell. **37**(9), 1751–1763 (2015)

Nighttime Haze Removal with Fusion Atmospheric Light and Improved Entropy

Xing Jin[1,2], Xiang Yang[1,2], Jingjing Zhang[1,2(✉)], and Zhicheng Li[1,2]

[1] School of Automation, China University of Geosciences,
Wuhan 430074, China
{jinxing,xiangyang,work.zhang,20121002364}@cug.edu.cn
[2] Hubei Key Laboratory of Advanced Control and Intelligent Automation
for Complex Systems, Wuhan 430074, China

Abstract. Haze removal is important for the normal work of computer vision system. However, most of the existing image dehazing methods are aimed at daytime haze images. These methods cannot always work well for night haze images since the spatially non-uniform environmental illumination are present at nighttime scenes that can generate glow. This makes nighttime haze removal from single image is an ill-posed problem with challenges. In this paper, we propose a novel algorithm for single nighttime image haze removal. We first remove the glow effects by decomposing the glow image from the nighttime haze image based on a nighttime haze imaging model which can account for spatially non-uniform environmental illumination and the glow effects in the image. Then, we estimate the atmospheric light by combining multiple patch sizes local atmospheric light using multiscale fusion algorithm. Transmission is estimated by maximizing the objective function which is designed by considering the image contrast and color distortion. Finally, haze is removed using the two estimated parameters. Experimental results show that the proposed algorithm can achieve haze-free results while removing the glow effects.

Keywords: Nighttime haze removal · Multi-scale fusion · Entropy

1 Introduction

The quality of images of outdoor scenes are usually degraded by adverse weather conditions like fog and haze. The degraded images often have low contrast and glow effects, which affect the normal performance of many computer vision systems since most of them assume that the input image is the haze-free scene radiance. Therefore, the effective haze removal of image is a work with great significance.

Many existing image dehazing algorithms are designed for daytime haze removal [1–4]. Almost all of these methods rely on the atmospheric scattering model [5] and the estimation of the parameters in the model. Based on this

© Springer Nature Singapore Pte Ltd. 2017
J. Yang et al. (Eds.): CCCV 2017, Part II, CCIS 772, pp. 323–333, 2017.
https://doi.org/10.1007/978-981-10-7302-1_27

model, Tan [1] proposed an image dehazing method by maximizing local image contrast. Fattal [2] estimated medium transmission using statistical independence between shading and albedo. He [3] restored a hazy image using the Dark Channel Prior which was derived from statistics of daytime haze-free images. Even though these methods are effective for daytime haze images, they show great defects under the nighttime hazy scenes. This is mainly due to the spatially non-uniform environmental illumination are present at nighttime scenes that can generate glow.

Recently, there have been some methods for nighttime haze removal [6–9]. Pei et al. [6] propose a color transfer method as a preprocessing step to transform the brightness and color of a nighttime haze image into a daytime haze image. Then the Dark Channel Prior method is used to estimate the transmission as well as bilateral filtering is applied as a post-processing step. Zhang et al. [7] proposed a new model to account for various non-uniform environmental illumination and removed the haze after light compensation and color correction. Ancuti et al. [8] proposed a nighttime haze removal algorithm based on image fusion. They estimated the atmospheric light component on image patch rather on the entire image by generating two inputs using multiple patch sizes and the Laplacian of the original image is defined to be the third input to reduce the glowing effect.

The above methods are effective for haze removal and reduce the effects of glow, but they do not really remove the glow since the models they adopted do not account for glow effects. Li et al. [9] modelled the glow effects by adding the atmospheric point spread function into the daytime haze imaging model. He decomposed the glow image from the original input image and restored the scene radiance using Dark Channel Prior. Although Li's method removed haze and glow effects, the resultant images may contain noise and blocking artifacts. In our work, we propose a novel method to solve these problems. First, we adopt Li's method to decompose the glow image. Then we estimate the atmospheric light using image fusion approach. Finally, transmission is estimated by maximizing the improved entropy.

The rest of paper is organized as follows. Section 2 introduces the nighttime imaging model. Section 3 describes the proposed approach. Section 4 provides the experimental results and comparisons with conventional works and the work is concluded in Sect. 5.

2   Nighttime Haze Imaging Model

In computer vision and computer graphics, the daytime haze imaging model which widely used in many image de-haze approaches is called the atmospheric scattering model [5]. Figure 1(a) shows a diagram of the daytime haze imaging model. The pixel intensity in the captured image is a linear combination of two parts: the direct transmission part and the airlight part. Mathematically, it can be expressed as:

$$I(x) = J(x)t(x) + A(1 - t(x)), \tag{1}$$

where $I(x)$ is the observed intensity at pixel x, $J(x)$ is the scene radiance when there is no haze or fog particles. A is the global atmospheric light, and $t(x)$ is the medium transmission describing the portion of the light that is not scattered and reaches the camera. $J(x)t(x)$ is called direct transmission, and $A(1 - t(x))$ is called atmospheric light. For a haze image $I(x)$, the goal of haze removal is to recover $J(x)$ from $I(x)$.

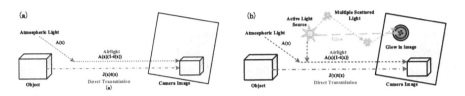

Fig. 1. Nighttime haze imaging model.

In the daytime imaging model, it assumes that the atmospheric light intensity is global constant. Nevertheless, nighttime scenes generally have active light sources that can generate glow when the presence of particles in the atmosphere is substantial. To express the active light sources and glow under the nighttime scenes, Li et al. [9] proposed a novel model for nighttime haze imaging by adding a glow model with an atmospheric point spread function ($APSF$) [10] into the daytime haze imaging model as follows [11]:

$$I(x) = J(x)t(x) + A(x)(1 - t(x)) + A_a(x) * APSF, \tag{2}$$

where $A(x)$ is the atmospheric light which is no longer globally uniform, but space varying. A_a is the active light sources, that the intensity is convolved with $APSF$. $J(x)t(x) + A(x)(1 - t(x))$ is called the nighttime haze image, $A_a * APSF$ is the glow image. Figure 1(b) shows diagram of the nighttime haze model. The active light sources is scattered as it travels through the suspended particles (haze and fog), resulting the glowing effects in the captured image. Given a nighttime haze image $I(x)$, the goal is to restore the scene radiance $J(x)$ by decomposing glow from $I(x)$ and estimating both the transmission $t(x)$ and the varying atmospheric light $A(x)$.

3   Nighttime Haze Removal

Our basic pipeline is illustrated in Fig. 2. It starts by decomposing the original input image into two images: nighttime haze image and glow image, as shown in Fig. 2(b), (c). Then, we estimate the atmospheric light by combining multiple patch sizes local atmospheric light using multiscale fusion algorithm as shown in Fig. 2(j). Transmission is estimated using the nighttime haze image and atmospheric light that has been estimated as shown in Fig. 2(k). Finally, the two estimated parameters are used to restore the scene radiance as shown in Fig. 2(l). In the following, the details of each step are discussed.

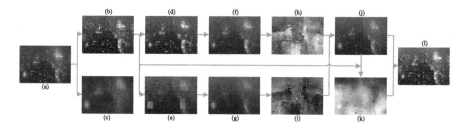

Fig. 2. The framework of the proposed method. (a) Original input image. (b) Night-time haze image. (c) Glow image. (d) Local atmospheric light with size 10×10. (e) Local atmospheric light with size 30×30. (f), (g) Refined results of (d), (e). (h), (i) Normalized weight maps of (f), (g). (j) Fusion atmospheric light. (k) Estimated transmission. (l) Output image.

3.1 Glow Decomposition

In the nighttime scenes, the brightness of the glow decreases gradually and smoothly away from the light sources due to the multiple scattering, it causes that the glow image has a "short tail" distribution in the gradient histogram [11]. Based on this fact, Li et al. proposed the glow decomposition method [11] by layer separation using relative smoothness. In this paper, we adopt Li's approach [9] to decompose the glow image from original input. The objective function for layer separation is defined as:

$$E(R) = \sum_x \left(\rho(R(x) * f_{1,2}) + \lambda((I(x) - R(x)) * f_3)^2 \right)$$
$$s.t.\ 0 \leq R(x) \leq I(x),$$
$$\sum_x R_r(x) = \sum_x R_g(x) = \sum_x R_b(x). \tag{3}$$

where $R(x) = J(x)t(x) + A(x)(1-t(x))$ is the nighttime haze image, $f_{1,2}$ is the two direction first order derivative filters, f_3 is the second order Laplacian filter and the operator $*$ denotes convolution. $\rho(s) = \min(s^2, \tau)$ is a robust function which preserve the large gradients of input image I in the remaining nighttime haze layer $R(x)$. λ is the parameter that controls the smoothness of the glow layer. The first inequality constraint is to ensure the solution is in a proper range, and the second constraint is to force the range of the intensity values for difference color channels to be balanced. The objective function in Eq. (3) can be solved efficiently using the half-quadratic splitting technique [11].

3.2 Atmospheric Light Estimation

After glow decomposition, we get the nighttime haze image. To obtain the haze-free scene radiance, we still need to estimate the atmospheric light and transmission. In a previous work [9], Li assumed that atmospheric light is locally constant and the brightest intensity in a local area is the atmospheric light of that area. He split the original input image into a grid of small patches (15×15)

and found the brightest pixel in each area as atmospheric light. However, Li's estimation of the varying atmospheric light is admittedly a rough approximation that suffers from two weaknesses. First, there may be of the boosting noise and blocking artifacts in the de-hazed results as shown by the sky region of Fig. 3(b). Second, it is very hard to select the size of the patches, since small patches are desirable to achieve fine spatial adaptation to the atmospheric light, but might also induce poor light estimates and reduced chance of capturing hazy pixels. By choosing a too large patch size, the haze is better removed, but the color might be shifted, the influence of the airlight might not be entirely removed and some details may remain poorly restored [8].

(a) Nighttime haze image   (b) Li's work   (c) Our work

Fig. 3. Dehazed results using different atmospheric light estimation methods.

To circumvent this problem, we employ a multiscale fusion approach to estimate the atmospheric light [8]. We assume that the atmospheric light is locally constant and define the local atmospheric light intensity to be:

$$A^c(x) = \max_{y \in \psi(x)} \left[\min_{z \in \Omega(y)} (I^c(z)) \right] = \max_{y \in \psi(x)} [I^c_{MIN}(z)], \tag{4}$$

where I is the observed intensity, A is the local atmospheric, $c \in \{r, g, b\}$ is the color channel index, and ψ, Ω are local patches. In this paper, all results have been generated using patches ψ twice the size of Ω.

Our fusion process approach is implemented in three main steps. First, based on the formula for estimating atmospheric light, we derive the two inputs of the fusion approach by using different sizes of the patches to estimate atmospheric light. As shown in Fig. 2(d), (e), for an original input image of size 500×300, we take the patch size 10×10 and 30×30, respectively. We then apply the guided image filter [12] as a post-process, the atmospheric light after refined is shown in Fig. 2(f), (g). In the second step, we calculate the corresponding normalized weight maps according to the contrast and saturation of the two inputs as shown in Fig. 2(h), (i). Finally, the derived inputs and the normalized weight maps are blended by means of multi-resolution fusion algorithm using a Laplacian pyramid decomposition of the inputs and a Gaussian pyramid of the normalized weights. By using these maps, we can get the fused atmospheric light as shown in Fig. 2(j). Figure 3 compares the final results of using the Li's method [9] and our proposed method, our result have a good visual effect without blocking artifacts at the sky region.

3.3   Transmission Estimation

As written in (2), once we estimate the Atmospheric light, the scene radiance depends on the transmission. In previous methods, transmission was estimated by using the Dark Channel Prior [3] even though they used different imaging models. As we all know, the Dark Channel Prior is a statistical result based on the observation on a great deal of daytime outdoor haze-free images. When it is applied to nighttime scenes, this assumption may fail and lead to inaccurate estimation of transmission.

Note that the entropy of haze-free image is bigger than that of hazy image at same scene, Park et al. [13] proposed a novel method to estimate transmission by maximizing the image entropy at non-overlapped sub-block regions without using any prior knowledge. The objective function he proposed is comprised of two functions. The first one is the image entropy $f_{entropy}$ as a contrast measure, which has been proven can be used to characterize texture of an image [14,15]. An image entropy can be expressed by the function of transmission t as follows:

$$f_{entropy}(t) = -\sum_{i=0}^{255} \frac{h_i(t)}{N} \log \frac{h_i(t)}{N}, \tag{5}$$

where N is the number of pixels in the image, $h_i(t)$ is the number of pixels that have intensity i in the gray-scaled image of scene radiance J calculated from (2), when the transmission is set to t. However, the de-hazed image may take values smaller than 0 or larger than 255 when transmission t is too small. Therefore, the second objective functions is designed to restrict excessive overflow and underflow as follows:

$$f_{fidelity}(t) = \min_{c \in \{r,g,b\}} s^c(t), \tag{6}$$

$$s^c(t) = \frac{1}{N}\sum_{p=1}^{N} \delta(p), \ \delta(p) = \begin{cases} 1, \ 0 \le J^c(p) \le 255 \\ 0, \ otherwise \end{cases} \tag{7}$$

where, $s^c(t)$ expresses the ratio of pixels between 0 and 255 at each color channel of the de-hazed image J when the transmission is set to t. Note that as the number of overflow and underflow pixels becomes fewer, $f_{fidelity}(t)$ becomes larger. The final objective function is defined as:

$$f_{objective}(t) = f_{entropy}(t) \cdot f_{fidelity}(t), \tag{8}$$

by maximizing the objective function in (8), the transmission t can be estimated from the daytime haze image. However, as shown in Fig. 4(b), maximizing this objective function of nighttime haze image may cause color distortion since the contrast has been over enhanced. In order to solve this problem, we introduce the third objective functions $f_{hue}(t)$ to indicate the color distortion before and after haze removal. It can be defined as:

$$f_{hue}(t) = \frac{h_J(t) - h_R(t)}{N}, \tag{9}$$

(a) Nighttime haze image   (b) Park's work   (c) Our work

Fig. 4. Dehazed results using transmission estimated by different objective functions.

$$h(t) = \arctan\left(\frac{O_{2x}(t)}{O_{1x}(t)}\right), \tag{10}$$

$$\begin{cases} O_{1x}(t) = \frac{R_x(t) - G_x(t)}{\sqrt{2}} \\ O_{2x}(t) = \frac{R_x(t) + G_x(t) - 2B_x(x)}{\sqrt{6}} \end{cases}, \tag{11}$$

where N is the number of pixels in the image, $O_{1x}(t)$ and $O_{2x}(t)$ are the two component of the opponent color space, $h(t)$ is corresponding to the hue of an image, $h_J(t)$ and $h_R(t)$ are the hue of scene radiance and nighttime haze image, respectively, when the transmission is set to t. In theory, $h_J(t)$ and $h_R(t)$ are equal [16]. However, the existing haze removal algorithms are difficult to achieve the desired effect, resulting the value of $h_J(t)$ and $h_R(t)$ are usually biased. Therefore, the color distortion after haze removal can be measured by calculating the degree of variation of $f_{hue}(t)$. Note that as the image has less color distortion after haze removal, $f_{hue}(t)$ becomes smaller. In other words, we can reduce the color distortion effectively by minimizing the proposed color distortion measure. Consequently, the improved objective function we propose is defined as:

$$f_{objective}(t) = \frac{f_{entropy}(t) \cdot f_{fidelity}(t)}{\max(f_{hue}(t), 0.01)}, \tag{12}$$

the minimum $f_{hue}(t)$ value is set by 0.01 to prevent division by zero. The objective function we proposed can well satisfy two quotas of contrast and color distortion. By maximizing the objective function in (12), we can estimate the transmission which provides good contrast and faithful dehazing results without distortions as shown in Fig. 4(c).

However, since the values of the transmission in outdoor hazy image is non-homogeneous and space-varying [17], local optimal transmission is estimated at each of the non-overlapped sub-block regions as follows:

$$t_k^{block} = \arg \max_{t \in \{0.01 \leq t \leq 1\}} f_{objective}(t), \tag{13}$$

where t_k^{block} is the k-th sub-block which is divided by pre-specified block size from a hazy image, the minimum t value is set by 0.01 to prevent division by zero. The size of the block has an important impact on the estimation of the t. If we choose a larger block, transmission may have different values in this region. Conversely, if the block we choose is too small, the estimated transmission may be inaccurate due to the insufficient number of pixels. We set a medium block size to 15×15 in this paper. We find the optimal solution by exhaustive searching of t with 0.01 step size within the range of 0.01 to 1.0, which achieves high computational efficiency than traditional optimization algorithms.

Since the transmission from (13) is estimated by non-overlapped local region, halo effect and block artifact may occur. We chose the guided filter [12] to alleviate those artifacts. Figure 2(k) shows the refined transmission map.

(a) Input images   (b) Pei's work   (c) Zhang's work   (d) Li's work   (e) our work

Fig. 5. The comparison of the proposed method with conventional methods.

4   Experimental Results

To demonstrate the effectiveness of the proposed algorithm, we evaluated our method compared our method with conventional methods [6,7,9]. Figure 5 shows the comparison of the proposed method with Pei's work [6], Zhang's work [7] and Li's work [9]. As can be seen from Fig. 5, Pei's method and Zhang's method can remove haze effectively, but there are color distortion, and glow effects are still visible since their model did not account for glow effects. Although Li's method removes the glow effects effectively, there are blocking artifacts at the sky area as well as areas with low brightness values look unnatural. The method we proposed can remove the glow effects effectively while provide better visual effect overall as shown in Fig. 5(e).

We also compared performance of the proposed method with other conventional methods [3,6,9] whose results are shown in Fig. 6 in terms of image entropy [15], structural similarity index (SSIM) [18] and peak signal to noise

(a) Input image   (b) He's work   (c) Pei's work   (d) Li's work   (e) Our work

Fig. 6. The comparison of the proposed method with other conventional methods.

ratio (PSNR) [19]. Image entropy is a measure of image contrast that can be used to characterize texture of an image. High entropy means that the image can present more detail and texture features. SSIM index can measure the structural similarity of the dehazed image to the original image. Table 1 shows the quantitative results in Fig. 6. From the statistical data in Table 1, we can see that the entropy and PSNR of our result improves obviously with the SSIM is very close to the others. The comparison results show that our result can provide more detailed texture features while maintaining high structural similarity.

Table 1. Quantitative measurements of results in Fig. 6.

Index	He	Pei	Li	Our
Entropy	17.6432	18.1718	20.7183	**21.8546**
SSIM	0.9991	0.9984	**0.9993**	0.9992
PSNR	28.3678	25.6779	28.8196	**29.7488**

5 Conclusion

In this paper, we propose an efficient algorithm for nighttime dehazing. We exploit a new nighttime haze imaging model which takes into account both the non-uniform illumination from artificial light sources and the scattering and attenuation effects of haze. Based on the new model, we first decompose the glow image from the nighttime haze image. Then, we estimate two parameters, the atmospheric light and transmission. Finally, scene radiance is restored using the two estimated parameters. Experimental results on nighttime hazy images show that the proposed algorithm can successfully enhance image visibility while provides a good visual effect.

Acknowledgments. This research was supported partially by the National Natural Science Foundation of China (NSFC) (61604135), the National Key Research and Development Program of China (2016YFF010020002), and the National Key Scientific Instrument and Equipment Development Projects of China (2012YQ0901670102).

References

1. Tan, R. T.: Visibility in bad weather from a single image. In: IEEE Conference on Computer Vision and Pattern Recognition (CVPR), pp. 1–8. IEEE Press, Anchorage (2008)
2. Fattal, R.: Single image dehazing. ACM Trans. Graph. (TOG) 27(3), 72 (2008)
3. He, K., Sun, J., Tang, X.: Single image haze removal using dark channel prior. IEEE Trans. Pattern Anal. Mach. Intell. 33(12), 2341–2353 (2011)
4. Tang, K., Yang, J., Wang, J.: Investigating haze-relevant features in a learning framework for image dehazing. In: IEEE Conference on Computer Vision and Pattern Recognition (CVPR), pp. 2995–3000 (2014)
5. Narasimhan, S.G., Nayar, S.K.: Vision and the atmosphere. Int. J. Comput. Vision 48(3), 233–254 (2002)
6. Pei, S. C., Lee, T. Y.: Nighttime haze removal using color transfer pre-processing and dark channel prior. In: IEEE International Conference on Image Processing (ICIP), pp. 957–960 (2012)
7. Zhang, J., Cao, Y., Wang, Z.: Nighttime haze removal with illumination correction. arXiv preprint arXiv:1606.01460 (2016)
8. Ancuti, C., Ancuti, C. O., Vleeschouwer, C. D., Bovik, A. C.: Nighttime dehazing by fusion. In: IEEE International Conference on Image Processing (ICIP) pp. 2256–2260 (2016)
9. Li, Y., Tan, R. T., Brown, M. S.: Nighttime haze removal with glow and multiple light colors. In: International Conference on Computer Vision (ICCV), pp. 226–234 (2015)
10. Narasimhan, S. G., Nayar, S. K.: Shedding light on the weather. In: IEEE Conference on Computer Vision and Pattern Recognition (CVPR), pp. I-665–I-672 (2003)
11. Li, Y., Brown, M.S.: Single image layer separation using relative smoothness. In: IEEE Conference on Computer Vision and Pattern Recognition (CVPR), pp. 2752–2759 (2014)
12. He, K., Sun, J., Tang, X.: Guided image filtering. In: Daniilidis, K., Maragos, P., Paragios, N. (eds.) ECCV 2010. LNCS, vol. 6311, pp. 1–14. Springer, Heidelberg (2010). https://doi.org/10.1007/978-3-642-15549-9_1
13. Park, D., Park, H., Han, D. K., Ko, H.: Single image dehazing with image entropy and information fidelity. In: IEEE International Conference on Image Processing (ICIP), pp. 4037–4041 (2015)
14. Deng, G.: An entropy interpretation of the logarithmic image processing model with application to contrast enhancement. IEEE Trans. Image Process. 18(5), 1135 (2009)
15. Gull, S.F., Skilling, J.: Maximum entropy method in image processing. Commun. Radar Sig. Process. IEE Proc. F 131(6), 646–659 (1984)
16. Chambah, M., Rizzi, A., Gatta, C., Besserer, B., Marini, D.: Perceptual approach for unsupervised digital color restoration of cinematographic archives. In: SPIE-IS&T Electronic Imaging, vol. 5008, pp. 138–149 (2003)
17. Kim, J.H., Sim, J.Y., Kim, C.S.: Single image dehazing based on contrast enhancement. In: IEEE International Conference on Acoustics, Speech and Signal Processing, vol. 7882, pp. 1273–1276 (2011)

18. Wang, Z., Bovik, A.C., Sheikh, H.R., Simoncelli, E.P.: Image quality assessment: from error visibility to structural similarity. IEEE Trans. Image Process. **13**(4), 600–12 (2004). A Publication of the IEEE Signal Processing Society
19. Hore, A., Ziou, D.: Image quality metrics: PSNR vs. SSIM. In: International Conference on Pattern Recognition, pp. 2366–2369 (2010)

Light Field Super-Resolution Using Cross-Resolution Input Based on PatchMatch and Learning Method

Mandan Zhao$^{(\boxtimes)}$, Xiangyang Hao, Chuanqi Cheng, and Jiansheng Li

Information Engineering University, Zhengzhou, China
mandanzhao@163.com, xiangyanghao2004@163.com, legeng3q@163.com,
ljszhx@163.com

Abstract. This paper addresses the problem of generating a super-resolution (SR) image from a low-resolution (LR) image assisted by nearby high-resolution (HR) image. The scaling factor of the super-resolved image is up to 8 times and even more, which is much larger than the ordinary super-resolution scaling factor. Combined patch match and learning based method for image super-resolution using a cross-resolution input. The method is used to super-resolve the images captured by a hybrid light field system consisting of a standard LF camera and a HR DSLR camera. We take the central high-resolution image as a reference, to deal with the around low-resolution images. Unlike other relative algorithm, the proposed method is exploited for the existence of large parallax between the captured images. The main process of our method is a combination of patch based (i.e., example based) algorithm and learning based (e.g., convolutional neural network) method, and does not require any calibration information. Experimental results show that our proposed method performs better than existing method on challenging scenes containing complex texture, specularity and large parallax. Both accuracy and visual improvements in our results are noticeable.

Keywords: Super resolution · Patchmatch based method
Light field

1 Introduction

In recent years, light field imaging [5] becomes one of the most extensively used method for capturing the 3D appearance of a scene. It measures the spatial and angular variations in the intensity of light [18]. In the early years, light field cameras required expensive hardware, such as multi-camera arrays [30]. Although recently commercial and industrial light field cameras such as Lytro [6] and RayTrix [1] are introduced, they still suffer from restricted sensor resolution which hamper their widespread appeal and adoption. Light field cameras must make a trade-off between spatial and angular resolution.

M. Zhao—Student author.

© Springer Nature Singapore Pte Ltd. 2017
J. Yang et al. (Eds.): CCCV 2017, Part II, CCIS 772, pp. 334–344, 2017.
https://doi.org/10.1007/978-981-10-7302-1_28

In this paper, we propose a highly accurate multi-view super-resolution method based on the light-field system introduced in [28]. The key idea of the proposed synthesis is a combination of CNN and patch-based method and the method fully uses the correspondence between images in different views. The patch-based method finds the textural similarity between the high resolution image and the low resolution image. However, when dealing with challenging scenes such as large parallax, the correspondence of the high resolution image and the low resolution image is much lower. As a result, the patch-based method is not effective enough to deal with the edges of the objects and the specularity and results in ghosting and blurring. Instead of pairing correspondence between images, VDSR (Very Deep convolutional networks for Super Resolution [19]) super-resolves images in a single view and fully extract the features of the low-resolution image, so it can cover the shortage of the patch-based method. Our method combines these two methods and takes advantage of both. On the one hand, the method explores the similarity between images. On the other hand, dealing with challenging scenes effectively becomes possible.

Our key technical contributions are: (a) the proposed method is able to dispose challenge scenes, especially large parallax; (b) the method performs better than existing methods in accuracy; (c) the super-resolution process is simple and effective and results in less time cost. Experimental result demonstrates that the proposed method obviously improves the quality of the reconstructed high resolution light field.

2   Related Work

Light fields [5] provide a new angular dimension allowing for various visual applications such as light field display [29] and light field microscopy [21]. Many recent works try to capture or synthesize high quality light fields form different types of input data. Light field super-resolution, which aims to improve the spatial resolution of the light fields, is a hotspot of these works.

2.1   Single Image Super-Resolution Method Using CNNs

Image super-resolution using deep convolutional networks is first introduced in [14]. The method differs fundamentally from existing external example-based approaches, in that the method does not explicitly learn the dictionaries [31] or manifolds [4] for modeling the patch space. SuperResolution Convolutional Neural Network (SRCNN) is a representative method for deep-learning-based SR approach and has been used with large improvements in accuracy. Kim *et al.* [19] has proposed a simple yet effective training procedure that learns residuals only and performs better than SRCNN in accuracy, which is called VDSR. We train our model based on the algorithm illustrated in [19].

We notice that VDSR utilizes contextual information spread over large image regions and performs better than the patch-based algorithms in the position of the edge of the object and the regions full of complex texture. However, VDSR is based on the single image and it could not fully use the information of different views.

2.2 Light Field Super-Resolution

For single camera light-field, the disadvantage is low spatial resolution. Recently there are several methods composed to restore high frequency information.

Spatial Super-Resolution. To increase the spatial resolution, Bishop *et al.* [7] proposed a method to estimate both high resolution depth map and light in the Bayesian framework under the prior of Lambertian textural. Another method, patch matching based techniques, are widely used in image processing, such as texture synthesis [15], image completion [25], denoising [8], deblurring [12] and image super-resolution [16,17]. Wanner and Goldluecke [28] introduced a hybrid imaging system using a patch-based algorithm. Cho *et al.* [11] explicitly model the calibration pipeline of Lytro cameras and propose a learning based interpolation method to obtain higher spatial resolution. However, the quality of the recovered light field images is not as good as that of the input high resolution images. The spatial high frequency details are lost in the super-resolution recovered images.

Angular Super-Resolution. To reconstruct novel views from sparse angular samples, some methods require the input to follow a specific pattern, or to be captured in a carefully designed way. For example, the work by Levin and Durand [20] takes in a 3D focal stack sequence and reconstructs the light field, using a prior based on the dimensionality gap. Shi *et al.* [24] leverage sparsity in the continuous Fourier spectrum to reconstruct a dense light field from 1D set of viewpoints. Marwah *et al.* [23] propose a dictionary-based approach to reconstruct light fields a coded 2D projection.

2.3 Hybrid Imaging

The idea of hybrid imaging was proposed in the context of motion deblurring [3], where a low resolution high speed video camera co-located with a high resolution still camera was used to deblur the blurred images. On the basis of that work, several examples of hybrid imaging have found utility in different applications. Cao *et al.* [9] propose a hybrid imaging system consisting of an RBG video camera and an LR multi-spectral camera to produce HR multi-spectral video using a co-located system. Another example of a hybrid imaging system is the virtual view synthesis system proposed by Tola *et al.* [26], where four regular video cameras and a time-of-flight sensor are used. They show that by adding the time-of-flight camera they could render better quality virtual views than just using a camera array with similar sparsity. Wang *et al.* [27] introduced another light-field attachment which combines a DSLR and 8 low-quality cameras around. They improve the accuracy of the super-resolved images but the algorithm of the synthesize is too complex and needs several times of iterations which limits the speed.

Accordingly, our method integrates patch-based method and VDSR, which makes use of the advantages of the two techniques.

3   Proposed Method

This section introduces the proposed patch-based method integrated convolutional networks for super-resolution of the side view images. The configuration is outlined in Fig. 1. Here is our basic idea: using patch-based method to fix the error of the image super-resolved by VDSR.

We consider an input of two images: a high-resolution image (the reference image which we denote as *Ref*), and a low-resolution image (we denote them as *Src*). The two images show the same scene in two different views, and the distance of the two views is 10 pixels in the light field.

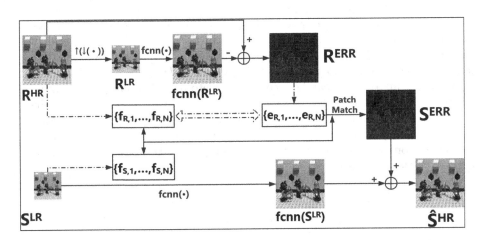

Fig. 1. The overview of the proposed patch based method integrated CNN.

3.1   Compute Initial Error

In this step, we aims to calculate an error map which presents the error of the image super-resolved by VDSR. As we known, the scaling factor is too large in our experiment's setup. For the single image super-resolution, it cannot handle under these too large factor. So we draw into the reference high-resolution image.

By down-sampling the *Ref* by a factor of N, we obtain the image R_{low} which is the same size as the *Src*. It is noted that the factor N is the result of the size of *Ref* divided by the size of *Src*. Then we super-resolve the R_{low} by a factor of N using VDSR, and denote it as R_{high}. The initial error map is obtained by subtracting *Ref* and R_{high}.

$$R_{error} = Ref - R_{high} \tag{1}$$

The very deep convolution network (VDSR) is inspired by [13]. This residual-learning network converges much faster than the standard CNN and gives a significant boost in performance.

In this paper, we note that the residual-learning network is not conflict with the error map between the high-resolution and low-resolution images. The residual-learning network is not the access to obtain the aim results. It is the more efficiency method.

3.2 Patch-Based Estimation

Now we get the error map between R_{high} and Ref from the first step. In this step, we use the patch-based method based on the error map (which denoted as R_{error}) at the view of Ref. We adopt the available patch match-based super-resolution method which improves the algorithm in [28]. In this step, we first build the dictionary D_{error} consisting of the extracted patches from the error map R_{error}. Then we extract patches from R_{high} to build dictionary D_{high}. Low resolution features are computed from each of the patches in D_{high} by down-sampling by a factor of N using the first and second order derivatives filters. The low resolution features are stored in dictionary D_{low}.

Gradient information can be incorporated into patch matching algorithms to improve accuracy when searching for similar patches. Chang *et al.* [10] use first- and second-order derivatives as features to facilitate matching. The PatchMatch based method also use first- and second-order gradients as the feature which is extracted from the low-resolution patches. The four 1-D gradient filters used to extract the features are:

$$g_1 = [-1, 0, 1], g_2 = g_1^T \tag{2}$$

$$g_3 = [1, 0, -2, 0, 1], g_4 = g_3^T \tag{3}$$

where the superscript "T" denotes transpose. For a low-resolution patch l, filter $\{g_1, g_2, g_3, g_4\}$ are applied and feature f_l is represented as concatenation of the vectorized filter outputs.

To super-resolve Src, the features f_j, which are calculated from each patch l_j of Src, are used to match. The 9 nearest neighbors in D_{low} with the smallest L_2 distance from f_j are computed. These 9 nearest neighbors in D_{low} (denoted as $\{f_{ref,k}^j\}_{k=1}^9$) correspond to 9 HR patches in D_{high} and these 9 HR patches maps 9 error patches in D_{error} (denoted as $\{e_{ref,k}^j\}_{k=1}^9$). Then the reconstruction weights motivated from [28] are calculated. The estimated error patches $\hat{e}_j$ corresponding to l_j is estimated by:

$$\hat{e}_j = \frac{\sum_{k=1}^9 w_k e_{ref,k}^j}{\sum_{k=1}^9 w_k}, w_k = exp\frac{-||f_j - f_{ref,k}^j||^2}{2\sigma^2} \tag{4}$$

So we get an error image (which denoted as S_{error}) at the view of Src with the sum of similarity weighted error patches from the dictionary D_{error}. We follow the same parameter setting in [28]. The patch size of high resolution patches is 64×64 and the patch size of low resolution patches is determined by the factor N. The S_{error}, which indicates the error of VDSR method at the view of Src, has the same size of the high-resolution image Ref.

3.3   Integrated Super-Resolution

We have got the error map S_{error} at the view of Src, which indicates the error of VDSR method. So in this step, we will integrate two proposed method to fully use the correspondence of the images of different views. Firstly, Src is super-resolved by VDSR and we denote the result as S_{cnn}. Then we add S_{cnn} and S_{error}. In this way, the defect of the VDSR super-resolved image is made up by the S_{error}. The final result is generated.

Here we explain why we compute S_{error} in the process of the synthesis. Patch-based method finds the textural similarity between the HR image and LR image, and the super-resolved image is a sum of the HR patches. At the scene of large parallax, the edges of the objects in image are quite different so the patch-based method results in blur. Also, the specularity cannot be restored well. VDSR can cover the shortage of the patch-based method due to the single view information. We combine two methods to take advantage of both.

4   Experimental Results

We evaluate the performance of our proposed method for side views and dense light field rendering on the Standford light field dataset [2] in several different scenes, including challenging scenes such as complex textures, specularity and large parallax.

Table 1. The superresolve scale is $\times 4$.

Dataset\method	Bicubic	VDSR	Patch based method	Our method
Cotton	40.5204	**43.6644**	39.9557	43.2467
ele	46.7392	48.5263	47.0081	**48.6014**
Flower	34.5007	37.0903	36.9175	**38.2336**
hippocampus	53.1375	50.5403	**53.3378**	51.0375
LegoBulldozer	31.479	34.6345	31.9685	**35.2196**
LegoGantry	28.8707	31.1338	31.2648	**32.544**
rectifiedsmall angular	26.0773	28.0312	27.4733	**29.2274**
Truck	33.8503	35.3746	37.1094	**38.0753**
Worm	39.7295	**44.3569**	38.8579	40.9949
JellyBeans	43.0806	45.5689	43.214	**45.9222**
Knights	30.9126	33.6056	31.5034	**34.0243**
Amethyst	32.6559	34.3501	34.8299	**35.7121**
TreasureChest	28.4255	29.6139	28.5932	**29.8761**
Average PSNR	36.1525	38.1916	37.0795	**38.6704**

Table 2. The superresolve scale is ×8.

Dataset\method	Bicubic	VDSR	Patch based method	Our method
Cotton	32.7439	35.2759	32.6775	**35.3286**
ele	41.7983	**43.3827**	40.606	43.0605
Flower	26.6311	27.7077	30.8944	**30.9321**
hippocampus	48.6558	48.0739	**50.7418**	48.6922
LegoBulldozer	27.048	29.261	27.173	**29.515**
LegoGantry	24.8678	26.6092	27.7943	**28.9813**
rectifiedsmall angular	22.6408	23.9477	22.8769	**24.3882**
Truck	30.4047	31.4772	33.4653	**33.8647**
Worm	34.7553	**39.1384**	34.8951	36.1801
JellyBeans	36.5531	40.4824	36.9044	**40.5603**
Knights	26.9367	28.9750	27.2989	**29.2336**
Amethyst	29.0107	30.1847	31.6159	**32.0057**
TreasureChest	25.5891	26.2591	25.5121	**26.2956**
Average PSNR	31.3566	33.1365	32.4966	**33.7722**

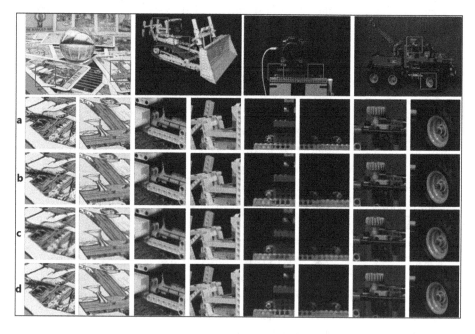

Fig. 2. Super-resolution results comparision between 3 methods. The super-resolve scale is ×4. From top to bottom: (a) ground truth, (b) VDSR, (c) patch based method, (d) our method.

4.1 Experiment Setup

For Standford data set, we select 9 views from each light field with similar layout to the light-field attachment. To make the scene challenging, we select the side view image with $d = 10$ in 8-adjacency distance from the central view. We evaluate our method in two different scales: $\times 4$, $\times 8$. The input low resolution side view images are obtained by down-sampling of each image with these two factors, and the original high resolution images can act as ground truth. For patch-based super-resolution, we follow the same setting up with [28]. For VDSR, we set the initial training parameters the same as [13]. In the end, we also text several microscope light field datasets, e.g. provided by Lin *et al.* [22].

4.2 Super-Resolution Results

We evaluate our method on all light fields in the dataset [2]. The PSNR values of the patch-based super resolution images, VDSR images and our super-resolution images of several listed scenes are shown in Tables 1 and 2. It is noticed that the PSNRs of our method are higher than those of patch-based method and VDSR method, reflecting in both two scales. It is due to the fact that our method fully use the correspondence of the images in different views and takes advantages of two kinds of synthesis.

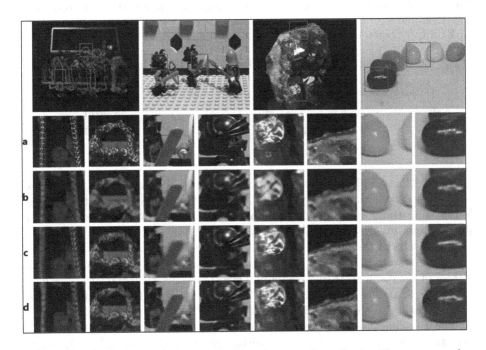

Fig. 3. Super-resolution results comparision between 3 methods. The super-resolve scale is $\times 8$. From top to bottom: (a) ground truth, (b) VDSR, (c) patch based method, (d) our method.

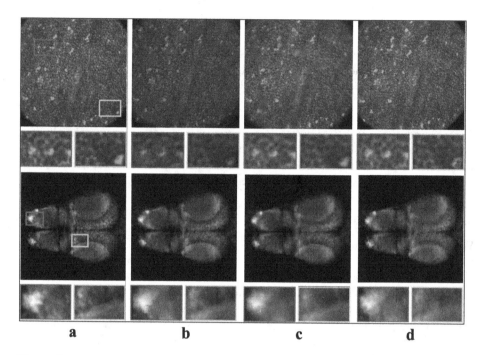

Fig. 4. Super-resolution results comparision between 3 methods in the microscope light field (Cells and Eye). The super-resolve scale is ×4. From left to right: (a) ground truth, (b) VDSR, (c) patch based method, (d) our method.

Table 3. The superresolve scale is ×4 in the microscope light field datasets.

Dataset\method	Bicubic	VDSR	Patch based method	Our method
Cells	32.1754	33.2491	33.4501	**35.9071**
Eye	33.1209	35.6751	35.6803	**38.8841**

Figures 2 and 3 illustrates some super-resolution patches cropped the simulations. It is obvious that the patches of out method contain better high frequency details than those of patch-based method. The patch-based method results in blurring and our method alleviates this mistake.

The results of the microscope light field are also presented in the Fig. 4. The top of the results is the Cells dataset, which be all in a muddle situation. The bottom of the results is the fly compound eye. These two typical microscope light field are unstructure and dusky. So the super-resolved results of our method are more similarity to the groundtruth (Table 3).

The run-time for our proposed algorithm is about 3 min per picture. The algorithm was implemented in C++ without optimization on an Intel i7 fourth generation processor with 32 GB of RAM. Compared to the synthesis in [27], the speed is much faster.

5   Conclusion

In this work, we proposed a highly accurate multi-view super-resolution method which is used to super-resolve the images captured by light field system. The main process of our method is a combination of patch based algorithm and convolutional neural network. Our method performs better in accuracy than existing method on challenging scenes containing complex texture, specularity and large parallax, while costing less time. Experimental result demonstrates that the proposed method obviously improved the quality of the reconstructed high resolution light field.

In the feature, we would like to utilize the natural property of the light field, which we will reach a better super-resolved results. Besides, some applications should be extended, such as the depth estimation, images sequence interpolation, and so on.

References

1. RayTrix: 3D light field camera technology. http://www.raytrix.de/
2. The (new) Standford light field archive. http://www.raytrix.de/
3. Ben-Ezra, M., Nayar, S.K.: Motion deblurring using hybrid imaging. In: Proceedings of the IEEE Conference on Computer Vision and Pattern Recognition (CVPR), pp. 657–664 (2003)
4. Bevilacqua, M., Roumy, A., Guillemot, C., Morel, A.: Low-complexity single-image super-resolution based on nonnegative neighbor embedding. In: BMVC (2012)
5. Bishop, T.E., Favaro, P.: The light field camera: extended depth of field, aliasing, and superresolution. IEEE Trans. Pattern Anal. Mach. Intell. **34**(5), 972–986 (2012)
6. Bishop, T.E., Favaro, P.: The light field camera: extended depth of field, aliasing, and superresolution. IEEE Trans. Pattern Anal. Mach. Intell. **34**(5), 972 (2012)
7. Bishop, T.E., Zanetti, S., Favaro, P.: Light field superresolution. In: 2009 IEEE International Conference on Computational Photography (ICCP), pp. 1–9. IEEE (2009)
8. Buades, A., Coll, B., Morel, J.M.: A non-local algorithm for image denoising. In: IEEE Computer Society Conference on Computer Vision and Pattern Recognition (CVPR 2005), vol. 2, pp. 60–65. IEEE (2005)
9. Cao, X., Tong, X., Dai, Q., Lin, S.: High resolution multispectral video capture with a hybrid camera system. In: Proceedings of the IEEE Conference on Computer Vision and Pattern Recognition (CVPR), pp. 297–304 (2011)
10. Chang, H., Yeung, D.Y., Xiong, Y.: Super-resolution through neighbor embedding. Vis. Pattern Recognit. **1**, 275–282 (2004)
11. Cho, D., Lee, M., Kim, S., Tai, Y.W.: Modeling the calibration pipeline of the Lytro camera for high quality light-field image reconstruction. In: ICCV (2013)
12. Cho, S., Wang, J., Lee, S.: Video deblurring for hand-held cameras using patch-based synthesis. ACM Trans. Graph. (TOG) **31**(4), 64 (2012)
13. Dong, C., Loy, C.C., He, K., Tang, X.: Learning a deep convolutional network for image super-resolution. In: Fleet, D., Pajdla, T., Schiele, B., Tuytelaars, T. (eds.) ECCV 2014. LNCS, vol. 8692, pp. 184–199. Springer, Cham (2014). https://doi.org/10.1007/978-3-319-10593-2_13

14. Dong, C., Loy, C.C., He, K., Tang, X.: Image super-resolution using deep convolutional networks. IEEE Trans. Pattern Anal. Mach. Intell. **38**(2), 295–307 (2016)
15. Efros, A.A., Freeman, W.T.: Image quilting for texture synthesis and transfer. In: Proceedings of the 28th Annual Conference on Computer Graphics and Interactive Techniques, pp. 341–346. ACM (2001)
16. Freedman, G., Fattal, R.: Image and video upscaling from local self-examples. ACM Trans. Graph. (TOG) **30**(2), 12 (2011)
17. Freeman, W.T., Jones, T.R., Pasztor, E.C.: Example-based super-resolution. IEEE Comput. Graph. Appl. **22**(2), 56–65 (2002)
18. Ihrke, I., Restrepo, J., Mignard-Debise, L.: Principles of light field imaging: briefly revisiting 25 years of research. IEEE Signal Process. Mag. **33**(5), 59–69 (2016)
19. Kim, J., Kwon Lee, S., Mu Lee, K.: Accurate image super-resolution using very deep convolutional networks. In: Proceedings of the IEEE Conference on Computer Vision and Pattern Recognition, pp. 1646–1654 (2016)
20. Levin, A., Durand, F.: Linear view synthesis using a dimensionality gap light field prior. In: IEEE Computer Society Conference on Computer Vision and Pattern Recognition (CVPR 2010), pp. 1831–1838. IEEE (2010)
21. Levoy, M., Ng, R., Adams, A., Footer, M., Horowitz, M.: Light field microscopy. ACM Trans. Graph. (TOG) **25**(3), 924–934 (2006)
22. Lin, X., Wu, J., Zheng, G., Dai, Q.: Camera array based light field microscopy. Biomed. Opt. Express **6**(9), 3179–3189 (2015)
23. Marwah, K., Wetzstein, G., Bando, Y., Raskar, R.: Compressive light field photography using overcomplete dictionaries and optimized projections. ACM Trans. Graph. (TOG) **32**(4), 46:1–46:12 (2013)
24. Shi, L., Hassanieh, H., Davis, A., Katabi, D., Durand, F.: Light field reconstruction using sparsity in the continuous fourier domain. ACM Trans. Graph. (TOG) **34**(1), 12:1–12:13 (2014)
25. Sun, J., Yuan, L., Jia, J., Shum, H.Y.: Image completion with structure propagation. ACM Trans. Graph. (ToG) **24**(3), 861–868 (2005)
26. Tola, E., Cai, Q., Zhang, Z., Zhang, C.: Virtual view generation with a hybrid camera array. Technical report (2009)
27. Wang, Y., Liu, Y., Heidrich, W., Dai, Q.: The light field attachment: turning a DSLR into a light field camera using a low budget camera ring. IEEE Trans. Vis. Comput. Graph. **23**, 2357–2364 (2016)
28. Wanner, S., Goldluecke, B.: Variational light field analysis for disparity estimation and super-resolution. IEEE TPAMI **36**(3), 606–619 (2014)
29. Wetzstein, G., Lanman, D., Hirsch, M., Heidrich, W., Raskar, R.: Compressive light field displays. IEEE Comput. Graph. Appl. **32**(5), 6–11 (2012)
30. Wilburn, B., Joshi, N., Vaish, V., Talvala, E.V., Antunez, E., Barth, A., Adams, A., Horowitz, M., Levoy, M.: High performance imaging using large camera arrays, pp. 765–776 (2005)
31. Yang, J., Wright, J., Huang, T., Ma, Y.: Image super-resolution as sparse representation of raw image patches. In: IEEE Conference on Computer Vision and Pattern Recognition (CVPR 2008), pp. 1–8. IEEE (2008)

A New Image Sparse Reconstruction Method for Mixed Gaussian-Poisson Noise with Multiple Constraints

Ziling Wu[1,2], Hongxia Gao[1,2], Yongfei Chen[1,2(✉)], and Hui Kang[3]

[1] School of Automation Science and Engineering, South China University
of Technology, Guangzhou 510641, China
{auzilingwu,c.yongfei}@mail.scut.edu.cn, hxgao@scut.edu.cn
[2] Engineering Research Centre for Manufacturing Equipment of Ministry
of Education, South China University of Technology, Guangzhou 510641, China
[3] Guangdong Polytechnic Normal University, Guangzhou 510665, China
spiritcherry@126.com

Abstract. The mixed Gaussian-Poisson noise is common in many systems. Sparse based methods are now considered state-of-the-art to reconstruct noisy images. Moreover, it's gaining increasing attention to improve the sparse reconstruction methods with more image priors like nonlocal similarity. But most related work is aimed at single noise. And because of the definition of sparse representation, the image can only lie in a low dimensional subspace. The cosparse model is then proposed to move the emphasis on the number of zeros in the representation, thus enlarges the subspace's dimensions. For the first time, we combine sparsity, nonlocal similarity and cosparsity to improve the reconstruction quality. Firstly, non local similarity is used as the melioration of sparse constraint. Then the data fidelity term and cosparsity constraint are added. The objective function is solved alternately and iteratively by IRLSM and GAP. Experimental results indicate that the proposed method can attain higher reconstruction quality.

Keywords: Mixed Gaussian-Poisson noise · Reconstruction
Sparsity Cosparsity · Nonlocal similarity

1 Introduction

Images are always degraded by noises in process of formulation, transmission and preservation. And in many applications, there is more than one kind of noises. For example, in the low-photon-counting systems like micro focus X-ray detection, astronomy and fluorescence microscopy, only limited photons are collected because of system requirement or physical constraints. Usually, the noise is modelled as Poisson distribution. And the electronic fluctuations and intrinsic thermal are always assumed as additive Gaussian noise [1]. Thus, Poisson noise and Gaussian noise co-exist in such systems. It's essential to remove the mixed

© Springer Nature Singapore Pte Ltd. 2017
J. Yang et al. (Eds.): CCCV 2017, Part II, CCIS 772, pp. 345–356, 2017.
https://doi.org/10.1007/978-981-10-7302-1_29

noise for better application. There are some but not many methods for mixed Gaussian-Poisson noise removal. PURE-LET [1], GAT+BM3D [2], GAT+BLS-GSM [3] are considered state-of-the-art. They are almost based on transform domain like wavelet.

The most popular noisy image reconstruction method should be the compressive sensing based methods. In the compressive sensing framework, the original clean image is sparse while the noisy one is not. Thus, the denoising process is actually to recover the sparsity of the image [4], i.e. to decompose the image into a linear combination of a small number of dictionary atoms [5]. It is also called synthesis sparsity. The synthesis sparsity explores the local information of the images and has shown great results. There are two main problems in sparse reconstruction. One is dictionary learning, the other is sparse decomposition [6]. The most representative sparse method should be K-SVD [7]. To improve the sparse reconstruction quality, it is gaining growing attractions to combine it with more image priors. Nonlocal similarity is the most common indicating that there are many similar patches in the images even if they are not in a local and adjacent area. Researches show that nonlocal similarity is good for reconstructing image structures [2,8]. Actually, clustering-based dictionary learning like K-LLD [9] is an indirect way to use the non local similarity in sparse reconstruction. And in recent years, the CSR [10] proposed by Dong et al. has been widely considered, which combines nonlocal similarity and sparsity to get the double sparse constraints that help to preserve more details in reconstruction. Another literature [11] combines ideas from nonlocal sparse models and the Gaussian scale mixture model to keep the sharpness and suppress undesirable artifacts. However, the methods mentioned above are aimed at single noise model. And also, sparsity itself is limited because images can only lie in a low dimensional subspace. Given this, a dual analysis viewpoint to sparse representations called cosparsity or analysis sparsity was proposed and has attracted more and more attentions. Cosparsity shifts its focus from the nonzero decomposition coefficients of synthesis dictionary to the zeros in analysis dictionary [12]. The differences between these two models were compared in detail in [12]. Nevertheless, unlike synthespis sparsity, the understanding of analysis sparsity remains shallow and scarce today. Among them, [13] introduced many greedy-like methods in cosparse frame-work including Greedy Analysis Pursuit (GAP), analysis CoSaMP (ACoSaMP), Analysis SP (ASP), etc. And Rubinstein et al. proposed the dual K-SVD based on cosparsity called AK-SVD [14] in 2013, which outperformed K-SVD in removing Gaussian noise. In [15] the cosparse model is cast into row-to-row optimizations and use quadratic programming to obtain sparseness maximization results. As the sparsity and cosparsity are complementary, it's reasonable to combine them together to promote reconstruction quality. In this paper, we propose a new reconstruction method for images degraded by mixed Gaussian-Poisson noise based on sparsity, cosparsity and nonlocal similarity. And in the follow-up to this article, we refer 'sparse/sparsity' to synthesis case and 'cosparse/cosparsity' to analysis one.

2   The Mixed Gaussian-Poisson Noise Model

Assume that $y \in \mathbb{R}$ is corrupted by mixed Gaussian-Poisson noise, then the observation model is as follow:

$$y = Poisson(u) + n, \tag{1}$$

where $Poisson(u)$ means that the original image $u \in \mathbb{R}^N$ is corrupted by Poisson noise; $n \sim \mathcal{N}(0, \sigma^2 I)$ is the additive Gaussian noise; N is the product of image length and width.

Usually, the probability density function (PDF) based on joint probability distribution for formula (1) is complicated and thus the corresponding objective function is difficult to solve. One of the common ways to tackle it is to transform the mixed noise into a Gaussian one using Generalized Anscombe Transform (GAT) [16]. Then the problem turns into the additive case that is easier to solve. While in paper [17], the authors built up the PDF based on independent probability distribution to simplify the objective function in a dual adaptive regularization (DAR) scheme to remove the mixed noise. In this paper, we adopt the same strategy as [17]. Under the independent probability distribution, the PDF changes into:

$$P_{mixed}(y|u) = \prod_{k=1}^{N} \frac{(u_k)^{y_k} e^{-u_k}}{y_k!} \frac{1}{\sqrt{2\pi\sigma^2}} e^{-\frac{\|y-u\|_2^2}{2\sigma^2}}, \tag{2}$$

where u_k and y_k is the k_{th} component of u and y.

Then the log-likelihood of formula (2) under MAP criterion is:

$$T_0(u) = \min_u -\sum_{k=1}^{N} \log \frac{(u_k)^{y_k} e^{-u_k}}{y_k!} - \log \frac{1}{\sqrt{2\pi\sigma^2}} e^{-\frac{\|y-u\|_2^2}{2\sigma^2}} \tag{3}$$

$$= \min_u \sum_{k=1}^{N}(u_k - y_k \log u_k) + \|y - u\|_2^2.$$

Motivated by [18], the Taylor's approximation is used to transform the logarithmic term into a quadratic term. Therefore, the objective function becomes: at i_{th} iteration:

$$T_0(u) = \min_u \|u - s^i\|_2^2 + \frac{2}{\eta_i} \|y - u\|_2^2, \tag{4}$$

$$s^i = u^i - \frac{1}{\eta^i} \nabla Poi(u^i), \tag{5}$$

where $\eta^i \in \mathbb{R}$ is the second-order coefficient and is calculated via Barzilai-Borwein [19]; ∇ is the gradient operator; $Poi(u) = \sum_{k=1}^{N}(u_k - y_k \log u_k)$ is the Poisson component.

3   The Proposed Method

We will formulate our objective function in this section on the basis of Sect. 2. Firstly, we use a sparse melioration based on nonlocal similarity to further constrain the synthesis sparse coefficients. Then, the mixed Gaussian-Poisson data fidelity term and cosparse constraint are combined with the modified synthesis model.

$$
T(u) = \min_{\alpha_m, u} \sum_{m=1} \|R_m u - D\alpha_m\|_2^2 + \mu \|\alpha_m - \beta_m\|_1
$$
$$
+ \tau(\|u - s^i\|_2^2 + \frac{2}{\eta_i} \|y - u\|_2^2) + \lambda \|\Omega u\|_0 . \tag{6}
$$

Thus, the objective function takes use of synthesis sparsity, analysis sparsity and nonlocal similarity and combines these priors together to improve the reconstruction quality.

To make it more intuitive, we rewrite the objective function (6) as:

$$
T(u) = \min_{\alpha, u} \|u - D\alpha\|_2^2 + \mu \|\alpha - \beta\|_1 + \tau(\|u - s^i\|_2^2 + \frac{2}{\eta_i} \|y - u\|_2^2) + \lambda \|\Omega u\|_0, \tag{7}
$$

where R_m in (6) is the operator to extract a patch in $\mathbb{R}^n$ at location m; η, τ and λ are parameters to balance the data fidelity term and constraint terms; $D = [D_1; D_2; \cdots D_c] \in \mathbb{R}^{n*c}$ is the synthesis dictionary. We used the same training strategy in [20] to get a clustering dictionary. $\Omega \in \mathbb{R}^{p*N}$ is the analysis dictionary updated by GAP; α/α_m is sparse coefficients; β/β_m is the melioration for sparse coefficients based on nonlocal similarity and decided via the following scheme:

Step1: Find out Q similar image patches that have smallest Euclidean distance with the input patch in the whole image;

Step2: Calculate the similarity via formula (9) [8], define the set of similar blocks θ':

$$
\theta' := \{j | w(u_{in}, u_j^{le}) > \varsigma\}, \tag{8}
$$

$$
w(u_{in}, u_j^{le}) = \frac{1}{Z(m)} e^{-\frac{\|u_m - u_n\|_{2,a}^2}{h^2}}, \tag{9}
$$

$$
Z(m) = \sum_n e^{-\frac{\|u_m - u_n\|_{2,a}^2}{h^2}}, \tag{10}
$$

where u_{in} means the input image to get β; u_j^{le} represents the j_{th} patch in Step 1; ς is the threshold; a is the standard deviation of the Gaussian kernel; h is the scaling parameter; $Z(m)$ is the normalized coefficient;

Step 3: Calculate the similar image patch:

$$
\tilde{u} = \sum w(u_{in}, u_j^{le}) * u_j^{le}, j \in \theta', \tag{11}
$$

Table 1. The proposed method SRMM.

SRMM method

Input: noisy observation with mixed Gaussian-Poisson noise, dictionary D;

(1) Initialize:
   Constant $u^0 = y$, β^0 is zero vector, $\eta^0 > 0$, $r > 0$, $w > 0$, $\epsilon > 0$, $\mu > 0$, $\tau > 0$,
   $V > 0$, $i = 0$, $t = 0$, total number of iterations T_{iter};
(2) Repeat:
   ① Update α^{t+1} using IRLSM: $\alpha^{t+1} = \arg\min_{\alpha} \left\| u^t - D\alpha \right\|_2^2 + \mu \left\| \alpha - \beta^t \right\|_1$;
   ② If $t < T_{iter}$, update: $u^{t+1} = D\alpha^{t+1}$;
            update β^{t+1} based on using the scheme in Section 3;
            update: $t = t + 1$; Go to (2)– ① ;
   ③ Else, output: $D\alpha$, Go to (3);
(3) Repeat:
   ① Calculate η^i based on Barzilai-Borwein;
   ② Calculate s^i : $s^i = u^i - \dfrac{1}{\eta^i} \nabla Poi(u^i)$;
   ③ Update u' and Ω using GAP:
   $$u' = \arg\min_{u} \left\| u - D\alpha \right\|_2^2 + \tau(\left\| u - s^i \right\|_2^2 + \frac{2}{\eta_i} \left\| y - u \right\|_2^2) + \lambda \left\| \Omega u \right\|_0$$
   ④ If $\ T_2(u') \le \max\limits_{v=\max(v-V,0),\cdots i} T_2(u^v) - \dfrac{w}{2}\eta^i \left\| u' - u^i \right\|_2^2$,
       update: $u^{i+1} = u, i = i+1$, go to (3);
   ⑤ Else, update: $\eta^i = r\eta^i$, go to (3)– ② ;
(4) Until: $\dfrac{\left\| u^i - u^{i-1} \right\|_2^2}{\left\| u^{i-1} \right\|_2^2} < \epsilon$;

Output: $u^* = u^i$.

Step 4: Calculate β:
$$\tilde{u} = D\beta. \tag{12}$$

To solve formula (7), we divide it into two sub-problems based on alternative optimization:
   Sub-problem 1: with u fixed, solve α:
$$\alpha = \arg\min_{\alpha} T_1(\alpha) = \min_{\alpha} \left\| u - D\alpha \right\|_2^2 + \mu \left\| \alpha - \beta \right\|_1, \tag{13}$$

Here we use the Iterative Re-weighted Least Squares Minimization (IRLSM) [21] for simplicity.
   Sub-problem 2: with α fixed, solve u:
$$u = \arg\min_{u} T_2(u) = \min_{u} \left\| u - D\alpha \right\|_2^2 + \tau(\left\| u - s^i \right\|_2^2 + \frac{2}{\eta_i} \left\| y - u \right\|_2^2) + \lambda \left\| \Omega u \right\|_0. \tag{14}$$

We choose GAP to solve (14) as it's the IRLSM method in cosparsity's framework.

The complete method called SRMM can be found in Table 1.

4   Experiments

In this section, experiments are displayed to demonstrate the effectiveness and superiority of the proposed method (SRMM). Firstly, experiments are conducted on commonly used natural images in Fig. 1(a)–(d). We add different degrees of mixed noise ($\sigma = 5$ and $\sigma = 25$) to test the availability. Examples are shown in Fig. 1(e)–(f). Moreover, two micro-focus X-ray noisy images are used as the real-data experimental subjects, see Fig. 1(g)–(h). Several methods are selected as comparisons: (1) mixed noise removal: PURE-LET (P-L) [1], GAT+BM3D (G+B) [2], GAT+BLS-GSM (G+B-G) [3], DAR [17]; (2) sparse-based: K-SVD [7] and AK-SVD [14]; (3) multiple-constraint-based: CSR [10].

Our proposed denoising approach has seven vital paramete6rs to be set: the patch size, the number of the atoms in learning dictionary, the number of the cluster centers, the iterations number T_{iter} and three regularization parameters μ, τ and λ. We set the patch size 7×7 and cluster centers number 8. Each cluster has 200 atoms. The regularization terms μ, τ and λ are set to be 1, 0.5 and 0.4, respectively. The maximum total iterations number T_{iter} is set 10. These parameters are selected empirically for higher performance in our experiment, but suitable variations to these parameters are acceptable depending on different image size and noise strength.

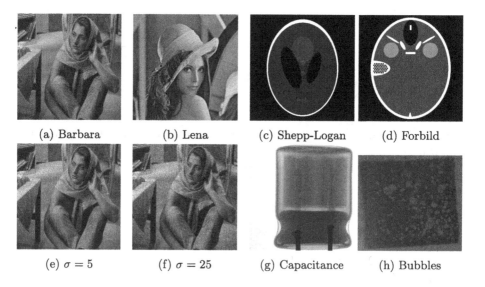

(a) Barbara	(b) Lena	(c) Shepp-Logan	(d) Forbild
(e) $\sigma = 5$	(f) $\sigma = 25$	(g) Capacitance	(h) Bubbles

Fig. 1. Original images and noisy images.

Due to the space limitation, we only show partial enlarged results of two natural images in Figs. 2 and 3. For micro-focus X-ray images, we demonstrate the edge-detection results in Figs. 4 and 5 as the micro-focus X-ray images are low-contrast and hard to compare directly. Objective indexes can be seen in Tables 2, 3 and 4. Full-Reference PSNR and MSSIM are used for natural images. And no reference MSR for smooth regions, LS for detail regions [17] and BRISQUE [22] are applied in micro-focus X-ray images since it's unable to get the noise-free micro-focus X-ray images.

For natural images, PURE-LET will not only blur details, but also bring artifacts that debase reconstruction quality. GAT+BLS-GSM suffers from over-smoothing, lots of details are lost. DAR suffers from details loss with slight noise remain. GAT+BM3D outperforms PURE-LET, GAT+BLS-GSM and DAR in visual effects. However, some details, especially weak edges are over-smoothed. K-SVD can preserve most details but there is noise left. Results of AK-SVD are not ideal for the reason that the edge transitions are not smooth. As the

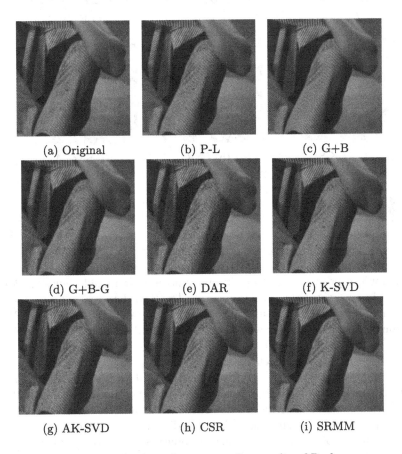

(a) Original   (b) P-L   (c) G+B

(d) G+B-G   (e) DAR   (f) K-SVD

(g) AK-SVD   (h) CSR   (i) SRMM

Fig. 2. Partial enlarged reconstruction results of Barbara.

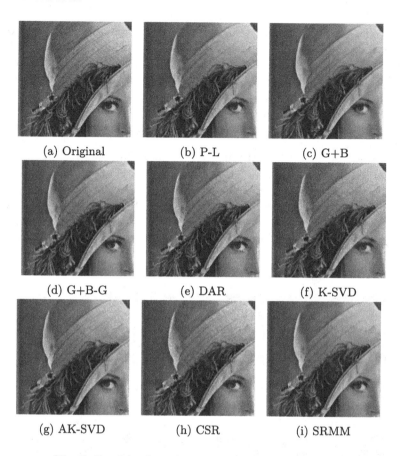

(a) Original   (b) P-L   (c) G+B

(d) G+B-G   (e) DAR   (f) K-SVD

(g) AK-SVD   (h) CSR   (i) SRMM

Fig. 3. Partial enlarged reconstruction results of Lena.

noise level especially the Poisson noise strength grows stronger, the noise in the non-smooth area is hard to be removed sufficiently. With two constraints, CSR reaches better results than the above methods. Nevertheless, some details are still lost. As for the proposed method, it can remove most mixed Gaussian-Poisson noise and in the meantime, preserve more edges and details.

For the reconstruction results of micro-focus X-ray images in Fig. 4, PURE-LET fails to reconstruct 'Capacitance'. GAT+BM3D, DAR, K-SVD, AK-SVD and CSR suffer from over-smoothness. The details of pins as indicated in Fig. 4 are less. The noise is not well eliminated in GAT+BLS-GSM. The proposed method SRMM, however, removes most noise and preserves more details in the pins. For 'Bubbles' in Fig. 5, the proposed method and GAT+BM3D can reconstruct most details including small and weak bubbles in the left side. But GAT+BM3D's result is smoother than SRMM. Other methods fail to reconstruct the shapes of the weak and small bubbles as much as SRMM does.

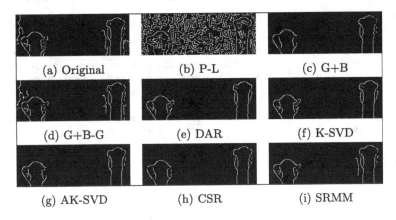

Fig. 4. Partial enlarged edge-detection results (Canny) of Capacitance.

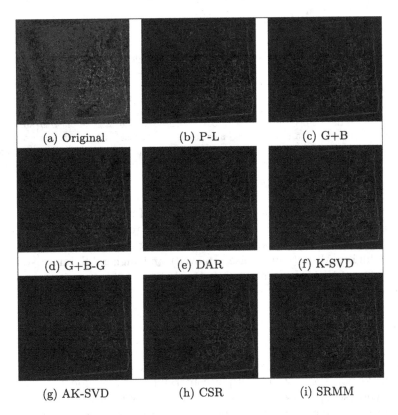

Fig. 5. Edge-detection results (local variance) of Bubbles.

Table 2. Objective indexes of natural images with $\sigma = 5$ in mixed noise.

		Noisy	P-L	G+B	G+B-G	DAR	K-SVD	AK-SVD	CSR	SRMM
Barbara	PSNR	26.767	33.583	28.813	28.813	30.478	33.193	31.463	33.568	**33.913**
	MSSIM	0.8762	0.9224	0.9540	0.9540	0.9516	0.9616	0.9479	0.9634	**0.9669**
Lena	PSNR	26.514	33.211	30.072	30.072	33.064	33.470	33.099	33.724	**33.893**
	MSSIM	0.8308	0.8770	0.9472	0.9472	0.9485	0.9479	0.9369	0.9496	**0.9504**
Shepp-Logan	PSNR	31.046	39.691	33.452	33.452	38.292	38.578	38.584	41.477	**42.233**
	MSSIM	0.7278	0.8154	0.8091	0.8091	0.9686	0.8243	0.8073	0.9204	**0.9958**
Forbild	PSNR	29.210	38.654	33.320	33.320	37.080	34.179	37.006	38.708	**38.897**
	MSSIM	0.7373	0.8659	0.8597	0.8597	0.8883	0.8497	0.8615	0.9379	**0.9944**

Table 3. Objective indexes of natural images with $\sigma = 25$ in mixed noise.

		Noisy	P-L	G+B	G+B-G	DAR	K-SVD	AK-SVD	CSR	SRMM
Barbara	PSNR	19.454	28.690	29.968	26.116	26.740	27.068	26.969	29.798	**29.971**
	MSSIM	0.6429	0.8221	**0.9224**	0.8737	0.8923	0.8601	0.8699	0.9203	0.9211
Lena	PSNR	19.414	30.490	31.210	27.828	28.792	31.012	29.726	31.027	**31.546**
	MSSIM	0.5711	0.7961	0.9148	0.8936	0.8971	0.9158	0.8871	0.9116	**0.9208**
Shepp-Logan	PSNR	20.001	28.622	29.793	27.395	28.665	28.436	28.440	31.876	**38.910**
	MSSIM	0.2732	0.3777	0.5293	0.5119	0.5141	0.4825	0.5063	0.5589	**0.9460**
Forbild	PSNR	19.816	27.770	29.634	27.150	28.498	27.770	27.759	29.378	**29.610**
	MSSIM	0.3534	0.6357	0.6583	0.6372	0.6158	0.6357	0.6365	0.6818	**0.9599**

Table 4. Objective indexes of micro-focus X-ray images.

		P-L	G+B	G+B-G	DAR	K-SVD	AK-SVD	CSR	SRMM
Capacitance	MSR	5.1978	6.4484	6.4264	6.4491	6.4413	6.4526	6.4452	**6.4530**
	LS	16.3957	14.7049	16.3197	13.0116	12.8816	13.8061	14.9567	**16.5292**
	BRISQUE	58.9075	48.2338	36.0041	39.4218	36.8020	35.2522	45.4149	**30.3488**
Bubbles	MSR	5.1586	5.1530	**6.9124**	4.2809	4.1368	5.2647	5.1626	5.1805
	LS	9.7317	8.3610	5.6096	11.0608	11.0619	9.4268	6.5187	**11.1155**
	BRISQUE	66.8818	69.6640	62.9114	55.0163	52.1887	46.2160	69.6345	**41.8585**

Tables 2, 3 and 4 show the objective indexes of all experiments. In Table 2, SRMM achieves higher PSNR and MSSIM in most results. Also, for micro-focus X-ray images, SRMM gets lowest BRISQUE indicating best overall quality. And relatively higher MSR (more noise removed) and higher LS (more structures kept) mean that SRMM can attain a better balance between noise-removal and edge-preservation.

5 Conclusions

In this paper, we propose a new image sparse reconstruction method for mixed Gaussian-Poisson noise with multiple constraints. In our model, three priors, namely synthesis sparsity, analysis sparsity and nonlocal similarity, are combined as the sparse constraints to obtain better results. Among them, the synthesis sparsity and analysis sparsity are complementary and reconstruct noisy images by means of re-covering the sparsity. And nonlocal similarity explores the structure information that helps to reconstruct details. The objective function is solved via IRLSM and GAP alternatively. Experimental results show that the proposed method outperforms the contrast methods both in visual effects and indexes, which will improve reconstruction quality.

Acknowledgments. This work was supported by Natural Science Foundation of China under Grant 61403146, Fundamental Research Funds for the Central Universities under Grant 2015ZM128, Science and Technology Program of Guangzhou, China under Grant 201707010054 and Science and Technology Program of Guangzhou, China under Grant 201704030072.

References

1. Luisier, F., Blu, T., Unser, M.: Image denoising in mixed Poisson-Gaussian noise. IEEE Trans. Image Process. **20**(3), 696–708 (2011)
2. Dabov, K., Foi, A., Katkovnik, V., Egiazarian, K.: Image denoising by sparse 3-D transform-domain collaborative filtering. IEEE Trans. Image Process. **16**(8), 2080–2095 (2007)
3. Portilla, J., Strela, V., Wainwright, M.J., Simoncelli, E.P.: Image denoising using Gaussian scale mixtures in the wavelet domain. IEEE Trans. Image Process. **12**(11), 1338–1351 (2003)
4. Lian, Q., Shi, B., Chen, S.Z.: Research advances on dictionary learning models, algorithms and applications. Acta Automatica Sin. **41**(2), 240–260 (2015)
5. Mallat, S.G., Zhang, Z.: Matching pursuits with time-frequency dictionaries. IEEE Trans. Signal Process. **41**(12), 3397–3415 (1993)
6. Elad, M., Aharon, M.: Image denoising via sparse and redundant representations over learned dictionaries. IEEE Trans. Image Process. **15**(12), 3736–3745 (2006)
7. Aharon, M., Elad, M., Bruckstein, A.: K-SVD: an algorithm for designing overcomplete dictionaries for sparse representation. IEEE Trans. Signal Process. **54**(11), 4311–4322 (2006)
8. Buades, A., Coll, B., Morel, J.M.: A non-local algorithm for image denoising. In: Proceedings of IEEE Computer Society Conference on Computer Vision and Pattern Recognition, pp. 60–65. IEEE Computer Society, Washington, DC (2005)
9. Chatterjee, P., Milanfar, P.: Clustering-based denoising with locally learned dictionaries. IEEE Trans. Image Process. **18**(7), 1438–1451 (2009)
10. Dong, W., Li, X., Zhang, L., et al.: Sparsity-based image denoising via dictionary learning and structural clustering. In: 24th Proceedings of IEEE Conference on Computer Vision and Pattern Recognition, pp. 457–464. IEEE Press, Colorado Springs (2011)
11. Dong, W., Shi, G., Ma, Y., Li, X.: Image restoration via simultaneous sparse coding: where structured sparsity meets Gaussian scale mixture. Springer Int. J. Comput. Vis. **114**(2), 217–232 (2015)
12. Nam, S., Davies, M.E., Elad, M., Gribonval, E.: The cosparse analysis model and algorithms. Appl. Comput. Harmonic Anal. **34**(1), 30–56 (2013)
13. Giryes, R., Nam, S., Elad, M., et al.: Greedy-like algorithms for the cosparse analysis model. Linear Algebra Appl. **441**(1), 22–60 (2014)
14. Rubinstein, R., Peleg, T., Elad, M.: Analysis K-SVD: a dictionary-learning algorithm for the analysis sparse model. IEEE Trans. Signal Process. **61**(3), 661–677 (2013)

15. Li, Y., Ding, S., Li, Z.: Dictionary learning with the cosparse analysis model based on summation of blocked determinants as the sparseness measure. Elsevier Dig. Sig. Process. **48**, 298–309 (2016)

16. Makitalo, M., Foi, A.: Optimal inversion of the generalized anscombe transformation for Poisson-Gaussian noise. IEEE Trans. Image Process. **22**(1), 91–103 (2013)

17. Wu, Z., Gao, H., Ma, G., Wan, Y.: A dual adaptive regularization method to remove mixed Gaussian-Poisson noise. In: Chen, C.-S., Lu, J., Ma, K.-K. (eds.) ACCV 2016. LNCS, vol. 10116, pp. 206–221. Springer, Cham (2017). https://doi.org/10.1007/978-3-319-54407-6_14

18. Harmany, Z.T., Marcia, R.F., Willett, R.M.: This is SPIRAL-TAP: sparse poisson intensity reconstruction algorithms theory and practice. IEEE Trans. Image Process. **21**(3), 1084–1096 (2012)

19. Barzilai, J., Borwein, J.M.: Two-point step size gradient methods. IMA J. Numer. Anal. **8**(1), 141–148 (1988)

20. Dong, W., Zhang, L., Shi, G., et al.: Image deblurring and super-resolution by adaptive sparse domain selection and adaptive regularization. IEEE Trans. Image Process. **20**(7), 1838–1857 (2011)

21. Daubechies, I., Devore, R., Fornasier, M.: Iteratively reweighted least squares minimization for sparse recovery. Commun. Pure Appl. Math. **63**(1), 1–38 (2008)

22. Mittal, A., Moorthy, A.K., Bovik, A.C.: No reference image quality assessment in the spatial domain. IEEE Trans. Image Process. **21**(12), 4695–4708 (2012)

Face Video Super-Resolution with Identity Guided Generative Adversarial Networks

Dingyi Li[1] and Zengfu Wang[1,2(✉)]

[1] Department of Automation, University of Science and Technology of China,
Hefei 230027, China
lidingyi@mail.ustc.edu.cn, zfwang@ustc.edu.cn
[2] Institute of Intelligent Machines, Chinese Academy of Sciences,
Hefei 230031, China

Abstract. Faces are of particular concerns in video surveillance systems. It is challenging to reconstruct clear faces from low-resolution (LR) videos. In this paper, we propose a new method for face video super-resolution (SR) based on identity guided generative adversarial networks (GANs). We establish a two-stage convolutional neural network (CNN) for face video SR, and employ identity guided GANs to recover high-resolution (HR) facial details. Extensive experiments validate the effectiveness of our proposed method from the following aspects: fidelity, visual quality and robustness to pose, expression and illuminance variations.

Keywords: Super-resolution · Face hallucination · Identity guidance
Generative adversarial networks (GANs)

1 Introduction

Video surveillance systems grow rapidly in recent years. They appear in many streets and buildings and play an important role in safety. However, their limited resolutions and scopes prevent them from providing more important information especially on human faces. Face super-resolution (SR), also called face hallucination [17], aims at reconstructing sharp face images from low-resolution (LR) observations. The study of face SR starts from the seminal work of Baker and Kanade [1]. They introduced an example-based SR method for face image SR. Example-based methods are very suitable for face SR since face images are of relatively fixed features. Traditional face SR methods use nearest neighbor [1], principal component analysis (PCA) [18], PCA and Markov network [12], sparse representation [23]. The powerful convolutional neural network, has also been introduced into face SR [28]. Perceptual loss [6,9] and generative adversarial networks (GANs) have been employed in face SR for visually pleasant SR results [3,16,24,25]. Although plenty of algorithms have been proposed, most of these algorithms conduct experiments on aligned frontal faces under restricted

© Springer Nature Singapore Pte Ltd. 2017
J. Yang et al. (Eds.): CCCV 2017, Part II, CCIS 772, pp. 357–369, 2017.
https://doi.org/10.1007/978-981-10-7302-1_30

illuminance. However, in real video surveillance systems, there might be pose, expression and illuminance variations. Plain learning based approaches may generate over-smoothed results when the scale factors are very large. Perceptual loss and GAN based methods produce sharp images but are difficult to guarantee high restoration fidelity.

In this paper, we introduce identity guidance to GANs. A new two-stage CNN for face video SR is proposed. Different from plain perceptual loss based methods which compute the Mean Square Error (MSE) loss between the perceptual feature maps, we employ discriminator net on the identity feature maps. Different from plain GAN based methods using raw SR and ground truth (GT) high-resolution (HR) images as the inputs of the discriminator net, we feed the identity feature maps into the discriminator net for better restoration of face elements. We compare our method with state-of-the-art SR methods and show the superiority of our method in terms of both restoration fidelity and visual quality.

We summarize our contributions as follows:

1. We propose a two-stage CNN as our generator net to facilitate visually appealing face video SR in our GAN framework.
2. We establish an identity guided GAN framework for face video SR for better fidelity and visual quality.
3. Our method is robust to pose, expression and illuminance variations in challenging videos. Our face SR results are superior compared with state-of-the-art SR methods.

The rest of the paper is organized as follows. In Sect. 2 we review related work. In Sect. 3, we illustrate the proposed method. In Sect. 4, we provide detailed experimental results and analysis. The paper is concluded in Sect. 5.

2   Related Work

Face SR has attracted much attention from researchers and companies. Baker and Kanade [1] introduced nearest neighbor in face SR. Liu et al. [12] employed a global linear model and a local Markov network to obtain good SR results. Ma et al. [13] used position patch to reconstruct faces. Jiang et al. [5] applied locality and sparsity constraint for least square inversion problem, leading to noise robust face SR. These methods perform well on aligned faces. When faces are not aligned, their performance drop dramatically. Yang et al. [22] extracted facial components in LR images and utilized component examples for learning. However, it is hard to detect facial components in extremely LR images. Deep CNNs were firstly applied in face SR by Zhou et al. [28]. Different from CNN based general image SR methods [6–10,14,26,27], face SR CNNs utilize the facial information. Zhu et al. [29] cascaded gated deep bi-network and dense face correspondence field for face SR.

GANs are also introduced in image processing tasks including face image SR to improve the visual quality of the super-resolved faces. Tuzel et al. [16]

proposed a deep global-local face SR net and introduced adversarial training to improve visual quality. Yu and Porikli [24] designed a discriminative generative net for face SR. Yu and Porikli [25] also presented a transformative discriminative neural network to super-resolve unaligned and very LR faces. Gulrajani et al. [3] compared plain GANs with Wasserstein GANs for face SR.

Although GANs have been employed in face SR, the inputs of discriminator net are simply the SR and HR images in previous methods. These methods utilize no person identity priors to distinguish one person from another, only aiming at obtaining sharper face images. In this paper, we utilize face identity features as the inputs of the discriminator net for improving the fidelity of the SR images and providing better results on facial components.

3 Our Video SR Approach

3.1 Overview of the Proposed Method

We illustrate our face video SR framework in Fig. 1(a). In our method, three networks are employed. They are the generator net, the identity net and the discriminator net. The generator net regards multiple LR images as the inputs and generates a SR image. The identity net is utilized to extract the identity feature maps of SR images and the original HR images. Our identity net applies the bottom layers of a face recognition CNN [20]. We employ the discriminator net to determine whether the SR identity feature maps look similar to the HR identity feature maps. Our method is different from plain MSE loss and perceptual loss [6,9] based SR methods as shown in Fig. 1(b) which simply employ the MSE between the SR feature maps and the HR feature maps of a pretrained VGG network [15]. Li et al. [11] uses identity loss to preserve identity and perceptual loss to enhance visual quality. However, these two losses are also MSE based. Apart from previous GAN based methods in Fig. 1(c) determining whether the SR images look similar to the HR images, our identity guided GAN determines whether the SR identity feature maps are similar to HR identity feature maps. Our method aims at obtaining more accurate identity features on super-resolved faces.

3.2 The Generator Net

We design a two-stage CNN as the generator net in our GAN framework for enhancing visual quality, as shown in Fig. 2. The proposed CNN is based on the state-of-the-art general video SR method named motion compensation and residual net (MCResNet) [10]. We interpolates LR inputs frames using the bicubic interpolation (Bicubic) and employs an optical flow method [2,7] for motion estimation and compensation. Then the motion compensation frames are fed into our residual video SR network. We employ two-stage deep residual learning by adding two skip-layer connections. The two stages both include a new convolutional path and an element-wise sum of the outputs of the convolutional

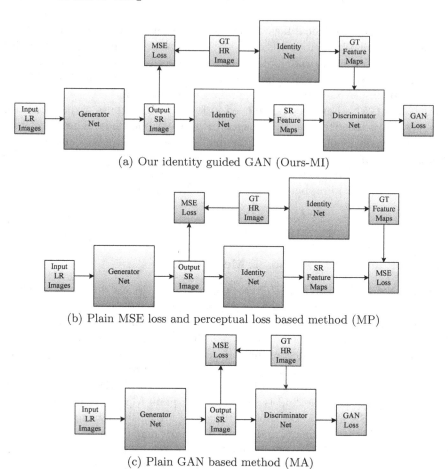

(a) Our identity guided GAN (Ours-MI)

(b) Plain MSE loss and perceptual loss based method (MP)

(c) Plain GAN based method (MA)

Fig. 1. Overview of our method and previous methods.

path and the interpolated centering input frame. The first stage restores high-frequency details. The blue components in Fig. 2 are the second stage of our network for denoising and illuminance adjustment. We find that although the proposed net obtains lower objective evaluation values than MCResNet using MSE loss, the new net provides better results than MCResNet in our GAN framework. Our proposed generator net adds the input centering frame twice so that the low-frequency details are better preserved in GAN training. However, the two-stage structure is a little bit worse than MCResNet with the MSE loss since the two-stage structure uses less filters.

Let $\mathbf{Y} = \{\mathbf{I}_{-T}^L, \mathbf{I}_{-T+1}^L, \cdots, \mathbf{I}_0^L, \cdots, \mathbf{I}_{T-1}^L, \mathbf{I}_T^L\}$ to be all the bicubically interpolated input frames where $\mathbf{I}_t^L$ is the t-th frame. Suppose we have N convolutional

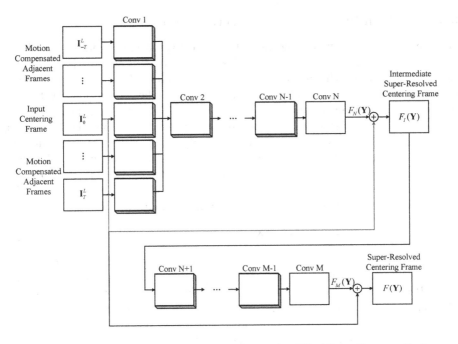

Fig. 2. The proposed generator net for face video SR. (Color figure online)

layers in the first stage and M convolutional layers in total. For the first stage we have

$$F_I(\mathbf{Y}) = F_N(\mathbf{Y}) + \mathbf{I}_0^L \tag{1}$$

where $F_I(\mathbf{Y})$ is the output of the first stage and $F_N(\mathbf{Y})$ is the output of the N-th convolutional layer. $\mathbf{I}_0^L$ is the centering frame. For the second stage we have

$$F(\mathbf{Y}) = F_M(\mathbf{Y}) + \mathbf{I}_0^L \tag{2}$$

where $F(\mathbf{Y})$ is the output of the whole network and $F_M(\mathbf{Y})$ is the output of the M-th convolutional layer. The interpolated input centering frame is added to the SR results twice for more accurate SR and easier optimization. The loss for the generator net in our GAN is

$$L_G = L_M + \lambda_A L_A \tag{3}$$

where L_M is the MSE loss and L_A is the adversarial loss. λ_A is the weight of the adversarial loss. We also have the MSE loss as

$$L_M = \frac{1}{2}||F(\mathbf{Y}, \Theta) - \mathbf{X}||_2^2 \tag{4}$$

where $F(\mathbf{Y}, \Theta)$ are the outputs of the generator net. Θ are the parameters of the generator net. $\mathbf{X}$ are the ground truth (GT) HR images. We will describe the definition of L_A in the following subsection.

3.3   The Identity Net and the Discriminator Net

For the identity net, we utilize the bottom 7 convolutional layers of one of the state-of-the-art face recognition CNN designed by Wen *et al.* [20]. The convolutional layer we use is the mid-level information extracted by the face recognition CNN. We find that high-level information at the top layers and low-level information at the bottom layers are not suitable for reconstructing facial details. Fully connected layers are not used so that the generator net is scalable to input image sizes. We use the pretrained model of [20]. The parameters of the identity net are not changed during training. The outputs of the identity net, which are the extracted identity feature maps, are directly fed into the discriminator net. For the discriminator net, we employ 4 convolutional layers with a stride of 2 and a fully connected layer. The loss of the discriminator net is

$$L_D = -log(D(I(\mathbf{X}))) - log(1 - D(I(F(\mathbf{Y})))) \qquad (5)$$

where L_D is the discriminative loss. I($\mathbf{X}$) and D(I($\mathbf{X}$)) are the outputs of the identity net and the discriminator net respectively when the inputs are the original HR images. F($\mathbf{Y}$) are the outputs of generator net when the inputs are the SR images. I(F($\mathbf{Y}$)) and D(I(F($\mathbf{Y}$))) are the outputs of the identity net and the discriminator net respectively. The adversarial loss, which is back-propagated through the discriminator net and the identity net to the generator net, is defined as

$$L_A = -log(D(I(F(\mathbf{Y})))) \qquad (6)$$

4   Experiments

4.1   Implementation Details

We conduct experiments on the challenging YouTube Faces Database [21]. The YouTube Faces Database contains 3,425 videos of 1,595 subjects. The videos are captured in different scenes with different illuminance. The faces are of various poses and expressions. We utilize the extracted face region videos of the first 1000 people for training, and the 3rd frame (the centering frame of the first 5 frames) of the first video of the last 95 people for testing. 181760 face images are used for training. We use Bicubic to resize all faces to a size of 80×80. Then we also apply Bicubic to downsample these 80×80 images with a scale factor of 4 to generate LR images. We conduct experiments on the Y channel in the YCbCr color space. The visual quality of the super-resolved images, which we mainly focus on, are inspected. Peak pixel-to-noise ratio (PSNR) and structural similarity (SSIM) [19] are utilized as the objective evaluations. 8 pixels on each border are eliminated when conducting the objective evaluations. We compare our method with the bicubic interpolation (Bicubic) and state-of-the-art general single image based SR methods very deep SR (VDSR) [8], denoising convolutional neural networks (DnCNN) [26] and very deep Residual Encoder-Decoder Network (RED-Net) [14], and general multi-frame based method MCResNet [10]. For VDSR, RED-Net and MCResNet, we stack 20, 30 and 20 layers respectively to achieve their best performance. All CNN based methods except for DnCNN [26] are trained on the same training dataset as ours for fair comparison using the Caffe platform [4] on a Linux workstation with an Intel i7-6700K CPU of 4.0 GHz, a Nvidia Titan X GPU and 64 GB memory. We also compare our method with face SR methods Ma's [13], structured face hallucination (SFH) [22] and Locality-constrained Representation (LcR) [5]. Since SFH is unable to detect faces on 20×20 low-resolution images, it provides no SR results.

In our generator net, T is 2, N is 10, M is 15. The first convolutional layer is of 64 feature maps for each frame. For layer 10 and layer 15, only 1 feature map is employed. The feature map size of the intermediate layers are 32. We also test our method with different configurations including MSE loss based (M), MSE loss and plain perceptual loss based (MP), MSE loss and plain adversarial loss based (MA), MCResNet with MSE loss and identity adversarial loss (MCRN-MI) and MSE and identity adversarial loss based (Ours-MI). For M, we employ the generator net in Fig. 2 with MSE loss. For MP, MSE loss between the SR and the HR images, and MSE loss between the SR and HR identity feature maps are computed in the structure of Fig. 1(b). For MA, we employ the framework in Fig. 1(c) as in plain GAN based SR methods. For Ous-MI we employ our identity guided GANs shown in Fig. 1(a). We use the MCResNet as the generator net for MCRN-MI. The other settings of MCRN-MI are the same as Ours-MI.

4.2 Results and Analysis

In Table 1, we show that MCResNet obtains the highest PSNR and SSIM values when trained with MSE loss only, better than our two-stage generator net which uses less filters. We also find that although our method trained with MSE loss gets slightly lower PSNR and SSIM values than MCResNet, it is more powerful to restore fine facial details when trained with both MSE loss and adversarial loss. In Table 2 we evaluate our method with different configurations objectively. MP, MA, MCRN-MI and Ours-MI get lower PSNR and SSIM values than M, but they achieve better visual quality. We find that for visual quality, Ours-MI outperforms Ma's, LcR, MP, MA and MCRN-MI. In Figs. 3, 4, 5, 6, 7, 8 and 9, we provide some of the face SR results. The results of MSE loss based methods VDSR, DnCNN, RED-Net, MCResNet, and M are over-smoothed.

Table 1. Objective evaluations of our generator net with MSE loss and other methods

Metric	Bicubic	VDSR [8]	DnCNN [26]	RED-Net [14]
PSNR	30.97	33.28	32.73	32.65
SSIM	0.8444	0.8989	0.8873	0.8868
Running time	0.001 s	0.101 s	0.004 s	0.161 s
Metric	MCResNet [10]	Ma [13]	LcR [5]	M
PSNR	34.67	23.08	27.86	34.50
SSIM	0.9241	0.5386	0.7425	0.9217
Running time	1.030 s	0.043 s	0.147 s	1.005 s

Table 2. Objective evaluations of our method with different configurations

Metric	M	MP	MA	MCRN-MI	Ours-MI
PSNR	34.50	33.14	33.18	32.90	32.81
SSIM	0.9217	0.9088	0.9083	0.8966	0.8968
Running time	1.005 s	1.007 s	1.000 s	1.017 s	0.997 s

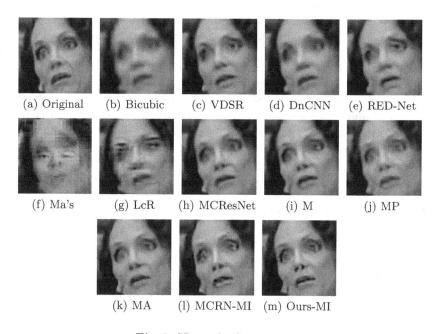

(a) Original   (b) Bicubic   (c) VDSR   (d) DnCNN   (e) RED-Net

(f) Ma's   (g) LcR   (h) MCResNet   (i) M   (j) MP

(k) MA   (l) MCRN-MI   (m) Ours-MI

Fig. 3. SR results for person 1.

The results of Ma's [13] and LcR [5] are of plenty of visual artifacts since they rely on careful face alignment. The results of MP is slightly sharper than M, but looks brighter than the original image. MA can super-resolve sharp edges but is still unable to restore fine facial details especially on eyes and noses. MCRN-MI may generate fine details, but is of lower quality than Ours-MI. Meanwhile, results of MCRN-MI may contain too much noise. It is hard for MCResNet to restore fine details in our framework since it employs only one stage and loses important facial information in the deep convolutional layers. Our two-stage generator net uses two skip-layer connections to better preserve low-frequency details. We can see that our MI super-resolves fine facial components such as eyes, noses and mouths. Our identity guided GAN based method is robust to pose, expression and illuminance changes. The computational cost of each methods are shown in Tables 1 and 2. For Bicubic, Ma's and LcR, only an Intel i7-6700K CPU is used. Other methods are tested with an Intel i7-6700K CPU and a Nvidia Titan X GPU. The running time of our method is about 1 s for super-resolving one facial image, which is acceptable in real applications.

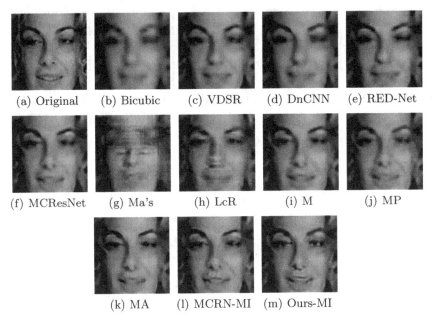

(a) Original   (b) Bicubic   (c) VDSR   (d) DnCNN   (e) RED-Net

(f) MCResNet   (g) Ma's   (h) LcR   (i) M   (j) MP

(k) MA   (l) MCRN-MI   (m) Ours-MI

Fig. 4. SR results for person 3.

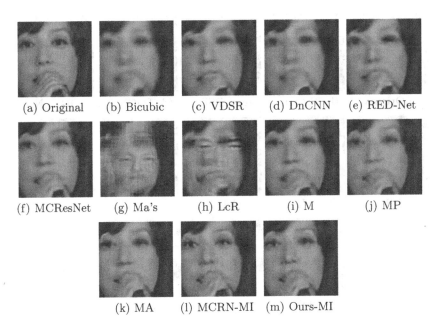

(a) Original   (b) Bicubic   (c) VDSR   (d) DnCNN   (e) RED-Net

(f) MCResNet   (g) Ma's   (h) LcR   (i) M   (j) MP

(k) MA   (l) MCRN-MI   (m) Ours-MI

Fig. 5. SR results for person 10.

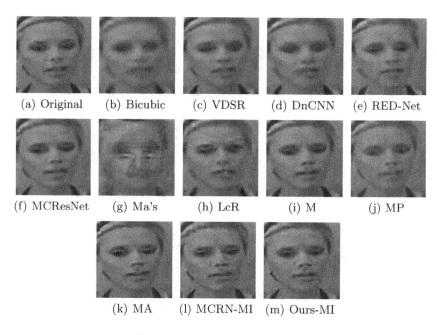

(a) Original   (b) Bicubic   (c) VDSR   (d) DnCNN   (e) RED-Net

(f) MCResNet   (g) Ma's   (h) LcR   (i) M   (j) MP

(k) MA   (l) MCRN-MI   (m) Ours-MI

Fig. 6. SR results for person 13.

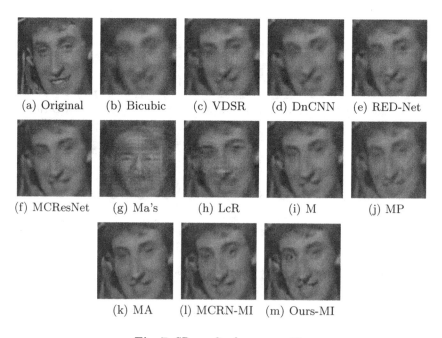

(a) Original   (b) Bicubic   (c) VDSR   (d) DnCNN   (e) RED-Net

(f) MCResNet   (g) Ma's   (h) LcR   (i) M   (j) MP

(k) MA   (l) MCRN-MI   (m) Ours-MI

Fig. 7. SR results for person 36.

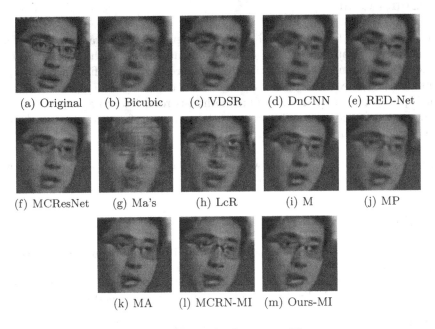

Fig. 8. SR results for person 65.

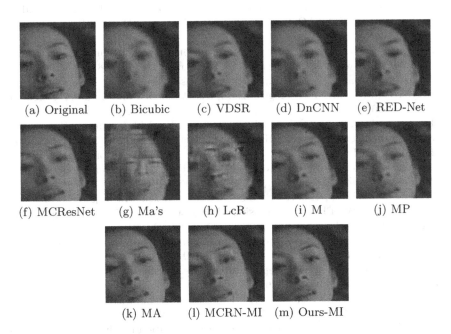

Fig. 9. SR results for person 90.

5 Conclusion

In this paper, we have proposed an identity guided GAN framework for face video SR. The proposed two-stage CNN generator net and the employment of the identity guidance for GANs facilitate the restoration of fine facial components. Extensive experiments have demonstrated the high fidelity and superior visual quality of our SR results. Our method is robust to pose, expression and illuminance variations. Our future work is combining face video SR with face video recognition in a joint manner.

Acknowledgements. This work was supported by the National Natural Science Foundation of China (no. 61472393).

References

1. Baker, S., Kanade, T.: Hallucinating faces. In: IEEE International Conference on Automatic Face and Gesture Recognition, pp. 83–88 (2000)
2. Drulea, M., Nedevschi, S.: Total variation regularization of local-global optical flow. In: Proceedings of the IEEE Conference on Intelligent Transportation Systems, pp. 318–323 (2011)
3. Gulrajani, I., Ahmed, F., Arjovsky, M., Dumoulin, V., Courville, A.: Improved training of wasserstein gans. arXiv preprint arXiv:1704.00028 (2017)
4. Jia, Y., Shelhamer, E., Donahue, J., Karayev, S., Long, J., Girshick, R., Guadarrama, S., Darrell, T.: Caffe: convolutional architecture for fast feature embedding. In: Proceedings of the ACM International Conference on Multimedia, pp. 675–678 (2014)
5. Jiang, J., Hu, R., Wang, Z., Han, Z.: Noise robust face hallucination via locality-constrained representation. IEEE Trans. Multimed. **16**(5), 1268–1281 (2014)
6. Johnson, J., Alahi, A., Li, F.F.: Perceptual losses for real-time style transfer and super-resolution. In: Proceedings of European Conference on Computer Vision, pp. 694–711 (2016)
7. Kappeler, A., Yoo, S., Dai, Q., Katsaggelos, A.K.: Video super-resolution with convolutional neural networks. IEEE Trans. Comput. Imaging **2**(2), 109–122 (2016)
8. Kim, J., Lee, J.K., Lee, K.M.: Accurate image super-resolution using very deep convolutional networks. In: Proceedings of the IEEE Conference on Computer Vision and Pattern Recognition, pp. 1646–1654 (2016)
9. Ledig, C., Theis, L., Huszar, F., Caballero, J., Aitken, A., Tejani, A., Totz, J., Wang, Z., Shi, W.: Photo-realistic single image super-resolution using a generative adversarial network. arXiv preprint arXiv:1609.04802 (2016)
10. Li, D., Wang, Z.: Video superresolution via motion compensation and deep residual learning. IEEE Trans. Comput. Imag. **3**(4), 749–762 (2017)
11. Li, M., Zuo, W., Zhang, D.: Deep identity-aware transfer of facial attributes. arXiv preprint arXiv:1610.05586 (2016)
12. Liu, C., Shum, H.Y., Freeman, W.T.: Face hallucination: theory and practice. Int. J. Comput. Vis. **75**(1), 115 (2007)
13. Ma, X., Zhang, J., Qi, C.: Hallucinating face by position-patch. Pattern Recognit. **43**(6), 2224–2236 (2010)

14. Mao, X.J., Shen, C., Yang, Y.B.: Image restoration using very deep convolutional encoder-decoder networks with symmetric skip connections. In: Proceedings of Advances in Neural Information Processing Systems, pp. 2802–2810 (2016)
15. Simonyan, K., Zisserman, A.: Very deep convolutional networks for large-scale image recognition. arXiv preprint arXiv:1409.1556 (2014)
16. Tuzel, O., Taguchi, Y., Hershey, J.R.: Global-local face upsampling network. arXiv preprint arXiv:1603.07235 (2016)
17. Wang, N., Tao, D., Gao, X., Li, X., Li, J.: A comprehensive survey to face hallucination. Int. J. Comput. Vis. **106**(1), 9–30 (2014)
18. Wang, X., Tang, X.: Hallucinating face by eigentransformation. IEEE Trans. Syst. Man Cybern. C Appl. Rev. **35**(3), 425–434 (2005)
19. Wang, Z., Bovik, A.C., Sheikh, H.R., Simoncelli, E.P.: Image quality assessment: from error visibility to structural similarity. IEEE Trans. Image Process. **13**(4), 600–612 (2004)
20. Wen, Y., Zhang, K., Li, Z., Qiao, Y.: A discriminative feature learning approach for deep face recognition. In: Proceedings of European Conference on Computer Vision, pp. 499–515 (2016)
21. Wolf, L., Hassner, T., Maoz, I.: Face recognition in unconstrained videos with matched background similarity. In: Proceedings of the IEEE Conference on Computer Vision and Pattern Recognition, pp. 529–534 (2011)
22. Yang, C.Y., Liu, S., Yang, M.H.: Structured face hallucination. In: Proceedings of the IEEE Conference on Computer Vision and Pattern Recognition, pp. 1099–1106 (2013)
23. Yang, J., Wright, J., Huang, T., Ma, Y.: Image super-resolution via sparse representation. IEEE Trans. Image Process. **19**(11), 2861–2873 (2010)
24. Yu, X., Porikli, F.: Ultra-resolving face images by discriminative generative networks. In: Proceedings of European Conference on Computer Vision, pp. 318–333 (2016)
25. Yu, X., Porikli, F.: Face hallucination with tiny unaligned images by transformative discriminative neural networks. Proceedings of AAAI Conference on Artificial Intelligence, pp. 4327–4333 (2017)
26. Zhang, K., Zuo, W., Chen, Y., Meng, D., Zhang, L.: Beyond a Gaussian denoiser: residual learning of deep CNN for image denoising. IEEE Trans. Image Process. **26**(7), 3142–3155 (2017)
27. Zhao, Y., Wang, R., Dong, W., Jia, W., Yang, J., Liu, X., Gao, W.: Gun: Gradual upsampling network for single image super-resolution. arXiv preprint arXiv:1703.04244 (2016)
28. Zhou, E., Fan, H., Cao, Z., Jiang, Y., Yin, Q.: Learning face hallucination in the wild. In: Proceedings of AAAI Conference on Artificial Intelligence, pp. 3871–3877 (2015)
29. Zhu, S., Liu, S., Loy, C.C., Tang, X.: Deep cascaded bi-network for face hallucination. In: Proceedings of European Conference on Computer Vision, pp. 614–630 (2016)

Exemplar-Based Pixel by Pixel Inpainting Based on Patch Shift

Zhenping Qiang[1,2], Libo He[1], and Dan Xu[1(✉)]

[1] School of Information Science and Engineering in Yunnan University,
#2, Cuihubei Road, Kunming 650091, People's Republic of China
22013000170@mail.ynu.edu.cn, danxu@ynu.edu.cn
[2] Department of Computer and Information Science,
Southwest Forestry University, Kunming, China
qzp@swfu.edu.cn

Abstract. This paper presents a novel exemplar-based image inpainting method. It performs patch-shift based information statistics and incorporates the result into the data filling process. The central idea to ensure the confidence pixel value to fill is based on patch shift that captures the statistic information from the expanded filling patches' most similar patches. Our approach outperforms the original exemplar-based inpainting in a large number of examples on removing large occluding objects and thin scratches in real and synthetic images, at the cost of some additional computation. It inherits well-known advantages of the exemplar-based inpainting method, such as the capability of maintaining the integrity and consistency of image structure information, and the adaptability to many application scenarios. Experiments on synthetic and natural images show the advantages of our approach, especially, in the big target removing task our method yields generally better results.

Keywords: Image inpainting · Patch shift
Exemplar-based image inpainting · Object removal

1 Introduction

Image inpainting, which is also known as "image completion", involves the issue of filling missing parts in images. This is a challenging task in computer vision [22]. First of all, the filling algorithm should be able to successfully complete complex natural images. Secondly, it should be able to handle incomplete images with large missing parts. Finally, the algorithm's execution should be in a fully automatic manner, i.e. without intervention from the user.

So far, many inpainting algorithms have been proposed to resolve these issues. We divide them into four categories with the respect of their capability of characterizing the redundancy of the image.

One category of image inpainting methods is diffusion-based [3,5,6,8,10,27]. This class of methods has been first introduced by Bertalmio et al. in [6]. These methods are naturally well suited for completing straight lines, curves, and

© Springer Nature Singapore Pte Ltd. 2017
J. Yang et al. (Eds.): CCCV 2017, Part II, CCIS 772, pp. 370–382, 2017.
https://doi.org/10.1007/978-981-10-7302-1_31

inpainting small regions. However, they are not well suited for recovering the large area textures because their incompetence of synthesizing semantic textures and structures [18].

The second category of methods that are more effective for inpainting large holes is exemplar-based, which is based on the seminal work on texture synthesis [14,28] and exploits image statistical and self-similarity priors. The authors in [19] further categorize these methods into two subcategories: MRF-based and matching-based. The MRF-based methods are realized by optimizing discrete Markov Random Fields (MRFs) [22,25]. These methods rearranged the patch or pixel locations to fill the image, and needed to solve a global optimization problem. This makes the methods more complex than matching-based methods. Matching-based methods [11,13,26,29] are inspired by local region-growing methods which grow the texture one patch at a time. Thanks to the proposed fast PatchMatch method [4], the computational complexity of the match-based inpainting methods have been relieved.

Methods in the third category do not involve any explicit interpolation in the image domain but rather in one or several transform spaces, e.g., applying sparse representations for decomposing the image into a texture and a geometry component and then using two dictionaries of different characteristics to complete the image [15]. In [20] Koh and Rodriguez-Marek proposed a method which modified the K-clustering with singular value decomposition (K-SVD) [1] inpainting method. This type of methods usually performs well in sparse missing data or thin domains image inpainting.

Finally, there are some hybrid methods, i.e. Bertalmio et al. [7] proposed a method to decompose an image into structure and texture components, the structure component is filled by using a PDE base method, while the texture component is filled by a texture synthesis method. Methods [9,21] used one unique energy function to combine different approaches. The paper [24] proposed the super-resolution method to retrieve high-frequency details of the inpainted areas. Arias et al. proposed a general variational framework for exemplar-based image inpainting [2,17] and achieved very good results. All of these methods are expected to achieve an effective balance between preserving edges or structures and restoring texture.

Due to these research works, image inpainting has made considerable progress. Especially in the aspect of image retouching, image inpainting has become a standard tool, e.g. the Content Aware Fill in Adobe Photoshop is implemented in [4,29] as reported in Adobe web site. However, there are numerous applications on image and video editing need to be enhanced in terms of quality and speed of the inpainting. There also have been a few research works on how to use image inpainting for super-resolution and zooming of images [16].

We notice that exemplar-based methods are effective in large area inpainting. However, these methods may not get ideal results for restoring complicated images. Firstly, in the patch matching, selected processing patches often centered at inpainting boundary pixels, then the patches possibly with a lot of unknown pixels have been processed earlier which will increase the matching error.

Besides, in the filling process, these methods filled all the unknown pixels in the unit of a patch, this may cause the error accumulation, and lead to unsatisfied inpainting results.

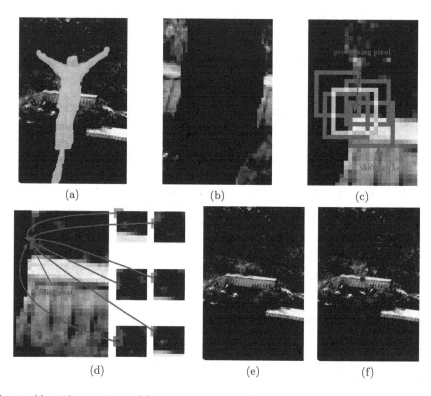

Fig. 1. Algorithm outline. (a) Masked image. (b) Selecting the top 10 pixels with the highest priority. (c) Shift patches of the processing pixel. (d) Combining a set of matched patches for each filling pixel. (e) Our result. (f) Result of [11].

To resolve these issues, we propose an exemplar-based inpainting method based on patch shift (as showed in Fig. 1). We select the pending pixels according to their structure priority (as in Fig. 1(b)). Then we select a series of neighboring patches which contain the selected pending pixels as patches to be processed (as in Fig. 1(c)). Moreover, we delete some patches that contain a large number of unknown pixels and patches that are completely known. Furthermore, we acquire the most similar patches for each remaining patch by matching it in the known region of the image according to the method proposed in paper [4]. Finally, for each unknown pixel in the processing patches, we get a series of pixels from matched patches which include the corresponding pixel to the filling pixel, and then fill the pixel according to the values of these corresponding pixels (as in Fig. 1(d)). A variety of experiments show our method is effective and performs well in large area repairing.

The rest of the article is organized as follows. In Sect. 2, we introduce the previous works which are related with our method. In Sect. 3, we present our exemplar-based inpainting method based on patch shift. The inpainting results and comparison in real nature image and synthetic image are discussed in Sect. 4, followed by the conclusion in Sect. 5.

2   Related Work

The original intention of exemplar-based inpainting methods are to filling-in large missing portions in an image (e.g., to remove an object) in a visually plausible way. For images which have the same patterns, we can use the "texture synthesis" algorithms to solve this problem. However, nature images often contains not only the same texture information, but also many foreground objects whose contour continuity is very sensitive to human eyes. Based on this, many exemplar-based inpainting methods are paying great attention to the structural information in valid pixels around the repaired boundary. A pioneering approach in this field was proposed by Criminisi et al. [11] that combined a structural reconstruction approach with the texture synthesis to utilize advantages of both approaches. They observed that it was important to complete an image starting by recovering the structures first, and based on this idea they proposed an order calculation method to ensure the patch which the isophotes flowing into it to be filled earlier.

Here, given an input image I, as well as a target region Ω and a source region S, generally $S = I - \Omega$. The goal of the exemplar-based inpainting is to fill Ω in a visually plausible way by copying patches or pixels from S. Also we use $\delta\Omega$ to represent the boundary of the target region, i.e. the fill front. In paper [11], the processing order is given by a patch priority measure defined as the product of two terms $(P(p) = C(p)D(p))$. The first term $C(p)$ is the confidence term accounts for the amount of known pixels versus unknown pixels in the processing patch, and the second term $D(p)$ is the data term to indicate the presence of some structures in the processing path.

Based on this initial priority calculation method, some methods are proposed to amend it. For instance, many methods [12,23,30] were proposed to define different patch processing orders in the way to ensure the structure information to be filled in first.

In these methods, on one hand, we noticed that there are many patches with large numbers of unknown pixels and large gradient information will be filled first. On the other hand, whole unknown pixels in the processing patch were filled at the same time in these methods. These two aspects may cause filling errors, and lead to the visual discontinuity of the repairing result eventually.

3   Model Description

In this section we describe the pipeline of our algorithm. The process of our algorithm has three steps: (i) selecting pixels; (ii) matching patches; (iii) filling pixels. This process iterates until the image is completely repaired.

3.1   Selecting Pixels

In this step, image product is used to calculate the repairing image, and mark the boundary of the repairing area based on the mask image. We only use data-term $D(p)$ as the priority term to determine which pixels to process first. An effective data-term $D(p)$ must tell about whether linear structures are presented or not at a point p and with which angle they cross the repaired region. Because of the incomplete information of the pixels $p \in \delta\Omega$, the gradient $\overrightarrow{\nabla I_q}$ cannot precisely estimated priority at this point. We use a better priority term defined as (1) which is proposed and validated in [12] to compute the priority of the marked boundary pixels.

$$D(p) = \|G_p \overrightarrow{n_p}\|_2. \tag{1}$$

where $\overrightarrow{n_p}$ is the local orientation to the mask front at the point p, and G_p is a weighted average of structure tensors estimated on non-masked parts of the target patch ψ_p, for color image, $I = (R, G, B)^T$, G_p can be defined as (2):

$$G_p = \sum_{q \in (\psi_p \cap (I-\Omega))} w_q \begin{pmatrix} R_x^2 + G_x^2 + B_x^2 & R_x R_y + G_x G_y + B_x B_y \\ R_x R_y + G_x G_y + B_x B_y & R_y^2 + G_y^2 + B_y^2 \end{pmatrix}. \tag{2}$$

and w_q is a normalized $2d$ Gaussian function centered at p. The subscripts x and y indicate spatial derivatives of different color channels.

At the end of pixel selecting step, we extract top N_1 pixels with the highest priority on the repairing boundary as candidate pixels to process in the subsequent step.

3.2   Matching Patches

According to selecting pixels step, pixels among the contour crosses Ω orthogonally are extracted. For each pixel p_i in these extracted pixels, we shift the patch ψ_{p_i}(the size of the path is $w_s \times w_s$) which centers at p_i to get all the patches which include the pixel p_i. The obtained set S_ψ of the patches is defined as (3):

$$S_\psi = \{\psi | p_i \in \psi, i = 1, 2, ...N_1\}. \tag{3}$$

Through this process, we extend the patches for further matching process. After shifting, the patches may contain a lot of unknown pixels or may be not containing any unknown pixel which are required to be filtered. For each patch $\psi_j \in S_\psi$, these filter condition can be formulized as (4):

$$0 < |p_k \in \psi_j, \ p_k \in \Omega| \leq \alpha_1 N. \tag{4}$$

where p_k is a pixel in patch ψ_j, N is the number of pixels in each patch (i.e. $w_s \times w_s$), α_1 is a conditional threshold.

After filtering, we obtained the processing patches set S_ψ', each patch in S_ψ' is matched using PatchMatch algorithm to obtain its most similar patch in the known region of the process image. As a result, we obtained a matched patches set M_ψ' correspond to the processing patches S_ψ'.

3.3   Filling Pixels

As discussed above, our algorithm extends the processing patches S'_ψ in matching patches step. This made the filling pixel may present in many different processing patches. Thus, when we want to fill pixel p_i, it is necessary to consider how to fill it based on the matched patches set M'_ψ.

In each iterative process of our algorithm, we first exact the pixel p_i belongs to the processing patch set S'_ψ. According to the location of p_i in each patch ψ_{p_i} in S'_ψ, we get its corresponding pixel value in the matched patch ψ_{p_i} from M'_ψ. For these corresponding pixels value of p_i, due to the weighted average method is more likely to cause fuzzy, in this paper, the median method is adopted to select the final filling value to fill p_i. Finally, after p_i is filled, our algorithm updates the repairing mask and continues to process the unprocessed pixels belongs to the processing patch. When all unprocessed pixels have been processed, our algorithm will enter the next iteration.

3.4   Confidence Term Processing

Since we have done patch shift of the selected pixel, we don't use the confidence information $C(p)$ of pixel p to compute the processing order. To ensure the patch that contains a larger amount of reliability information to be filled in a prior order, we adopt a new strategy to use and update the confidence term $C(p)$. During initialization, the value of $C(p)$ is set to $C(p) = 0 \ \forall p \in \Omega$, and $C(p) = 1 \ for \ \forall p \in S$. Our algorithm don't update $C(p)$ immediately after fill a pixel. Before we compute the data term $D(p)$, if $C(p) = 0 \ for \ \forall p \in \Omega$, we set $D(p) = 0$. Until $D(p) = 0 \ for \ \forall p \in \delta\Omega$, we update $C(p)$ and set $C(p) = 1$ for all filled pixels.

With this strategy, we can ensure that the repair is done from the structure area on the contour $\delta\Omega$ of the target region Ω, and layer by layer to repair the target region Ω. Figure 2 shows an example of the our algorithm.

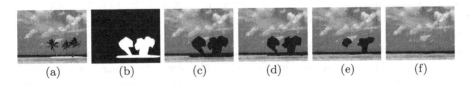

(a)	(b)	(c)	(d)	(e)	(f)

Fig. 2. An example of our algorithm. (a) Input image. (b) Mask image. (c) Masked image. (d) Intermediate result image. (e) Intermediate result image. (f) Result.

4   Result and Comparisons

Here we apply our proposed exemplar-based pixel by pixel inpainting algorithm to a variety of applications, ranging from purely synthetic images to full-color photographs that contain complex textures. At the same time, we take some side-by-side comparisons to some previously proposed methods.

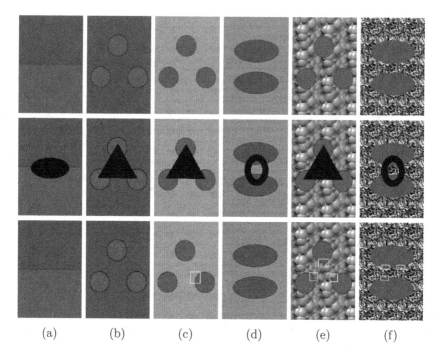

(a)   (b)   (c)   (d)   (e)   (f)

Fig. 3. Comparisons on synthetic images. The first row shows the original synthetic images. The second row shows the unknown region (black areas). The third row shows the results and defective areas (in the yellow boxes). (Color figure online)

4.1 Comparisons on Synthetic Images

Figure 3 presents six examples for synthetic images inpainting. The first row presents the original synthetic images; the second row shows the unknown region (black areas), and the third row shows the results and the defective areas (in the yellow boxes). The (a) and (b) columns show the inpainting of the synthetic images without texture, as the results showed, for such synthetic images, the proposed method is very effective and there are no obvious flaws in the results. (c) and (d) columns show the inpainting of the synthetic images with regular texture, as the results showed, for this type of synthetic images, there are some minor flaws in the results, but the repaired effect of the regular texture effect is very good. (e) and (f) columns show the inpainting results of the synthetic images with irregular texture. The results show that our method has the ability to repair the overall structure of composite images with the complex background, but there are more flaws in the results. This mainly due to the strong gradient area existed in the background texture, and these areas have a higher priority order to process, the foreground object's boundary may be undermined in background texture's repaired.

Fig. 4. Comparisons for object removal. The first row shows five original images. In the remaining five rows, the first to the fifth columns show the masked images, results of [11], results of [24], results of [17] and our results.

4.2 Comparisons on Full-Color Photographs

To verify the effect of our method to full-color photograph inpainting applications, we have conducted a lot of experimental comparisons to classical methods

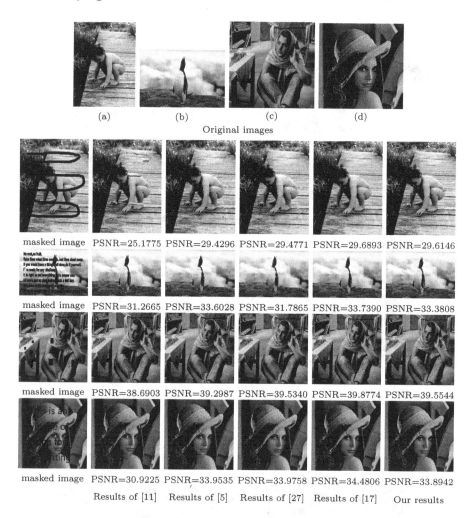

Fig. 5. Comparisons for scratch and text removal. The first row shows four original images. In the remaining four rows, the first to the sixth columns show the masked images, results of [11], results of [5], results of [27], results of [17] and our results.

on commonly used image inpainting datasets. These applications include large target removal, scratch repair and texture removal.

Figure 4 presents the comparison between our algorithm and related exemplar-based inpainting algorithms. The first row presents the five original images, and the others rows, from the first to the fifth columns show masked images and the results of the methods in [11,17,24] and ours. For the image with regular background texture (Fig. 4(a)), the methods for comparison and our method can repair the obvious boundary (the coastline in Fig. 4(a)) very well. For the image with irregular background texture (Fig. 4(b)), our method can be a good way to repair the boundary of images, while the repaired area does not have any obvious texture garbage. In the image Fig. 4(c) and (d), there are prominent structure need to be restored, our method achieves the best repairing results too. The Criminisi's method has obviously repairing failures for inpainting images in Fig. 4(c) and (d). The pole is clearly disconnected in the result of approach [17]. Although the expected structures were repaired in the results of method [24], there are serrated edge on the poles in the repaired result of Fig. 4(c) and there are a lot of blurring in the repaired result of Fig. 4(d). For the inpainting image (Fig. 4(e)), which has obvious boundaries and different types of textures, our method also achieved the best results. On one hand, our method ensures the continuity of the contour, on the other hand, there is no distinct blur.

Furthermore, Fig. 5 shows a lot of examples we have done for scratch and text removal. For the scratch repairing in Fig. 5(a) and the text removal in Fig. 5(b), the PSNR of our algorithm is similar to the algorithm [17] and is superior some other algorithms [5,11,27]. For repairing of damaged texture areas, even algorithms [5,17,27] nearly achieve the similar PSNR as our method, but they still have some differences. Figure 6 shows that when we zoom in the repaired pictures, we can find our repaired picture is better than algorithms [5,27] and close to the best scratch repainting algorithm [17]. The PSNR values of repairing algorithm whose result is presented in Fig. 5(d) of algorithms [5,17,27] are little better than our method. However, when we enlarge parts detail of repaired pictures (presented in the Fig. 7.), we find repaired pictures of algorithms [5,17,27] still have obvious outline of 'of'. Thus, our algorithm is still superior to the traditional algorithms [5,27] that are good at scratch repairing.

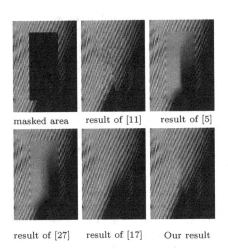

Fig. 6. Comparisons of the inpainting details of Fig. 5(c).

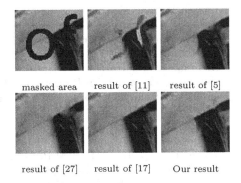

Fig. 7. Comparisons of the inpainting details of Fig. 5(d).

4.3   Implementation Details and Parameters

All experiments are running on a PC with an Intel Core i7 2.3 GHz CPU and 16 G RAM. Parameters of the algorithm are kept as constant for testing are presented in this paper, and the size of patch is set to 9×9, the number $N_1 = 10$ in Eq. (3), $\alpha_1 = 0.1$ in Eq. (4).

Limitations. Our method may fail when there is a strong structure in both the whole and the part of the image. This shortcoming is actually existed in all exemplar-based inpainting methods.

5   Conclusion

In this paper, through a lot of experiments, we prove patch shift is effective for exemplar-based inpainting algorithm and our method is effective for scratch or text removal, object removal and missing block completion.

Acknowledgments. This work is supported by the projects of National Natural Science Foundation of China (11603016, 61540062), the Key Project of Yunnan Applied Basic Research (2014fa021) and project of Research Center of Kunming Forestry Information Engineering Technology (2015FBI06).

References

1. Aharon, M., Elad, M., Bruckstein, A.: K-SVD: an algorithm for designing overcomplete dictionaries for sparse representation. IEEE Trans. Signal Process. **54**(11), 4311–4322 (2006)
2. Arias, P., Facciolo, G., Caselles, V., Sapiro, G.: A variational framework for exemplar-based image inpainting. Int. J. Comput. Vision **93**(3), 319–347 (2011)
3. Ballester, C., Bertalmio, M., Caselles, V., Sapiro, G., Verdera, J.: Filling-in by joint interpolation of vector fields and gray levels. IEEE Trans. Image Process. **10**(8), 1200–1211 (2001)
4. Barnes, C., Shechtman, E., Finkelstein, A., Goldman, D.: Patchmatch: a randomized correspondence algorithm for structural image editing. ACM Trans. Graph.-TOG **28**(3), 24 (2009)
5. Bertalmio, M., Bertozzi, A.L., Sapiro, G.: Navier-stokes, fluid dynamics, and image and video inpainting. In: Proceedings of the 2001 IEEE Computer Society Conference on Computer Vision and Pattern Recognition, CVPR 2001, vol. 1, pp. I-355–I-362 (2001)
6. Bertalmio, M., Sapiro, G., Caselles, V., Ballester, C.: Image inpainting. In: Proceedings of the 27th Annual Conference on Computer Graphics and Interactive Techniques, pp. 417–424. ACM Press/Addison-Wesley Publishing Co. (2000)
7. Bertalmio, M., Vese, L., Sapiro, G., Osher, S.: Simultaneous structure and texture image inpainting. IEEE Trans. Image Process. **12**(8), 882–889 (2003)
8. Bornemann, F., März, T.: Fast image inpainting based on coherence transport. J. Math. Imaging Vision **28**(3), 259–278 (2007)
9. Bugeau, A., Bertalmío, M., Caselles, V., Sapiro, G.: A comprehensive framework for image inpainting. IEEE Trans. Image Process. **19**(10), 2634–2645 (2010)

10. Chan, T.F., Shen, J.: Nontexture inpainting by curvature-driven diffusions. J. Vis. Commun. Image Represent. 12(4), 436–449 (2001)
11. Criminisi, A., Perez, P., Toyama, K.: Object removal by exemplar-based inpainting. In: Proceedings of the 2003 IEEE Computer Society Conference on Computer Vision and Pattern Recognition, vol. 2, pp. II-721–II-728. IEEE (2003)
12. Daisy, M., Buyssens, P., Tschumperlé, D., Lézoray, O.: A smarter exemplar-based inpainting algorithm using local and global heuristics for more geometric coherence. In: 2014 IEEE International Conference on Image Processing (ICIP), pp. 4622–4626. IEEE (2014)
13. Drori, I., Cohen-Or, D., Yeshurun, H.: Fragment-based image completion. ACM Trans. Graph. (TOG) 22, 303–312 (2003). ACM
14. Efros, A.A., Leung, T.K.: Texture·synthesis by non-parametric sampling. In: The Proceedings of the Seventh IEEE International Conference on Computer Vision, vol. 2, pp. 1033–1038 (1999)
15. Elad, M., Starck, J.L., Querre, P., Donoho, D.L.: Simultaneous cartoon and texture image inpainting using morphological component analysis (MCA). Appl. Comput. Harmon. Anal. 19(3), 340–358 (2005)
16. Fadili, M.J., Starck, J.L., Murtagh, F.: Inpainting and zooming using sparse representations. Comput. J. 52(1), 64–79 (2009)
17. Fedorov, V., Facciolo, G., Arias, P.: Variational framework for non-local inpainting. Image Process. On Line 5, 362–386 (2015)
18. Guillemot, C., Le Meur, O.: Image inpainting: overview and recent advances. IEEE Signal Process. Mag. 31(1), 127–144 (2014)
19. He, K., Sun, J.: Image completion approaches using the statistics of similar patches. IEEE Trans. Pattern Anal. Mach. Intell. 36(12), 2423–2435 (2014)
20. Koh, M.S., Rodriguez-Marek, E.: Turbo inpainting: iterative K-SVD with a new dictionary. In: IEEE International Workshop on Multimedia Signal Processing, MMSP 2009, pp. 1–6. IEEE (2009)
21. Komodakis, N.: Image completion using global optimization. In: 2006 IEEE Computer Society Conference on Computer Vision and Pattern Recognition, vol. 1, pp. 442–452. IEEE (2006)
22. Komodakis, N., Tziritas, G.: Image completion using efficient belief propagation via priority scheduling and dynamic pruning. IEEE Trans. Image Process. 16(11), 2649–2661 (2007)
23. Le Meur, O., Gautier, J., Guillemot, C.: Examplar-based inpainting based on local geometry. In: 2011 18th IEEE International Conference on Image Processing (ICIP), pp. 3401–3404. IEEE (2011)
24. Le Meur, O., Guillemot, C.: Super-resolution-based inpainting. In: Fitzgibbon, A., Lazebnik, S., Perona, P., Sato, Y., Schmid, C. (eds.) ECCV 2012. LNCS, vol. 7577, pp. 554–567. Springer, Heidelberg (2012). https://doi.org/10.1007/978-3-642-33783-3_40
25. Pritch, Y., Kav-Venaki, E., Peleg, S.: Shift-map image editing. In: 2009 IEEE 12th International Conference on Computer Vision, pp. 151–158. IEEE (2009)
26. Sun, J., Yuan, L., Jia, J., Shum, H.Y.: Image completion with structure propagation. ACM Trans. Graph. (ToG) 24(3), 861–868 (2005)
27. Telea, A.: An image inpainting technique based on the fast marching method. J. Graph. Tools 9(1), 23–34 (2004)
28. Wei, L.Y., Levoy, M.: Fast texture synthesis using tree-structured vector quantization. In: Proceedings of the 27th Annual Conference on Computer Graphics and Interactive Techniques, pp. 479–488. ACM Press/Addison-Wesley Publishing Co. (2000)

29. Wexler, Y., Shechtman, E., Irani, M.: Space-time video completion. In: Proceedings of the 2004 IEEE Computer Society Conference on Computer Vision and Pattern Recognition, CVPR 2004, vol. 1, pp. I-120–I-127. IEEE (2004)
30. Xu, Z., Sun, J.: Image inpainting by patch propagation using patch sparsity. IEEE Trans. Image Process. **19**(5), 1153–1165 (2010)

GAN Based Sample Simulation for SEM-Image Super Resolution

Maoke Yang[1,2], Guoqing Li[1(✉)], Chang Shu[1,2], Pan Zhao[1,4], and Hua Han[1,2,3]

[1] Institute of Automation, Chinese Academy of Sciences,
95 Zhongguancun East Road, Beijing 100190, China
guoqing.li@ia.ac.cn
[2] University of Chinese Academy of Sciences, Beijing, China
[3] The Center for Excellence in Brain Science and Intelligence Technology,
Beijing, China
[4] Harbin University of Science and Technology, Harbin, China

Abstract. We propose to employ image super resolution to accelerate collection speed of scanning electric microscopes (SEM). This process can be done by collecting images in lower resolution, and then upscale the collected images with image super-resolution algorithms. However, because of physical factors, SEM-images collected in different resolution changed not only in their scale, but also with noise level and physical distortion. Consequently, it is hard to obtain training dataset. In order to solve this problem, we designed a generative adversarial network (GAN) to fit the noise of SEM images, and then generate realistic training samples from high resolution SEM data. Finally, a fully convolutional network have been designed to perform image super-resolution and image denoise at the same time. This pipeline works well on our SEM-image dataset.

Keywords: Image super resolution · Generative adversarial network
Scanning electric microscope

1 Introduction

Recent needs [1,2,24] to reconstruct the neural connection of animal-brain-tissue calls for effective methods to collect numerous microscopic images of tissue slices. A efficient way [3] to collect microscopic images is to first cut biological samples into thousands ultra-thin slices, which have a thickness of about 30 nm to 70 nm, then put these slices on several silicon wafers, and use several SEMs to take images for these slices in parallel. Our laboratory collect microscopic images in this way too. This method did accelerate the data collection process, however, the speed is still not as fast as we expected. For example, in a typical high resolution image of an ultra-thin slice, each pixel correspond to a physical size of 2 nm, and it take 2 μs to collect each pixel. Therefore, a single tissue slice of size 1 mm^2 needs 2.5×10^{11} pixels to represent, which will spend a single SEM at least 138.8 h to collect. This means that taking images for ten thousand of

© Springer Nature Singapore Pte Ltd. 2017
J. Yang et al. (Eds.): CCCV 2017, Part II, CCIS 772, pp. 383–393, 2017.
https://doi.org/10.1007/978-981-10-7302-1_32

ultra-thin slices needs one SEM continuously working at least 6.6 years! Using 6 SEMs working in parallel still needs more than a year. Undoubtably, further accelerating the collection process is of great significance, and obstructed by expensive cost, it is not feasible to increase the collection speed only by increase the number of microscopes. This work explore the way to speed up the collection of every single microscope by using the technology of image super-resolution and image denoise at the same time.

Single image super resolution (SISR) tries to increase the image resolution, and recover the detailed information as much as possible. A basic assumption of image super resolution is that high resolution images contain much redundant information, and thus can be recovered from low resolution images. According to [4], the linear degradation model of SISR is formulated by:

$$\mathbf{z} = D_s H \mathbf{x} \qquad (1)$$

where $\mathbf{z} \in \mathbf{R}^{M \times N}$ is the input low resolution (LR) image, $\mathbf{x} \in \mathbf{R}^{Ms \times Ns}$ is the unknown high resolution (HR) image, linear operation $H \in \mathbf{R}^{MNs^2 \times MNs^2}$ blurs the image, and operation $D_s \in \mathbf{R}^{MN \times MNs^2}$ decrease the image scale by a factor s. Many recent work trying to learn a mapping from low resolution images to high resolution images with extra data set, and the state-of-the-art mapping function usually parametrized with deep neural networks.

It is possible to use the technology of image super-resolution and image denoise to accelerate the collection process of SEM-images. Automatic analysis of biological-tissue-image require the image contains enough pixel for each organelle, however, as discussed before, the collection process of high resolution SEM-images is really time-consuming. In practice, an effective way to accelerate collection process is to take images in a relatively lower resolution, and then increase the image resolution by image super-resolution algorithms. Reducing half of the imaging resolution can accelerate four times of the imaging speed. In order to balance the conflict between collection speed and final image quality, this work only consider to upscale the collected image with factor of 2.

The most difficult problem in our application is to get training dataset. Traditional image super-resolution problem do not consider the disturb of random noise, while SEM-images usually be notoriously noised, which means we cannot just use pretrained image-super-resolution-model to do our job. Besides, traditional SISR algorithms usually down-sample HR images to create training sample, but SEM-images that collected with difference parameter would have difference noise level, which means we cannot create the training dataset in a similar way. A Possible solution is to take images in the same place with different resolutions to create the training samples. However, as observed in the experiment, electron beam will heat the tissue slices in the collecting process, and cause some physical deformation to the biological sample. As a result, even collected in the same place with same parameters, two images would have huge difference on pixel level. This change does not affect biological meaning, but really obstruct the training of image-super resolution model.

Considering the SEM-image super resolution problem, the noise usually come from irregular scattering electrons, thus it is okay to think the noise is "adding" into the true image. Then we can formulate the degradation model as:

$$\mathbf{z} = D_s H(\mathbf{x} + N) \tag{2}$$

Where $\mathbf{z} \in \mathbf{R}^{M \times N}$ is the collected low resolution (LR) image; $\mathbf{x} \in \mathbf{R}^{Ms \times Ns}$ is the unknown high resolution (HR) image, which we want to restore from the collected LR image $\mathbf{z}$. Linear operation $H \in \mathbf{R}^{MNs^2 \times MNs^2}$ and operation $D_s \in \mathbf{R}^{MN \times MNs^2}$ represent the degradation process as Eq. (1). The $N \in \mathbf{R}^{M \times N}$ is the added noise. This formulation can be further written as:

$$\mathbf{z} = D_s H \mathbf{x} + G \tag{3}$$

In which $G = D_s H N$. Following this formulation, we can add some noise to the down-sampled HR images to create our training image pairs. This method can be easily applied to bridge the gaps between the noisy and clean images, at the same time, the distortion between low resolution images and high resolution images is no longer a problem. In this work, we trained a generative adversarial network [6] to fit the noise of low resolution SEM-images, and then use the generator to add the learned noise to down-sampled high resolution image patches. This methods can help us build suitable dataset for our training and validation.

Overall, the contributions of this work are mainly in three aspects:

1. A pipeline has been proposed to accelerate the SEM-image collection of large scale biological-tissue imaging.
2. A GAN has been designed to produce high quality training samples for the super-resolution and denoise problem of SEM-images.
3. We trained an end-to-end fully convolutional network to solve the SEM-image super-resolution problem.

2   Related Work

2.1   Example-Based Image Super Resolution

Example-based image super resolution method learns a mapping from LR image patches to HR image patches. Traditionally, as summarized in [9], example-based SISR methods employ hand-designed feature, hand-designed upscale, hand-designed nonlinear mapping to reconstruct the HR image from the input LR image. SRCNN [10] merged the process of feature extraction, nonlinear mapping, and reconstruction into a single model, training a simple three-layer fully convolutional network to perform feature extractor, nonlinear mapping function, and HR image reconstructor at the same time. In their work, they need to first upscale LR image into desired scale, and then restore detailed information with the three-layer model. Later works such as VDSR [11] build a deeper convolutional network to learn a better mapping. Recent works try to merge the upscale process into the super resolution model, and train an end-to-end model to solve the SISR problem. EEDS [9] employed transposed convolutional layer to increase the image scale, while ESPCN [12] use self-designed sub-pixel convolutional layer to bridge the gap between different scale. In a further development, SRGAN [13] solve the SISR problem with generative adversarial network, using a discriminator to push the generator learn a nature image manifold, which can produce more vivid HR images.

2.2   Image Denoise

Image denoise is a low-level image processing problem, which plays a fundamental role in many computer vision applications. BM3D [14] exploited a nonlocal image modeling by a 3D grouping, followed by a collaborative filtering through transform-domain shrinkage of the 3D-array. Later, Burger et al. [5] employed a simple multi-layer perceptron to learn the mapping from noisy images to a noise-free images. Recently, Chang et al. [15] found that when the data is enough, one single network can do multi-tasks such as image denoise, image super resolution, and many other tasks separately, which illustrated that data-driven methods can solve many image recovery problems well.

2.3   Generative Adversarial Networks

Generative Adversarial Networks (GAN) [6] is a technique to training generative models that can approximate image distribution via a two player adversarial game, and has been shown to generate high quality images [7,16–18]. In this framework, the generator (G) tries to generate realistic-looking images to deceive the discriminator (D), while the D learns to discriminate real images and the generated images. Actually, the D in this framework plays a role of loss function, and tell the G how to generate more lifelike pictures. The training process of GAN can be formulated as follows:

$$\min_G \max_D V(D,G) = \mathbf{E}_{x \sim p_{data}(x)}[log D(x)] + \mathbf{E}_{z \sim p_z(z)}[log(1 - D(G(z)))] \quad (4)$$

where $p_{data}(x)$ is the distribution of nature image, and z is some random noise. Typically, a generator take a fix dimension random noise as input, which then projected to realistic-looking images.

One of the most important property of GAN is that this deep generation model do not need explicit formula. Thus, it is possible to use GANs to estimate the noise distribution of SEM image, and then reproduce realistic training samples for our model.

3   Sample Simulation and SEM-Image Super Resolution

3.1   Create High Quality Training Samples with GAN

We designed a Generative Adversarial network (GAN) to estimate the noise distribution of SEM-images, and then the generator is used to add the learned noise to down-sampled HR image patches and generate LR patches. We refer this model as Noise-GAN in later discussion. It may be possible to add random noise to the down-sampled HR image patches to produce the fake LR image patches, but it is hard to precisely estimate the distribution of noise. Thus, this work explore to fit the noise distribution with GAN.

In the practice, our work down-sample the HR image patches at first, and then feed these patches into the generator to add the generated noise, which makes these patches look real. In the training process, the discriminator tries to distinguish the true LR patches and the fake LR patches, while the generator tries to cheat the discriminator. After 30 epochs' training, the generator is good enough to produce realistic-looking LR patches, and even human cannot distinguish the difference between true LR patches and fake LR patches.

The structure of Noise-GAN follows the principle of DCGAN [7] and W-GAN [8]. The generator projects a 128-d random noise to a noise map of 64×64, which are additional added with the input image patch to finally generate the fake LR patch. The generated fake LR patches and the truly collected LR patched then be used to train the discriminator. A little detail need to notice is that in every training batch, the fake LR image patches and the true LR image patches were collected in the same place, thus they have the same semantic information. Besides, these patches need to subtract their mean value before feed into the discriminator to avoid the disturb of luminance. With these method, the only difference between the fake patches and the true patches is the noise level, and will prevent the discriminator classify these patches with other information. This process will make sure the GAN learns to only fit the noise. The tanh activation rather than relu activation was used in the generator, and we use RMSProp optimizer with learning rate of 0.00001 to train this model. The network is illustrate as Fig. 1.

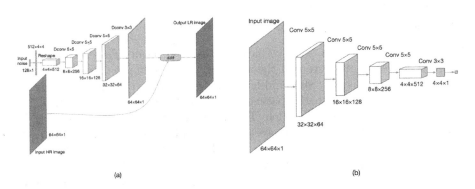

(a)   (b)

Fig. 1. The Noise-GAN structure: (a) The generator, which add generated noise to the HR image patches; (b) The discriminator, which is training to distinguish the true LR image patches and the generated LR patches.

In Fig. 2, we compare the generated LR patches with true LR patches and the down-sampled HR patches. These patches was cropped from a big picture of our SEM-image data set. We will introduce this data set in Sect. 4.

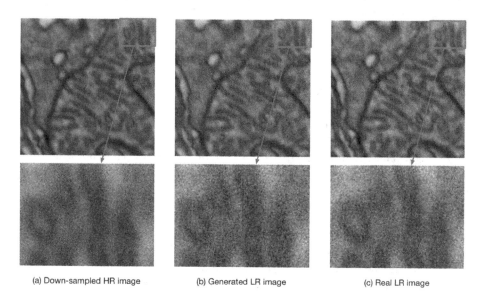

(a) Down-sampled HR image   (b) Generated LR image   (c) Real LR image

Fig. 2. The comparison between generated patch and real patch: (a) Real HR image patch, which have a pixel size of 1 nm; The HR patch was down-sampled to have the same scale with (c). (b) Generated fake LR patch, which use (a) as input; (c) Real LR patch, which was collected with pixel size of 2 nm.

3.2   End-to-End SEM-Image Super Resolution

We designed a deep network to verification the final result. This network may not works as well as other state-of-the-art image super resolution methods, but it is simple to build, easy to train, and have enough capacity to learn the mapping from noised LR patches to HR patches. We call this network as SESR, which means SEM-images super resolution. This network can easily get average PSNR of 36.038 dB on Set5 after 13000 interactions' training on the 91-image dataset.

The architecture of our network described in Fig. 3(c) was inspired by inception model [19–21], ResNet [22], and EEDS [9]. This model have structures to capture multi-scale information, and use skip connect to transfer low-level information to the upper layer. Block1 as illustrated in Fig. 3(a) have four sub branches, and each branch have different receptive filed changed from 1 to 7 to capture features in different scale. These extracted features then be concated and processed by a convolutional layer. Block2 works in a similar way, big convolution kernels are decomposed into small asymmetric convolutions for computational efficiency. Block1 and Block2 can be seen as assemble models, and the features extracted with difference scale are assembled together for a further prediction. A 5×5 transposed convolutional layer are used to increase the image scale. Besides, as discussed in [21,22], skip connection used in this model can also ensure the network converges faster. The res-block in this model is the same with ResNet [22].

In the training process, we use mean squared error as the object function, and use Adam [23] optimizer to train this network, with a learning rate of 0.0003. 10 thousands of interactions is enough to train this network to convergence. The network structure is illustrated as Fig. 3.

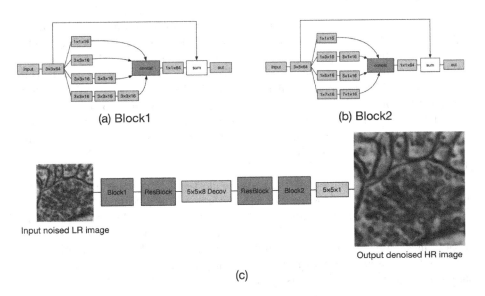

(a) Block1   (b) Block2

(c)

Fig. 3. The structure of SESR. The four sub branches in (a) and (b) are used to capture features in different scale, and the skip connection in these blocks are used to pass low frequency information to upper layer in forward propagation, and speed up the training in back propagation.

Compared with two-stage methods such as SRCNN and VDSR, this model can proposes images end to end, and compared with other one-stage methods such as EEDS and ESPCN, this methods is light-weighted. The storage size of the parameter is less than 1 MB.

4 Experiments and Results

The SESR model was trained with tree group of training data in this experiment to verification the effect of the generated training set. To the best of our knowledge, there are no previous work trying to deal with the SEM-image super resolution problem, so we compare the result with other two possible methods.

4.1   SEM-Image Dataset

The SEM-images used in this work was collected form the Zeiss Super55 Electrical microscope of our laboratory. We take images in the same place with pixel size changed from 1 to 8 nm, corresponding a image size changed from 8192 × 8192 to 1024 × 1024. Because of the large size of these images, 8 images of each size are enough for training, validation, and testing. In the experiment, we crop 3 images for training, 1 image for validation, and 4 images for testing.

4.2   Training and Testing

The SESR model has been trained with three group of training data to illustrate the importance of the generated training data in our application. The fist group is the 91 images as mentioned before, we denote this dataset as "-NT". The second dataset is the real collected SEM dataset, we denote this dataset as "-SEM". The image which have a pixel size of 1 nm was down-sampled to the correspond scale to act as the ground truth, and then down-sample the ground truth image to generate the LR image patches as input. The third dataset is also the SEM dataset, but the difference with the second dataset is that this dataset use the generated LR image patches as input. We use "-GAN" to denote this dataset.

We testing these model with the real collected LR patches as input. It is important to notice that because of the distortion and deformation in the collection process, the output result need to be registered with the ground truth to be measured quantitatively. In order to avoid the difference made by the registration algorithm, for every input image, the output result of each method are registered to the ground truth with the same transformation. Then, the peak signal to noise ratio (PSNR) was used to measure the final result.

Table 1. Average results on test set. Because of the results are measured after registration, the final PSNR are a little lower than many other application.

Method	Bicubic	BM3D+Bicubic	Bicubic+BM3D	SESR-NT	SESR-SEM	SESR-GAN
2 nm × 2	20.421	21.2400	20.9200	20.383	20.5900	**21.2527**
4 nm × 2	21.191	22.2647	21.8899	21.225	21.5382	**22.4889**
8 nm × 2	20.031	20.7056	20.4631	20.077	21.5382	**21.2368**

We compare the final result with other two possible methods. One is to remove noise with BM3D at fist, and then enlarge the image size with Bicubic algorithm, the other method is to upscale the image with Bicubic algorithm at first, and then remove the noise with BM3D. Because we do not know the noise pattern precisely, we estimate the $\sigma = 25$ for the BM3D algorithm. This two kind of methods illustrated different results in our dataset. We illustrated the average PSNR of these results as Table 1.

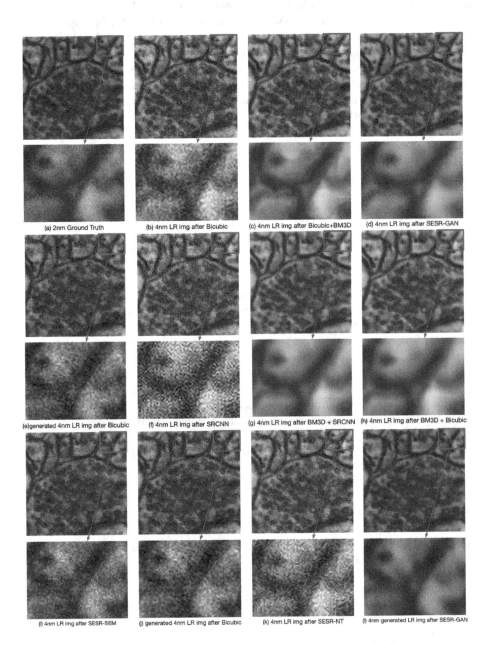

Fig. 4. Final results and comparison with other methods

A group of results is illustrated as Fig. 4. From Fig. 4(f), (i) and (k), we can see that methods trained with nature images are disturbed by the noise. From Fig. 4(d) we can see that using BM3D to remove the noise after bicubic destroy some detailed structure. From Fig. 4(g) and (h) we can see that denoising with BM3D at first can inhibit the noise well, however blurred the final result. And the similar appearance between Fig. 4(d) and (i) in return illustrate that the generated training dataset matters. In all, SESR model trained with the generated training set performs better than other methods, and the final result looks more similar than others. The images predicted by BM3D+Bicubic, although have high PSNR on the test set, blurs the detailed structure of the image, which is not expected in the real application. Although the results of our model also looks a little blurred, this pipeline do illustrated a promising way to really solve our problem.

5   Conclusion

This work proposed a pipeline to deal with the SEM-images super resolution problem. SEM-image super-resolution algorithms have a high potential to accelerate the collection speed of SEMs, and finally break through the bottleneck of obtaining high volume biological tissue image sequence. However, since the noise in SEM-images can not be neglected, SEM-image super resolution problem cannot be treated as the way of ordinary SISR problem. The most difficult part in this application is to get suitable training data to train our model. This work first analyzed the difference between SEM images and ordinary images, and then designed a GAN to fit the distribution of the noise in SEM images. After that, the generator can be used to generate training samples for the final training process. Comparing with other possible solution, this training-sample-generation method performs better, and still have much potential to improve.

Acknowledgments. This paper is supported by the Scientific Instrument Developing Project of the Chinese Academy of Sciences, Grant No. YZ201671, and Special Program of Beijing Municipal Science and Technology Commission (No. Z161100000216146).

References

1. Lichtman, J.W., Winfried, D.: The big and the small: challenges of imaging the brains circuits. Science **334**(6056), 618–623 (2011)
2. Kasthuri, N., Hayworth, K.J., Berger, D.R., et al.: Saturated reconstruction of a volume of neocortex. Cell **162**(3), 648–661 (2015)
3. Hayworth, K.J., Morgan, J.L., Schalek, R., et al.: Imaging ATUM ultrathin section libraries with WaferMapper: a multi-scale approach to EM reconstruction of neural circuits (2014)
4. Romano, Y., Isidoro, J., Milanfar, P.: RAISR: Rapid and Accurate Image Super Resolution. IEEE Trans. Comput. Imaging **3**(1), 110–125 (2017)
5. Burger, H.C., Schuler, C.J., Harmeling, S.: Image denoising: can plain neural networks compete with BM3D? In: 2012 IEEE Conference on Computer Vision and Pattern Recognition (CVPR), pp. 2392–2399. IEEE (2012)

6. Goodfellow, I., Pouget-Abadie, J., Mirza, M., et al.: Generative adversarial nets. In: Advances in Neural Information Processing Systems, pp. 2672–2680 (2014)

7. Radford, A., Metz, L., Chintala, S.: Unsupervised representation learning with deep convolutional generative adversarial networks. arXiv preprint arXiv:1511.06434 (2015)

8. Arjovsky, M., Chintala, S., Bottou, L.: Wasserstein GAN. arXiv preprint arXiv:1701.07875 (2017)

9. Wang, Y., Wang, L., Wang, H., et al.: End-to-end image super-resolution via deep and shallow convolutional networks. arXiv preprint arXiv:1607.07680 (2016)

10. Dong, C., Loy, C.C., He, K., Tang, X.: Learning a deep convolutional network for image super-resolution. In: Fleet, D., Pajdla, T., Schiele, B., Tuytelaars, T. (eds.) ECCV 2014. LNCS, vol. 8692, pp. 184–199. Springer, Cham (2014). https://doi.org/10.1007/978-3-319-10593-2_13

11. Kim, J., Lee, J.K., Lee, K.M.: Accurate image super-resolution using very deep convolutional networks. In: Proceedings of the IEEE Conference on Computer Vision and Pattern Recognition, pp. 1646–1654 (2016)

12. Shi, W., Caballero, J., Huszr, F., et al.: Real-time single image and video super-resolution using an efficient sub-pixel convolutional neural network. In: Proceedings of the IEEE Conference on Computer Vision and Pattern Recognition, pp. 1874–1883 (2016)

13. Ledig, C., Theis, L., Huszr, F., et al.: Photo-realistic single image super-resolution using a generative adversarial network. arXiv preprint arXiv:1609.04802 (2016)

14. Dabov, K., Foi, A., Katkovnik, V., et al.: BM3D image denoising with shape-adaptive principal component analysis. In: Signal Processing with Adaptive Sparse Structured Representations, SPARS 2009 (2009)

15. Chang, J.H., Li, C.L., Poczos, B., et al.: One network to solve them all-solving linear inverse problems using deep projection models. arXiv preprint arXiv:1703.09912 (2017)

16. Berthelot, D., Schumm, T., Metz, L.: BEGAN: Boundary Equilibrium Generative Adversarial Networks. arXiv preprint arXiv:1703.10717 (2017)

17. Zhao, J., Mathieu, M., LeCun, Y.: Energy-based generative adversarial network. arXiv preprint arXiv:1609.03126 (2016)

18. Denton, E.L., Chintala, S., Fergus, R.: Deep generative image models using a Laplacian pyramid of adversarial networks. In: Advances in Neural Information Processing Systems, pp. 1486–1494 (2015)

19. Szegedy, C., Liu, W., Jia, Y., et al.: Going deeper with convolutions. In: Proceedings of the IEEE Conference on Computer Vision and Pattern Recognition, pp. 1–9 (2015)

20. Szegedy, C., Vanhoucke, V., Ioffe, S., et al.: Rethinking the inception architecture for computer vision. In: Proceedings of the IEEE Conference on Computer Vision and Pattern Recognition, pp. 2818–2826 (2016)

21. Szegedy, C., Ioffe, S., Vanhoucke, V., et al.: Inception-v4, inception-ResNet and the impact of residual connections on learning. arXiv preprint arXiv:1602.07261 (2016)

22. He, K., Zhang, X., Ren, S., et al.: Deep residual learning for image recognition. In: Proceedings of the IEEE Conference on Computer Vision and Pattern Recognition, pp. 770–778 (2016)

23. Kingma, D., Ba, J.: Adam: a method for stochastic optimization. arXiv preprint arXiv:1412.6980 (2014)

24. Berning, M., Boergens, K.M., Helmstaedter, M.: SegEM: efficient image analysis for high-resolution connectomics. Neuron **87**(6), 1193–1206 (2015)

PSO-Based Single Image Defogging

Fan Guo, Lijue Liu$^{(\boxtimes)}$, and Jin Tang

School of Information Science and Engineering,
Central South University, Changsha 410083, Hunan, China
guofancsu@163.com

Abstract. This paper proposes a novel defogging algorithm based on particle swarm optimization (PSO) to adaptively and automatically select parameter values. Owing to the lack of enough information to solve the equation of image degradation model, existing defogging methods generally introduce some parameters and set these values fixed. Inappropriate parameter setting leads to difficulty in obtaining the best defogging results for different input foggy images. In this paper, we mainly focus on the way to select optimal parameter values for image defogging. The proposed method is applied to two representative defogging algorithms by selecting the two main parameters and optimizing them using the PSO algorithm. A comparative study and qualitative evaluation demonstrate that the better quality results are obtained by using the proposed method.

Keywords: Single image defogging · Particle Swarm Optimization Parameter selection · Defogging effect assessment

1 Introduction

Most automatic systems assume that the input images have clear visibility, therefore removing the effects of bad weather from these images is an inevitable task. In the past decades, extensive research efforts have been conducted to remove fog or haze from a single input image. Most of these methods [1–4] intend to recover scene radiance using the image degradation model that describe the formation of a foggy image. Tan [1] removed fog by maximizing the local contrast of the restored image. Nishino *et al.* [2] proposed a Bayesian probabilistic method that estimates the scene albedo and depth from a foggy image with energy minimization of a factorial Markov random field. He *et al.* [3] estimated the transmission map and the airlight of the degradation model using the dark channel prior. Tarel *et al.* [4] introduced an atmospheric veil to restore image visibility based on the fast median filter. However, these methods are controlled by a few parameters with fixed values that cannot be automatically adjusted for different foggy images.

In recent years, people are quite interested in automatic fog removal, which is useful in applications such as surveillance video [5], intelligent vehicles [6], and

© Springer Nature Singapore Pte Ltd. 2017
J. Yang et al. (Eds.): CCCV 2017, Part II, CCIS 772, pp. 394–406, 2017.
https://doi.org/10.1007/978-981-10-7302-1_33

outdoor object recognition [7]. In this paper, we thus focus on the PSO-based adaptive parameter adjustment for single image defogging.

It is widely agreed that PSO algorithm can be used for digital image processing, such as image retrieval [8], image segmentation [9], etc. However, due to the lack of proper objective criterion of defogging effect as the fitness function, present defogging algorithms seldom use genetic algorithm to effectively remove fog from a single image. Though the objective image quality evaluation methods have achieved some promising results, they are just applied to assess the quality of degraded image, such as image denosing results and image deblurring results. The aim of defogging algorithm is to recover color and details of the scene from input foggy image. Unlike image quality assessment, the fog can not be addressed like a classic image noise or degradation which may be added and then removed. Meanwhile, there is no easy way to have a reference no-fog image, and the quality evaluation criteria of degraded image, such as the structural similarity (SSIM) [10], the peak signal-to-noise ratio (PSNR) [11], and the mean square error (MSE), are not suitable for assessing image defogging effects. This makes the problem of adaptive parameter adjustment of defogging algorithm not straightforward to solve.

In this paper, we mainly focus on the way to select parameter values for single image defogging. Here, the defogging effect assessment index presented in our previous work [12] is taken as the fitness function of the proposed PSO-based method to adaptively adjust the parameter values for different input foggy images. In Sect. 2 of the paper, the limitation of existing methods are described. Section 3 introduces the proposed parameter value selection approach to single image defogging. Experimental results and conclusions are presented in Sects. 4 and 5, respectively.

2   Limitation of the Existing Defogging Methods

Most current defogging methods recover the scene radiance by solving the image degradation model. Since the model contains three unknown parameters and the solving process is an ill-posed inverse problem, it is thus inevitable to introduce many application-based parameters that used in various assumptions for image defogging. A large quantity of experimental results shows that the selection of the algorithm parameters has direct influence on the final defogging effect. However, there exists a major problem for the parameter setting in most defogging algorithms, i.e. the parameters always have fixed values in the defogging algorithms.

In our experiments, we find that the fixed parameter values caused that the fog removal algorithms just have good defogging effect for a certain kind of foggy image, and the algorithms may not work well for the images captured under other foggy conditions. For example, He's algorithm [3] has mainly three parameters to control: ω which alters the amount of haze kept at all depths, c the patch size for estimating transmission map, and t_0 restrict the transmission to a lower bound to make a small amount of fog preserve in very dense fog regions.

All these parameters have fixed value suggested by the authors, such as the fog parameter ω , which is set to be 0.95 in the algorithm [3]. Our experimental results given in Fig. 1 using He's algorithm [3] show that, if ω is adjusted downward, more fog will be kept, and vice versa. Using $\omega = 0.95$ keeps a slight amount of fog effect around at all depths. However, the experiments show that ω sometimes needs to be decreased when an image contains substantial sky regions, otherwise the sky region may wind up having artifacts. An example showing the need to decrease ω is presented in Fig. 1. The defogging result with $\omega = 0.95$ is shown in Fig. 1(b). One can clearly see that the sky looks contoured since the fog removed by He's algorithm was too strong in this region, and the defogging result has no obvious visibility improvement with $\omega = 0.65$. If setting $\omega = 0.12$, the sky region becomes brighter and smoother, which makes the whole image look more natural.

| (a) | (b) | (c) | (d) |

Fig. 1. Fog removal results with fixed parameter value. (a) Original image. (b) Unpleasing contour effect with $\omega = 0.95$. (c) No obvious visibility improvement with $\omega = 0.65$. (d) Smoother sky region using $\omega = 0.12$

Most defogging algorithms have introduced some parameters, which lead to user interaction and make the final defogging effect hard to control as well. For example, Tarel's algorithm [4] is controlled by five parameters [4] in which p is the percentage of removed atmospheric veil, s_v the assumed maximum size of white objects in the image, b the white balance control for global or local process, s_i the maximum size of adapted smoothing to soften the noise amplified by the restoration, and g an extra factor during final gamma correction. Figure 2 shows the defogging results obtained with different parameter values using Tarel's algorithm [4]. One can clearly see that the restoration is too light with $p = 0.7$, $b = 0.5$, $s_v = 19$, $s_i = 1$, $g = 1.3$ and too strong with $p = 0.99$, $b = 0$, $s_v = 19$, $s_i = 1$, $g = 1$. It seems better with $p = 0.96$, $b = 0.5$, $s_v = 19$, $s_i = 1$, $g = 1.3$. On the right, most of the colors are removed due to a too large value of $b = 1$, and $b = 0.5$ leads to better results. It is obvious that the controllability can be greatly improved and the user-interaction can be also largely reduced if defogging algorithm has no more than two parameters. Therefore, distinguishing between main parameters which directly affect the results and other less important parameters which can be considered as fixed values, and then automatically adjusting their values are very important for the defogging algorithms.

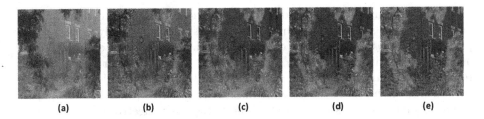

Fig. 2. Fog removal results with many fixed parameters. (a) Original image. (b) Defogging result with $p = 0.7$, $b = 0.5$, $s_v = 19$, $s_i = 1$, $g = 1.3$. (c) Defogging result with $p = 0.99$, $b = 0$, $s_v = 19$, $s_i = 1$, $g = 1$. (d) More pleasing result with $p = 0.96$, $b = 0.5$, $s_v = 19$, $s_i = 1$, $g = 1.3$. (e) Less pleasing result with $p = 0.99$, $b = 1$, $s_v = 19$, $s_i = 1$, $g = 1$

In order to solve the "fixed parameter value" problem, on the one hand we should choose two main parameters and optimize them for different defogging algorithms, and on the other hand we must use an effective defogging effect assessment index to automatically determine the parameter values for different input image. However, the defogging evaluation results obtained using image quality evaluation criteria is often inconsistent with human visual conception, and the existing enhancement assessment methods are mainly from image contrast and do not consider the color restoration effect of the defogging results.

It can thus be seen that "fixed parameter value" is the most common problem existing in most fog removal methods. However, it is hard to realize adaptive adjustment of the algorithm parameters by the defogging results due to the lack of an effective assessment index. Therefore, a PSO-based parameter value selection algorithm using proper defogging effect assessment index is proposed in this paper to select the parameter values for single image defogging.

3   The Proposed Parameter Value Selection Approach

Since most defogging algorithms may produce oversaturated or undersaturated colors with fixed parameters, the two main parameters that directly affect the defogging results are first selected. And then a PSO algorithm combined with a proper defogging assessment index is used here to enable these defogging methods to adaptively set the required two parameters for different input images, and enhance the visibility of scenes with natural color and the best defogging effect. In this section, the detailed descriptions of the two main parameters selection approach to various representative defogging algorithms are first given in Subsect. 3.1. Then, the assessment index for measuring defogging effect is reported in Subsect. 3.2. Finally, the PSO-based parameter value selection method is presented in Subsect. 3.3.

3.1   Optimal Parameter Selection

The task of optimal parameter selection is to distinguish between two main parameters which directly affect the results and other less important parameters which can be considered as fixed values for various defogging methods. There are two ways to select the two main parameters: one is to analyze the related parameters from the perspective of physical mechanism, and the other is to tune one of algorithm parameters by fixing the rest to see whether the defogging results have significant change. Two representative fog removal methods [3,4] are taken as examples to describe the parameter selection process in this paper. Since He's method [3] is recognized as one of the most effective ways to remove fog, and Tarel's method [4] is regarded as one of the fastest defogging algorithms at present.

For He's method [3], a small amount of haze for distance objects is kept to make the final defogging results seem more natural and preserve the feeling of depth as well. There are two key parameters for the fog preservation purpose: ω $(0 < \omega < 1)$ and t_0 $(0 < t_0 < 1)$. Other parameters have less influence on the final defogging effect, and can thus be regarded as less important parameters. For example, the influence of the large patch size c can be effectively reduced by the soft matting used in He's method [13]. Therefore, the patch size suggested by the author is reasonable. Figure 1 shows He's defogging results using different ω. The fog removal results using different t_0 or patch size c are shown in Fig. 3. One can clearly see that t_0 has significant influence on the final results, while the patch size c on the contrary has little influence.

Fig. 3. He's defogging results with different parameter values. (a) The influence of two main parameters on final results with $\omega = 0.95$, $c = 7$. From left to right: original foggy image with size of 400×300, the results obtained with $t_0 = 0.1, 0.5, 1$, respectively. (b) The influence of parameter c on the final results with $\omega = 0.95$, $t_0 = 0.5$. From left to right: the results obtained with $c = 30, 15, 9, 3$, respectively

For Tarel's method [4], the most important two steps that determine the final defogging effect are atmospheric veil remove and gamma correction. Since the value of $p(0 < p < 1)$ controls the amount of atmospheric veil that can be removed, this parameter is useful to compromise between highly restored visibility where colors may appear too dark, and less restored visibility where colors are clearer. The parameter $g(0 < g < 10)$ is used to perform gamma correction to achieve more colorful result. Experimental results show that the larger the value of g, the clearer the defogging result is. Compared with the two parameters p and g, other parameters b, s_v and s_i have less effect on the final defogging results, as shown in Fig. 4.

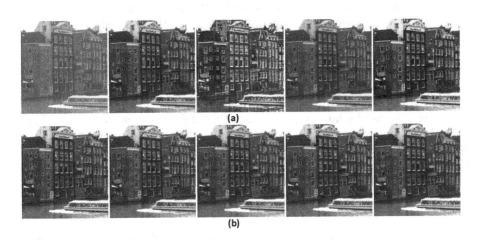

Fig. 4. Tarel's defogging results with different parameter values. (a) The influence of two main parameters on the final results with $b = 0.5$, $sv = 17$, $s_i = 1$. From left to right: original foggy image with size of 370 × 400, the results obtained with ($p = 0.1$, $g = 3$), ($p = 0.95$, $g = 3$), ($p = 0.1$, $g = 3$), ($p = 0.1$, $g = 7$), respectively. (b) The influence of other three parameters on final results with $p = 0.1$, $g = 5$. From left to right: the results obtained with ($b = 0$, $sv = 17$, $s_i = 1$), ($b = 1$, $sv = 17$, $s_i = 1$), ($b = 0.5$, $sv = 5$, $s_i = 1$), ($b = 0.5$, $sv = 30$, $s_i = 1$), ($b = 0.5$, $sv = 17$, $s_i = 10$), respectively

From the two examples showed in Figs. 3 and 4, we can deduce that the two main parameters generally have significant influence on the final results, and also have definite physical meanings. While other parameters, such as the patch size or the window size of smoothing, have much less effect on the results and can thus be considered as fixed values. The conclusions can be applied to other defogging algorithms to select the key parameters for them.

3.2 Measurement of Defogging Effect

The CNC index, an effective defogging evaluation indicator proposed in our previous work [12] is used here to guide the parameter adjustment process. For the input foggy image **x** and its corresponding fog removal image **y**, the CNC index is obtained after carrying out the following steps: (i) compute the ratee of visible edges after and before fog removal, (ii) calculate the image color naturalness index (CNI) and color colourfulness index (CCI) to measure the color naturalness of the defogging image $\mathbf{y}$, and (iii) Combine the three components e, CNI and CCI to yield an overall defogging effect measure:

$$\text{CNC}(\mathbf{x}, \mathbf{y}) = h\left(e(\mathbf{x}, \mathbf{y}),\ \text{CNI}(\mathbf{y}),\ \text{CCI}(\mathbf{y})\right) \tag{1}$$

For the overall variation trend of the three indexes, the statistical results show that the peak of CNI curve stands for the most natural result, but it is not necessarily the best defogging effect. However, the best effect must have good naturalness (high CNI value). When the image is overenhanced, the color is distorted, and CNI goes down rapidly. For e and CCI, they achieve the best effect before reaching their peaks. When the image is overenhanced, the curves continue ascending. After reaching their peaks, these curves begin to go down. Therefore, if the uptrend of e and CCI (from their best effect points to their curve's peaks) can be largely counteracted by the downtrend of CNI, and the peak of CNC curve can be more close to the real best effect point. Meanwhile, the value variation of CNI is small, while that of e and CCI is relatively big. Thus, the effect of e and CCI on the CNC index needs to be weakened. The CNC index between image $\mathbf{x}$ and $\mathbf{y}$, *i.e.* the function h in (1) can be defined as

$$\text{CNC}(\mathbf{x}, \mathbf{y}) = e(\mathbf{x}, \mathbf{y})^{1/5} \cdot \text{CNI}(\mathbf{y}) + \text{CCI}(\mathbf{y})^{1/5} \cdot \text{CNI}(\mathbf{y}) \tag{2}$$

As explained above, a good result is described by the large value of CNC. Therefore, the optimal value of the two main parameters of defogging algorithm scan be obtained when the CNC index (2) achieves the largest value.

3.3 Parameter Value Selection Using Particle Swam Optimization

The particle swarm optimization (PSO) is one of the most important swarm intelligence paradigms [14], and its searching is based on the simulations of social behaviors such as animals herding, fish schooling, and birds flocking where the swarm search for food in a collaborative manner. The PSO is easy to implement, and therefore, it is employed to solve the optimization problems in many applications. Using PSO, the two main parameters of various defogging algorithms can be automatically determined on the basis of the CNC index. The simple but effective PSO-based parameter selection consists in the following steps:

Parameter Initialization. Initialize the beginning parameters, such as swarm populations, and the number of training iterations. Also, the particles are randomly located and the movement vector is randomly assigned. To prevent the blind search of particles, the particle's position and velocity are constrained in a certain range.

The task using the PSO is to find the best combination of the two main parameters according to an objective criterion. Therefore, the initial position of a particle can be expressed as an 1-by-2 matrix of random value RV, and each element of $RV \in (0, 1)$. $X_{\min}$ and X_{max} are the minimum and maximum range values for the position of the two parameters. The $X_{\min}$ is set to 0, and the value of $X_{\max}$

depends on the value ranges of the two parameters. For example, the value ranges of parameters p and g in Tarel's method are $(0, 1)$ and $(0, 10)$, respectively, so the 2 parameter values can be written as $p = RV$ and $g = 10 \times RV$.

Fitness Function. Let *Gbest* be the best known position of the entire swarm and let *Pbest* be the best known position of particle i. Store *Gbest* and all *Pbest* locations at the current iteration by using an evaluation process employing the fitness function for all particles.

For image defogging, the fitness is measured by the CNC index [see Eq. (2)] of a defogged image obtained by the method presented above, because a color image with good defogging effect includes many visible edges and its color must be natural and colorful. Thus, the best fitness value is the one with the largest CNC value. Therefore, using the CNC index, the *Gbest* and *Pbest* can be obtained by selecting the highest fitness.

Termination Condition. If the number of training iterations is terminated or the accuracy is satisfied, then output *Gbest* and *Pbest* locations, and the algorithm terminates. Otherwise, go to Step (4). When the number of training iterations is reached, the PSO computation is terminated.

In our experiment, the swarm populations *sizepop* is set to 20 and the number *maxgen* of training iterations is set to 35, the total number of parameter p or g can thus be calculated as $2 \times sizepop + (maxgen - 1) \times sizepop = 2 \times 20 + (35 - 1) \times 20 = 720$. The final optimal value of p or g corresponds to the maximum value of the 720 candidate particle values indicates the selected value for the two main parameters. The enhanced image with the two final parameter values is our final fog removal result.

Iterative Optimization. Calculate the movement vectors in Eq. (3) for all particles. Next, modify the locations of all particles utilizing Eq. (4) and then go to Step (2). The movement vector (location) is specified as follows:

$$V_i(t + 1) = wV_i(t) + c_1 \times r_1 \times (Pbest_i - X_i(t)) + c_2 \times r_2 \times (Gbest - X_i(t)) \quad (3)$$

where $V_i = (V_{i1}, V_{i2}, ..., V_{im}) \in \Re^m$, and the particle's initial velocity is $0.5 \times RV$. $V_i(t + 1)$ represents the movement vector of particle i at the $(t+1)$th iteration, w indicates the inertia weight ($w = 1$ in our experiments), c_1 and c_2 denote the acceleration coefficients which are random numbers in $[0,1]$, we set $c_1 = c_2 = 0.5$ for all results reported in this paper. r_1 and r_2 are also two randomly generated values in $[0,1]$. Moreover, in Eq. (3), the first term $wV_i(t)$ denotes the particle's inertia, the second term $c_1 \times r_1 \times (Pbest_i - X_i(t))$ indicates the particle's cognition-only model, and the third term $c_2 \times r_2 \times (Gbest - X_i(t))$ stands for the particle's social-only model. The location of particle i is modified by Eq. (4).

$$X_i(t + 1) = X_i(t) + V_i(t + 1) \quad (4)$$

where $X_i(t + 1)$ represents the location of particle i at the $(t + 1)$th iteration, which is used to estimate the optimal value of the two main parameters. $V_i(t + 1)$ denotes the movement vector of particle i at the $(t+1)$th iteration. Hence, the new location of particle i is to add its current location vector to its movement vector. Therefore, we can deduce that the iterative optimization operation is designed to suit our needs regarding good image defogging effect with relatively large CNC index value and maintaining the diversity at the same time.

4   Experimental Results

The publicly available dataset frida2 [14] is used to evaluate image defogging methods. Although the testing images in this dataset are very similar in view structure, the reason why we use the this dataset images as our testing images is that the dataset contains synthetic no-fog images and associated foggy images for 66 diverse road scenes, while capturing the fog and no-fog natural images with the same scenes is very hard in most cases. The absolute difference (AD) on the images between defogged images and target images without fog is used as performance metric, and good results are described by small value of AD. To verify the effectiveness and validity of the proposed parameter value selection method, three criteria have been considered: (i) generation number influence, (ii) qualitative comparison, and (iii) quantitative evaluation. In the experiments, all the results are obtained by executing Matlab R2008a on a PC with 3.10 GHz Intel® CoreTM i5-2400 CPU.

4.1   Generation Number Evaluation

The final condition of the proposed PSO-based algorithm is to reach the given number *maxgen* of training iterations. To evaluate the influence of the iteration number *maxgen* used in the proposed method, some group experiments are performed by varying the iteration number *maxgen* from 20 to 50, the parameter values and the resulting AD metics of a test image are presented in Table 1. One can clearly see that the results are visually and statistically close (the value range of AD are [60.7332, 61.9503] for He's method, and [42.2918, 43.7579] for Tarel's method) when varying *maxgen* from 20 to 50. It demonstrates that the influence of the iteration number is very limited in the proposed method. The experiments on a large quantity of other test images also confirm the observations.

Table 1. AD metics of the two representative defogging methods under *maxgen* from 20 to 50

	Gen.	20	25	30	35	40	45	50
He's method	AD	61.8255	61.3164	61.0950	60.7332	61.4351	61.6160	61.9503
Tarel's method	AD	43.7579	43.4142	43.1007	42.8579	42.2918	42.6439	43.1403

4.2 Qualitative Comparison

For further evaluation of the PSO-based method for selecting the most proper parameter values, the image dataset provide by Tarel et al. [15] is used to validate the accuracy of the parameter selection, since this dataset provides the original image without fog and the image with fog for 66 road scenes simultaneously. For each image, we obtain the fog removal results with the default and the auto-adaptive parameter values for He's and Tarel's algorithms. We also give the corresponding fog-free images as the reference images for comparison. An illustrative example is shown in Fig. 5. One can clearly see that the defogging results obtained using the auto-adaptive values achieve a better enhancement effect compared to the results obtained using the default values for both defogging algorithms. Note that the color of He's results seems much dark than that of no-fog reference images, and there are some halo artifacts in Tarel's results. All these problems are caused by the defogging algorithm itself, not the proposed parameter selection method. The accuracy of the PSO-based parameter selection can be validated by determining the most proper parameter values and producing the best results for each defogging method.

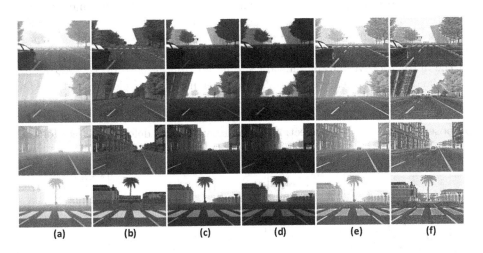

Fig. 5. Visual comparison of defogging results for public database frida2 [14]. (a) Foggy images. (b) Fog-free images. (c) He's results obtained using default parameter values ($\omega = 0.95$, $t_0 = 0.1$, $c = 3$). (d) He's results obtained using auto-adaptive parameter values. (e) Tarel's results obtained using default parameter values ($p = 0.95$, $b = 0.5$, $s_v = 9$, $s_i = 1$, $g = 1.3$). (f) Tarel's results obtained using auto-adaptive parameter values

4.3 Quantitative Evaluation

To quantitatively assess the proposed parameter value selection method, we compute the AD index value for the images in Fig. 5, and the statistical results are shown in Table 2. One can notice that the AD value obtained by the auto-adaptive parameter is smaller than that of default parameters for both defogging algorithms, which means that the better defogging effect can be obtained by using the proposed method. This confirms our observations in Fig. 5.

Table 2. AD index between enhanced images and no-fog images for the testing images in Fig. 5

	Foggy image	He's method (default value)	He's method (adaptive value)	Tarel's method (default value)	Tarel's method (adaptive value)
Fig. 5(Row#1)	69.7350	43.1087	**41.8408**	51.8163	**43.3998**
Fig. 5(Row#2)	77.8368	50.3951	**40.3007**	56.4112	**47.9156**
Fig. 5(Row#3)	78.0347	51.9249	**42.5308**	54.0593	**49.2353**
Fig. 5(Row#4)	66.4877	52.5317	**41.8013**	47.9604	**37.4284**

The AD index is also tested for more test images in public database frida2 (66 images). Figure 6(a) shows the statistical results of the AD for He's method and Fig. 6(b) shows the AD results for Tarel's method. In Fig. 6, circle "o" stands for foggy image, circle "●" stands for the defogging image obtained using the default parameter values, and circle "●" stands for the defogging image obtained by the auto-adaptive parameter values. The horizontal axes are the AD index values and vertical axes are the image number index. It is clear that the ADs of adaptive parameter results are smaller than that of other results for both defogging methods. This indicates that the fog removal results obtained by the proposed parameter value selection method have better defogging effect for the

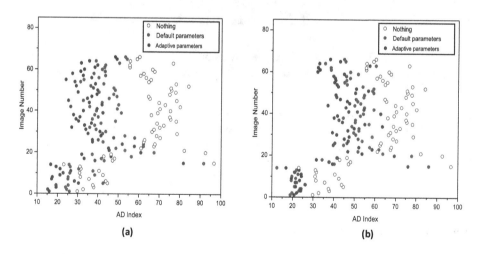

Fig. 6. AD index results for the 66 test images in public database frida2 [15]. (a) He's statistical results. (b) Tarel's statistical results (Color figure online)

public image database compared to the other results. This is also consistent with the assessment results of AD and human visual perception.

5   Conclusions

In this paper, a novel PSO-based parameter value selection method was proposed. Different from the most defogging methods which generally fix the parameter values, the proposed approach can help defogging methods automatically select optimal parameter values for different foggy images. In the proposed method, the two main parameters which directly affect the results are first distinguished from other less important parameters which can be considered as fixed values. Then, the two parameter values are adaptively determined by using the PSO algorithm. The proposed parameter selection method has been applied to two representative defogging algorithms, which demonstrated the superior performance of the proposed scheme in both qualitatively and quantitatively. Although the proposed method provided a new way to solve the parameter adjustment problem for single image defogging, in the future, we will try to investigate the parameter value selection issue based on more advanced assessment index, since the CNC index may not be the best one to measure image defogging effect.

Acknowledgements. This work was supported by the National Natural Science Foundation of China (61573380, 61502537), and the Postdoctoral Science Foundation of Central South University (No. 126648).

References

1. Tan, R.T.: Visibility in bad weather from a single image. In: Proceedings of IEEE Conference on Computer Vision and Pattern Recognition, pp. 1–8 (2008)
2. Nishino, K., Kratz, L., Lombardi, S.: Bayesian defogging. Int. J. Comput. Vis. **98**, 263–278 (2012)
3. He, K.M., Sun, J., Tang, X.O.: Single image haze removal using dark channel prior. IEEE Trans. Pattern Anal. Mach. Intell. **33**, 2341–2353 (2011)
4. Tarel, J.P., Hautiere, N.: Fast visibility restoration from a single color or gray level image. In: Proceedings of IEEE International Conference on Computer Vision, pp. 2201–2208 (2009)
5. Lagorio, A., Grosso, E., Tistarelli, M.: Automatic detection of adverse weather conditions in traffic scenes. In: Proceedings of IEEE Fifth International Conference on Advanced Video and Signal Based Surveillance, pp. 273–279 (2008)
6. Hautiere, N., Tarel, J.-P., Aubert, D.: Towards fog-free in-vehicle vision systems through contrast restoration. In: Proceedings of IEEE Conference on Computer Vision and Pattern Recognition, pp. 2374–2381 (2007)
7. Hautiere, N., Tarel, J.-P., Halmaoui, H., Bremond, R., Aubert, D.: Enhanced fog detection and free-space segmentation for car navigation. Mach. Vis. Appl. **25**, 667–679 (2014)

8. Jiji, G.W., DuraiRaj, P.J.: Content-based image retrieval techniques for the analysis of dermatological lesions using particle swarm optimization technique. Appl. Soft Comput. **30**, 650–662 (2015)

9. Li, Y.Y., Jiao, L.C., Shang, R.H., Stolkin, R.: Dynamic-context cooperative quantum-behaved particle swarm optimization based on multilevel thresholding applied to medical image segmentation. Inf. Sci. **294**, 408–422 (2015)

10. Wang, Z., Bovik, A.C., Sheikh, H.R., Simoncelli, E.P.: Image quality assessment: from error visibility to structural similarity. IEEE Trans. Image Process. **13**, 600–612 (2004)

11. Ji, Z.X., Chen, Q., Sun, Q.S., Xia, D.D.: A moment-based nonlocal-means algorithm for image denoising. Inf. Process. Lett. **109**, 1238–1244 (2009)

12. Guo, F., Tang, J., Cai, Z.X.: Objective measurement for image defogging algorithms. J. Cent. South Univ. **21**, 272–286 (2014)

13. He, K.M.: Single image haze removal using dark channel prior, Ph.D. dissertation, The Chinese University of Hong Kong (2011)

14. Kennedy, J., Eberhart, R.C.: Particle swarm optimization. In: IEEE Proceedings of International Conference Neural Network, pp. 1942–1948 (1995)

15. Tarel, J.-P., Hautiere, N., Caraffa, L., Cord, A., Halmaoui, H., Gruyer, D.: Vision enhancement in homogeneous and heterogeneous fog. IEEE Intell. Transp. Syst. Mag. **4**, 6–20 (2012)

Adaptive Measurement Network for CS Image Reconstruction

Xuemei Xie[✉], Yuxiang Wang, Guangming Shi, Chenye Wang, Jiang Du, and Xiao Han

Xidian University, Xi'an, China
xmxie@mail.xidian.edu.cn

Abstract. Conventional compressive sensing (CS) reconstruction is very slow for its characteristic of solving an optimization problem. Convolutional neural network can realize fast processing while achieving comparable results. While CS image recovery with high quality not only depends on good reconstruction algorithms, but also good measurements. In this paper, we propose an adaptive measurement network in which measurement is obtained by learning. The new network consists of a fully-connected layer and ReconNet. The fully-connected layer which has low-dimension output acts as measurement. We train the fully-connected layer and ReconNet simultaneously and obtain adaptive measurement. Because the adaptive measurement fits dataset better, in contrast with random Gaussian measurement matrix, under the same measurement rate, it can extract the information of scene more efficiently and get better reconstruction results. Experiments show that the new network outperforms the original one.

Keywords: Compressive sensing · Image reconstruction
Deep learning · Adaptive measurement

1 Introduction

Compressive sensing (CS) theory [1–3] is able to acquire measurements of signal at sub-Nyquist rates and recover signal with high probability when the signal is sparse in a certain domain. Random Gaussian matrix is often used as the measurement matrix because we must ensure that the basis of sparse domain is incoherent with measurement. When it comes to reconstruction, there are two main kinds of reconstruction methods: conventional reconstruction methods [4–9] and deep learning reconstruction methods [10–13].

A large amount of compressive sensing reconstruction methods have been proposed. But almost all of them get the reconstruction result by solving optimization, which makes them slow. In recent years, some deep learning

---

This paper was presented in part at the CCF Chinese Conference on Computer Vision, Tianjin, 2017. This paper was recommended by the program committee.

J. Yang et al. (Eds.): CCCV 2017, Part II, CCIS 772, pp. 407–417, 2017.
https://doi.org/10.1007/978-981-10-7302-1_34

approaches have been proposed. With its characteristic of off-line training and online test, the speed of reconstruction has been greatly improved.

The first paper [11] applying deep learning approach to solve the CS recovery problem used stacked denoising autoencoders (SDA) to recover signals from undersampled measurements. SDA consists of fully-connected layers, which means larger network with the signal size growing. This imposes a large computational complexity and can lead to overfitting. DeepInverse [13], utilizing convolutional neuron network (CNN) layers, works with arbitrary measurement, which means that the whole image can be reconstructed by it. But with the signal size growing, the cost of measurement grows too. ReconNet [10] used a fully-connected layer along with convolutional layers to recover signals from compressive measurements block-wise. It can reduce the network complexity and the training time while ensuring a good reconstruction quality. However, ReconNet used fixed random Gaussian measurement, which is not optimally designed for signal.

In this paper, we take a fully-connected layer which has low-dimension output as measurement. The fully-connected layer and reconstruction network ReconNet are put together to be an adaptive measurement network. It can be proved the adaptive measurement network performs better than ReconNet with fixed measurement by comparing the trained weights. Experiment shows that the adaptive measurement matrix can obtain more information of images than fixed random Gaussian measurement matrix. And the images reconstructed from adaptive measurement have larger value of PSNR.

The structure of the paper is organized as follows. Section 2 introduces the fixed random Gaussian measurement. And the description of adaptive measurement network is introduced in Sect. 3. Section 4 conducts the experiments, and Sect. 5 concludes the paper.

2   Fixed Random Gaussian Measurement

ReconNet is a deep learning CS reconstruction approach, which can recover rich semantic content even at a low measurement rate of 1%. It is a convolutional neural network and the process of training and testing ReconNet are shown in Fig. 1.

ReconNet consists of one fully-connected layer and six convolutional layers in which the first three layers and last three layers are identical. The function of these layers is described as follows. The fully-connected layer takes CS measurements as input and outputs a feature map of size 33×33. The first/last three layers are inspired by SRCNN [14], which is a CNN-based approach for image super-resolution. Except for the last convolutional layer, all the other convolutional layers followed by ReLU. Only when the input of CNN has structure information can CNN work. So the fully-connected layer plays a role of recovering some structure information from CS measurements and then convolutional layers enhance output of full-connected layer to a high-resolution image.

The training dataset consists of input data and ground truth. All the images in dataset are 33×33 size patches extracted from original images. Input data of

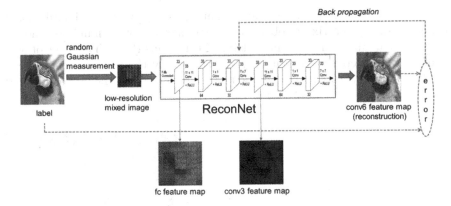

Fig. 1. The process of training and testing ReconNet with fixed random Gaussian measurement. Feature maps of fully connected layer (fc feature map), third convolutional layer (conv3 feature map) and sixth convolutional layer (conv6 feature map) at measurement rate 25%.

ReconNet for training is obtained by measuring each of the extracted patches using a random Gaussian matrix Φ. For a given measurement rate, a random Gaussian matrix of appropriate size is firstly generated and then its rows are orthonormalized to get Φ. The input of testing and training is obtained by using the same random Gaussian matrix. Before being measured, the 33×33 size block should be reshaped into a 1089-dimension column vector.

The loss function is given by

$$L(\{W\}) = \frac{1}{T} \sum_{i}^{T} \|f(y_i, \{W\}) - x_i\|^2. \tag{1}$$

$f(y_i, \{W\})$ is the i–th reconstruction image of ReconNet, x_i is the i–th original signal as well as the i–th label, W means all parameters in ReconNet. T is the total number of image blocks in the training dataset. The loss function is minimized by adjusting W using backpropagation. For each measurement rate, two networks are trained, one with random Gaussian initialization for the fully connected layer, and the other with a deterministic initialization, in each case, weights of all convolutional layers are initialized using a random Gaussian with a fixed standard deviation. The network which provides the lower loss on a validation test will be chosen.

The test process does not include the dotted line part in Fig. 1. The high-resolution scene image is divided into non-overlapping blocks of size 33×33 and each of them is reconstructed by feeding in the corresponding CS measurements to ReconNet. The reconstructed blocks are arranged appropriately to form a reconstruction of the image.

It can be proved that fully-connected layer can recover some structure information. The block of high-resolution scene image is firstly measured by random Gaussian matrix and then multiply parameters of fully-connected layer, this process equals to multiplying a square matrix as Fig. 2 shows. The diagonal numbers of square matrix are obviously larger than the other numbers, and we can see it as an approximate unit matrix, which means the i-th element of output of fully-connected layer is mainly determined by the i-th element of high resolution block. Feature maps can also be used to prove that the fully-connected layer can recover some information.

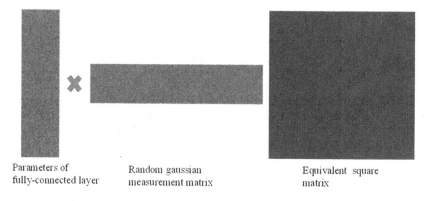

Parameters of   Random gaussian   Equivalent square
fully-connected layer   measurement matrix   matrix

Fig. 2. The equivalent square matrix at measurement rate 10%.

The feature map of fully-connected layer (fc feature map in Fig. 1) can be obtained after image parrot (label in Fig. 1) is sent into the trained ReconNet at measurement rate 10%. It can be seen that the fully-connected layer can recover some structure information. The feature maps of fully-connected layer (fc feature map in Fig. 1), third convolutional layer (conv3 feature map in Fig. 1) and sixth convolutional layer (conv6 feature map in Fig. 1) at measurement rate 25% are also shown in Fig. 1. We can see that it is a process from low-resolution to high-resolution.

The main drawback of random measurement is that they are not optimally designed for signal. Therefore, the adaptive measurement is possibly a promising approach.

3   Adaptive Measurement Network

We put a fully-connected layer and ReconNet together to form the adaptive measurement network as Fig. 3 shows. The fully-connected layer which has low-dimension output is considered as measurement. The measurement rate is determined by dimension of network input and fully-connected layer output.

The whole Fig. 3 shows the process of training adaptive network (including the dashed line part). In the training stage, ground truth still consists of 33×33 size patches extracted from the original images. Different from the random Gaussian measurement network, input data of training set is the same with ground truth instead of the output of Gaussian measurements.

When it comes to test, the parameter of fully-connected layer is taken as measurement while ReconNet still works as reconstruction network. Figure 3 without dashed line part shows the process of testing.

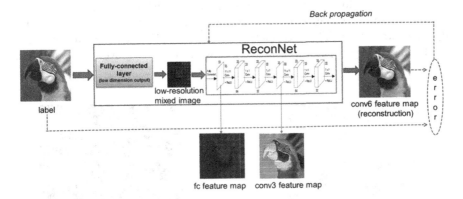

Fig. 3. The process of training and testing with adaptive measurement. Feature maps of fully connected layer (fc feature map), third convolutional layer (conv3 feature map) and sixth convolutional layer (conv6 feature map) at measurement rate 10%.

Accordingly, the loss function of new network is given by

$$L(\{W\}) = \frac{1}{T} \sum_{i}^{T} \|f(x_i, \{W, K\}) - x_i\|^2. \tag{2}$$

where K is the parameter of new added fully-connected layer. Difference between (1) and (2) is that in (2) reconstruction image is determined by x_i and $\{W, K\}$, but y_i and $\{W\}$ in (1).

Compared to the original one, the new network has more parameter to train. The initial value of the network is random Gaussian. There is a high probability that a better measurement more adaptive to data set can be obtained.

It can be proved the new ReconNet of adaptive measurement network performs better than the original one. The fully-connected layer can recover more structure information. The equivalent process is shown in Fig. 4(b). Compared with Fig. 4(a), the value of square matrix in Fig. 4(b) is more dispersing. The i–th element of output of fully-connected layer is mainly determined by the i–th element of high resolution block and its neighboring elements. So, the measurement can acquire information more effectively. As shown in Fig. 5, in equivalent square matrix, the value of white part is larger than black part. A high-resolution block is firstly reshaped to a high-resolution vector. The output of fully-connected layer is obtained by multiplying the high-resolution vector by the equivalent square matrix. We take the 1–th element as an example. The 1–th element of output is obtained by multiplying the high-resolution vector by 1–th row vector of square matrix. The elements of 1–th row vector can be seen as the weights of column vector. It is obvious that the red elements of the high-resolution vector have larger weights, which means the 1–th element of output of fully-connected layer is mainly determined by those red elements of high-resolution vector. In high-resolution block, those red elements correspond to the 1–th element and its neighboring elements. Since the 1–th element and its neighboring elements are relevant, the values of them are approximate, which means the 1-th element of output of fully-connected layer is more determined by the 1–th element of high-resolution block. The i–th element can be explained the same way. It is proved that the fully-connected layer of new ReconNet recover more structure information.

(a)   (b)

Fig. 4. The equivalent square matrix (a) Fixed random Gaussian measurement with MR $= 10\%$ and (b) adaptive measurement with MR $= 10\%$.

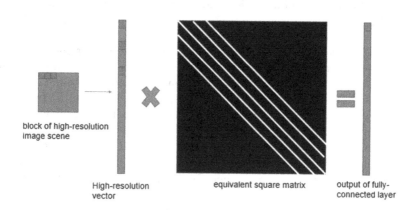

block of high-resolution image scene

High-resolution vector   equivalent square matrix   output of fully-connected layer

Fig. 5. Simplified explanation.

Adaptive measurement network's feature maps of fully-connected layer (fc feature map), third convolutional layer (conv3 feature map) and sixth convolutional layer (conv6 feature map) at measurement rate 10% are shown in Fig. 3.

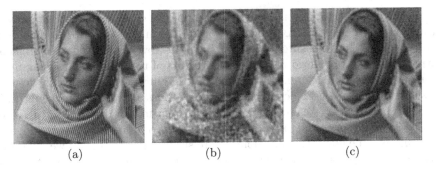

Fig. 6. Test image and reconstruction results of Barbara. (a) the original image (b) the reconstruction result of random Gaussian measurement network (c) the reconstruction result of adaptive measurement network.

In contrast to Fig. 1, the feature maps of adaptive network are obviously better even at measurement 10%.

Figure 6 shows an example of reconstruction results at two kinds of measurement. The measurement rate is 10%. Figure 6(a) is the original image. Figure 6(b) is the reconstruction result of random Gaussian measurement network. Figure 6(c) is the reconstruction result of adaptive network. The adaptive reconstruction result is more attractive visually.

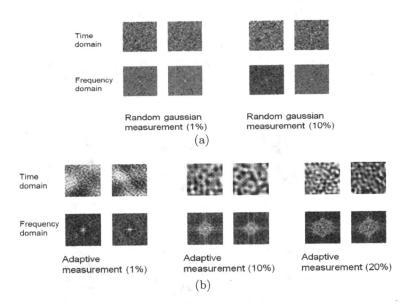

Fig. 7. Measurement matrix. (a) is random Gaussian measurement matrix at measurement rate 1%, 10% in time and frequency domain. (b) is adaptive measurement matrix at measurement rate 1%, 10%, 20% in time and frequency domain.

Adaptive measurement network's better performance can also be proved through measurement matrix. Since the original signal is reshaped to a column vector before being measured, we reshape some row vectors of measurement matrix to size 33×33. Two reshaped row vectors of the random Gaussian measurement matrix at measurement rate 1% and 10% in time and frequency domain are shown in Fig. 7(a). The content of random Gaussian measurement matrix is obviously irregular. We cannot get any useful information from Fig. 7(a). Two reshaped row vectors of adaptive measurement matrix at measurement rate 1%, 10%, 20% in time and frequency domain are shown in Fig. 7(b). As we all know, most of the energy of an image is concentrated in the low frequency part. When the measurement rate is low, some high frequency information must be discarded to reconstruct the contours of the image as fully as possible. However, with the increase in measurement rate, the ability of measurement is enhanced. The high-frequency information in adaptive measurement increases gradually. We can also know it from frequency domain image. So, the reconstructed image will become clearer.

4   Results

In this section, we conduct reconstruction experiments at both fixed random Gaussian measurement and adaptive measurement.

We use the caffe framework for network training on the MATLAB platform. Our computer is equipped with Intel Core i7-6700 CPU with frequency of 3.4 GHz, NVidia GeForce GTX 980 GPU, 64 GB RAM, and the framework runs on the Ubuntu 14.04 operating system.

The dataset consists of 21760 33×33 size patches extracted from 91 images in [14] with a stride equal to 14. It is worthy to mention that because Recon-Net reconstruct image block-wise and the size of block is fixed, zero-padding

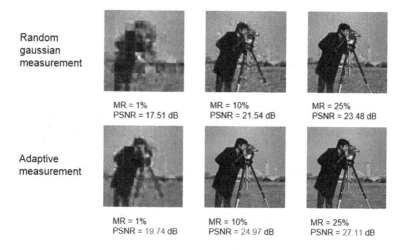

Fig. 8. The reconstruction results of image cameraman at different measurement rates.

operation is applied to input images of different size. But we find symmetric padding acts better than zero-padding, so all the experiment results are based on symmetric padding instead of zero-padding.

We use cameraman image to test both the networks, and the result of reconstruction is shown as follows.

It is shown in Fig. 8 that the reconstruction results of image cameraman at different measurement rates with different measurements. Our results are more attractive visually.

The reconstruction results for 11 test images at measurement rate 1%, 10%, 25% with different measurements are shown in Table 1. All results show that adaptive measurement outperforms random Gaussian measurement.

Table 1. The reconstruction results for 11 test images at measurement rate 1%, 10%, 25% with different measurements.

Image	Measurement	Rate 25%	Rate 10%	Rate 1%
Monarch	Adaptive	29.25 dB	26.65 dB	17.70 dB
	Gaussian	24.95 dB	21.49 dB	15.61 dB
Parrots	Adaptive	30.51 dB	27.59 dB	21.67 dB
	Gaussian	26.66 dB	23.36 dB	18.93 dB
Barbara	Adaptive	27.40 dB	24.28 dB	21.36 dB
	Gaussian	23.58 dB	22.17 dB	19.08 dB
Boats	Adaptive	32.47 dB	28.80 dB	21.09 dB
	Gaussian	27.83 dB	24.56 dB	18.82 dB
Cameraman	Adaptive	27.11 dB	24.97 dB	19.74 dB
	Gaussian	23.48 dB	21.54 dB	17.51 dB
Fingerprint	Adaptive	32.31 dB	26.55 dB	16.22 dB
	Gaussian	26.15 dB	20.99 dB	15.01 dB
Flinstones	Adaptive	27.94 dB	23.83 dB	16.12 dB
	Gaussian	22.74 dB	19.04 dB	14.14 dB
Foreman	Adaptive	36.18 dB	33.51 dB	25.53 dB
	Gaussian	32.08 dB	29.02 dB	22.03 dB
House	Adaptive	34.38 dB	31.43 dB	22.93 dB
	Gaussian	29.96 dB	26.74 dB	20.30 dB
Lena	Adaptive	31.63 dB	28.50 dB	21.49 dB
	Gaussian	27.47 dB	24.48 dB	18.51 dB
Peppers	Adaptive	29.65 dB	26.67 dB	19.75 dB
	Gaussian	25.74 dB	22.72 dB	17.39 dB
MeanPSNR	Adaptive	30.80 dB	27.53 dB	20.33 dB
	Gaussian	26.42 dB	23.28 dB	17.94 dB

5 Conclusion

We have presented an adaptive measurement obtained by learning. We showed that the adaptive measurement provides better reconstruction results than the fixed random Gaussian measurement. It is shown that the learned measurement matrix is more regular in time domain. It is clear that the learned measurement matrix is more adaptive to data set than the fixed one. That's an important reason why adaptive measurement works better. What's more, our network is universal, which can be applied to all kinds of images.

Acknowledgements. This work is supported by the National Natural Science Foundation of China (Grant No. 61472301, 61632019) and the Foundation for Innovative Research Groups of the National Natural Science Foundation of China (No. 61621005).

References

1. Kašin, B.S.: The widths of certain finite-dimensional sets and classes of smooth functions. Izv. Akad. Nauk SSSR Ser. Mat. **41**(2), 334–351 (1977)
2. Candès, E., Romberg, J.: Sparsity and incoherence in compressive sampling. Inverse Prob. **23**(3), 969–985 (2007)
3. Rauhut, H.: Random sampling of sparse trigonometric polynomials. Found. Comput. Math. **22**(6), 737–763 (2008)
4. Candes, E.J., Romberg, J., Tao, T.: Robust uncertainty principles: exact signal reconstruction from highly incomplete frequency information. IEEE Trans. Inf. Theory **52**(2), 489–509 (2004)
5. Duarte, M.F., Wakin, M.B., Baraniuk, R.G.: Wavelet-domain compressive signal reconstruction using a Hidden Markov Tree model. In: IEEE International Conference on Acoustics, Speech and Signal Processing, pp. 5137–5140. IEEE (2008)
6. Peyré, G., Bougleux, S., Cohen, L.: Non-local regularization of inverse problems. In: Forsyth, D., Torr, P., Zisserman, A. (eds.) ECCV 2008. LNCS, vol. 5304, pp. 57–68. Springer, Heidelberg (2008). https://doi.org/10.1007/978-3-540-88690-7_5
7. Baraniuk, R.G., Cevher, V., Duarte, M.F., et al.: Model-based compressive sensing. IEEE Trans. Inf. Theory **56**(4), 1982–2001 (2010)
8. Li, C., Yin, W., Jiang, H., et al.: An efficient augmented Lagrangian method with applications to total variation minimization. Comput. Optim. Appl. **56**(3), 507–530 (2013)
9. Kim, Y., Nadar, M.S., Bilgin, A.: Compressed sensing using a Gaussian Scale Mixtures model in wavelet domain. In: IEEE International Conference on Image Processing, pp. 3365–3368. IEEE (2010)
10. Kulkarni, K., Lohit, S., Turaga, P., Kerviche, R., Ashok, A.: Reconnet: non-iterative reconstruction of images from compressively sensed measurements. In: Proceedings of the IEEE Conference on Computer Vision and Pattern Recognition, pp. 449–458 (2016)
11. Mousavi, A., Patel, A.B., Baraniuk, R.G.: A deep learning approach to structured signal recovery. In: 2015 53rd Annual Allerton Conference on Communication, Control, and Computing (Allerton), pp. 1336–1343. IEEE, September 2015
12. Yao, H., Dai, F., Zhang, D., Ma, Y., Zhang, S., Zhang, Y.: DR^2-Net: Deep Residual Reconstruction Network for Image Compressive Sensing. arXiv preprint arXiv:1702.05743 (2017)

13. Mousavi, A., Baraniuk, R.G.: Learning to invert: signal recovery via deep convolutional networks. arXiv preprint arXiv:1701.03891 (2017)
14. Dong, C., Loy, C.C., He, K., Tang, X.: Learning a deep convolutional network for image super-resolution. In: Fleet, D., Pajdla, T., Schiele, B., Tuytelaars, T. (eds.) ECCV 2014. LNCS, vol. 8692, pp. 184–199. Springer, Cham (2014). https://doi.org/10.1007/978-3-319-10593-2_13

Image Segmentation and Classification

A Band Grouping Based LSTM Algorithm for Hyperspectral Image Classification

Yonghao Xu[1], Bo Du[2(✉)], Liangpei Zhang[1], and Fan Zhang[1]

[1] State Key Laboratory of Information Engineering in Survey Mapping,
and Remote Sensing, Wuhan University,
Wuhan 430079, People's Republic of China
[2] School of Computer, Wuhan University,
Wuhan 430072, People's Republic of China
gunspace@163.com

Abstract. In recent years, many deep learning based methods were proposed to deal with the hyperspectral image (HSI) classification task. So far, most of these methods focus on the spectral integrality but neglect the contextual information among adjacent bands. In this paper, we propose to use the long short-term memory (LSTM) model with an end-to-end architecture for the HSI classification. Moreover, considering the high dimensionality of hyperspectral data, two novel grouping strategies are proposed to better learn the contextual features among adjacent bands. Compared with the traditional band-by-band strategy, the proposed methods prevent a very deep network for the HSI. In the experiments, two benchmark HSIs are utilized to evaluate the performance of proposed methods. The experimental results demonstrate that the proposed methods can yield a competitive performance compared with existing methods.

Keywords: Long Short-Term Memory (LSTM)
Recurrent Neural Network (RNN) · Deep learning
Feature extraction · Hyperspectral image classification

1 Introduction

Different from the conventional image which only covers the human visual spectral range with RGB bands, hyperspectral image (HSI) can cover a much larger spectral range with hundreds of narrow spectral bands. It has been widely used in urban mapping, forest monitoring, environmental management and precision agriculture [1]. For most of these applications, HSI classification is a fundamental task, which predicts the class label of each pixel in the image.

Compared with traditional classification problems [12–14, 16, 17], HSI classification is more challenging due to the curse of dimensionality [7], which is also known as the Hughes phenomenon [9]. In order to alleviate this problem, dimensionality reduction methods are proposed, which can be divided into feature selection [3] and feature extraction (FE) [2] methods. The main purpose of

© Springer Nature Singapore Pte Ltd. 2017
J. Yang et al. (Eds.): CCCV 2017, Part II, CCIS 772, pp. 421–432, 2017.
https://doi.org/10.1007/978-981-10-7302-1_35

feature selection is to preserve the most representative and crucial bands from the original dataset and discard those making no contribution to the classification. By designing suitable criteria, feature selection methods can eliminate redundancies among adjacent bands and improve the discriminability of different targets. FE, on the other hand, is used to find an appropriate feature mapping to transform the original high-dimensional feature space into a low-dimensional one, where different objects tend to be more separable.

Witnessing the achievements by deep learning methods in the fields of computer vision and artificial intelligence [10,11], a promising way to extract deep features for hyperspectral data has become a feasible option. In [5], Chen et al. introduce the concept of deep learning into HSI classification for the first time, using a multilayer stacked autoencoders (SAEs) to extract deep features. After the pre-training stage, the deep networks are then fine-tuned with the reference data through a logistic regression classifier. In a like manner, a deep belief networks (DBNs) based spectral-spatial classification method for HSI is proposed in [6], where both the single layer restricted Boltzmann machine and multilayer DBN framework are analyzed in detail.

Despite the fact that these deep learning based methods possess better generalization ability compared with shallow methods, they mainly focus on the spectral integrality and directly classify the image in the whole feature space. However, the pixel vectors in the HSI are also sequential data, where the contextual information among adjacent bands is discriminative for the recognition of different objects.

To overcome the aforementioned drawbacks, we propose to use the long short-term memory (LSTM) model to extract spectral features, which is an updated version of recurrent neural networks (RNNs). Considering the high dimensionality of hyperspectral data, two novel grouping strategies are proposed to better learn the contextual features among adjacent bands. The major contributions of this study are summarized as follow.

1. As far as we know, it is the first time that an end-to-end architecture for the HSI classification with the LSTM is proposed, which takes the contextual information among adjacent bands into consideration.
2. Two novel grouping strategies are proposed to better learn the contextual features among adjacent bands for the LSTM. Compared with the traditional band-by-band strategy, proposed methods prevent a very deep network for the HSI.

The rest of this paper is organized as follows. Section 2 describes the spectral classification with the LSTM in detail. The information of data sets used in this study and the experimental results are given in Sect. 3. Conclusions and other discussions are summarized in Sect. 4.

2 Spectral Classification with LSTM

In this Section, we will first make a brief introduction to RNNs and the LSTM. Then, strategies for processing the spectral information with the LSTM are presented.

2.1 RNNs

RNNs [15] are important systems for processing sequential data, which allow cyclical connections between neural activations at different time steps.

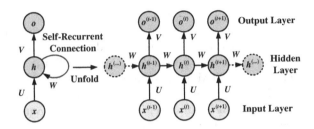

Fig. 1. The architecture of a recurrent neural network.

The architecture of a recurrent neural network is shown as Fig. 1 Given a sequence of values $x^{(1)}, x^{(2)}, \ldots, x^{(\tau)}$, apply the following update equations for each time step from $t = 1$ to $t = \tau$.

$$h^{(t)} = g\left(b_a + Wh^{(t-1)} + Ux^{(t)}\right) \tag{1}$$

$$o^{(t)} = b_o + Vh^{(t)} \tag{2}$$

$$g(x) = \tanh(x) = \frac{e^x - e^{-x}}{e^x + e^{-x}} \tag{3}$$

where b_a and b_o denote bias vectors. U, V and W are the weight matrices for input-to-hidden, hidden-to-output and hidden-to-hidden connections, respectively. $x^{(t)}$, $h^{(t)}$ and $o^{(t)}$ are the input value, hidden value and output value at time t, respectively. The initialization of $h^{(0)}$ in (1) is specified with Gaussian values.

From (1) we can see that the hidden value of RNNs is determined by both the input signal at the current time step and the hidden value at the previous time step. In this manner, both the contextual information and the underlying pattern of the sequential data can be discovered. For the classification task, the softmax function can be added at the last time step to calculate the probability that the input data belongs to the ith category.

$$P(y = i|\theta, b) = s\left(o^{(\tau)}\right) = \frac{e^{\theta_i o^{(\tau)}} + b_i}{\sum_{j=1}^{k} e^{\theta_j o^{(\tau)}} + b_j} \tag{4}$$

where θ and b are the weight matrix and bias vector, respectively. k is the number of classes. The loss function of the whole network can be defined as

$$\mathcal{L} = -\frac{1}{m}\sum_{i=1}^{m}[y_i\log\left(\hat{y}_i\right) + (1-y_i)\log\left(1-\hat{y}_i\right)] \tag{5}$$

where y_i and $\hat{y}_i$ denote the label and predicted label of the ith data, respectively. m is number of training samples. The optimization of a RNN can be accomplished by the mini-batch stochastic gradient descent with the back-propagation through time (BPTT) algorithm [19].

2.2 LSTM

The main challenge when training the RNN is the long-term dependencies that gradients tend to either vanish or explode during the back-propagation phase. To mitigate this problem, a gated RNN called LSTM is proposed in [8]. The core component of the LSTM is the memory cell which replaces the hidden unit in traditional RNNs. As shown in Fig. 2, there are four main elements in the memory cell, including an input gate, a forget gate, an output gate and a self-recurrent connection. The forward propagation of the LSTM for time step t is defined as follows.

Input gate:

$$i^{(t)} = \sigma\left(W_i x^{(t)} + U_i h^{(t-1)} + b_i\right) \tag{6}$$

Forget gate:

$$f^{(t)} = \sigma\left(W_f x^{(t)} + U_f h^{(t-1)} + b_f\right) \tag{7}$$

Output gate:

$$o^{(t)} = \sigma\left(W_o x^{(t)} + U_o h^{(t-1)} + b_o\right) \tag{8}$$

Cell state:

$$c^{(t)} = i^{(t)} \odot g\left(W_c x^{(t)} + U_c h^{(t-1)} + b_c\right) + f^{(t)} \odot c^{(t-1)} \tag{9}$$

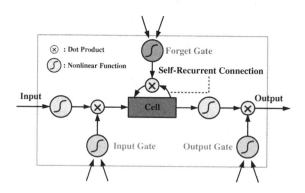

Fig. 2. Illustration of an LSTM model.

LSTM output:

$$h^{(t)} = o^{(t)} \odot g\left(c^{(t)}\right) \tag{10}$$

where W_i, W_f, W_o, W_c, U_i, U_f, U_o and U_c are weight matrices. b_i, b_f, b_o and b_c are bias vectors. $\sigma(x) = 1/(1 + \exp(-x))$ is the sigmoid function and $\odot$ denotes the dot product.

Similar to traditional RNNs, the LSTM network can be trained by the mini-batch stochastic gradient descent with the BPTT algorithm. Refer to [8] for more detailed descriptions.

2.3   The Proposed Band Grouping Based LSTM Algorithm

Previous literatures have shown that the deep architecture possesses better generalization ability when dealing with the complicated spectral structure [5,6]. While existing methods focus on the integrality of spectra, the LSTM network pays more attention to the contextual information among adjacent sequential data. Therefore, how to divide the hyperspectral vector into different sequences in a proper way is crucial to the performance of the network. A natural idea is to consider each band as a time step and input one band at a time. However, hyperspectral data usually has hundreds of bands, making the LSTM network too deep to train in such a circumstance. Thus, a suitable grouping strategy is needed.

Let n be the number of bands and τ be the number of time steps in the LSTM. Then the sequence length of each time step is defined as $m = floor(n/\tau)$, where $floor(x)$ denotes rounding down x. For each pixel in the hyperspectral image, let $z = [z_1, z_2, \ldots, z_i, \ldots, z_n]$ be the spectral vector, where z_i is the reflectance of the ith band. The transformed sequences are then denoted by $x = [x_1, x_2, \ldots, x_i, \ldots, x_\tau]$, where x_i is the sequence at the ith time step. In what follows, we introduce two grouping strategies proposed in this paper.

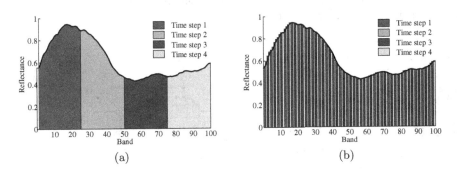

Fig. 3. (a) Grouping strategy 1. Adjacent bands are divided into the same sequence according to the spectral orders. (b) Grouping strategy 2. Every group in this case will cover a large spectral range. The bands marked with the same color will be fed into the LSTM network at each step time.

Grouping Strategy 1: Divide the spectral vector into different sequences according to the spectral order:

$$x^{(1)} = [z_1, z_2, \ldots, z_m]$$

$$\ldots$$

$$x^{(i)} = \left[z_{(i-1)m+1}, z_{(i-1)m+2}, \ldots, z_{im}\right] \qquad (11)$$

$$\ldots$$

$$x^{(\tau)} = \left[z_{(\tau-1)m+1}, z_{(\tau-1)m+2}, \ldots, z_{\tau m}\right]$$

where $x^{(i)}$ is the sequence at time i. As shown in Fig. 3(a), strategy 1 makes the signals inside a group continuous without any intervals and each group concentrates on a narrow spectral range. The spectral distance between different time steps will be relatively longer under such circumstances.

Grouping Strategy 2: Divide the spectral vector with a short interval:

$$x^{(1)} = \left[z_1, z_{1+\tau}, \ldots, z_{1+\tau(m-1)}\right]$$

$$\ldots$$

$$x^{(i)} = \left[z_i, z_{i+\tau}, \ldots, z_{i+\tau(m-1)}\right] \qquad (12)$$

$$\ldots$$

$$x^{(\tau)} = [z_\tau, z_{2\tau}, \ldots, z_{\tau m}]$$

Compared to strategy 1, each group in this case will cover a larger spectral range and the spectral distance between different time steps will be much shorter, as shown in Fig. 3(b).

After grouping the spectral vector z into different sequences $x^{(1)}, \ldots, x^{(\tau)}$, the LSTM network can be utilized to extract the contextual features among adjacent spectra. A fully connected (FC) layer and a softmax layer are added following the LSTM to accomplish the image classification. The complete spectral classification framework is illustrated as Fig. 4.

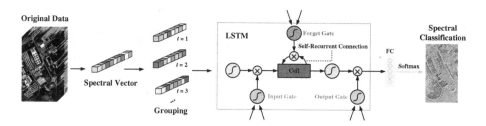

Fig. 4. Spectral classification with the proposed LSTM network. FC means the fully connected layer. The spectral vector of each pixel is divided into several groups. At each time step, a group of spectra are fed into the LSTM network.

3 Experiment Results and Analysis

3.1 Data Description

In our experiments, two benchmark hyperspectral data sets, including the Pavia University and Indian Pines, are utilized to evaluate the performance of the proposed method.

The first data set is acquired by the Reflective Optics Systems Imaging Spectrometer (ROSIS) sensor over the Pavia University, northern Italy. This image consists of 103 spectral bands with 610×340 pixels and it has a spectral coverage from $0.43\,\mu m$ to $0.86\,\mu m$ and a spatial resolution of $1.3\,m$. The training and test set are listed in Table 1.

Table 1. Number of training and test samples used in the Pavia University data set.

Class number	Class name	Training	Test
1	Asphalt	548	6083
2	Meadows	540	18109
3	Gravel	392	1707
4	Trees	542	2522
5	Metal sheets	256	1089
6	Bare soil	532	4497
7	Bitumen	375	955
8	Bricks	514	3168
9	Shadows	231	716
Total		3930	38846

The second data set is gathered by the Airborne Visible/Infrared Imaging Spectrometer (AVIRIS) sensor over the Indian Pines test site in Northwestern Indiana. After the removal of the water absorption bands, the image consists of 200 spectral bands with 145×145 pixels. It has a spectral coverage from $0.4\,\mu m$ to $2.5\,\mu m$ and a spatial resolution of $20\,m$. The training and test set are listed in Table 2.

All the experiments in this paper are randomly repeated 30 times with different random training data. The overall accuracy (OA) and Kappa coefficient [1] are utilized to quantitatively estimate different methods. Both the average value and the standard deviation are reported. The experiments in this paper are implemented with an Intel i7-5820K 3.30-GHz processor with 32 GB of RAM and a NVIDIA GTX1080 graphic card.

3.2 Analysis About the LSTM

In this subsection, we first evaluate two grouping strategies proposed in this paper with different number of time steps. As shown in Fig. 5, strategy 2 outperforms strategy 1 on both data sets. The reason behind this phenomenon lies

Table 2. Number of training and test samples used in the Indian Pines data set.

Class number	Class name	Training	Test
1	Alfalfa	30	16
2	Corn-notill	150	1278
3	Corn-mintill	150	680
4	Corn	100	137
5	Grass-pasture	150	333
6	Grass-trees	150	580
7	Grass-pasture-mowed	20	8
8	Hay-windrowed	150	328
9	Oats	15	5
10	Soybean-notill	150	822
11	Soybean-mintill	150	2305
12	Soybean-clean	150	443
13	Wheat	150	55
14	Woods	150	1115
15	Buildings-Grass-Trees-Drives	50	336
16	Stone-Steel-Towers	50	43
Total		1765	8484

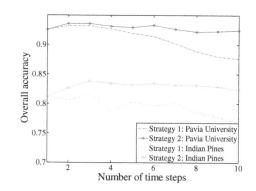

Fig. 5. The performance of the LSTM with different number of time steps.

on two aspects. First, the sequence divided by strategy 2 covers a wider spectral range compared with strategy 1, which means more abundant spectral information is fed into the LSTM cell at each time step in the case of strategy 2. Second, the spectral distance between different time steps is much shorter in strategy 2. Under such a circumstance, it is easier for the LSTM to learn the contextual features among adjacent spectral bands. Besides, as the number of time steps increases, the OA has a trend of rising first then getting steady or decreasing. This result shows that a too deep architecture may not be suitable for the LSTM to extract spectral features. For all the data sets in this paper, we set the number of time steps as 3. The number of neurons in the FC layer is set as 128.

3.3   Classification Results

In this subsection, we will report the classification results of the proposed methods along with other approaches, including raw (classification with original spectral features using RBF-SVM), PCA (classification with first 20 PCs using

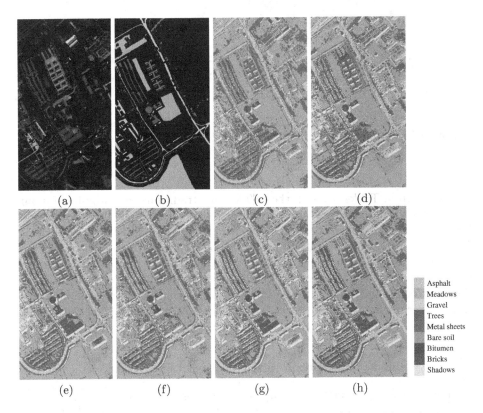

Fig. 6. Classification maps for the Pavia University data set. (a) The false color image. (b) Ground-truth map. (c) Raw. (d) PCA. (e) SAE. (f) LSTM-band-by-band. (g) LSTM-strategy 1. (h) LSTM-strategy 2.

RBF-SVM) and SAE [5] (spectral classification with SAE). We use the LibSVM [4,18] for the SVM classification in our experiments. The range of the regularization parameters for the five-fold cross-validation is from 2^8 to 2^{10}. The LSTM is implemented under Theano 0.8.2 and other experiments in this paper are carried out under MATLAB 2012b. The classification maps of different methods are shown in Figs. 6 and 7 and the quantitative assessment is shown in Tables 3 and 4.

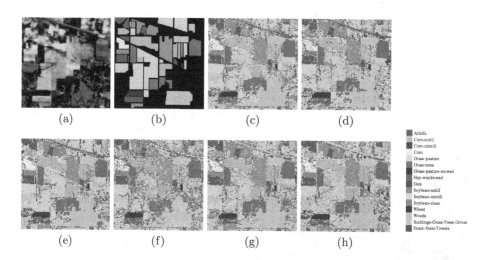

Fig. 7. Classification maps for the Indian Pines data set. (a) The false color image. (b) Ground-truth map. (c) Raw. (d) PCA. (e) SAE. (f) LSTM-band-by-band. (g) LSTM-strategy 1. (h) LSTM-strategy 2.

As shown in Tables 3 and 4, the LSTM with traditional band-by-band input fails to get a high accuracy and performs even worse than shallow methods in both data sets. By contrast, the LSTM with proposed grouping strategies yields better results and the overall accuracy can be improved about 5% to 13% on different data sets. The main reason here is that the band-by-band strategy would generate a too deep network and may result in information loss during the recurrent connection. Take the Pavia University data set for example. Since the number of time steps in this case will reach 103, after unfolding the LSTM, the depth of the network will also be 103, making it very hard for the network's

Table 3. Classification results of the Pavia University data set.

Method	Raw	PCA	SAE	LSTM-band-by-band	LSTM-strategy 1	LSTM-strategy 2
OA	90.80 ± 0.37	89.46 ± 0.36	91.97 ± 1.34	88.04 ± 1.52	92.86 ± 0.64	$\mathbf{93.14 \pm 0.72}$
Kappa × 100	87.62 ± 0.48	85.82 ± 0.46	89.21 ± 1.70	84.03 ± 1.91	90.36 ± 0.83	$\mathbf{90.71 \pm 0.95}$

Table 4. Classification results of the Indian Pines data set.

Method	Raw	PCA	SAE	LSTM-band-by-band	LSTM-strategy 1	LSTM-strategy 2
OA	79.68 ± 0.87	74.72 ± 0.99	83.25 ± 0.98	71.35 ± 1.19	80.68 ± 1.08	**84.11 ± 1.11**
Kappa × 100	76.68 ± 0.97	71.05 ± 1.18	80.72 ± 1.11	67.22 ± 1.43	77.77 ± 1.23	**81.57 ± 1.24**

training. In general, the LSTM with proposed strategy 2 achieves the best results compared with other spectral FE methods.

4 Conclusions

In this paper, we have proposed a band grouping based LSTM algorithm for the HSI classification. The proposed method has the following characteristics. (1) Our method takes the contextual information among adjacent bands into consideration, which is ignored by existing methods. (2) Two novel grouping strategies are proposed to better train the LSTM. Compared with the traditional band-by-band strategy, proposed methods prevent a very deep network for the HSI and yield better results.

Since the proposed method mainly focuses on the deep spectral FE, our future work will try to take the deep spatial features into consideration.

Acknowledgments. This work was supported in part by the National Natural Science Foundation of China under Grants U1536204, 60473023, 61471274, and 41431175, China Postdoctoral Science Foundation under Grant No.2015M580753.

References

1. Bioucas-Dias, J.M., Plaza, A., Camps-Valls, G., Scheunders, P., Nasrabadi, N., Chanussot, J.: Hyperspectral remote sensing data analysis and future challenges. IEEE Geosci. Remote Sens. Mag. **1**(2), 6–36 (2013)
2. Bruce, L.M., Koger, C.H., Li, J.: Dimensionality reduction of hyperspectral data using discrete wavelet transform feature extraction. IEEE Trans. Geosci. Remote Sens. **40**(10), 2331–2338 (2002)
3. Chang, C.I., Du, Q., Sun, T.L., Althouse, M.L.: A joint band prioritization and band-decorrelation approach to band selection for hyperspectral image classification. IEEE Trans. Geosci. Remote Sens. **37**(6), 2631–2641 (1999)
4. Chang, C.C., Lin, C.J.: LIBSVM: a library for support vector machines. ACM Trans. Intell. Syst. Technol. **2**(3), 389–396 (2011)
5. Chen, Y., Lin, Z., Zhao, X., Wang, G., Gu, Y.: Deep learning-based classification of hyperspectral data. IEEE J. Sel. Top. Appl. Earth Obs. Remote Sens. **7**(6), 2094–2107 (2014)
6. Chen, Y., Zhao, X., Jia, X.: Spectral-spatial classification of hyperspectral data based on deep belief network. IEEE J. Sel. Top. Appl. Earth Obs. Remote Sens. **8**(6), 1–12 (2015)
7. Donoho, D.L.: High-dimensional data analysis: the curses and blessings of dimensionality. AMS Math Chall. Lect. **1**, 32 (2000)

8. Hochreiter, S., Schmidhuber, J.: Long short-term memory. Neural Comput. **9**(8), 1735–1780 (1997)
9. Hughes, G.: On the mean accuracy of statistical pattern recognizers. IEEE Trans. Inf. Theory **14**(1), 55–63 (1968)
10. Ji, S., Xu, W., Yang, M., Yu, K.: 3D convolutional neural networks for human action recognition. IEEE Trans. Pattern Anal. Mach. Intell. **35**(1), 221–31 (2013)
11. Krizhevsky, A., Sutskever, I., Hinton, G.E.: ImageNet classification with deep convolutional neural networks. Adv. Neural Inf. Process. Syst. **25**(2), 2012 (2012)
12. Liu, T., Gong, M., Tao, D.: Large-cone nonnegative matrix factorization. IEEE Trans. Neural Netw. Learn. Syst. **28**(9), 2129–2142 (2016)
13. Liu, T., Tao, D.: Classification with noisy labels by importance reweighting. IEEE Trans. Pattern Anal. Mach. Intell. **38**(3), 447–461 (2016)
14. Liu, T., Tao, D., Song, M., Maybank, S.J.: Algorithm-dependent generalization bounds for multi-task learning. IEEE Trans. Pattern Anal. Mach. Intell. **39**(2), 227–241 (2017)
15. Rumelhart, D.E., Hinton, G.E., Williams, R.J.: Learning representations by back-propagating errors. Nature **323**, 533–536 (1986)
16. Tao, D., Li, X., Wu, X., Maybank, S.J.: General tensor discriminant analysis and gabor features for gait recognition. IEEE Trans. Pattern Anal. Mach. Intell. **29**(10), 1700–1715 (2007)
17. Tao, D., Li, X., Wu, X., Maybank, S.J.: Geometric mean for subspace selection. IEEE Trans. Pattern Anal. Mach. Intell. **31**(2), 260–274 (2009)
18. Tao, D., Tang, X., Li, X., Wu, X.: Asymmetric bagging and random subspace for support vector machines-based relevance feedback in image retrieval. IEEE Trans. Pattern Anal. Mach. Intell. **28**(7), 1088–1099 (2006)
19. Werbos, P.J.: Backpropagation through time: what it does and how to do it. Proc. IEEE **78**(10), 1550–1560 (1990)

A Self-adaptive Cascade ConvNets Model Based on Three-Way Decision Theory

Wen Shen, Zhihua Wei$^{(\boxtimes)}$, Cairong Zhao, and Duoqian Miao

Department of Computer Science and Technology, Tongji University,
Shanghai 201804, China
zhihua_wei@tongji.edu.cn

Abstract. Convolutional Neural Networks (ConvNets) have a great improvement on the classification performance compared to traditional image classification technologies and become one of the leaders in computer vision. In this paper, we present a Correcting Reliability Level (CRL) supervised three-way decision (3WD) cascade model to implement image classification of mass commodity data. Our model simulates the human decision process by using 3WD to determine "accepted" or "unsure" for the classification result. When judged as "unsure", CRL will supervise the 3WD and learn more information to make the final prediction. In addition, we introduce a Class Grouping algorithm based on feedback to learn the similarity between classes, which help us to train several expert ConvNets for different types of commodity images. Experimental results show that our model can effectively reduce the classification error rate compared with the base classifier.

Keywords: Three-way decisions · Cascade model
Convolutional Neural Networks · Image classification
Correcting Reliability Level

1 Introduction

Conventional image classification techniques were limited in their ability to process natural image data in their raw form. It required careful engineering and considerable domain expertise to design a feature extractor that transformed the raw data (the pixel values of an image) into a suitable internal representation or feature vector from which the classifier could classify patterns in the input [6]. On the contrary, deep-learning methods like ConvNets can learn representation directly from the data.

ConvNets are now the most commonly used large-scale image classification models. As early as 1990, ConvNets were trained for the task of classifying low-resolution images of handwritten digits [1]. However, with limited computation and

---

W.Shen—student.

© Springer Nature Singapore Pte Ltd. 2017
J. Yang et al. (Eds.): CCCV 2017, Part II, CCIS 772, pp. 433–444, 2017.
https://doi.org/10.1007/978-981-10-7302-1_36

training data, ConvNets were not popular until recent years. Krizhevsky et al. [4] trained a large, deep ConvNet (named AlexNet) to win over other contestants in the ILSVRC-2012 competition which was the first time a model performed so well on ImageNet dataset. Zeiler and Fergus [17] explained a lot of the intuition behind ConvNets and showed how to use a multi-layered deconvnet [18] to visualize the filters and weights correctly. Simonyan & Zisserman [12] created a simple and deep (19 layers) VGG Net, which reinforced the notion that ConvNets have to have a deep network of layers. Szegedy et al. [13] introduced the Inception module and built GoogLeNet. Coming up with the Inception module, a creative structuring of layers can lead to improved performance and computationally efficiency. He et al. [3] presented a residual learning framework to ease the training of deep networks and built a 152 layer ResNet, which won ILSVRC 2015 with an incredible error rate of 3.6% (lower than the error rate of humans, around 5–10%). Larsson et al. [5] advanced ResNet and built FractalNet which shows that explicit residual learning is not a requirement for building ultra-deep neural networks. All in all, the development of ConvNets is amazing.

In this paper, we use ConvNets for mass commodity image classification. However, a single ConvNet is limited in its ability of treating all the classes fairly. Combining the predictions of many different models is a very successful way to reduce test errors. Therefore, we propose a Class Grouping algorithm based on feedback to learn the similarity between classes and train several deep ConvNets become experts of different types of commodity images.

Schmidhuber et al. [10] declared that only winner neurons are trained. They trained several deep neural columns become experts on inputs preprocessed in different ways and average their predictions. Simonyan and Zisserman [11] proposed a two-stream ConvNet architecture which incorporates spatial and temporal networks. Each stream was implemented using a deep ConvNet, softmax scores of which were combined by late fusion.

Inspired by works above, we build a cascade ConvNets model to implement large-scale image classification. However, more innovative than those works, our model introduces the theory of three-way decisions to simulate human decision process. We build a 3WD-based cascade model including 3 layers, the first layer is a base ConvNet for all images and the third layer consists of several expert ConvNets trained for similar classes. The most important second layer is a three-way decision layer, which controls whether the data flows from the first layer to the third layer.

The notion of three-way decisions was originally introduced by the needs to explain the three regions of probabilistic rough sets. A theory of three-way decisions is constructed based on the notions of acceptance, rejection and non-commitment [15], whenever it is impossible to make an acceptance or a rejection decision, the third noncommitment decision is made [14]. Three-way decisions play a key role in everyday decision-making and have been widely used in many fields and disciplines. Three-way spam filtering systems [19], for example, add a suspected folder to allow users make further examinations of suspicious emails, thereby reducing the chances of misclassification. Three-way decisions are also commonly used in medical decision making [8,9]. In the threshold approach to clinical decision making proposed in [9], by comparing the probability of disease with a pair of a "testing" threshold and a "test-treatment" threshold, doctors make one of three decisions: (a) no treatment no further testing; (b) no treatment but further testing; (c) treatment without further testing.

This paper extends the application of three-way decisions to the image classification. When there is doubt about the classification result(the first layer), our model will make a noncommitment decision(the second layer) and learn more information from expert classifiers(the third layer) to make the final prediction.

The structure of this paper is as follows. In Sect. 2 we describe our model. Next, in Sect. 3 we show the experiments and analyze the results. Last, we summarize our work in Sect. 4.

2   Model

We use GoogLeNet model throughout the paper. GoogLeNet, as defined in [13], uses 9 Inception modules in the whole architecture, which is a network in the network structure [7]. By using the Inception models, the network is deeper and wider, and is better than previous models.

In Sect. 2.1, we define some symbols that will be referred later. Next, in Sect. 2.2, we introduce the Class Grouping algorithm that based on feedback. Later, in Sect. 2.3 we introduce the three-way decision cascade model. Last, Sect. 2.4 shows the ultimate form of our model, CRL-supervised 3WD cascade model.

2.1   Symbols Definition

Before going further into our model, we define some symbols.

Img, the input image.

$CAT = \{c_1, c_2, \cdots, c_i, \cdots, c_C\}$, the class set, including C classes.

$P = \{p_1, p_2, \cdots, p_i, \cdots, p_C\}$, the classification result of a ConvNet model, where p_i is the probability that Img is classified as class c_i.

$Conf = (n_{ij})_{C \times C}$, the confusion matrix of ConvNet test result, where n_{ij} is the number of images of class c_i being classified as class c_j. The bigger the n_{ij}, the easier that images of class c_i are classified as class c_j.

$Threshold\ of\ possible\ classes\ (Th\text{-}pos)$. Obviously, we do not need to consider the situation that Img belonging to class c_i if p_i is very small. Therefore, we need a threshold to determine the possible classes of Img. If p_i is no less than $Th\text{-}pos$, we think that Img may belong to class c_i. We stipulate that $Th\text{-}pos$ is no less than $\frac{1}{C}$.

$Top\text{-}1\ class$ (referred to as c_{top}), the class considered the most probable by the model, the probability is p_{top}.

2.2   Class Grouping Algorithm Based on Feedback

For commodity images of web-based platforms, many commodity classes are similar to each other, which is difficult for both humans and machines to distinguish them (see Fig. 1). Therefore, a classifier trained for all the classes is not enough for distinguishing those similar classes, we need some more specified classifiers trained for certain similar classes. In this paper, we propose a Class Grouping (CG) algorithm (see Algorithm 1) based on the feedback of the classification results.

Fig. 1. Experimental data, 4 samples each class. Classes c_1, c_5, c_{10} and c_{14} are of similar features, they are all kinds of sweaters. And classes c_3, c_{18}, c_{29}, c_{30}, c_{31} and c_{32} are kinds of trousers.

Algorithm 1. Class Grouping

Input: CAT; $S = (s_{ij})_{C \times C}$; cluster number K;
Output: K clusters;
Make each class in CAT a cluster;
Compute pair-wise distance of all clusters, $d_{mn} = min(s_{ij}), \forall c_i \in clt_m \,\&\, \forall c_j \in clt_n$;
repeat
    find two clusters that are closest to each other;
    merge the two clusters form a new cluster clt_{new};
    compute the distance form clt_{new} to all other clusters.
until there are only K clusters

We define s_{ij} the similarity between class c_i and class c_j(see Formula (1)).

$$s_{ij} = \frac{n_{ij}}{\sum_{t=1}^{C} n_{it}} \times \frac{n_{ji}}{\sum_{t=1}^{C} n_{jt}} \tag{1}$$

After running CG algorithm, we get K clusters: *cat-1*, *cat-2*, $\cdots$, *cat-K*. We train an expert ConvNet for each cluster(see Sect. 3.2). These ExpConvNets will be used to build the third layer of cascade model.

Based on class grouping experimental results, we introduce *similar-classes*. If class c_i and c_j are belong to the same subset *cat-k*, we call that c_i and c_j are similar-classes. Similar-classes are of high probability to be wrongly classified to each other and therefore require expert judgments.

2.3   Three-Way Decision Based Cascade Model

A theory of 3WD is constructed based on the notions of acceptance, rejection and noncommitment. Inspired by 3WD theory, we no longer directly accept the classification result of the base classifier. Instead, we make one of two decisions: (a) accept it if it is reliable; (b) opt for a noncommitment if it is not reliable. Since this is not a binary-decision problem with two options, but a multiclass classification problem, there is no "reject" option. We judge the classification result is not reliable if meeting two conditions: (i) existing class $c_a (c_a \neq c_{top})$ and $p_a \geq Th\text{-}pos$; (ii) c_a and c_{top} belonging to the same subset $cat\text{-}k$. The condition (i) guarantees that Img may belong to class c_a, condition (ii) guarantees that c_a and c_{top} are similar-classes. We define these conditions under the hypothesis that meeting these two conditions means that the classifier is confused. The classifier considers that Img can be predicted as c_a or c_{top}. Thus, we need to put the image into the expert classifier $ExpConvNet\text{-}k$ for further judgment. Formula (2) is the 3WD process.

$$3WD = \begin{cases} \text{delay,} & \text{satisfiying condition (i) and (ii)} \\ \text{accept,} & \text{otherwise} \end{cases} \tag{2}$$

The ability of a single classifier is limited. Therefore, combining several different models is a way to reduce the error rate [4]. Cascade [2] is a special case of ensemble learning. The basic idea of cascade is the connection of multiple classifiers. The information is passed between layers and the output information of the upper classifier is used as the additional information of the next classifier.

Under the guidance of 3WD theory, we establish a self-adaptive cascade ConvNets model, including 3 layers (see Algorithm 2). The first layer is a base classifier (a base ConvNet), the second layer is a 3WD layer and the third layer is expert layer, including several expert classifiers (ExpConvNets). We put Img into the first layer (a base classifier) and send the classification result P^1 into the 3WD layer. Next, the 3WD layer will make one of two decisions: (a) accept P^1 as the final P; (b) opt for a noncommitment and put Img into $ExpConvNet\text{-}k$ (the classification result is P^2). Finally, we calculate probability P based on P^1 and P^2, see Formula (3).

$$p_i = \begin{cases} p_{c_i}^2, & \text{if } c_i \in cat\text{-}k \\ p_{c_i}^1, & \text{otherwise} \end{cases} \tag{3}$$

where $p_{c_i}^1$ represents the probability of the image being predicted as class c_i by the base classifier and $p_{c_i}^2$ represents the probability of the image being predicted as class c_i by the expert classifier.

Algorithm 2. 3WD-CM

Input: image Img,size $n * n$
Output: prediction P,size C
Input Img into the first layer (a base ConvNet), get P^1;
Send the classification result P^1 to the 3WD layer;
if $\exists\ c_a$, & $p_a \geq$ *Th-pos* **then**
  if $c_a \in$ *cat-k* & $c_{top} \in$ *cat-k* **then**
    put Img into *ExpConvNet-k*, get P^2;
    calculate the final P with Formula (3)
  else
    $P = P^1$
  end if
else
  $P = P^1$
end if

2.4 CRL-supervised 3WD Cascade Model

3WD decides which images may need expert judgments. But the experimental result (see Table 2) tells us that blindly following the decision of 3WD is not a good idea. Suppose that the base classifier considers Img as class c_i while 3WD layer delays this result and after expert judgment, the cascade model finally judge Img as class c_j. Experimental experience tells us that there are two situations: (1) in most cases, c_i is the correct class while c_j is wrong; (2) on the contrary, in most cases, c_i is wrong while c_j is the correct class. Situation (1) tells us that the 3WD makes the classification result from right to wrong; while situation (2) tells us that 3WD makes the classification result from wrong to right. Obviously, we welcome the latter situation. Thus, we need to supervise the 3WD process.

We define *Correcting Reliability Level (CRL)* to supervise the 3WD process. A high CRL means that the 3WD has a high probability of making right decision. We define TF_i (Truth to False) to describe the situation (1) above, and FT_i (False to Truth) to describe the situation (2) above. CRL computing see Formula (4).

$$CRL_i = \begin{cases} \frac{N_{FT_i} - N_{TF_i}}{N_{FT_i} + N_{TF_i}} & \text{while } N_{FT_i} > N_{TF_i}\ \&\ N_{FT_i} > 0 \\ 0 & \text{otherwise} \end{cases} \tag{4}$$

We use a random function

$$R(p) = binomial(1, p) \tag{5}$$

to move CRL value to a Boolean value, "True" means following the 3WD and putting Img into the expert classifier, while "False" means "accept" the result of base classifier. p is the probability of return "True". $R(CRL_i)$, for instance, has a probability of CRL_i to return "True". Therefore, the larger the CRL value is, the more likely that Img will be passed into the expert classifier.

On the basis of 3WD-CM, we add a CRL table after the 3WD layer. CRL table is used to determine whether *Img* is worthy of expert judgment (see Algorithm 3).

Algorithm 3. CRL-CM

  Input: image *Img*,size $n * n$
  Output: prediction P,size C
  Input *Img* into the first layer (a base ConvNet), get P^1;
  Send the classification result P^1 to the 3WD layer;
  if $\exists\ c_a$, & $p_a \geq$ *Th-pos* **then**
    if $c_a \in$ *cat-k* & $c_{top} \in$ *cat-k* **then**
      Check CRL table and get $crl = CRL_{top}$
      if R(crl) is TRUE **then**
        put *Img* into *ExpConvNet-k*, get P^2;
        calculate the final P with Formula (3)
      else
        $P = P^1$
      end if
    else
      $P = P^1$
    end if
  else
    $P = P^1$
  end if

---

3 Experiments

In this section, we show the experiment details. In Sect. 3.1, we introduce the experimental dataset. Next, in Sect. 3.2 we introduce the class grouping process and ExpConvNets training. Then, in Sect. 3.3 we show the results of 3WD-based Cascade Model. At last, in Sect. 3.4 we analyze the results of CRL-supervised 3WD Cascade Model.

3.1 JD Clothing Dataset

The experimental data of this paper is JD clothing dataset, examples see Fig. 1. JD is one of the most famous B2C shopping site in China and the first large-scale integrated business platform to be listed in the United States. JD has a strong market share; therefore, it has accumulated a large number of commodity image data, which provides researchers with a lot of resources. Our experimental dataset has about 400,000 clothing images, including 37 classes. The dataset is divided into training set, validation set and test set at a ratio of 8: 1: 1.

In this paper, we report two error rates: top-1 and top-5, where the top-5 error rate is the fraction of test images for which the correct label is not among the five labels considered most probable by the model [4].

3.2   Class Grouping and ExpConvNets Training

In this paper, we do experiments on a deep learning framework named Caffe. Yosinski et al. [16] pointed out that fine-tuning is better than randomly initialize parameters. Our experimental results also confirm this. We start from a pre-trained model (GoogLeNet on ImageNet LSVRC-2014) and fine-tune it. The top-1 error rate is 44.59% and the top-5 error rate is 9.40%, which are better than randomly initializing (the top-1 error rate is 57.11% and the top-5 error rate is 21.16%). We call the fine-tuned model Base Model (BM) below and the latter experiments will fine-tune models on the basis of it.

Test results confirm that images of many classes are easily to be misclassified to each other, like class c_1 and c_{10} (see Fig. 2). Thus, we group those similar classes into the same subset with CG algorithm introduced in Sect. 2.2.

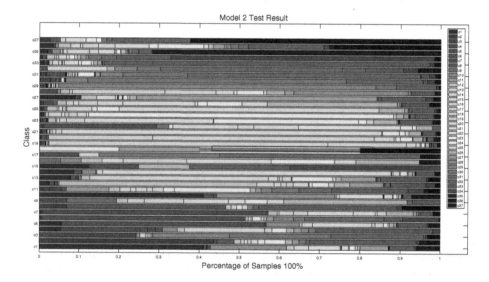

Fig. 2. Stacked bar chart of test result of BM. Take class c_1 as example, there is about 20% of test images misclassified as class c_{10}. Similarly, there is about 40% of test images of class c_{10} misclassified as class c_1.

We set $K = 5$ and divide CAT into 5 subsets. After class grouping, we train an expert ConvNet for each subset. We fine-tune the BM and adapt most of the architecture (only change the output number of the last layer), and resume training from the BM weights. Table 1 shows the class grouping result and the ExpConvNets error rates.

Table 1. Class grouping results and ExpConvNets error rates.

Subset	Classes	ExpConvNet	Top-1 error rate(%)
cat-1	c_2, c_8, c_{15}, c_{27}, c_{33}, c_{37}	ExpConvNet-1	9.16
cat-2	c_1, c_5, c_{10}, c_{14}, c_{25}, c_{35}	ExpConvNet-2	41.17
cat-3	c_3, c_6, c_{17}, c_{18}, c_{21}, c_{29}, c_{30}, c_{31}, c_{32}	ExpConvNet-3	23.50
cat-4	c_4, c_{12}, c_{13}, c_{34}	ExpConvNet-4	16.67
cat-5	c_7, c_9, c_{11}, c_{16}, c_{19}, c_{20}, c_{22}, c_{23}, c_{24}, c_{26}, c_{28}, c_{36}	cat-5ExpConvNet-5	33.48

3.3   3WD-based Cascade Model

In order to get the CRL table, we first need to test images with 3WD-CM. We set Th-pos 0.1 in this experiment. The top-1 error rate of 3WD-CM is 44.401%, reducing by 0.189% compared with BM; top-5 error rate is 9.475%, increasing by 0.075% compared with BM. The results are bad. The top-1 error rate reduces a little bit, and the top-5 error rate does not drop but increase.

Table 2 shows the counting results of TF and FT. We can see that there are totally 1195 cases of situation (1) and 1,273 cases of situation (2). Therefore, in fact only 78 samples are modified correctly by 3WD-CM, the accuracy increases by only 0.189%. When c_1 is considered as c_{top} by the base classifier, there are totally 613 images considered needing expert judgement by 3WD, wherein, 135 images are modified correctly and 478 images are modified incorrectly. Thus, there are 343 images being modified incorrectly in total. This shows that when c_1 is considered as c_{top}, we should better ignore the decision of 3WD that *Img* needing expert judgment. We should better accept c_1 as the prediction result. On the contrary, if the base classifier considers c_{10} as c_{top}, we would better follow the 3WD and do an expert judgment for *Img*. Because, for c_{10}, the number of images which are modified correctly is greater than the number of images which are modified incorrectly. Therefore, we use CRL to supervise 3WD process.

Table 2. Classification results of 3WD-CM

Class	N_{FT}	N_{TF}	NET	Class	N_{FT}	N_{TF}	NET	Class	N_{FT}	N_{TF}	NET	Class	N_{FT}	N_{TF}	NET
c_1	135	478	-343	c_{11}	1	2	-1	c_{21}	0	22	-22	c_{31}	64	52	12
c_2	4	7	-3	c_{12}	9	9	0	c_{22}	3	8	-5	c_{32}	1	1	0
c_3	11	3	8	c_{13}	3	13	-10	c_{23}	22	10	12	c_{33}	15	42	-27
c_4	1	1	0	c_{14}	73	29	44	c_{24}	37	32	5	c_{34}	0	2	-2
c_5	33	18	15	c_{15}	0	0	0	c_{25}	3	8	-5	c_{35}	17	43	-26
c_6	16	13	3	c_{16}	0	0	0	c_{26}	11	19	-8	c_{36}	12	10	2
c_7	5	2	3	c_{17}	0	0	0	c_{27}	6	20	-14	c_{37}	3	3	0
c_8	5	2	3	c_{18}	0	0	0	c_{28}	9	1	8	Total	1273	1195	78
c_9	3	2	1	c_{19}	60	58	2	c_{29}	21	26	-5				
c_{10}	602	150	452	c_{20}	58	54	4	c_{30}	30	55	-25				

Note: NET($= N_{FT} - N_{TF}$) means the net number of samples modified correctly by 3WD-CM.

3.4 CRL-supervised 3WD Cascade Model

Now, we calculate CRL of each class, see Table 3. We establish CRL-supervised 3WD cascade model (CRL-CM). Table 4 shows the classification performance of CRL-CM under different $Th\text{-}pos$ values (0.1, 0.2, 0.3 and 0.4). We test 30 times for each $Th\text{-}pos$ value and take the average error rate. Compared with BM, the top-1 error rate reduces by about 1.09% when $Th\text{-}pos = 0.1$. Top-5 error rate does not reduce obviously. With the increasement of $Th\text{-}pos$, the error rate reduces less, because the greater the $Th\text{-}pos$ is, the more harsh that 3WD determines a sample being "unsure", thus, those misclassified samples lose the chance of being modified correctly. The experimental results show that the CRL-CM can effectively reduce the classification error rate compared with a single base ConvNet.

Table 3. CRL table.

Class	CRL	Class	CRL	Class	CRL	Class	CRL
c_1	0	c_{11}	0	c_{21}	0	c_{31}	0.103
c_2	0	c_{12}	0	c_{22}	0	c_{32}	0
c_3	0.571	c_{13}	0	c_{23}	0.375	c_{33}	0
c_4	0	c_{14}	0.431	c_{24}	0.072	c_{34}	0
c_5	0.294	c_{15}	0	c_{25}	0	c_{35}	0
c_6	0.103	c_{16}	0	c_{26}	0	c_{36}	0.091
c_7	0.429	c_{17}	0	c_{27}	0	c_{37}	0
c_8	0.429	c_{18}	0	c_{28}	0.8		
c_9	0.2	c_{19}	0.017	c_{29}	0		
c_{10}	0.601	c_{20}	0.036	c_{30}	0		

Table 4. Average error rates of CRL-CM under different $Th\text{-}pos$.

$Th\text{-}pos$	Top-1 error rate(%)	Top-5 error rate(%)
0.1	43.50 (1.09)	9.386 (0.017)
0.2	43.62 (0.97)	9.398 (0.005)
0.3	43.68 (0.91)	9.415 (-0.013)
0.4	43.77 (0.82)	9.413 (-0.011)

Note: The numbers in parentheses are reduced error rates (%) compared with Model 2.

4 Conclusion

In this paper we integrate several different ConvNets to build a CRL-supervised 3WD cascade model. Experimental results show that our model can effectively reduce the error rate compared with a single ConvNet. The contributions of this paper are: (i) Simulating the human decision process by using 3WD to construct a cascade model with several ExpConvNets which become experts on inputs preprocessed in different ways; (ii) introducing CRL to supervise 3WD process which reduces error rate effectively. In future work, we will do more experiments with public datasets like ImageNet to prove the validity of our model.

Acknowledgments. The work is partially supported by the National Natural Science Foundation of China (No. 61573259, 61673301, 61573255 and 61673299), the program of Further Accelerating the Development of Chinese Medicine Three Year Action of Shanghai (No. ZY3-CCCX-3-6002) and the Natural Science Foundation of Shanghai (NO. 15ZR1443800).

References

1. Cun, Y.L., Boser, B., Denker, J.S., Howard, R.E., Habbard, W., Jackel, L.D., Henderson, D.: Handwritten digit recognition with a back-propagation network. In: Advances in Neural Information Processing Systems, pp. 396–404 (1990)
2. Gama, J., Brazdil, P.: Cascade generalization. Mach. Learn. **41**(3), 315–343 (2000)
3. He, K., Zhang, X., Ren, S., Sun, J.: Deep residual learning for image recognition, pp. 770–778 (2016)
4. Krizhevsky, A., Sutskever, I., Hinton, G.E.: Imagenet classification with deep convolutional neural networks. In: International Conference on Neural Information Processing Systems, pp. 1097–1105 (2012)
5. Larsson, G., Maire, M., Shakhnarovich, G.: Fractalnet: Ultra-deep neural networks without residuals (2016)
6. Lecun, Y., Bengio, Y., Hinton, G.: Deep learning. Nature **521**(7553), 436–444 (2015)
7. Lin, M., Chen, Q., Yan, S.: Network in network. Computer Science (2014)
8. Lurie, J.D., Sox, H.C.: Principles of medical decision making. Spine **24**(5), 493 (1999)
9. Pauker, S.G., Kassirer, J.P.: The threshold approach to clinical decision making. New Engl. J. Med. **302**(20), 1109 (1980)
10. Schmidhuber, J., Meier, U., Ciresan, D.: Multi-column deep neural networks for image classification. In: Computer Vision and Pattern Recognition, pp. 3642–3649 (2012)
11. Simonyan, K., Zisserman, A.: Two-stream convolutional networks for action recognition in videos. Adv. Neural Inf. Process. Syst. **1**(4), 568–576 (2014)
12. Simonyan, K., Zisserman, A.: Very deep convolutional networks for large-scale image recognition. Computer Science (2015)
13. Szegedy, C., Liu, W., Jia, Y., Sermanet, P.: Going deeper with convolutions. In: IEEE Conference on Computer Vision and Pattern Recognition, pp. 1–9 (2015)
14. Yao, Y.: Three-way decisions with probabilistic rough sets. Inf. Sci. **180**(3), 341–353 (2011)
15. Yao, Y.: An outline of a theory of three-way decisions. In: Yao, J.T., Yang, Y., Słowiński, R., Greco, S., Li, H., Mitra, S., Polkowski, L. (eds.) RSCTC 2012. LNCS (LNAI), vol. 7413, pp. 1–17. Springer, Heidelberg (2012). https://doi.org/10.1007/978-3-642-32115-3_1

16. Yosinski, J., Clune, J., Bengio, Y., Lipson, H.: How transferable are features in deep neural networks? Eprint Arxiv 27, pp. 3320–3328 (2014)
17. Zeiler, M.D., Fergus, R.: Visualizing and understanding convolutional networks. In: Fleet, D., Pajdla, T., Schiele, B., Tuytelaars, T. (eds.) ECCV 2014. LNCS, vol. 8689, pp. 818–833. Springer, Cham (2014). https://doi.org/10.1007/978-3-319-10590-1_53
18. Zeiler, M.D., Taylor, G.W., Fergus, R.: Adaptive deconvolutional networks for mid and high level feature learning. In: IEEE International Conference on Computer Vision, pp. 2018–2025 (2011)
19. Zhou, B., Yao, Y., Luo, J.: A three-way decision approach to email spam filtering. In: Farzindar, A., Kešelj, V. (eds.) AI 2010. LNCS (LNAI), vol. 6085, pp. 28–39. Springer, Heidelberg (2010). https://doi.org/10.1007/978-3-642-13059-5_6

Saliency Detection via CNN Coarse Learning and Compactness Based ELM Refinement

Ruirui Li[✉], Shihao Sun, Lei Yang, and Wei Hu

Beijing University of Chemical Technology,
No. 15, Beisanhuandong Road, Chaoyang District, Beijing, China
ilydouble@gmail.com, 472527311@qq.com, ylxx@live.com,
huwei@mail.buct.edu.cn

Abstract. Salient object detection has attracted a lot of research in computer vision. It plays a vital role in image retrieval, object recognition and other image processing tasks. Although varieties of methods have been proposed, most of them heavily depend on feature selection and fail in the case of complex scenes. We propose a processing framework for saliency detection which contains two main steps. It uses deep convolutional neural networks (CNNs) to find a coarse saliency region map that includes semantic clues. Then it refines the coarse saliency map by training an extreme learning machine (ELM) on a group of color and texture compactness features. To get final saliency objects, it synthesizes the coarse saliency region map and several multiscale saliency maps that are obtained by refining the coarse one together. The method achieves good experimental results and can be used to improve the existing salient object detection methods as well.

Keywords: CNNs · ELM · Saliency · Compactness · Multiscale

1 Introduction

Visual saliency aims at detecting salient attention-grabbing parts in an image. It has received increasing interest in recent years. Though early research primarily focus on predicting eye-xations in images, it has shown that salient object detection, which segments entire objects from images, is more useful and has been successfully applied in object recognition [22], image classification [29], object tracking [28], image compression [23], and image resizing [2]. Despite recent progress in deep learning, salient object detection remains a challenging problem that calls for more accurate solutions.

Without a rigorous definition of image saliency, traditional saliency detection methods rely on several saliency priors. The contrast prior is the most popular one, which can be further categorized as local contrast and global contrast [7,31]. Conventionally, the contrast based methods use hand-crafted features based on human knowledge on visual attention and thus they may not generalize well in different scenarios.

© Springer Nature Singapore Pte Ltd. 2017
J. Yang et al. (Eds.): CCCV 2017, Part II, CCIS 772, pp. 445–460, 2017.
https://doi.org/10.1007/978-981-10-7302-1_37

Several researchers [19,26,33] propose CNN based approaches for saliency detection. Though better performance has been achieved, there are still two major issues of prior CNN based saliency detection methods [27]. Firstly, CNNs use limited sizes of local image patches as input. To consider the spatial consistency, the CNNs networks require carefully designing and become extremely complex. Secondly, saliency priors, which are shown to be elective in previous work, are completely discarded by most CNN based methods.

Because of this, in our work, we present a progressive framework for image saliency computation called CELM. We leverage the advantages of high-level semantically meaningful features from deep learning as well as hand-crafted features when inferring saliency maps. Specifically, the framework has two procedures and learns the saliency map in a coarse-to-fine manner. The coarse-level image saliency semantically identities rough regions for salient objects by the CNN. The CNN takes the whole images as input and trains a global model to measure the saliency score of each pixel in an image, generating a coarse-level saliency map in a lower resolution. The fine-level image saliency is achieved by an ELM-based classification. This step is guided by the coarse-level saliency map and the input RGB image from which we fetch compactness features. After that, the coarse-level saliency map and the refined saliency map are synthesis together to get the final result. Figure 1 shows some saliency results generated by our approach. Extensive experiments on the standard benchmarks of image saliency detection demonstrate that the proposed CELM has better performance compared with state-of-the-art approaches. In summary, this paper makes the following main contributions to the community:

- A progressive saliency framework is developed by integrating CNN and ELM, taking advantage of both semantic and hand-crafted features. This model is general to be extended to improve the current saliency detection method.
- A heuristic learning method based on ELM is proposed to refine the coarse saliency image. It utilizes a new compactness hypothesis on superpixels to find subtle structures which has the same color and texture distribution as the positive samples.
- A saliency image synthesis algorithm based on saliency priors is proposed to fusion multiple leveled saliency maps.

The remainder of the paper is organized as follows. Section 2 reviews related work and differentiates our method from such work. Section 3 introduces our proposed method. Extensive experimental results and comparisons are presented in Sect. 4. And Sect. 5 concludes this paper.

2   Related Works

2.1   Image Saliency Detection

Image saliency detection approaches can be roughly categorized into two groups: bottom-up and top-down. Bottom-up methods focus on the low level features

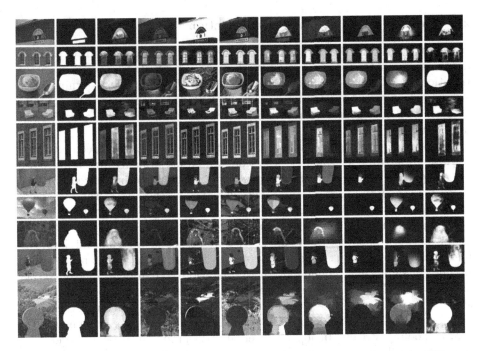

Fig. 1. Results from HKU-IS by different methods, from left to right: input, GT, our CELM, FT, GC, HC, DRFI, GMR, QCUT, PISA, DISC

(e.g. orientation, color, intensity, etc.). And the low level based methods can be further divided into local methods, global methods and hybrid of previous two according to the spatial scope of saliency computation. Local methods design saliency features by considering the contrast in a small neighborhood. As an example, in [16], multiscale image features (colors, intensity and orientations) are combined to generate saliency map. However, if there is a lot of high frequency noise in an image, local methods may result in a very poor performance. On the contrary, global methods compute saliency of an image region using its contrast over the entire image, which can tackle aforementioned problems. For example, Cheng et al. [9] used Histogram based Contrast (HC) and spatial information-enhanced Region based Contrast (RC) to measure saliency. But on the other hand, global methods ignore the details of local regions, leading to the blurring of the edges of the saliency. In order to combine the advantages of the complementary pair, Chen et al. [5] simultaneously integrate local and global structure information by designing a structure-aware descriptor based on the intrinsic bi-harmonic distance metric. Top-down methods move attention to high level features (e.g., faces, humans, cars, etc.), and are usually task-dependent. Yang et al. [32] proposed a top-down visual saliency model which incorporates a layered structure from top to bottom: CRF, sparse coding and image patches. Considering both importance of low and high features, Borji et al. [4] proposed a boosting model by integrating bottom-up features and top-down features.

2.2 Deep Learning for Saliency Detection

In recent years, with the development of deep learning, the methods of saliency detection based on deep learning have become a hotspot. Compared with the traditional manual extraction of features, the ones based on deep learning (such as convolution neural network) not only have better robustness, but also contains a higher level of semantic information, which is very important for salient object detection.

In [33], local and global context are integrated into a multi-context deep learning framework for saliency detection, whose performance was improved a lot, compared to many conventional approaches. Instead of using a fixed size local context, Li et al. [19] used a spatially varying one, which relies on the actual size of the surrounding regions. Furthermore, in order to dig more valuable information hidden inside the concatenated multiscale deep features, Li et al. [33] used neural network architecture at integrating stage. Though significant improvements have been made, the efficiency of deep feature extraction is not satisfied because of significant redundancy in computation and storage. In [10], rather than treat each region as an independent unit in feature extraction without any shared computation, Li et al. proposed an end-to-end deep contrast network which consists of a pixel-level fully convolutional stream and a segment-wise spatial pooling stream. In [6], Chen et al. also proposed an end-to-end deep hierarchical saliency network, whose architecture works in a global to local and coarse to fine manner. Our method also learns the saliency map in a progressive way. But different from Chens work, we use extreme learning machine to get fine saliency maps. While our approach leverages the advantages of high-level semantically meaningful features from deep learning, it also integrates hand-crafted features when inferring saliency maps.

3 Proposed Method

3.1 Progressive Framework

The pipeline of the proposed method is summarized in Fig. 2. We train the CNN to get the coarse-level saliency map. It can be found that Pixels in the coarse-level saliency map are probably unconnected to form coherent regions. To maintain the spatial structure, we divided the original image into a number of superpixels. Then we statistically compute the labels of superpixels through the coarse-level saliency map. Combined with a group of extracted features of the superpixels, an ELM classier for the given image is trained, and its condense output for each super-pixel is used as a measure of saliency. This procedure is carried out in a multiscale way. Finally, the detected results of different scales and the coarse saliency map are synthesized to form a strong and fine saliency map result.

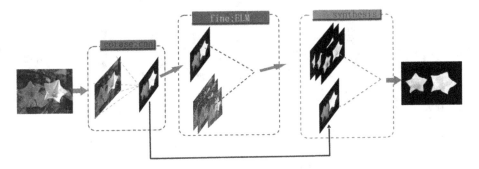

Fig. 2. Saliency detection framework

3.2 Coarse Saliency Map

We use all the images in the MSRA10K dataset and their labeled ground truth to train the coarse-level CNN. Its architecture is similar to the general AlexNet [18] proposed by Krizhevsky et al. Fig. 3 depicts the overall architecture of our CNN which contains six layers. The first five are convolutional and the last one is fully connected. All of them contain learnable parameters. The input of our CNN contains three RGB channels of whole images and one channel of average map as Fig. 4 illustrated. They are all resized to 256 × 256 and then filtered by the first convolutional layer (COV1) with 96 kernels of size 11 × 11 × 4 with a padding of 5 pixels. The result 62 × 62 × 96 feature maps are then sequentially given to a rectified linear unit (ReLU1) followed by a LRN and a max-pooling layer (MAXP1) which performs max pooling over 3 × 3 spatial neighborhoods with a stride of 2 pixels. The output of the MAXP1 is 31 × 31 × 96 features and then passed to the second convolutional layer (COV2). The number of filter kernels is changed from COV2 to COV5. They are set to 5, 3, 3, and 3 respectively. According to the parameter configuration of each layer, the architecture of the CNN can be described concisely by layer notations with layer sizes:

COV1(62 × 62 × 96)→RELU1→MAXP1→
COV2(31 × 31 × 96)→RELU2→MAXP2→
COV3(15 × 15 × 256)→RELU3→
COV4(15 × 15 × 384)→RELU4→
COV5(15 × 15 × 256)→RELU5→MAXP5→
FC1(7 × 7 × 256)→4096.

The last fully connected layer serves like a SVM. It computes the linear transformations of the feature vector and outputs 4096 saliency scores. The 4096 values are later re-arranged to a 64 × 64 coarse-level saliency map.

3.3 Saliency Refinement

The coarse-level saliency detection takes the whole image as input. It mainly considers the global saliency region in the image. Since it pays less attention

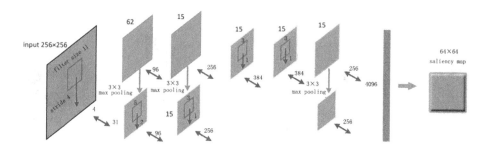

Fig. 3. CNN architecture

Fig. 4. Four input channels for CNN

to local context information, the salient pixels in coarse saliency map may be unconnected and may mistakenly lose subtle salient structures. To further refine the saliency result, we train a binary classier through Extreme Learning Machine [13] (ELM). The extreme learning machine (ELM) is proposed by Huang et al. [13,14]. It is a kind of machine learning algorithm for the single-layer feed-forward neural network (SLFN) [15] and its architecture is given in Fig. 5. The ELM contains three layers: the input layer, the hidden layer, and the output layer. The ELM randomly initializes the weights between input layer and hidden layer as well as the bias of hidden neurons. It analytically determines the weights between the hidden layer and the output layer using the least-squares method. Compared with Neural networks (NN), support vector machines (SVM) and other popular learning methods, the ELM has several significant advantages, such as real-time learning, high accuracy, least user intervention. In our experiments, sigmoid neurons are chosen for the training.

We use both the coarse saliency image and the original given image as input and propose a multiscale superpixel-based statistic method to label the saliency image. The procedure of saliency refinement is shown as Fig. 6. The original given RGB image is converted into the CIE LAB color space first, and is efficiently segmented into multi-leveled sets of superpixels by the SLIC algorithm [1]. We use two thresholds, T_h and T_l ($T_h > T_l$) to collect the training samples from the coarse saliency map. If average saliency value of the superpixel is higher than T_h, the superpixel is labeled as positive sample. If average saliency value of the superpixel is lower than T_l, the superpixel is labeled as negative sample.

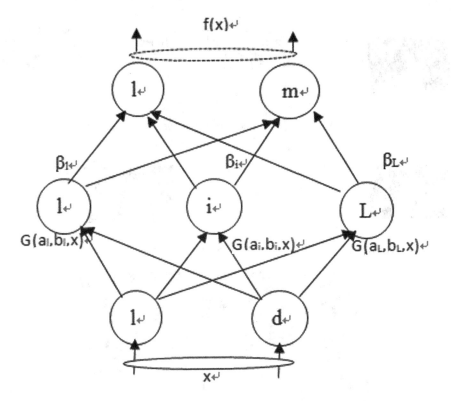

Fig. 5. ELM architecture: single layer feed forward neural network

Those whose values are between T_h and T_l are discarded. The thresholds are statically computed by the formula as:

$$\begin{cases} T_h = \min[(1+a)*(M+b), 0.9] \\ T_l = \min[(1-a)*(M+b), 0.1] \end{cases} \tag{1}$$

where M is the mean grayscale value of the coarse saliency map; a is the scope ratio parameter; b is the saliency ratio parameter. In our experiments, a is generally set to be 0.8 and b is computed by 0.5 subtraction of the mean gray scale value of average map.

The classification is based on a group of 16-dimensional feature of compactness. According with boundary prior assumption, we depart the superpixels into two groups, the center group and the boundary group. As the Fig. 7 illustrated, the areas which are masked by the transparent red color are the boundary group, and the left areas are the center group. For both groups, we compute compactness values on both color and texture.

The method to compute the compactness features is similar to the work [12]. For a superpixel v_i, the compactness degree Θ are computed by the Eq. 2 where W is a weight factor, θ is the scatter degree, c is an element in the set of { L, a, b, Lab}. The compactness degree is the summation of weighted scatter degree.

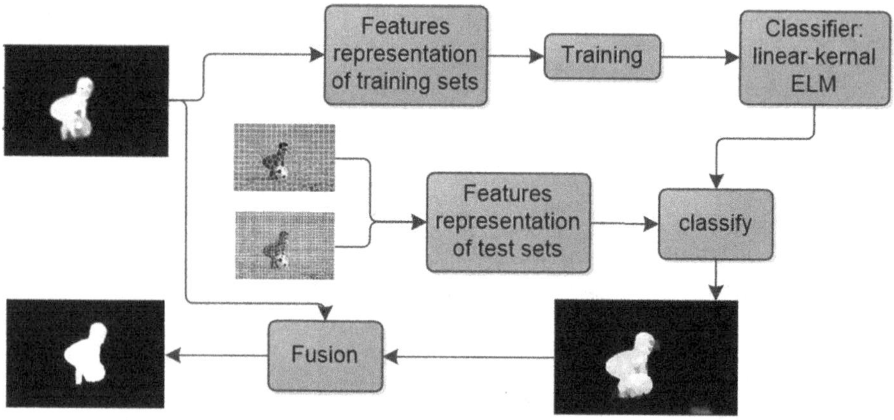

Fig. 6. Procedure of saliency refinement

Fig. 7. The Center group and the boundary group

The weight value is computed according to the pattern of Gaussian functions by Eq. 4. The scatter degree for v_i is computed by Eq. 3 which is the reciprocal form of weighted linear combination of spatial distance factor $D(p_i, p_j)$. We use the Euclidean distance to measure the spatial distance.

$$\Theta^c(v_i) = \sum_{j=1}^{N} W^c(v_i, v_j) \cdot \theta^c(v_j) \tag{2}$$

$$\theta^c(v_j) = \frac{1}{\sum_{j=1}^{N} D(p_i, p_j) \cdot W^c(v_i, v_j)} \tag{3}$$

$$W^c(v_i, v_i) = \frac{1}{\Omega_i} \exp(-\frac{(D(c_i, c_j)^2}{2\sigma^2}) \tag{4}$$

The algorithm computes the weight values for individual color channel and the whole color space. For color weights computation, we simply use the mean color value of the superpixel to represent v_i and v_j. For texture weights computation, we use a uniform LBP histogram stated on the superpixel. It is commonly a 59-dimensional vector.

The 16 features of superpixels and their saliency labels gotten from the coarse saliency map are all sent to the ELM as input. According to our approach, an ELM classier is trained. The output of ELM is confidence value of each superpixel and it is used as a measure of saliency. We perform the process in multiple scales of superpixels. In our experiments, most of the images in dataset are in about 400 × 400 pixels resolution. Thus we choose three numbers for superpixels to control the scales; they are [150, 250, 750].

3.4   Image Synthesis

The refined saliency map is generated by integrating several rened saliency maps. For each saliency map, we first perform a smoothing operation through the GrabCut algorithm. The GrabCut algorithm [24] was designed by Rother et al. from Microsoft Research Cambridge, UK. It extracts foreground using iterated graph-cuts. The GrabCut requires a mask map in which all pixel values are set with a value in 0, 1, 2, 3 which means foreground, background, probably foreground, and probably background. We compute two thresholds by Eq. 1 to set the mask values using the following rules:

$$M(i,j) = \begin{cases} 0, & M(i,j) < T_l \\ 1, & M(i,j) > T_h \\ 2, & T_l < M(i,j) < (T_h + T_l)/2 \\ 3, & T_h > M(i,j) > (T_h + T_l)/2 \end{cases} \tag{5}$$

The quality of the saliency map should be evaluated before synthesis. This step is carried out based on two hypotheses: the compactness and the variances. It is observed that salient regions are usually distributed close and the pixels in saliency map probably have high variance. Thus we compute the assessment weight $\psi(S_i)$ by

$$\psi(S_i) = norm(\frac{1}{\sum_{i=1}^{H} \sum_{j=1}^{W} S(i,j)D(S(i,j), S_{GC})^2}) \tag{6}$$

where $S(i,j)$ is the pixel with position(i,j), S_{GC} is the saliency gravity, and $norm()$ is the normalization operation. Multiple saliency maps then are integrated in a linear weighted sum form of involved saliency map according to its corresponding assessment score.

4 Experiment Result and Analysis

4.1 Experiments Setup

To evaluate the effectiveness of the proposed method, we perform experiments on five different types of datasets publicly available, which are MSRA10K [9], ECSSD [30], PASCAL-S [21], DUT-OMRON [31], and HKU-IS [19] datasets. All of the dataset are annotated by people and have pixel-wise groundtruth. The feature of each dataset is listed in Table 1.

Table 1. The description of the five datasets

Dataset	Size	Source	Description
MSRA10K	10000	MSRA	Only one salient object
ECSSD	1000	Internet	Structurally complex images
DUT-OMRON	5168	Dalian University of Technology	Controversial annotations
PASCAL-S	850	VOC2010	Contain 12 subjects
HKU-IS	4447	Hong Kong University	Contain multiple salient objects

4.2 Evaluation Metrics

In the comparison experiments, we use Precision-Recall (PR) curves, $F_{0.3}$ metric, AUC, and Mean Absolute Error (MAE) to evaluate the proposed method. With saliency value in the range [0,255], the P-R curve is obtained by generating the binary map when the threshold varies from 0 to 255, and comparing the binary result with the ground-truth. The F-measure is defined as,

$$F = \frac{(1 + \beta)^2 \cdot Precision \cdot Recall}{\beta^2 \cdot Precision + Recall} \tag{7}$$

where β^2 is set to be 0.3 as suggested in [1]. AUC is the area under ROC. As indicated in [11], PR curves, AUC and F metric provide a quantitative evaluation, while MAE provides a better estimate of the dissimilarity between the saliency map and binary ground truth. The MAE computes the average pixel-wise difference between saliency map S and the binary ground truth G.

$$MAE = \frac{1}{W \times H} \sum_{x=1}^{W} \sum_{y=1}^{H} |S(x, y) - G(x, y)| \tag{8}$$

4.3 Performance Comparison

In this subsection, we evaluate the proposed method on MSRA10K, ECSSD, OMRON, PASCAL-S, HKU-IS dataset and compare the performance with 7 state-of-the-art algorithms, including HC [9], GC [8], GMR [31], PISA [25],

Table 2. Comparison between our CELM and other methods

Dataset	Metrics	HC	GC	GMR	PISA	QCUT	DRFI	DISC	Ours
MSRA10K	MAE	0.215	0.150	0.126	0.101	0.116	0.126	0.044	0.039
	F_β	0.677	0.766	0.846	0.861	0.873	0.877	0.940	0.952
ECSSD	MAE	0.331	0.233	0.187	0.150	0.173	0.170	0.119	0.121
	F_β	0.455	0.598	0.738	0.766	0.766	0.782	0.799	0.838
OMRON	MAE	0.310	0.217	0.189	0.141	0.126	0.150	0.119	0.095
	F_β	0.381	0.496	0.612	0.631	0.684	0.664	0.659	0.711
PASCAL-S	MAE	0.354	0.264	0.233	0.192	0.198	0.207	0.156	0.151
	F_β	0.423	0.536	0.643	0.657	0.683	0.688	0.725	0.744
HKU-IS	MAE	0.292	0.215	0.175	0.128	0.143	0.145	0.103	0.087
	F_β	0.493	0.580	0.712	0.752	0.766	0.776	0.784	0.834

QCUT [3], DRFI [17], and DISC [6]. For fair comparison, we use the original source code provided by authors or the detection results provided by the corresponding literatures. The quantitative comparisons are shown in Figs. 8 and 9, and Table 1. We train the coarse CNN model on MSRA10K and test the model on other datasets to prove the generalization performance (Table 2).

The existing CNN based model DISC uses two CNN to train an end-to-end saliency detection model in a coarse to fine manner. Different from it, our model choose the ELM to refine the saliency map combining deep features and handicraft features. Our CELM based model improves the F-measure achieved by the DISC [6] by 1.3%, 4.9%. 7.9%, 2.6% and 6.8% respectively on MSRA10K,

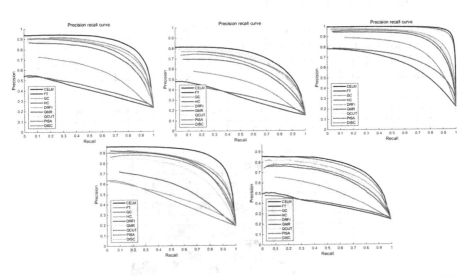

Fig. 8. Precision-Recall Curves on datasets, from left to right, up to down: ECSSD, OMRON, MSRA10K, HKU-IS, PASCAL-S

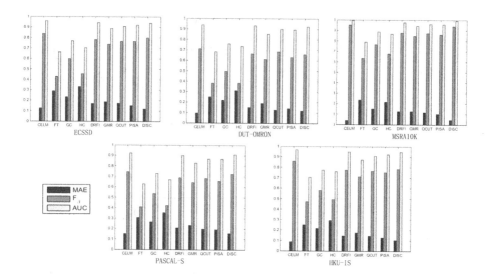

Fig. 9. MAE, F-measure and AUC values of compared methods on five datasets, from left to right, up to down: ECSSD, OMRON, MSRA10K, PASCAL-S, HKU-IS

ECSSD, OMRON, PASCAL-S, HKU-IS. At the same time, our CELM based model lowers the MAE by 11.4%, 20.2%, 3.2%, 15.5% on MSRA10K, OMRON, PASCAL-S and HKU-IS. Our method outperforms all the seven previous methods based on the three evaluation metrics.

4.4 Analysis of Propose Method

For most images, the approach achieves good salient results as Fig. 1 shows. However, the final saliency map heavily depended on the results gotten from CNN. Our CELM based model may fail if the coarse CNN output totally wrong region as Fig. 10 illustrated. Because the saliency map generated by CNN indicates the location of the salient object, if this saliency map indicates the wrong location of the salient map, the detection will failed. To further improve the performance, we could replace the AlexNet CNN with the state-of-the-art CNN network,

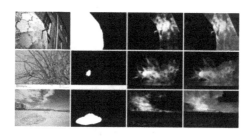

Fig. 10. Failed cases, from left to right are Input, groundtruth, coarse map, CELM

i.e. the DCL network [20]. The DCL network is proposed in CVPR 2016. It is one of the best models to detect salient object.

We use the same refining and synthesis framework as Subsects. 3.3 and 3.4 described. We compare our proposed CELM-DCL method with the original CELM, the DCL, and the other seven state of the art methods on three datasets: HKU-IS(1446), ECSSD and OMRON as Fig. 11 illustrated. This is because the DCL only provides results on these three datasets, and only 1446 results are provided in the HKU-IS dataset. For quantitative evaluation, we show comparison results with PR curves and F-measure scores in Table 3 and Fig. 12.

Table 3. Comparison between CELM-DCL and other methods

Dataset	Metrics	GMR	PISA	QCUT	DRFI	DISC	CELM	DCL	CELM-DCL
ECSSD	MAE	0.187	0.150	0.173	0.170	0.119	0.126	0.068	0.067
	F_β	0.738	0.766	0.766	0.782	0.799	0.838	0.902	0.907
OMRON	MAE	0.189	0.141	0.126	0.150	0.119	0.116	0.080	0.075
	F_β	0.612	0.631	0.684	0.664	0.659	0.711	0.756	0.768
HKU-IS(1446)	MAE	0.175	0.128	0.143	0.145	0.103	0.102	0.048	0.057
	F_β	0.712	0.752	0.766	0.776	0.784	0.834	0.907	0.913

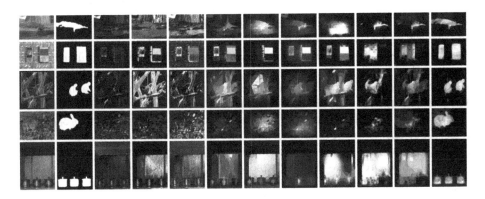

Fig. 11. Results from HKU-IS dataset by different methods, from left to right: input, groundtruth, FT, GC, HC, DRFI, GMR, QCUT, PISA, DISC, CELM, CELM-DCL

In Table 3, the best results are marked red and the second ones are marked green. Our CELM-DCL based model improves the F-measure achieved by the DCL by 0.5%, 1.6% and 0.7% respectively on ECSSD, OMRON, and HKU-IS. At the mean-time, the CELM-DCL based model lowers the MAE by 1.5% and 6.25% on ECSSD and OMRON.

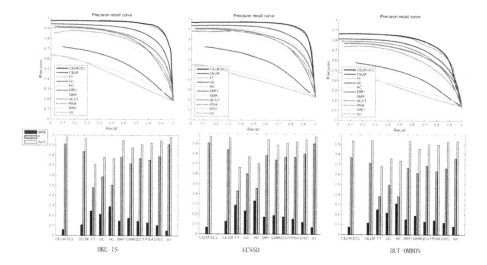

Fig. 12. Comparison of GMR, PISA, QCUT, DRFI, DISC, CELM, CDL, CELM-CDL on HKU-IS, ECSSD and DUT-OMRON; up: Precision-Recall curves; down: MAE, F-measure, and AUC

5   Conclusions

In this paper, we propose a saliency detection framework through the combination of CNN and ELM. To carry it out, we statically label the coarse map, extract the compactness features on two groups, and synthesis the multiple saliency map based on their qualities. Experiments show that the CELM get excellent salient results on all five datasets. Further improvements are also made by replacing the AlexNet with the network that the DCL uses. Extent experiments prove that the CELM-DCL outperforms the state of the art. It is proved that our approach can be used not only as a complete method, but also as a lifting method for current CNN based method. Future works include extending our work to a pixel-wise and accuracy approach as well as exploring better CNN networks.

References

1. Achanta, R., Shaji, A., Smith, K., Lucchi, A., Fua, P., Susstrunk, S.: Slic superpixels compared to state-of-the-art superpixel methods. IEEE Trans. Patt. Anal. Mach. Intell. **34**(11), 2274 (2012)
2. Avidan, S., Shamir, A.: Seam carving for content-aware image resizing. ACM Trans. Graph. **26**(3), 10 (2007)
3. Aytekin, Ç., Ozan, E.C., Kiranyaz, S., Gabbouj, M.: Visual saliency by extended quantum cuts. In: IEEE International Conference on Image Processing (2015)
4. Borji, A.: Boosting bottom-up and top-down visual features for saliency estimation. In: IEEE Conference on Computer Vision and Pattern Recognition, pp. 438–445 (2012)

5. Chen, C., Li, S., Qin, H., Hao, A.: Structure-sensitive saliency detection via multilevel rank analysis in intrinsic feature space. IEEE Trans. Image Process. **24**(8), 2303–16 (2015)
6. Chen, T., Lin, L., Liu, L., Luo, X., Li, X.: DISC: deep image saliency computing via progressive representation learning. IEEE Trans. Neural Netw. Learn. Syst. **27**(6), 1135 (2016)
7. Cheng, M.M., Mitra, N.J., Huang, X., Torr, P.H., Hu, S.M.: Global contrast based salient region detection. IEEE Trans. Patt. Anal. Mach. Intell. **37**(3), 569–582 (2015)
8. Cheng, M.M., Warrell, J., Lin, W.Y., Zheng, S., Vineet, V., Crook, N.: Efficient salient region detection with soft image abstraction, pp. 1529–1536 (2013)
9. Cheng, M.M., Zhang, G.X., Mitra, N.J., Huang, X., Hu, S.M.: Global contrast based salient region detection. In: IEEE Conference on Computer Vision and Pattern Recognition, pp. 409–416 (2011)
10. He, S., Lau, R.W., Liu, W., Huang, Z., Yang, Q.: SuperCNN: a superpixelwise convolutional neural network for salient object detection. Int. J. Comput. Vis. **115**, 330–344 (2015)
11. Hornung, A., Pritch, Y., Krahenbuhl, P., Perazzi, F.: Saliency filters: contrast based filtering for salient region detection. In: Computer Vision and Pattern Recognition, pp. 733–740 (2012)
12. Hu, P., Wang, W., Zhang, C., Lu, K.: Detecting salient objects via color and texture compactness hypotheses. IEEE Trans. Image Process. **25**(10), 4653–4664 (2016)
13. Huang, G.B., Zhou, H., Ding, X., Zhang, R.: Extreme learning machine for regression and multiclass classification. IEEE Trans. Syst. Man Cybern. Part B **42**(2), 513–529 (2012)
14. Huang, G.B., Ding, X., Zhou, H.: Optimization Method Based Extreme Learning Machine for Classification. Elsevier Science Publishers B.V., Amsterdam (2010)
15. Huang, G.B., Zhu, Q.Y., Siew, C.K.: Extreme learning machine: theory and applications. Neurocomputing **70**(1), 489–501 (2006)
16. Itti, L., Koch, C.: Computational modelling of visual attention. Nat. Rev. Neurosci. **2**(3), 194 (2001)
17. Jiang, H., Wang, J., Yuan, Z., Wu, Y., Zheng, N., Li, S.: Salient object detection: a discriminative regional feature integration approach. In: IEEE Conference on Computer Vision and Pattern Recognition, pp. 2083–2090 (2013)
18. Krizhevsky, A., Sutskever, I., Hinton, G.E.: ImageNet classification with deep convolutional neural networks. In: International Conference on Neural Information Processing Systems, pp. 1097–1105 (2012)
19. Li, G., Yu, Y.: Visual saliency based on multiscale deep features. In: Computer Vision and Pattern Recognition, pp. 5455–5463 (2015)
20. Li, G., Yu, Y.: Deep contrast learning for salient object detection, pp. 478–487 (2016)
21. Li, Y., Hou, X., Koch, C., Rehg, J.M., Yuille, A.L.: The secrets of salient object segmentation. In: IEEE Conference on Computer Vision and Pattern Recognition, pp. 280–287 (2014)
22. Lin, L., Wang, X., Yang, W., Lai, J.H.: Discriminatively trained and-or graph models for object shape detection. IEEE Trans. Patt. Anal. Mach. Intell. **37**(5), 959–72 (2015)
23. Ma, Y.F., Lu, L., Zhang, H.J., Li, M.: A user attention model for video summarization. In: Tenth ACM International Conference on Multimedia, pp. 533–542 (2002)

24. Rother, C., Kolmogorov, V., Blake, A.: GrabCut: interactive foreground extraction using iterated graph cuts. ACM Trans. Graph. (TOG) **23**(3), 309–314 (2004)
25. Wang, K., Lin, L., Lu, J., Li, C., Shi, K.: PISA: pixelwise image saliency by aggregating complementary appearance contrast measures with edge-preserving coherence. IEEE Trans. Image Process. **24**(10), 3019–3033 (2015)
26. Wang, L., Lu, H., Xiang, R., Yang, M.H.: Deep networks for saliency detection via local estimation and global search. In: IEEE Conference on Computer Vision and Pattern Recognition, pp. 3183–3192 (2015)
27. Wang, L., Wang, L., Lu, H., Zhang, P., Ruan, X.: Saliency detection with recurrent fully convolutional networks. In: Leibe, B., Matas, J., Sebe, N., Welling, M. (eds.) ECCV 2016. LNCS, vol. 9908, pp. 825–841. Springer, Cham (2016). https://doi.org/10.1007/978-3-319-46493-0_50
28. Wu, H., Li, G., Luo, X.: Weighted attentional blocks for probabilistic object tracking. Vis. Comput. **30**(2), 229–243 (2014)
29. Wu, R., Yu, Y., Wang, W.: SCaLE: supervised and cascaded laplacian eigenmaps for visual object recognition based on nearest neighbors. In: Computer Vision and Pattern Recognition, pp. 867–874 (2013)
30. Yan, Q., Xu, L., Shi, J., Jia, J.: Hierarchical saliency detection. In: Computer Vision and Pattern Recognition, pp. 1155–1162 (2013)
31. Yang, C., Zhang, L., Lu, H., Xiang, R., Yang, M.H.: Saliency detection via graph-based manifold ranking. In: Computer Vision and Pattern Recognition, pp. 3166–3173 (2013)
32. Yang, J.: Top-down visual saliency via joint CRF and dictionary learning. In: IEEE Conference on Computer Vision and Pattern Recognition, pp. 2296–2303 (2012)
33. Zhao, R., Ouyang, W., Li, H., Wang, X.: Saliency detection by multi-context deep learning. In: Computer Vision and Pattern Recognition, pp. 1265–1274 (2015)

PreNet: Parallel Recurrent Neural Networks for Image Classification

Junbo Wang[1,2], Wei Wang[1], Liang Wang[1,2], and Tieniu Tan[1(✉)]

[1] Center for Research on Intelligent Perception and Computing,
National Laboratory of Pattern Recognition, Institute of Automation,
Chinese Academy of Sciences, Beijing, China
{junbo.wang,wangwei,wangliang,tnt}@nlpr.ia.ac.cn
[2] University of Chinese Academy of Sciences (UCAS), Beijing, China

Abstract. Convolutional Neural Networks (CNNs) have made outstanding achievements in computer vision, e.g., image classification and object detection, by modelling the receptive field of visual cortex with convolution and pooling operations. However, CNNs have ignored to model the long-range spatial contextual information in images. It has long been believed that recurrent neural networks (RNNs) can model temporal sequences well by virtue of horizontal connections, and have been successfully applied in speech recognition and language modelling. In this paper, we propose a hierarchical parallel recurrent neural network (PreNet) to model spatial context for image classification. In this network, when transforming the whole image into sequences in four directions, we adopt the way of row-by-row/column-by-column scanning instead of traditional pixel-by-pixel scanning, for the convenience of fast convolution implementation. Following the recurrent network, a max-pooling operation is used to reduce the dimensionality of the obtained feature maps. The resulting PreNet can be easily paralleled on GPUs. We evaluate the proposed PreNet on two public benchmark datasets: MNIST and CIFAR-10. The proposed model can achieve the state-of-the-art classification performance, which demonstrates the advantage of PreNet over many comparative CNN structures.

Keywords: Convolutional neural networks · Image classification Recurrent neural networks

1 Introduction

CNNs have become very popular in the field of deep learning, and have been successfully applied in computer vision, e.g., image classification [22], object detection [6], and semantic segmentation [26]. These successes are generally considered to be attributed to modelling the receptive field of visual cortex with local convolution operation and achieving translation invariance with max-pooling operation. However, several researchers argue that variant poolings may lose

© Springer Nature Singapore Pte Ltd. 2017
J. Yang et al. (Eds.): CCCV 2017, Part II, CCIS 772, pp. 461–473, 2017.
https://doi.org/10.1007/978-981-10-7302-1_38

important location information [33], and local convolution can not capture the global spatial context in an image.

Different from the feedforward connections of convolutional neural networks, RNNs have horizontal feedback connections, which can model the long-range dependency in sequential data. With the capability of sequence modelling, RNNs with long short-term memory [12] have achieved great success in machine translation [3], question and answering [19], video super-resolution [16,17] and speech recognition [10]. Recently, more studies have been devoted to improving the memory capacity of RNNs for modelling longer-range contextual information, e.g., neural turing machine [14] and associative long short-term memory [5].

In this paper, we propose a hierarchical parallel recurrent neural network to model spatial context for image classification. Instead of using traditional recurrent neural networks in a pixel-by-pixel scanning order which leads to very long sequences (e.g., 10,000-length sequence for 100 × 100 pixel image), we adopt parallel recurrent neural networks to scan images in the way of row-by-row/column-by-column. Parallel recurrent neural networks not only shorten the sequence length to 100 for a 100 × 100 pixel image, but also use fast convolution operation to propagate information. Following parallel recurrent neural networks, a max-pooling operation is used to reduce the dimensionality of the obtained features maps. Through stacking parallel recurrent neural network layer and max-pooing layer, a hierarchical parallel recurrent neural network (PreNet) is proposed for image classification.

The most similar work to our PreNet is ReNet [35] which also replaces the convolution+pooling layer with four recurrent neural networks. However, our PreNet is much different from ReNet in many aspects. First, ReNet vertically sweeps the obtained feature maps based on the result of horizontal sweep, while PreNet recurrently handles the image sequences in four directions simultaneously. Second, ReNet sweeps the input from patch to patch in a pixel-by-pixel way, while PreNet sweeps the input in the way of row-by-row/column-by-column. Third, ReNet utilizes a fully connected operation at each recurrent time, while PreNet employs a efficient convolutional operation for fast parallel execution.

Finally, we evaluate the proposed PreNet on two widely-used benchmark datasets: MNIST and CIFAR-10. The experimental results demonstrate that the proposed model can achieve the state-of-the art classification performance on these datasets.

2   Model Description

In this part, we will first introduce traditional recurrent neural networks [11] and parallel recurrent neural networks, and then describe our proposed PreNet. Finally, we will compare the model structure between PreNet and the other two (ReNet and ConvNet).

2.1 Traditional Recurrent Neural Network

Different from common multilayer perception, RNN propagates information via horizontal feedback connections which can model the long-range dependency in sequential data, while multilayer perception passes on the activations of the neurons to the other neurons via feedforward connection. The input of the hidden layer in RNN includes both the input at current timestep and the hidden activations at previous timestep. Given a T length input sequence $\{x_t\}_{t=1,\cdots,T}$, where x_t is an input vector at timestep t, the hidden activations h_t and the output activations o_t at time $t(t = 1, \cdots, T)$ are computed as follows:

$$h_t = \sigma\left(x_t w_{xh} + h_{t-1} w_{hh} + b\right) \tag{1}$$

$$o_t = h_t w_{ho} \tag{2}$$

where w_{xh}, w_{hh}, and w_{ho} denote the network weights, b denotes the bias term, and σ denotes the activation function $\sigma\left(x\right) = \frac{1}{1+e^{-x}}$.

2.2 Parallel Recurrent Neural Network

Stollenga et al. [32] proposed a PyraMiD-LSTM model for biomedical volumetric image segmentation, which employs an elegant scanning way to considering each pixels entire spatio-temporal context. Besides, since the PyraMiD-LSTM uses a rather different scanning way, which is different from the traditional cuboid order of computations in MD-LSTM [13], it becomes very easy to parallelize.

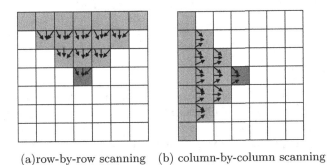

(a)row-by-row scanning   (b) column-by-column scanning

Fig. 1. The pyramidal connection topology of parallel recurrent neural network.

As we can see from Fig. 1, all inputs to all units are from four directions: left, right, up, or down. In Fig. 1, we adopt the way of row-by-row/column-by-column scanning instead of traditional pixel-by-pixel scanning, for the convenience of fast convolution implementation. For example, when the convolutional kernel size is 3, the stride is 1, and the elements in the second row (column) are the results of convolution of the first row (column) elements with the learned linear filter at the stride. Figure 1 just shows the scanning ways of from left to right (from up to down), and the computation from other two directions can be operated in the same way. Since all the elements in a whole grid row (or column) can be computed independently, the proposed model can be easy to parallelize.

Assuming that $X = \{x_{i,j}\}$ denotes the input image or the feature map from a previous layer, where $X \in R^{c \times h \times w}$, and c denotes the number of channels or the feature dimensionality, h denotes the height, and w denotes the width. Specifically, we split X into h rows, where $x_t \in R^{c \times 1 \times w}$ $(t = 1, 2, \ldots, h)$ is the t-th row of the input image. The input sequence length is h or w, the input vector at time t is x_t, the weights are w_{xh} and w_{hh}, the activations of the hidden layer are h^i ($i \in \{lr, rl, ud, du\}$), where i denotes the sweep direction, i.e., left-to-right, right-to-left, up-to-down, down-to-up, respectively. When the recurrent model computes the hidden output of the direction of up-to-down, the hidden output of other three directions can be gained simultaneously in the same way. After the results of four directions are gained, the final output can be obtained by summing them. As we use the padding operation in the convolution operation, the output feature maps can keep the same size as the input. Now, each vector in final output feature maps can represent corresponding features in the context of the whole image.

$$h_t^{lr} = \sigma \left(x_t^{lr} * w_{xh}^{lr} + h_{t-1}^{lr} * w_{hh}^{lr} + b^{lr} \right) \tag{3}$$

$$h_t^{rl} = \sigma \left(x_t^{rl} * w_{xh}^{rl} + h_{t-1}^{rl} * w_{hh}^{rl} + b^{rl} \right) \tag{4}$$

$$h_t^{ud} = \sigma \left(x_t^{ud} * w_{xh}^{ud} + h_{t-1}^{ud} * w_{hh}^{ud} + b^{ud} \right) \tag{5}$$

$$h_t^{du} = \sigma \left(x_t^{du} * w_{xh}^{du} + h_{t-1}^{du} * w_{hh}^{du} + b^{du} \right) \tag{6}$$

$$h = h^{lr} + h^{rl} + h^{ud} + h^{du} \tag{7}$$

2.3   Our PreNet

Based on the fact that parallel recurrent neural network employs an elegant scanning way to consider each pixels entire spatio-temporal context and it is easy to parallelize, we propose a hierarchical parallel recurrent neural network (PreNet) to model spatial context for image classification. As we can see from the above section, the PreNet architecture can map an input image to an output feature map. Therefore, we can stack multiple PreNet layers to make our network deeper, which can extract more complex and useful features of the input image. In our experiment, we first directly stack multiple PreNet layers followed by fully connected layers to form a deep network for image classification. However, the computational cost is high because the size of the feature map in all PreNet layers is constant. To reduce the dimensionality of the obtained features maps, a max-pooling operation is added after the PreNet layers. The proposed hierarchy PreNet architecture is shown in Fig. 2, and the main process is as follows.

Given input image, it first enters one PreNet layer which comprises of four recurrent neural networks from four directions respectively, and every recurrent

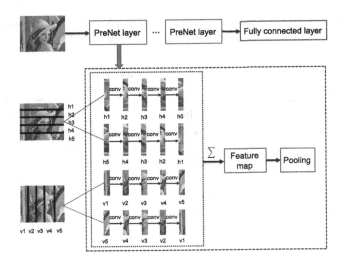

Fig. 2. The proposed Parallel Recurrent Neural Network (PreNet).

neural network handles the input image by the way of row-by-row/column-by-column scanning. Then it enters a pooling layer which maps the input feature map to a smaller size than the input. Let us denote function φ by the combination of the PreNet layer and the pooling layer, so we can stack multiple φ to form a hierarchy PreNet network. Finally, it is followed by two fully connected layers whose activation functions are rectify and softmax, respectively.

2.4   Model Compariation

In this section, we will compare the proposed PreNet with conventional convolutional neural networks (ConvNet) and ReNet. Since the comparison between ReNet and ConvNet is described in [35], we just explain the similarities and differences between PreNet and the other two here.

The comparison between PreNet and ReNet

- Both networks sweep the input image or feature map from four same directions. ReNet handles the sequences in the horizontal two directions and vertical two directions sequentially, while PreNet handles the sequences in four directions simultaneously. PreNet is obviously easier to parallelize than ReNet.
- ReNet sweeps the input from patch to patch in a pixel-by-pixel way, while PreNet handles the input in the way of row-by-row/column-by-column scanning. Both the way of scanning can ensure that each feature activation gains contextual information with respect to the whole image.
- Both networks can reduce the dimensionality of the feature map, which results in lower computational cost. However, ReNet reduces the dimension of the input feature map by dividing it into many non-overlapping patches and the

existence of learned lateral connections, but PreNet reduces the dimension of the input feature map by a max-pooling operation.

The comparison between PreNet and ConvNet

- Both architectures employ a set of filters in the input image or the feature map from the layer below. Nevertheless, the information of PreNet is transmitted via lateral connections which propagate across the whole image, while ConvNet only captures local information in a fixed input size. The lateral connections make the model extract more contextual information at each layer.
- They both have the max-pooling layer. For ConvNet, pooling operations are responsible for the translation invariant property. But for PreNet, pooling operations are mainly responsible for reducing dimensionality.

3   Experiments

To demonstrate the effectiveness of the proposed PreNet, we apply it to image classification on two publicly available datasets.

3.1   Datasets

The two evaluation datasets and corresponding experimental protocols are described as follows.

MNIST. The dataset [23] is composed of binary images of handwritten digits, and has been widely used for training various image processing systems. The images are all gray with a size of 28 × 28, each of which represents a handwritten digit from 0 to 9. In this experiment, we follow the standard split and use 50,000, 10,000 and 10,000 images for training, validation and test, respectively.

CIFAR-10. The dataset [21] consists of 60000 color images. These images have a size of 32 × 32, each of which belongs to one of 10 classes including airplane, automobile, bird, cat, deer, dog, frog, horse, ship and truck, totally 6000 images per class. Following the standard procedure, we split the dataset into 40,000 training, 10,000 validation and 10,000 test images. We normalize each pixel to have zero-mean and unit-variance across all training samples.

3.2   Pre-processing and Data Augmentation

It is believed that data preprocessing can usually have a significant impact on the final performance of a model [20]. To make fair comparisons, we exploit different pre-processing strategies for different datasets. For the MNIST dataset, we normalize each pixel of an image to the range of $[0, 1]$. For the CIFAR-10 dataset, we perform z-score standardization to make image pixels have zero-mean and unit-variance.

According to [22], data augmentation can always reduce overfitting and improve final performance. To extensively compare our model with others, we conduct our experiment using non-augmentation and augmentation, respectively. In the context of data augmentation, we mainly exploit three kinds of transformations [24,35] including flipping, shifting, random cropping.

3.3   Model Parameter Setup

In Table 1, we illustrate parameter settings of the proposed PreNet on the MNIST and CIFAR-10 datasets. The main parameters include the number of PreNet layers N_{pre}, their corresponding feature dimensionality d_{pre} and kernel size k_{pre}, the number of pooling layers N_{pool} and their corresponding kernel size k_{pool}, the number of fully connected layers N_{fc} and their corresponding feature dimensionality d_{fc}, and the activation function f_{fc}. All our experiments are performed on a ubuntu computer with an NVIDIA Tesla K40 GPU.

Table 1. Model parameters used in the experiments.

Parameters	MNIST	CIFAR-10
N_{pre}	2	2
d_{pre}	32	48
k_{pre}	7	5
N_{pool}	2	2
k_{pool}	2	2
N_{fc}	1	1
d_{fc}	128	256
f_{fc}	$max(0,x)$	$max(0,x)$
$Flipping$	No	Yes
$Shifting$	Yes	Yes
$RandomCropping$	Yes	Yes

3.4   Training

We apply stochastic gradient descent (SGD) with nesterov momentum [34] to train the networks. At the same time, to release the exploding gradient problem [2], we employ the gradient norm clipping strategy [28]. To avoid overfitting during training, in addition to the data augmentation, we perform dropout after all PreNet layers and fully connected layers. In addition, we use batch normalization [18] to accelerate the convergence of the proposed model. During training, we choose the best model which minimizes the classification loss function on the validation set. We set the learning rate 0.01 at start and drop it by a factor of 10 after every 500 epochs. The momentum term is 0.9, the batch size is 100, and all weights are initialized according to a uniform distribution.

3.5   Results and Analysis

We compare the proposed PreNet with several state-of-the-art models on the MNIST and CIFAR-10 datasets in Tables 2 and 3, respectively. All results to be compared in Tables 2 and 3 are from [35]. On the MNIST dataset, our proposed PreNet model can outperform most state-of-the-art models. On the CIFAR-10 dataset, the PreNet model performs comparably to most state-of-the-art models. It is worth mentioning that PreNet performs 0.16% and 1.35% better than ReNet on MNIST and CIFAR-10, respectively. Although the performance of our model doesn't outperform some state-of-the-art models on CIFAR-10, there are some reasons behind it. As we all know, the digital structure on MNIST is simple, while the intra-class spatial context variation on CIFAR-10 is complicated.

To verify the advantage of PreNet further, we compare PreNet with ConvNet on the parameters of the same order of magnitude on the CIFAR-10 dataset. Specifically, we use the same number of layers for PreNet and ConvNet, and compare their results in Table 4, where the ConvNet-one means one layer convolution, and PreNet-one means one layer PreNet, and the meanings of "-two" are similar. In this experiment, we set the number of fully connected layers as 1, and the number of neurons as 128. All the results in Table 4 are computed in the same computing environment, and data augmentation is not used in these experiments. From the experimental results, we can find that PreNets can achieve

Table 2. Comparison with existing models on MNIST.

Method	Test error
DropCNN [36]	0.28%
PreNet	**0.29%**
S-SCNN [9]	0.31%
DBSN [4]	0.35%
CKN [27]	0.39%
DSN [24]	0.39%
SCNN [29]	0.40%
FMP [8]	0.44%
Maxout [7]	0.45%
ReNet [35]	0.45%
NIN [25]	0.47%
COSFIRE [1]	0.52%

Table 3. Comparison with existing models on CIFAR-10.

Method	Test error
FMP [8]	4.5%
S-SCNN [9]	6.28%
NIN [25]	8.8%
Maxout [7]	9.35%
PV-Maxout [31]	9.39%
BO [30]	9.5%
PreNet	**11%**
DCNN [22]	11%
DropCNN [36]	11.10%
ReNet	12.35%
SP-CNN [37]	15.13%
PCFD [15]	15.6%

Table 4. Comparison with ConvNet models on CIFAR-10 (without data augmentation).

Method	Test accuracy
ConvNet-one	74.66%
PreNet-one	**77.5%**
ConvNet-two	80.92%
PreNet-two	**83.57%**

Table 5. The experimental results of PreNet on the CIFAR-10 by setting different parameters.

Parameter	Test accuracy
$N_{pre} = 1$	71.39%
$N_{pre} = 2$	78.15%
$d_{pre} = 16$	69.23%
$d_{pre} = 32$	71.39%
$d_{pre} = 48$	72.73%
$d_{pre} = 64$	72.80%
$k_{pre} = 3$	70.76%
$k_{pre} = 5$	71.39%
$k_{pre} = 7$	71.41%
$d_{fc} = 32$	70.12%
$d_{fc} = 64$	71.39%
$d_{fc} = 128$	72.98%
$d_{fc} = 256$	73.79%
$d_{fc} = 512$	74.28%

about 3% better than ConvNets. These results demonstrate the advantage of PreNet compared with convolutional neural networks further.

In the following, we testify how the parameters of the proposed PreNet affect the final performance. In a nutshell, we study a parameter by changing its value while fixing other parameters. In Table 5, we present the experimental results of different parameter setups on the CIFAR-10 dataset. Similarly, all the results in Table 5 are gained in the same computing environment, and data augmentation is not used in these experiments. The meaning of these parameters are explained in the previous section. The first column in table indicates four kinds of parameters respectively: the number of PreNet layers, the feature dimensionality of each PreNet layer, the kernel size of each PreNet layer, the feature dimensionality of each fully connected layer. The second column is their corresponding results.

In our experiments, we set different values for these parameters respectively. By increasing the values of N_{pre}, d_{pre}, k_{pre} and d_{fc}, the performance of the model gradually gets better. The situation is obviously consistent with our thought. As we do not fine tuning the learning rate, the model performance does not increase obviously when changing the values of some parameters such as k_{pre}. What calls for special attention is that the model performance cannot increase too much when the value of the parameter becomes too large because of overfitting.

Finally, we show the feature maps of the convolutional layers in PreNet and ConvNet, respectively. In Figs. 3 and 4, there are some feature maps of PreNet and ConvNet corresponding to a bird and a horse. The layer1 means the first

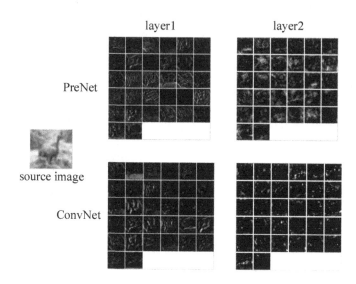

Fig. 3. The feature maps of PreNet and ConvNet on a bird image.

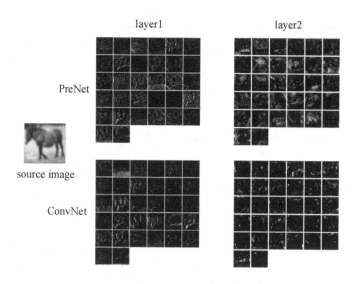

Fig. 4. The feature maps of PreNet and ConvNet on a horse image.

layer (PreNet or ConvNet) map, and the layer2 means the second layer (PreNet or ConvNet) map. As we can see from the Figs. 3 and 4, PreNet can capture richer context information compared to the little contour information of ConvNet. This is also consistent with our original motivation.

4 Conclusion

In this paper, we have proposed a hierarchical parallel recurrent neural network (PreNet) to model spatial context for image classification. The main idea is to handle the input image in the way of row-by-row/column-by-column scanning from four different directions by recurrent convolution operation. On the one hand, this structure makes the model easily capture global context information of the object in the image. On the other hand, it naturally facilitates parallel computing of the model. Our experimental results have shown the advantage of PreNet compared to common convolution neural networks and ReNet.

To further extend our work, there are several things worth exploring: First, the recurrent unit of PreNet is just conventional RNN instead of LSTM. It is a fact that LSTM has a powerful ability of modeling a long sequence, which can capture long-range memory about contextual information. It can be used in more complicated datasets if LSTM is applied to the PreNet model. Second, given that ReNet sweeps the input from patch to patch in a pixel-by-pixel way, a variant of PreNet can be explored in this respect. That is to say, a variant of PreNet can scan the input from patch to patch (e.g., slice row to slice row) in the same way as the original work principle.

Acknowledgements. This work is jointly supported by National Key Research and Development Program of China (2016YFB1001000), and National Natural Science Foundation of China (61525306, 61633021).

References

1. Azzopardi, G., Petkov, N.: Trainable cosfire filters for keypoint detection and pattern recognition. IEEE Trans. Pattern Anal. Mach. Intell. 35(2), 490–503 (2013)
2. Bengio, Y., Simard, P., Frasconi, P.: Learning long-term dependencies with gradient descent is difficult. IEEE Trans. Neural Netw. 5(2), 157–166 (1994)
3. Cho, K., Van Merriënboer, B., Gulcehre, C., Bahdanau, D., Bougares, F., Schwenk, H., Bengio, Y.: Learning phrase representations using RNN encoder-decoder for statistical machine translation. arXiv preprint arXiv:1406.1078 (2014)
4. Ciresan, D.C., Meier, U., Gambardella, L.M., Schmidhuber, J.: Deep big simple neural nets excel on handwritten digit recognition. CoRR abs/1003.0358 (2010). http://arxiv.org/abs/1003.0358
5. Danihelka, I., Wayne, G., Uria, B., Kalchbrenner, N., Graves, A.: Associative long short-term memory. arXiv preprint arXiv:1602.03032 (2016)
6. Girshick, R., Donahue, J., Darrell, T., Malik, J.: Rich feature hierarchies for accurate object detection and semantic segmentation. In: Proceedings of the IEEE Conference on Computer Vision and Pattern Recognition, pp. 580–587 (2014)

7. Goodfellow, I.J., Warde-Farley, D., Mirza, M., Courville, A., Bengio, Y.: Maxout networks. arXiv preprint arXiv:1302.4389 (2013)
8. Graham, B.: Fractional max-pooling. arXiv preprint arXiv:1412.6071 (2014)
9. Graham, B.: Spatially-sparse convolutional neural networks. arXiv preprint arXiv:1409.6070 (2014)
10. Graves, A., Mohamed, A.R., Hinton, G.: Speech recognition with deep recurrent neural networks. In: 2013 IEEE International Conference on Acoustics, Speech and Signal Processing (ICASSP), pp. 6645–6649. IEEE (2013)
11. Graves, A.: Supervised sequence labelling. In: Graves, A. (ed.) Supervised Sequence Labelling with Recurrent Neural Networks. SCI, vol. 385, pp. 5–13. Springer, Heidelberg (2012). https://doi.org/10.1007/978-3-642-24797-2_2
12. Graves, A., Schmidhuber, J.: Framewise phoneme classification with bidirectional LSTM and other neural network architectures. Neural Netw. **18**(5), 602–610 (2005)
13. Graves, A., Schmidhuber, J.: Offline handwriting recognition with multidimensional recurrent neural networks. In: Advances in Neural Information Processing Systems, pp. 545–552 (2009)
14. Graves, A., Wayne, G., Danihelka, I.: Neural turing machines. arXiv preprint arXiv:1410.5401 (2014)
15. Hinton, G.E., Srivastava, N., Krizhevsky, A., Sutskever, I., Salakhutdinov, R.R.: Improving neural networks by preventing co-adaptation of feature detectors. arXiv preprint arXiv:1207.0580 (2012)
16. Huang, Y., Wang, W., Wang, L.: Bidirectional recurrent convolutional networks for multi-frame super-resolution. In: Advances in Neural Information Processing Systems, pp. 235–243 (2015)
17. Huang, Y., Wang, W., Wang, L.: Video super-resolution via bidirectional recurrent convolutional networks. IEEE Trans. Pattern Anal. Mach. Intell. (2017)
18. Ioffe, S., Szegedy, C.: Batch normalization: accelerating deep network training by reducing internal covariate shift. arXiv preprint arXiv:1502.03167 (2015)
19. Iyyer, M., Boyd-Graber, J.L., Claudino, L.M.B., Socher, R., Daume III, H.: A neural network for factoid question answering over paragraphs. In: EMNLP, pp. 633–644 (2014)
20. Kotsiantis, S., Kanellopoulos, D., Pintelas, P.: Data preprocessing for supervised leaning. Int. J. Comput. Sci. **1**(2), 111–117 (2006)
21. Krizhevsky, A., Hinton, G.: Learning multiple layers of features from tiny images (2009)
22. Krizhevsky, A., Sutskever, I., Hinton, G.E.: ImageNet classification with deep convolutional neural networks. In: Advances in Neural Information Processing Systems, pp. 1097–1105 (2012)
23. LeCun, Y., Bottou, L., Bengio, Y., Haner, P.: Gradient-based learning applied to document recognition. Proc. IEEE **86**(11), 2278–2324 (1998)
24. Lee, C.Y., Xie, S., Gallagher, P., Zhang, Z., Tu, Z.: Deeply-supervised nets. arXiv preprint arXiv:1409.5185 (2014)
25. Lin, M., Chen, Q., Yan, S.: Network in network. arXiv preprint arXiv:1312.4400 (2013)
26. Long, J., Shelhamer, E., Darrell, T.: Fully convolutional networks for semantic segmentation. In: Proceedings of the IEEE Conference on Computer Vision and Pattern Recognition, pp. 3431–3440 (2015)
27. Mairal, J., Koniusz, P., Harchaoui, Z., Schmid, C.: Convolutional kernel networks. In: Advances in Neural Information Processing Systems, pp. 2627–2635 (2014)
28. Pascanu, R., Mikolov, T., Bengio, Y.: On the difficulty of training recurrent neural networks. arXiv preprint arXiv:1211.5063 (2012)

29. Simard, P.Y., Steinkraus, D., Platt, J.C.: Best practices for convolutional neural networks applied to visual document analysis. In: Document Analysis and Recognition, p. 958. IEEE (2003)
30. Snoek, J., Larochelle, H., Adams, R.P.: Practical Bayesian optimization of machine learning algorithms. In: Advances in Neural Information Processing Systems, pp. 2951–2959 (2012)
31. Springenberg, J.T., Riedmiller, M.: Improving deep neural networks with probabilistic maxout units. arXiv preprint arXiv:1312.6116 (2013)
32. Stollenga, M.F., Byeon, W., Liwicki, M., Schmidhuber, J.: Parallel multidimensional LSTM, with application to fast biomedical volumetric image segmentation. In: Advances in Neural Information Processing Systems, pp. 2980–2988 (2015)
33. Su, H., Liu, F., Xie, Y., Xing, F., Meyyappan, S., Yang, L.: Region segmentation in histopathological breast cancer images using deep convolutional neural network. In: 2015 IEEE 12th International Symposium on Biomedical Imaging (ISBI), pp. 55–58. IEEE (2015)
34. Sutskever, I., Martens, J., Dahl, G., Hinton, G.: On the importance of initialization and momentum in deep learning. In: Proceedings of the 30th International Conference on Machine Learning (ICML 2013), pp. 1139–1147 (2013)
35. Visin, F., Kastner, K., Cho, K., Matteucci, M., Courville, A., Bengio, Y.: ReNet: a recurrent neural network based alternative to convolutional networks. arXiv preprint arXiv:1505.00393 (2015)
36. Wan, L., Zeiler, M., Zhang, S., Cun, Y.L., Fergus, R.: Regularization of neural networks using dropconnect. In: Proceedings of the 30th International Conference on Machine Learning (ICML 2013), pp. 1058–1066 (2013)
37. Zeiler, M.D., Fergus, R.: Stochastic pooling for regularization of deep convolutional neural networks. arXiv preprint arXiv:1301.3557 (2013)

Person Re-identification Based on Body Segmentation

Hua Jiang and Liang Zhang[(✉)]

Tianjin Key Lab of Advanced Signal Processing,
Civil Aviation University of China, Tianjin, China
l-zhang@cauc.edu.cn

Abstract. Person re-identification is a difficult problem to solve in the process of video analysis of non-overlapping multi-camera surveillance system. A new algorithm of person re-identification is proposed in the base of the human segmentation parts in this article. First, the human body segmentation is achieved based on the depth of bone points. Second, the optimal key frame is selected by using the scoring strategy for all parts of the same human multi-frame images segmentation. Different weights for the global color feature and the HOG feature is assigned. Third, all the characteristics are combined to establish a human target model, and EMD (Earth Movers Distance) distance is used to determine the similarity between the targets. The Kinect REID and BIWI RGBD-ID databases are used in the experiments. The results show that the proposed method has stronger robustness and a higher recognition rate.

Keywords: Person re-identification · Human division
Depth information · Color characteristics · HOG characteristics

1 Introduction

The human body re-recognition is the basic research work of pedestrian gesture, action and behavior recognition which goal is to correlate pedestrian images obtained from multiple cameras, and then judge these from the different people who are estimated images of the human body are the same person [1]. Due to its non-contact characteristics, this technique has broad application prospects in monitoring video data processing, automatic photo annotation and image retrieval.

Early re-identification studies focused on two directions: one is based on the feature representation of the method. The other is based on the distance measurement method [2]. The feature class method aims at designing strong differentiation and stability features. This method improves the accuracy of re-recognition to a certain extent, but it only considers the human target from the overall characteristics, and lacks the spatial binding information about the human target. The measurement method requires a low design requirement for the feature, from the perspective of the measurement distance. Its performance

© Springer Nature Singapore Pte Ltd. 2017
J. Yang et al. (Eds.): CCCV 2017, Part II, CCIS 772, pp. 474–485, 2017.
https://doi.org/10.1007/978-981-10-7302-1_39

largely depends on the selection of the sample. When sufficient samples are available, the distance function learned can be generally applied to the recognition problem in a variety of environments. And when the number of samples are small, there will be a fitting phenomenon. In addition, the training data samples need to manually annotate and consume a lot of labor and time costs [3].

In the traditional pedestrian re-recognition methods, pedestrian appearance is usually viewed as a whole for modeling and matching. In order to improve the accuracy of human re-recognition, in the identification of the use of local matching based on the use of pedestrian global appearance processing architecture. The method can better cope with the impact of pedestrian appearance of the deformation. To this end, the researchers proposed a variety of block based pedestrian re-recognition method. Alahi [4] et al. Used a set of rectangles range large from small to divide the human body into multiple rectangular regions, and then combined each local feature to construct a new appearance model. Marimon [5] et al. extracted the characteristics of the overall image of the pedestrian, and then divided it into four equal parts, nine equal points, sixteen equal parts, and then calculate the characteristics. Bak [6] et al. Maximized the distance between the major color sets of the upper and lower parts of the body, dividing the body into the upper and lower parts, and using the body position detector to find a meaningful sub-region. SDALF (Symmetry-Driven Accumulation of Local Features) method [1]. First, the human body is divided into three parts of the head, trunk and legs, and then use the body's symmetry were the trunk and legs which were divided, according to the symmetry axis extraction weighted HSV color histogram, the MSCR (Maximified Stable Color Regions) feature, and the RHSP (Recurrent High-Structured Patches) feature. But this division does not guarantee accurate part color. For example, if the person's shirt is longer and into the leg portion, then the color characteristics of the leg will be contaminated by unwanted shirt colors.

This paper presents a method of human re-identification based on body segmentation. First, the human body segmentation is performed that it is based on the depth of bone points. The optimal frame is selected for all parts of the multi-frame images using the position scoring strategy. And then fused the global color histogram and HOG characteristics of the body parts of the characterization. The experimental results show that the pedestrian re-recognition method based on block matching has higher robustness and higher recognition precision.

2   Partial Segmentation and Optimal Frame Selection

Body segmentation means that the human image is divided into several meaningful parts according to the human body structure. Proper and accurate segmentation not only can improve the recognition accuracy, but also reduce the complexity of the subsequent image processing. In this paper, the segmentation process is shown in Fig. 1.

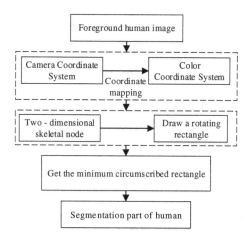

Fig. 1. Flow chart of human division

2.1   Segmentation Based on Depth of the Skeletal Point

For the foreground image that has been denoised, the segmentation of this paper is based on the position information of 20 skeletal nodes corresponding to the human body in the Kinect Camera coordinate system [7], as shown in Fig. 2.

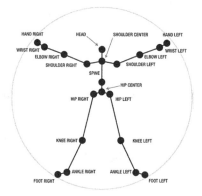

Fig. 2. Marked with 20 skeletal nodes of the human body

2.1.1   Draw a Rotating Rectangle

Select the location of the two ends of the skeleton nodes, the distance between the nodes are long of the rectangle. According to the experiment and experience. In addition to the width of the trunk part of the long two-thirds, other parts of the width are set to a long one-half. The midpoint of the coordinates of the two skeletal nodes are set as the rotation center of the rotating rectangle, and the rotation rectangle is drawn with the angle between the connection of the two skeletal nodes and the horizontal plane, as shown in Fig. 3.

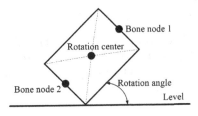

Fig. 3. Draw rotating rectangle

2.1.2   Partial Segmentation

Modern clothes have many styles and colors, the color of the sleeves and the torso is often part of the pattern of color which is not the same, the pants of the thigh and calf at the color may not be the same. And the head area is relatively subtle, not easy to extract effective information, feet area is easily blocked. Based on the above considerations, except to the head and the foot. The whole body is divided into 9 parts which are trunk, right upper arm, right lower arm, left upper arm, left lower arm, right thigh, right calf, left thigh, and left calf. There are 9 pictures in the human body on the rectangle inside the box is part of the split out of the human body, as shown in Fig. 4.

Fig. 4. 9 Parts after the human body segmentation

2.2   Optimal Frame Selection Based on Site Scoring

Because of the human always moving in the shooting process. The angle and gesture of the Kinect relative to the moving target are different greatly during the period of time. Different parts may be blocked at different times.

So the integrated multi-frame image, the block or other reasons for the loss of the use of other parts of the frame image to complement the corresponding parts. The use of site scoring strategy, the body of the 9 parts was selected optimal image preservation.

Among them, the site score using the following three indicators

(1) The ratio of the non-background area to the entire pair of images
    Traverse the entire image, the pixels are as white as the background part, the sum is represented by Δ, The foreground is indicated by sum_2, set the foreground part of the ratio of the whole image is Δ, which is

$$\Delta = \frac{sum_2}{sum_1 + sum_2} \tag{1}$$

(2) Angle difference of human body
    The human target in the Kinect of Camera coordinate system is shown in Fig. 5.
    The x, y, z direction of the angle difference were recorded as α, β, γ that set the total offset of χ, in order to illustrate the effectiveness of the algorithm, the two images of the same human body in different sets are selected under different illumination and angle conditions, and two different images of the same human body have a viewing angle of 90 to 180°, So the query set and the candidate set of the human body image selection have the biggest difference in the shooting angle, which is

$$\chi = \begin{cases} \frac{180-\alpha}{180} + \frac{180-\beta}{180} + \frac{180-\gamma}{180}, & Query\ set \\ \frac{\alpha}{180} + \frac{\beta}{180} + \frac{\gamma}{180}, & Candidate\ set \end{cases} \tag{2}$$

(3) Image quality evaluation

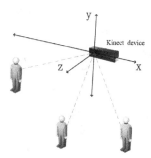

Fig. 5. The angle of the human body under the Kinect perspec-tive

There are two image quality evaluation in the general idea. One is subjective evaluation: subjective evaluation of the image quality by the observer; the other is objective evaluation: the use of algorithms to assess the image quality. Subjective evaluation methods are unavoidable and consistent with the subjective feelings, but they are also subjects to the subjective factors such as the professional background, psychological and mood of the observer. Objective evaluation method is accurate and fast. It has a unique assessment value and is more suitable for use in the actual project, but it and the subjective feelings of people have a certain access.

This paper uses a gradient-based common image sharpness evaluation function Tenengrad [8]. The function employs operators to extract gradient values in both horizontal and vertical directions respectively, Defined as followed

$$Ten = \frac{1}{n} * \sum_x \sum_y S(x,y)^2 \tag{3}$$

Where $S(x,y) = \sqrt{G_x * I(x,y) + G_y * I(x,y)}$ is the gradient of the image I at (x,y), G_x, G_y for *Sobel* convolutions, Respectively as

$$G_x = \begin{bmatrix} -1 & 0 & +1 \\ -2 & 0 & +2 \\ -1 & 0 & +1 \end{bmatrix}, G_y = \begin{bmatrix} -1 & -2 & -1 \\ 0 & 0 & 0 \\ +1 & +2 & +1 \end{bmatrix} \tag{4}$$

n is the total number of pixels in the image.

Based on the above indicators and assigned different weights, the overall score for each position is K

$$K = \frac{1}{2}\Delta + \frac{3}{10}Ten + \frac{1}{5}\chi \tag{5}$$

Each site of the human body takes the highest rated image as a sample.

3   Feature Design

The process of human re-recognition in this article is shown in Fig. 6.

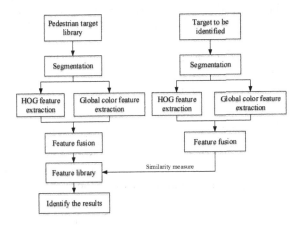

Fig. 6. Receptor flow chart of human target

3.1   Global Color Characteristics

The color feature is an important attribute to the image. RGB color space is composed of three primary colors that are mixed, the physical meaning is clear and suitable for the work of the picture tube, but the RGB color space does not have the light either the deformation and spectral invariance, and HSV color space is relative to these advantages. So the database image from RGB space to HSV space to experiment. Although the color histogram have the advantages which are simple to calculate, it has features that are insensitive to scale. But loses the spatial relationship between colors. Aibing Rao proposed the use of circular color histogram [9] to characterize the spatial characteristics of the color of the method to solve the problem.

Make $A_{ij} = |R_{ij}|$, for $i = 1, 2, \ldots, M$ and $j = 1, 2, \ldots, N$, so get an matrix $M \times N : A = (A_{ij})_{M \times N}$. This matrix is a circular color histogram that represents the number of colors in a ring, the row represents the color value, the column represents the number of rings, $|R_{ij}|$ indicates the number of color values i in the j rings. The measure of the circular color histogram is defined as

$$d(I, J) = \sqrt{\sum_{i=1}^{M} \sum_{j=1}^{N} (A_{ij} - B_{ij})^2} \qquad (6)$$

3.2   Directional Gradient Histogram

The so-called histogram of oriented gradients (HOG) refers to a local region descriptor based on the gradient direction [10]. It constructs local and global surface features by statistically localized gradient histograms. Due to the normalization of local and global gradient histograms, it has a strong anti-interference ability for slight changes in the surface of the human body caused by changes in light factors. The main idea of the HOG feature is to describe the shape of the local target in the foreground of the human body using the gradient of the gradient or edge.

Set the gradient of pixel (x, y) in the input image be

$$G_x(x, y) = H(x + 1, y) - H(x - 1, y) \qquad (7)$$

$$G_y(x, y) = H(x, y + 1) - H(x, y - 1) \qquad (8)$$

Where $G_x(x, y)$, $G_y(x, y)$, $H(x, y)$ respectively represent the horizontal gradient, vertical gradient, and pixel values at pixel (x, y) in the input image. The gradient amplitude $G(x, y)$ and the gradient direction $\alpha(x, y)$ at pixel (x, y) are

$$G(x, y) = \sqrt{G_x(x, y)^2 + G_y(x, y)^2} \qquad (9)$$

$$\alpha(x, y) = \tan^{-1}[\frac{G_x(x, y)}{G_y(x, y)}] \qquad (10)$$

3.3   Similarity Measure

Bhattacharyya distance has been widely used in image processing and computer vision due to the faster speed of operation. But because of the real scene at different times, the location of the same human body targets often exist posture, angle changes, easily lead to HOG characteristics of the shift. If the use of only two goals corresponding to the gradient histogram comparison of the Bhattacharyya distance method, the mismatch rate is bound to increase. The EMD cross distance [11] avoids this situation, which is often used to measure the similarity of a set.

The HOG feature P of the human body is represented as a set of multiple feature sets, $P = ((\alpha_1, \omega_{\alpha 1}), (\alpha_2, \omega_{\alpha 2}), \ldots, (\alpha_m, \omega_{\alpha m}))$, α_i represents a directional gradient histogram vector, and $\omega_{\alpha i}$ represents the weight of the vector α_i. Then the EMD distance of the HOG feature $P_A = ((a_1, \omega_{a 1}), (a_2, \omega_{a 2}), \ldots, (a_m, \omega_{am}))$ of the target A and the HOG feature $P_B = ((b_1, \omega_{b 1}), (b_2, \omega_{b 2}), \ldots, (b_m, \omega_{bm}))$ of the target B are defined as

$$D_{EMD}(A, B) = \min_{f_{ij}} \frac{\sum_{i=1}^{m} \sum_{j=1}^{n} d_{ij} f_{ij}}{\sum_{i=1}^{m} \sum_{j=1}^{n} f_{ij}}, i = 1 \ldots m; j = 1 \ldots n \qquad (11)$$

Where d_{ij} is the European distance of vector a_i and vector b_j, and f_{ij} is the transport stream.

Fusing the color feature and the HOG feature matching result, and then measuring the similarity among all the human body in the target body and the candidate set. Set γ, μ be the weight of the color feature and HOG, and the integrated distance between the two targets A and B is

$$D(A, B) = \gamma D_{color}(A, B) + \mu D_{EMD}(A, B) \qquad (12)$$

Where $D_{color}(A, B) = \sqrt{1 - \sum_i \frac{I_A(i) \cdot I_B(i)}{\sum_i I_A(i) \cdot I_B(i)}}$.

Several experiments were done to set the weight to $\gamma = 0.65, \mu = 0.35$.

4   Experimental Results and Analysis

Generally speaking, the body re-recognition will have two sets of data sets, query set and candidate set. This experiment is carried out in the Kinect REID database and the BIWI RGBD-ID database. These two databases are based on the depth of information on the human body to re-identify the common database. General re-recognition of the human body as a similarity sorting problem. The mainstream of target identification criteria is CMC curve(cumulative matching curve) [12]. The ranking k in the CMC curve represents the search for the target body in the candidate set. In the first k search. Results find the ratio of the target body to be queried. In this paper, the use of the evaluation criteria, and only use of color features, only use of HOG features and projection method of three methods for comparison.

4.1   Test Results and Analysis of Database Kinect REID

Kinect REID database have total of 71 human bodies, each body have about 120 frames, Fig. 7 are parts of the data example. 120 frames data of each human body randomly selected 60 as the query set, the remaining 60 frames as a candidate set, the two sets of each human body target segmentation, and then select the optimal score through the part of the frame.

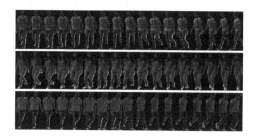

Fig. 7. Parts of the data in the database Kinect REID

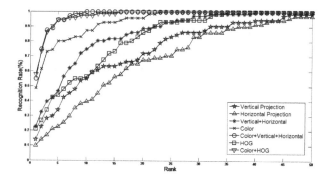

Fig. 8. The algorithm results in the database Kinect REID

As can be seen from Fig. 8 combining the color feature with the HOG feature significantly improves the recognition rate. Table 1 shows the comparison of the results of this algorithm with the SDALF, MCMimpl, SGLTrP3 and ED + SKL methods. It can be seen that the SGLTrP3 recognition rate is 66%, better than this and other methods at the first matching rate. But from the fifth recognition rate, this method is 94.3%, much higher than the other four methods.

4.2   Test Results and Analysis of Database BIWI RGBD-ID

Database BIWI RGBD-ID consists of two parts. Namely, Still dataset and Walking dataset, each part also contains training set of 28 human bodies and test set

Table 1. The algorithm is compared with other algorithms in the Kinect REID database

Method	Rec[a] rate 1	Rec rate 5	Rec rate 10	Rec rate 30	Rec rate 50
SDALF [13]	41%	70%	82%	98%	100%
MCMimpl [13]	51%	78%	87%	99%	100%
SGLTrP3 [14]	66%	82%	91%	100%	100%
ED + SKL [15]	56%	86%	94%	100%	100%
The method of this article	57.7%	94.3%	97.2%	100%	100%

[a]Notes: Rec is logogram of Recognition.

Table 2. The algorithm in this section is compared with other algorithms in the Still dataset

Method	Rec[a] rate 1	Rec rate 5	Rec rate 10	Rec rate 30	Rec rate 50
Face + Skeleton (SVM) [16]	52%	81%	90%	98%	100%
Nearest Neighbor [17]	27%	45%	82%	97%	100%
ED + SKL [15]	31%	68%	81%	100%	100%
The method of this article	58.9%	91.0%	98.7%	100%	100%

[a]Notes: Rec is logogram of Recognition.

of 50 human bodies. Taking into account the lack of training part of this article and if only a separate use of the query set or candidate set of experimental samples less, so mix the training set and test set together, there are total 78 people for re-recognition experiments (Table 2).

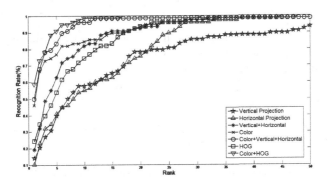

Fig. 9. The experimental results of this algorithm in the Still dataset section

Figure 9 shows the recognition results of the seven curves using the color feature, the typical method, the HOG feature, and the combination of the two in the Still dataset section. When the HOG feature is used only, the recognition rate of the top 10 is lower, but after the combination of the two of then, the recognition rate has been significantly improved, the first matching rate reach at 58.9%, the 10th matching rate at 98.7% which is higher than other similar algorithms (Fig. 10).

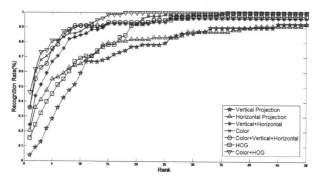

Fig. 10. The experimental results of this algorithm in the Walking dataset section

Table 3. The algorithm in this section is compared with other algorithms in the Walking dataset

Method	Rec[a] rate 1	Rec rate 5	Rec rate 10	Rec rate 30	Rec rate 50
Face + Skeleton (SVM) [16]	43.9%	74%	85%	96.5%	100%
Nearest Neighbor [17]	21%	43%	77%	97%	100%
ED + SKL [15]	26%	61%	78%	100%	100%
The method of this article	46.2%	80.7%	91.0%	100%	100%

[a]Notes: Rec is logogram of Recognition.

Table 3 compares this algorithm with the methods of machine learning algorithms such as Face + Skeleton (SVM), Nearest Neighbor and ED + SKL, except that the method is slightly 2.3% higher than the SVM method at the first matching rate. The rate of the beginning of the other matching rate, the algorithm has achieved the highest recognition rate.

5   Conclusion

In this paper, we propose an accurate segmentation of the human body according to the depth information of the human joints. The optimal frame is selected as the experimental sample for the multi-frame images of a human. It is proved that the human body segmentation based on deep bone nodes can enhance the robustness to occlusion problem, angle of view change and attitude change. In the re-recognition stage, a new object recognition algorithm combining color feature and HOG feature is proposed to improve the recognition rate. The next step is to improve the HOG characteristics and find better description features to further improve the recognition rate.

References

1. Farenzena, M., Bazzani, L., Perina, A., et al.: Person re-identification by symmetry-driven accumulation of local features. In: Computer Vision and Pattern Recognition, pp. 2360–2367. IEEE (2010)
2. Liu, C., Gong, S., Loy, C.C., Lin, X.: Person re-identification: what features are important? In: Fusiello, A., Murino, V., Cucchiara, R. (eds.) ECCV 2012. LNCS, vol. 7583, pp. 391–401. Springer, Heidelberg (2012). https://doi.org/10. 1007/978-3-642-33863-2_39
3. Qin, H.K., et al.: Summary of intelligent video surveillance technology. J. Comput. Sci. **38**(6), 1093–1118 (2015). In chinese
4. Alahi, A., Vandergheynst, P., Bierlaire, M., et al.: Cascade of descriptors to detect and track objects across any network of cameras. Comput. Vis. Image Underst. **114**(6), 624–640 (2010)
5. Alahi, A., Marimon, D., Bierlaire, M., et al.: A master-slave approach for object detection and matching with fixed and mobile cameras. In: IEEE International Conference on Image Processing, pp. 1712–1715. IEEE (2008)
6. Bak, S., Corvee, E., Brmond, F., et al.: Person re-identification using spatial covariance regions of human body parts. In: IEEE International Conference on Advanced Video and Signal Based Surveillance, pp. 435–440. IEEE (2010)
7. Shotton, J., Kipman, A., Kipman, A., et al.: Real-time human pose recognition in parts from single depth images. Commun. ACM **56**(1), 116–124 (2013)
8. Arun, R., Nair, M.S., Vrinthavani, R., et al.: An alpha rooting based hybrid technique for image enhancement. Eng. Lett. **19**(3), 159–168 (2011)
9. Rao, A., Srihari, R.K., Zhang, Z.: Spatial color histograms for content-based image retrieval. In: IEEE International Conference on TOOLS with Artificial Intelligence, p. 183. IEEE Computer Society (1999)
10. Dalal, N., Triggs, B., Triggs, B.: Histograms of oriented gradients for human detection. CVPR **1**(12), 886–893 (2005)
11. Fu, A.Y., Liu, W., Deng, X.: Detecting phishing web pages with visual similarity assessment based on earth mover's distance (EMD). IEEE Trans. Dependable Secure Comput. **3**(4), 301–311 (2006)
12. Bolle, R.M., Connell, J.H., Pankanti, S., et al.: The Relation between the ROC curve and the CMC. In: Fourth IEEE Workshop on Automatic Identification Advanced Technologies, vol. 2005, pp. 15–20. IEEE (2005)
13. Pala, F., Satta, R., Fumera, G., et al.: Multimodal person reidentification using RGB-D cameras. IEEE Trans. Circ. Syst. Video Technol. **26**(4), 788–799 (2016)
14. Imani, Z., Soltanizadeh, H.: Person reidentification using local pattern descriptors and anthropometric measures from videos of kinect sensor. IEEE Sens. J. **16**(16), 6227–6238 (2016)
15. Wu, A., Zheng, W.S., Lai, J.H.: Robust Depth-Based Person Re-Identification. IEEE Press (2017)
16. Munaro, M., Basso, A., Fossati, A., et al.: 3D reconstruction of freely moving persons for re-identification with a depth sensor. In: IEEE International Conference on Robotics and Automation, pp. 4512–4519. IEEE (2014)
17. Munaro, M., Fossati, A., Basso, A., Menegatti, E., Van Gool, L.: One-shot person re-identification with a consumer depth camera. In: Gong, S., Cristani, M., Yan, S., Loy, C.C. (eds.) Person Re-Identification. ACVPR, pp. 161–181. Springer, London (2014). https://doi.org/10.1007/978-1-4471-6296-4_8

Bidirectional Adaptive Feature Fusion for Remote Sensing Scene Classification

Weijun Ji[1,2], Xuelong Li[1], and Xiaoqiang Lu[1(✉)]

[1] Center for OPTical IMagery Analysis and Learning (OPTIMAL),
State Key Laboratory of Transient Optics and Photonics,
Xi'an Institute of Optics and Precision Mechanics, Chinese Academy of Sciences,
Xi'an 710119, Shaanxi, People's Republic of China
jiweijun2015@opt.cn, {xuelong_li,luxiaoqiang}@opt.ac.cn
[2] University of Chinese Academy of Sciences, 19A Yuquanlu, Beijing 100049,
People's Republic of China

Abstract. Convolutional neural networks (CNN) have been excellent for scene classification in nature scene. However, directly using the pretrained deep models on the aerial image is not proper, because of the spatial scale variability and rotation variability of the HSR remote sensing images. In this paper, a bidirectional adaptive feature fusion strategy is investigated to deal with the remote sensing scene classification. The deep learning feature and the SIFT feature are fused together to get a discriminative image presentation. The fused feature can not only describe the scenes effectively by employing deep learning feature but also overcome the scale and rotation variability with the usage of the SIFT feature. By fusing both SIFT feature and global CNN feature, our method achieves state-of-the-art scene classification performance on the UCM and the AID datasets.

Keywords: Feature fusion · Remote sensing scene classification

1 Introduction

High Spatial Resolution (HSR) remote sensing image scene classification aims to automatically label an aerial image with the specific category. It is the basis of land-use object detection and image understanding which are widely used in the field of military and civilian. Remote sensing image scene classification is a fundamental problem which has attracted much attention. The most vital and challenging task of the scene classification is to develop an effective holistic representation of the aerial image to directly model an image scene.

To develop the effective holistic representation, many methods have been proposed in recent years. Primitively, a bottom-up scheme was proposed to model

---

W. Ji—His research interests include remote sensing scene classification and computer vision.

© Springer Nature Singapore Pte Ltd. 2017
J. Yang et al. (Eds.): CCCV 2017, Part II, CCIS 772, pp. 486–497, 2017.
https://doi.org/10.1007/978-981-10-7302-1_40

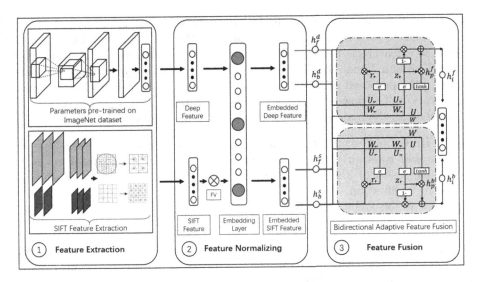

Fig. 1. The flowchart of the proposed method. Part 1 shows as the process of extracting features. Part 2 shows as the process of feature normalizing. Part 3 shows as the process of the bidirectional feature fusion.

a HSR remote sensing image scene by three "pixel-region-scene" steps. To further express the scene, many researchers tried to represent the scene directly without classifying pixels and regions. For example, Bag-of-the-Visual-Words (BoVW) [20] and some extensions of BoVW were proposed to improve the classification accuracy. Besides, the family of latent generative topic models [1] have also been applied in HSR image scene classification.

All these methods have better performance, they cannot satisfy the demand of higher accuracy. One main reason is that these methods may lack the flexibility and adaptivity to different scenes [19]. Recently, it has been proven that deep learning methods can adaptively learn image features which are suitable for specific scene classification tasks and achieve far better classification performances. However, there are two major issues that seriously influence the use of the deep learning method. (1) The deep learning method needs large training data to train the model and is time consuming. (2) The pre-trained models take little consideration of the HSR remote sensing image characteristics [13]. The spatial scale variability and rotation variability of the HSR remote sensing images cannot be expressed precisely by pre-trained models.

In order to relieve the above issues, a feature fusion strategy is proposed in the paper. Many works have proved that SIFT feature has better performance on overcoming the scale variability and rotation variability [16,17]. The proposed method can be divided into three steps. First, the deep features and the SIFT features are extracted by CNN and SIFT filter respectively. Second, the two features are normalized by normalization layer to get the same dimension. Finally, the normalized features are adaptively assigned with optimal weights to get the fused feature. Specially, the fusion strategy is to assign average weights to confidence scores from the deep feature and the SIFT feature. The optimal weight of a feature is trained by the model and is optimized by Back Propagation Through Time (BPTT). With the help of above adaptive feature fusion strategy, the proposed method can get a discriminative representation for scene classification. The flowchart of the proposed method is shown as Fig. 1.

In general, the major contributions of this paper are as follows:

1. For HSR image scene classification, deep learning features are sensitive to scale and rotation variant. To address the aforementioned problem, deep learning features and SIFT features are together exploited remote sensing scene classification.
2. A new fusion architecture is investigated to take full advantage of the information of features.

The rest of the paper is organized as follows: In Sect. 2, the related works on HSR scene classification are reviewed. Section 3 gives the detailed description of our proposed method. The experiments of our method on two data sets are shown in Sect. 4. In the last Section, we conclude this paper.

2   Related Work

In this section we provide previous work on remote sensing scene classification and feature fusion.

Recently, several methods have been proposed for remote sensing scene classification. Many of these methods are based on deep learning [3,10,13–15,26]. The deep learning method becomes the mainstream approach, because it uses a multi-stage global feature learning strategy to adaptively learn image features and cast the aerial scene classification as an end-to-end problem. To further improve the classification accuracy, many researchers have proposed different methods to overcome the issues which pre-trained models taking little consideration of the HSR remote sensing image characteristics. Some of them have addressed the problem from datasets, they have enlarged the datasets to increase the scale and rotation invariant information. Y. Liu et al. [11] added random structure noise i.e. random-scale stretching to capture the essential feature robust to scale change. G. Cheng et al. [4,5] rotated the original data to enlarge the dataset and increased the rotation information. Others focused on the deep learning architecture. G. Cheng et al. [4] added two fully connected layers and changed the loss function to ensure all the rotation data has the same representation. Moreover, some researchers cared about the classifier. F. Zhang et al. [23] adapted the boosting method to merge some weak networks to get more accurate results.

There has been a long line of previous work incorporating the idea of fusion for scene classification. For example, [18] designed sparse coding based multiple feature fusion (SCMF) for HSR image scene classification. SCMF sets the fused result as the connection of the probability images obtained by the sparse codes of SIFT, the local ternary pattern histogram Fourier, and the color histogram features. [24] also used four features and concatenated the quantized vector by k-means clustering for each feature to form a vocabulary. All these fusion methods proposed for HSR image scene classification obtained good results to some degree. Nevertheless, all of these methods were limited because the fused features were not effective enough to represent the scene. In this paper, the strategy fuses deep learning features and SIFT features together to achieve comparable performance.

3   Proposed Method

The proposed method involves three parts: (1) feature extraction. (2) feature normalizing. (3) feature fusion. The proposed method fuses the two features to get the final classification result.

3.1   Feature Extraction

This Section of feature extraction is divided into two parts, the first part is the deep feature extraction and the second part is the SIFT feature extraction. Deep features are extremely effective features for scene classification, and they are the guarantee of classification accuracy. Considering the limitation of the training data in the satellite image and the high computation complexity of tuning convolutional neural networks, we use the pre-trained convolutional neural networks. [19] compares three representative high-level deep learned scene classification methods. The comparison shows that VGG-VD-16 gets the best result.

VGG gives a thorough evaluation of networks by increasing depth using an architecture with very small (33) convolution filters, which shows a significant improvement on the accuracies, and can be generalised well to wide range of tasks and datasets. In our work, we use one of its best-performance models named VGG-VD-16, because of its simpler architecture and slightly better results. It is composed of 13 convolutional layers and followed by 3 fully connected layers, thus results in 16 layers. The parameters of the model are pre-trained on approximately 1.2 million RGB images in the ImageNet ILSVRC 2012. We carry on the experiment to verify that the first fully connected layer perform best in the experiment. So the first fully connected layer is chosn as the feature vectors of the images [19].

The SIFT feature [12] describes a patch by the histograms of gradients computed over a (4×4) spatial grid. The gradients are then quantized into eight bins, so the dimension of the final feature vector is 128 $(4 \times 4 \times 8)$.

The SIFT feature searches over all scales and image locations. It is implemented effectively by using a difference-of Gaussian function to identify potential interest points that are invariant to scale and orientation. The initial image is incrementally convolved with Gaussians to produce images separated by a constant factor k in scale space. Adjacent images scales are subtracted to produce the difference-of Gaussian images. This step detect the scale invariance feature from different scale space. And the date will experience key point localization, orientation assignment and key point descriptor. The features with scale invariant and rotation invariant are extracted.

3.2   Feature Normalizing

After feature extraction, the dimension of the deep feature is 4096 and the dimension of the SIFT feature is $128 \times N$ (where N indicates the number of the points of the interest). The dimensions of the two features are different so we need to utilize the normalizing method to reshape the two features in the same dimension.

We firstly use the FV [16] to encode the SIFT feature. In essence, the SIFT features generated from FV encoding method is a gradient vector of the log-likelihood. By computing and concatenating the partial derivatives, the mean and variance of Gaussian functions, the dimension of the final feature vector is $2 \times K \times F$ (where F indicates the dimension of the local feature descriptors and K denotes the size of the dictionary). And then, we resize the size of the dictionary to control the dimension of the output feature. Finally, we use the encoding method to express SIFT feature more effectively.

Then the deep feature and the SIFT feature are together to get the same dimension by a normalizing layer. The normalizing layer is composed of two fully connected layers. The dimensions of the fully connected layers and the number of the connection layers are verified by the experiments. The number of the fully connection layer is set to 2 and the dimensions of the fully connection layers are 4096 dimensions and 2048 dimensions. The parameters of the fully connection layer are trained together with the below proposed fusion strategy.

3.3   Feature Fusion

We propose a novel bidirectional adaptive feature fusion method inspired by recurrent neural networks [7,8]. The process of the feature fusion can be divided into three procedures. The input features pass through the primary fusion node to get a primary fusion feature. Then the primary feature and the first input node feature are fused with different weights to get the intermediate feature. After getting the two intermediate features from two directions, the two intermediate features are summed together with different weights and with the bias to get the final fusion feature. The reset ratio and the update ratio should be calculated firstly before the feature fusion. The details of the feature fusion is described as below.

In forward direction, the framework of the feature fusion is shown as Fig. 2. There are two input nodes in the framework, the first input feature is the deep learning feature marked as h_f^d and the second input feature is the SIFT feature marked as h_f^s. The reset ratio and the update ratio are calculated respectively by Eqs. 1 and 2. The primary fusion feature marked as h_f^p is calculated by Eq. 3. The primary fusion feature and the deep learning feature are as the two input features to calculate the intermediate fusion feature (marked as h_f^i) Eq. 4.

$$z_f = sigmoid(W_z h_f^s + U_z h_f^d) \tag{1}$$

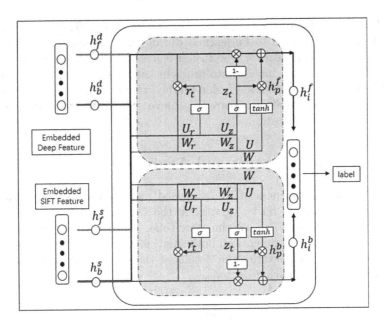

Fig. 2. The details of the feature fusion strategy, the input features are the embedded deep feature and embedded SIFT feature. The output is the label of the image. The two nodes are described as the bidirectional feature fusion strategy.

$$r_f = sigmoid(W_r h_f^s + U_r h_f^d) \tag{2}$$

$$h_f^p = tanh(W h_f^s + r_f \circ U h_f^d) \tag{3}$$

$$h_f^i = z_f * h_f^d + (1 - z_f) * h_f^p \tag{4}$$

In backward direction, the first input feature is the SIFT feature marked as h_b^s and the second input feature is the deep learning feature marked as h_b^d. The reset ratio and the update ratio are also calculated respectively by Eqs. 5 and 6. The primary fusion feature marked as h_b^p is calculated by Eq. 7. The primary fusion feature and the SIFT feature are as the two input features to calculate the intermediate fusion feature (marked as h_b^i) by Eq. 8.

$$z_b = sigmoid(W_z h_b^d + U_z h_b^s) \tag{5}$$

$$r_b = sigmoid(W_r h_b^d + U_r h_b^s) \tag{6}$$

$$h_b^p = tanh(W h_b^d + r_b \circ U h_b^s) \tag{7}$$

$$h_b^i = z_b * h_b^s + (1 - z_b) * h_b^p \tag{8}$$

In bidirectional feature fusion, h_b^i and h_f^i are calculated together to get the final fusion feature y_t. The formula is show as (9).

$$y = W_f h_f^i + W_b h_b^i + b_y \tag{9}$$

In all these formula Eqs. 1 to 9. $W_z, W_r, U_z, U_r, W, U, W_f, W_b, b_y$ are all the weight parameters needed to be learned during the training the model. After getting the final fusion feature, the fusion feature passes through the softmax layer to classify the HSR images into different categories. The model is trained by BPTT method. The proposed fusion method takes full advantage of the deep learning feature and the SIFT feature to achieve comparable performance in the HSR image scene classification.

4 Experimental Results and Analysis

To evaluate the effectiveness of the proposed method for scene classification, we perform experiments on two datasets: the UC Merced Land Use dataset [20] and the AID dataset [19]. At the same time, in order to evaluate the fusion strategy, we perform experiments on both single features (deep feature and SIFT feature) and the fused feature. The details of the experiments and the results are described in the following sections.

4.1 Dataset and Experiment Set Up

UC Merced Land Use dataset: It contains 21 scene categories. Each category contains 100 images. Each scene image consists of 256 × 256 pixels, with a spatial resolution of one foot per pixel. The example images are shown as Fig. 3. In the experiment, we randomly choose 80 images from each category as the training set, the rest images are chosen as the testing set. Some of the scenes are similar to each other causing the scene classification to be challenging.

Fig. 3. Example images associated with 21 land-use categories in the UC-Merced data set: (1) agricultural, (2) airplane, (3) baseball diamond, (4) beach, (5) buildings, (6) chaparral, (7) dense residential, (8) forest, (9) freeway, (10) golf course, (11) harbor, (12) intersection, (13) medium residential, (14) mobile home park, (15) overpass, (16) parking lot, (17) river, (18) runway, (19) sparse residential, (20) storage tanks and (21) tennis court

AID: It is a new dataset containing 10000 images which make it the biggest dataset of high spatial resolution remote sensing images. Thirty scene categories are included in this dataset. The number of different scenes is different from each other changing from 220 to 420. The shape of the scene image consists of 600 × 600 pixels. We follow the parameters setting in the [19]. Fifty images of each category are chosen as the training set and the rest of the images are set as the testing set. The dataset is famous for its images with high intra-class diversity and low inter-class dissimilarity. The example images are shown as Fig. 4.

Table 1. Comparison with the previous reported accuracies with the UC merced data set

Method	OA(%)	Method	OA(%)
BOVW	72.05	S-UFL	82.30
pLSA	80.71	SAL-PTM	88.33
SPCK++	76.05	GBRCN	94.53
LDA	81.92	SRSCNN	95.10
SPM+SIFT	82.30	CNN	93.10
SIFT+FC	81.67	PROPOSED	95.48

4.2 Experiment on UC Merced Dataset

To evaluate the proposed method, we compare our method with the state-of-the-art methods in scene classification. The comparative scene classification methods include BOVW [20], pLSA [2], LDA [1], SPM [9], SPCK [21], SIFT+SC [6], S-UFL [22], GBRCN [23], SAL-PTM [25] and SRSCNN [11]. The results are showed in Table 1. From Table 1, it can be concluded that the proposed method performs better than other methods. Compared with GBRCN which combines many CNN for scene classification, our method is 0.91% better than it. Our method is 0.38% better than SRSCNN which randomly selects patches from image and stretch to the specific scale as input to train CNN. The experiment results indicate the effectiveness of the proposed method.

4.3 Experiment on AID Dataset

To evaluate the proposed method on AID, the compared methods are chosen to code SIFT feature with models such as BOVW, pLSA, LDA, SPM, VLAD, FV and LLC. The deep models are chosen as VGG-VD-16 and the proposed method. Table 2 gives the comparison, which also shows the effectiveness of the proposed method.

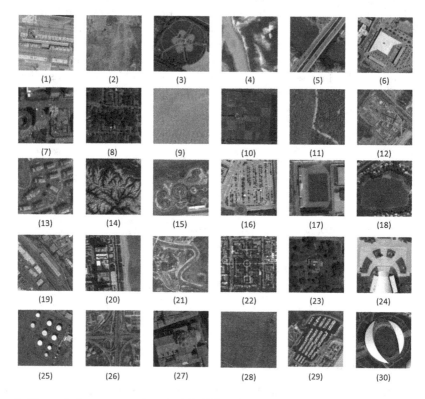

Fig. 4. Example images associated with 30 land-use categories in the UC-Merced data set: (1) airport, (2) bare land, (3) baseball field, (4) beach, (5) bridge, (6) center, (7) church, (8) commercial, (9) dense residential, (10) desert, (11) farmland, (12) forest, (13) industrial, (14) meadow, (15) medium residential, (16) mountain, (17) park, (18) parking, (19) play ground, (20) pond, (21) port, (22) railway station, (23) resort, (24) river, (25) school, (26) sparse residential, (27) square, (28) stadium, (28) storage tanks, (29) viaduct and (30) viaduct

Table 2. Comparison with the feature fusion with the aid data set

Method	BOVW	FV	LLC	pLSA
OA(%)	68.37	78.99	63.24	63.07
Method	SPM	VGG-VD-16	VLAD	PROPOSED
OA(%)	45.52	89.64	68.96	93.56

4.4 Experiment on Feature Fusion

To evaluate the proposed fusion strategy, we compare the classification accuracy in experiments with single deep feature and experiments with single SIFT feature encoded with fisher vector. Moreover, we compare the fusion strategy with sum of the features and the average of the features. The result shows that the single SIFT feature encoded with fisher vector performs worst in the experiment. At the same time, the deep feature and the fusion feature all get the classification accuracy exceeding 90%. The proposed fusion strategy obtains the best performance. The sum of feature gets the accuracy of 94.01% and the average of the features gets the accuracy of 93.05%. The classification accuracy of the proposed fusion strategy is shown as the best result compared with other fusion strategys. The final result is shown in Table 3. Our experiment is conducted on the UC Merced data set.

Table 3. Comparison with the feature fusion with the UC merced data set

Method	FV	VGG-VD-16	SUM	AVG	Proposed
OA(%)	85.11	93.09	94.01	93.05	95.48

5 Conclusion

The proposed method aims to solve the scale variance in high spatial resolution remote sensing images scene classification. Fusion of deep feature and SIFT feature is evaluated to get the state-of-the-art performance. The experiments on UCM and AID demonstrate the effectiveness of the proposed method.

References

1. Blei, D.M., Ng, A.Y., Jordan, M.I.: Latent dirichlet allocation. J. Mach. Learn. Res. **3**, 993–1022 (2003)
2. Bosch, A., Zisserman, A., Muñoz, X.: Scene classification via pLSA. In: Leonardis, A., Bischof, H., Pinz, A. (eds.) ECCV 2006. LNCS, vol. 3954, pp. 517–530. Springer, Heidelberg (2006). https://doi.org/10.1007/11744085_40
3. Castelluccio, M., Poggi, G., Sansone, C., Verdoliva, L.: Land use classification in remote sensing images by convolutional neural networks. J. Mol. Struct. Theochem **537**(1), 163–172 (2015)
4. Cheng, G., Ma, C., Zhou, P., Yao, X., Han, J.: Scene classification of high resolution remote sensing images using convolutional neural networks. In: Proceedings of IEEE International Geoscience and Remote Sensing Symposium, pp. 767–770 (2016)
5. Cheng, G., Zhou, P., Han, J.: Learning rotation-invariant convolutional neural networks for object detection in VHR optical remote sensing images. IEEE Trans. Geosci. Remote Sens. **54**(12), 7405–7415 (2016)

6. Cheriyadat, A.M.: Unsupervised feature learning for aerial scene classification. IEEE Trans. Geosci. Remote Sens. **52**(1), 439–451 (2014)

7. Chung, J., Gulcehre, C., Cho, K.H., Bengio, Y.: Empirical evaluation of gated recurrent neural networks on sequence modeling. Eprint Arxiv (2014)

8. Graves, A., Mohamed, A., Hinton, G.: Speech recognition with deep recurrent neural networks. In: Proceedings of IEEE International Conference on Acoustics, Speech and Signal Processing, pp. 6645–6649 (2013)

9. Lazebnik, S., Schmid, C., Ponce, J.: Beyond bags of features: spatial pyramid matching for recognizing natural scene categories. In: Proceedings of IEEE Conference on Computer Vision and Pattern Recognition, vol. 2, pp. 2169–2178 (2006)

10. Li, X., Lu, Q., Dong, Y., Tao, D.: SCE: a manifold regularized set-covering method for data partitioning. IEEE Trans. Neural Netw. Learn. Syst. (2017)

11. Liu, Y., Zhong, Y., Fei, F., Zhang, L.: Scene semantic classification based on random-scale stretched convolutional neural network for high-spatial resolution remote sensing imagery. In: Proceedings of IEEE International Geoscience and Remote Sensing Symposium, pp. 763–766 (2016)

12. Lowe, D.G.: Distinctive image features from scale-invariant keypoints. Int. J. Comput. Vis. **60**(2), 91–110 (2004)

13. Lu, X., Zheng, X., Yuan, Y.: Remote sensing scene classification by unsupervised representation learning. IEEE Trans. Geosci. Remote Sens. **55**, 5185–5197 (2017)

14. Luus, F.P.S., Salmon, B.P., Bergh, F.V.D., Maharaj, B.T.J.: Multiview deep learning for land-use classification. IEEE Geosci. Remote Sens. Lett. **12**(12), 1–5 (2015)

15. Nogueira, K., Penatti, O.A.B., Santos, J.A.D.: Towards better exploiting convolutional neural networks for remote sensing scene classification. Patt. Recogn. **61**, 539–556 (2017)

16. Perronnin, F., Dance, C.: Fisher kernels on visual vocabularies for image categorization. In: Proceedings of IEEE Conference on Computer Vision and Pattern Recognition, pp. 1–8 (2007)

17. Perronnin, F., Sánchez, J., Mensink, T.: Improving the fisher kernel for large-scale image classification. In: Daniilidis, K., Maragos, P., Paragios, N. (eds.) ECCV 2010. LNCS, vol. 6314, pp. 143–156. Springer, Heidelberg (2010). https://doi.org/10.1007/978-3-642-15561-1_11

18. Sheng, G., Yang, W., Xu, T., Sun, H.: High-resolution satellite scene classification using a sparse coding based multiple feature combination. Int. J. Remote Sens. **33**(8), 2395–2412 (2012)

19. Xia, G.S., Hu, J., Hu, F., Shi, B., Bai, X., Zhong, Y., Zhang, L., Lu, X.: AID: a benchmark data set for performance evaluation of aerial scene classification. IEEE Trans. Geosci. Remote Sens. **55**, 3965–3981 (2017)

20. Yang, Y., Newsam, S.: Bag-of-visual-words and spatial extensions for land-use classification. In: Proceedings of Sigspatial International Conference on Advances in Geographic Information Systems, pp. 270–279 (2010)

21. Yang, Y., Newsam, S.: Spatial pyramid co-occurrence for image classification. In: Proceedings of International Conference on Computer Vision, pp. 1465–1472 (2011)

22. Zhang, F., Du, B., Zhang, L.: Saliency-guided unsupervised feature learning for scene classification. IEEE Trans. Geosci. Remote Sens. **53**(4), 2175–2184 (2015)

23. Zhang, F., Du, B., Zhang, L.: Scene classification via a gradient boosting random convolutional network framework. IEEE Trans. Geosci. Remote Sens. **54**(3), 1793–1802 (2016)

24. Zheng, X., Sun, X., Fu, K., Wang, H.: Automatic annotation of satellite images via multifeature joint sparse coding with spatial relation constraint. IEEE Geosci. Remote Sens. Lett. **10**(4), 652–656 (2012)

25. Zhong, Y., Zhu, Q., Zhang, L.: Scene classification based on multifeature probabilistic latent semantic analysis for high spatial resolution remote sensing images. IEEE Trans. Geosci. Remote Sens. **53**(11), 6207–6222 (2015)
26. Zou, Q., Ni, L., Zhang, T., Wang, Q.: Deep learning based feature selection for remote sensing scene classification. IEEE Geosci. Remote Sens. Lett. **12**(11), 2321–2325 (2015)

Local Features Based Level Set Method for Segmentation of Images with Intensity Inhomogeneity

Hai Min[1], Li Xia[2,3], Qianqian Pan[2,4], Hao Fu[2,4], Hongzhi Wang[3], and Hai Li[2,3(✉)]

[1] School of Computer and Information, Hefei University of Technology, Hefei 230009, China
[2] Anhui Province Key Laboratory of Medical Physics and Technology, Center of Medical Physics and Technology, Hefei Institutes of Physical Science, Chinese Academy of Sciences, Hefei 230031, China
hli@cmpt.ac.cn
[3] Cancer Hospital, Chinese Academy of Sciences, Hefei 230031, China
[4] University of Science and Technology of China, Hefei 230027, China

Abstract. Local region-based level set models have recently been recognized as promising methods to segment images with intensity inhomogeneity. In these models, local intensity information in a neighborhood of predetermined size is extracted and then embedded into the energy function, where the local neighborhood intensities are assumed to be rather constant. Complex image characteristics, such as variation in degree of intensity inhomogeneity and noise levels, can lead to severe challenges for accurate image segmentation when using only a fixed scale parameter for local regions. In this paper, we propose a new multi-scale local feature-based level set method based on previous studies of multi-scale image filtering methods. Our novel method can adaptively determine the optimal scale parameter for each pixel during contour evolution, alleviating the challenges caused by severe intensity inhomogeneity. Our experimental results illustrate the good performance of the proposed level set method.

Keywords: Intensity inhomogeneity · Level set
Local maximum description difference · Local region descriptor
Multi-scale

1 Introduction

Local region-based level set models [1,14,20], such as local region descriptors (LRDs) method [2], localizing region-based (LRB) active contours [4], local binary fitting (LBF) model [7,8], local image fitting (LIF) model [19], and local Chan-Vese (LCV) model [16], have been recognized as effective methods to segment images with intensity inhomogeneity. In these models, local intensity information in a neighborhood of a predetermined size is extracted and then

© Springer Nature Singapore Pte Ltd. 2017
J. Yang et al. (Eds.): CCCV 2017, Part II, CCIS 772, pp. 498–508, 2017.
https://doi.org/10.1007/978-981-10-7302-1_41

embedded into the energy functional with CV-like structure, thus guiding the evolution of deformable contour used to identify object boundaries. Though they do not assume homogeneous intensity in the whole object being segmented, these methods are limited by the assumption that the intensities in each local region are rather constant. Therefore, selection of an optimal value of the scale parameter is a critical factor for segmenting images with intensity inhomogeneity. To determine the optimal size of local region, a trial-and-error solution, along with visual assessment of segmentation accuracy, is usually employed in the traditional procedure. Complex image characteristics, such as variations in the degree of intensity inhomogeneity and noise levels for different regions or images, lead to severe challenges for accurate image segmentation using only a fixed scale parameter for all local regions.

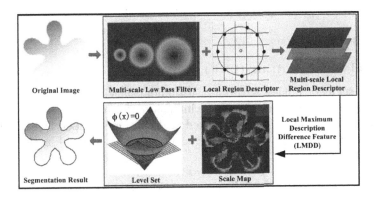

Fig. 1. The schematic diagram of the proposed multi-scale local region-based level set method.

In recent years, multi-scale level set methods have been explored to overcome the difficulties caused by single scale methods. For example, Lin et al. [9] presented a multi-scale level set framework to segment echocardiographic images, where a coarse scale was first used to extract image boundaries and fine scale was then adopted to refine the results. A similar scheme was used by Kim et al. [3] to track non-rigid object boundaries. However, these methods are essentially traditional multi-scale image processing approaches with a predetermined scale parameter in each step; rather than an adaptive approach for scale parameter selection.

Motivated by previous studies on multi-scale image processing [5, 10, 12, 13], we propose a new multi-scale local feature-based level set method for image segmentation. Our new method can adaptively determine the optimal scale parameter for each pixel during contour evolution. Figure 1 is a schematic diagram of the proposed method. First, by using the multi-scale low pass filters, we construct multi-scale local region descriptors. Based on the descriptors, a local maximum description difference feature (LMDD) is defined, which is associated with the maximum response of multi-scale high-pass filters. Since intensity inhomogeneity is believed to be primarily located in the low-frequency band [11, 15], the LMDD feature is expected to significantly reduce the influence of intensity inhomogeneity on image segmentation. Meanwhile, the optimal scale value is determined automatically. The LMDD feature is then incorporated into one typical local-region based level set model with Chan-Vese (CV)-like structures, namely LBF model [8], to construct the energy function. Finally, minimization of this energy completes the segmentation. It should be noted that the proposed method can easily be incorporated into other typical local region based level set models, such as LIF, LCV etc.

The rest of the paper is organized as follows: In Sect. 2, we will introduce the related works and the way to construct the novel multi-scale local region-based level set model. Experimental results and associated performance analysis are illustrated in Sect. 3.

2   Method

2.1   Local Binary Fitting (LBF) Model

In order to segment images with intensity inhomogeneity, Li et al. [8] proposed the LBF model, which draws upon the intensity information in local regions by using the kernel function with one fixed scale parameter. Let $\Omega \in R^2$ be the image domain, and $I : \Omega \to R$ the given image. The energy functional of the LBF model is defined as:

$$
\begin{aligned}
E = \lambda_1 &\int_\Omega \int_{in(c)} K_\sigma(x-y)(I(y)-f_1(x))^2 dy dx \\
+ \lambda_2 &\int_\Omega \int_{out(c)} K_\sigma(x-y)(I(y)-f_2(x))^2 dy dx \\
+ &\mu l + \upsilon p(\phi)
\end{aligned}
\tag{1}
$$

where I(y) denotes the intensity configuration of point $y \in \Omega$. The segmenting curve c is represented by the zero level set, i.e. $c = \{x \in \Omega | \phi(x) = 0\}$. $in(c)$ and $out(c)$ represent the inside and outside region of evolving contour c, respectively. l denotes the length of c. K_σ is the Gaussian kernel with standard deviation σ. $\lambda_1, \lambda_2, \mu$ and υ are fixed parameters. $p(\phi)$ is used to avoid the re-initialization step. f_1 and f_2 are smooth functions approximating the local image intensities inside and outside the contour c, respectively. Obviously, the energy functional (1) is region-scalable, and σ plays a key role to control the size of local regions [8]. However, in the classical LBF model, only one fixed scale parameter σ is applied for each image and there is no general guideline for LBF model to choose suitable scale parameters for different images.

2.2   Local Maximum Description Difference Feature (LMDD)

In this section, we will introduce how to construct the multi-scale local region descriptor and the LMDD feature.

Multi-scale Local Region Descriptor. The most common model to describe intensity inhomogeneity [6,18] can be written as:

$$I = bJ + n \tag{2}$$

where $J : \Omega \rightarrow R$ is the true image to be restored, $b : \Omega \rightarrow R$ denotes the intensity inhomogeneity field, and $n : \Omega \rightarrow R$ is the noise.

Based on the assumption that the spectrum of intensity inhomogeneity is mainly concentrated in the lower frequency band, the local region descriptors can be constructed by using multi-scale low-pass filters, e.g. the Gaussian filter, the mean filter or median filter, etc. We take Gaussian filter as an example to elucidate how the multi-scale low-pass filters are used and embedded into the local region-based level set model. The multi-scale Gaussian filter is given by:

$$K_{\sigma_k}(x - y) = \frac{1}{\sqrt{2\pi}\sigma_k} e^{\frac{-|x-y|^2}{2\sigma_k^2}}, \quad k = 1, 2...m, \tag{3}$$

where x is the center pixel and y denotes the pixel in the neighborhood. Neighborhood scale is controlled by $\sigma_k = 2k + 1$. After determining the filters, multi-scale local region descriptors LI_k^{LBF} for LBF model is given by:

$$LI_k^{LBF} = \frac{\int_\Omega K_{\sigma_k}(x - y)I(y)dy}{\int_\Omega K_{\sigma_k}(x - y)dy} \tag{4}$$

It can be seen that LI_k^{LBF} denotes the Gaussian weighting mean in local regions with different scale.

LMDD and Optimal Scale Value. After obtaining the multi-scale local region descriptor, we can calculate the LMDD feature. First, the multi-scale local region description difference d_k is defined as:

$$d_k(x) = (I(x) - LI_k^{LBF}(x))^2 \tag{5}$$

Then, the LMDD feature $M(x)$ is given by:

$$M(x) = max_k(d_k(x)) \tag{6}$$

The optimal scale value for local region is obtained as follows:

$$s(x) = \arg max_k(d_k(x)) \tag{7}$$

It can be seen that, d_k indicates the approximation degree between LI_k and the original image I. Since LI_k^{LBF} is constructed by using the low-pass filters, d_k is actually a high-pass filtering operator. The LMDD feature, which will be embedded into the level set energy functional, is the maximum response of multi-scale high-pass filters.

The advantages to extract the LMDD feature are as follows: First, through LMDD feature, i.e. the maximum response of multiple high-pass filters with different scale, the intensity inhomogeneity located in low frequency band can be greatly restrained. Meanwhile, image details, as well as object boundaries, in high frequency are well preserved. Second, with the segmenting contour evolving, the LMDD feature for a pixel near the boundary and the corresponding optimal scale of local region will also be adjusted. Figure 2 shows an example of the optimal scale update.

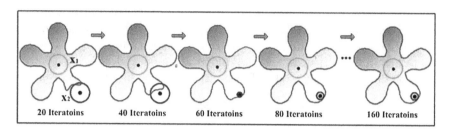

Fig. 2. Illustrates the evolution of the segmenting contour (red boundary) at different iteration and the corresponding optimal scale (denoted by the diameter of green circle and blue circle) for local regions centered at x_1 and x_2. (Color figure online)

Feature Incorporation and Multi-scale Local LBF Models (MS-LBF). In this section, we will introduce how to incorporate the LMDD feature into the traditional level set energy functional with CV-like structure:

$$E = E_D + E_R = E_{in} + E_{out} + E_R \qquad (8)$$

where E_D is the data term, which consists of two items, E_{in} and E_{out}, corresponding to the inside and outside regions of the evolving contour, respectively. E_R is the regularization term with the aim to smooth the evolving contour and avoid the reinitialization step. In this paper, E_R includes the arc length penalty term [18] and the re-initialization penalty term:

$$
\begin{aligned}
E_R &= \mu l + \upsilon p(\phi) \\
&= \mu \int_\Omega |\nabla H(\phi)| \mathrm{d}x + \upsilon \int_\Omega (\nabla \phi - 1)^2 \mathrm{d}x
\end{aligned} \qquad (9)
$$

Formula (5) can be divided into two parts $d_{k,in}^{MS-LBF}$ and $d_{out,k}^{MS-LBF}$, corresponding to the inside and outside region of evolving contour:

$$d_{k,in}^{MS-LBF} = (I - LI_{k,in}^{MS-LBF})^2 \qquad (10)$$

$$d_{out,k}^{MS-LBF} = (I - LI_{k,out}^{MS-LBF})^2 \qquad (11)$$

where

$$LI_{k,in}^{MS-LBF} = \frac{\int_\Omega K(x-y,\sigma_k)I(y)H(\phi)\mathrm{d}y}{\int_\Omega K(x-y,\sigma_k)H(\phi)\mathrm{d}y} \qquad (12)$$

$$LI_{k,out}^{MS-LBF} = \frac{\int_{\Omega} K(x-y,\sigma_k)I(y)(1-H(\phi))dy}{\int_{\Omega} K(x-y,\sigma_k)(1-H(\phi))dy} \tag{13}$$

$LI_{k,in}^{MS-LBF}$ and $LI_{k,out}^{MS-LBF}$ are the multi-scale local region descriptors with scale σ_k of inside and outside regions of evolving contour, respectively. $d_{k,in}^{MS-LBF}$ and $d_{out,k}^{MS-LBF}$, represent the multi-scale local region description difference with scale σ_k of inside and outside regions of evolving contour. Then, the LMDD feature inside and outside of contour at image pixel I(x) can be computed as:

$$M_{in}^{MS-LBF}(x) = max_k(d_{k,in,i,j}^{MS-LBF}(x)) \tag{14}$$

$$M_{out}^{MS-LBF}(x) = max_k(d_{k,out,i,j}^{MS-LBF}(x)) \tag{15}$$

According to (8), the multi-scale data term E_D^{MS-LBF} of MS-LBF model is obtained by:

$$\begin{aligned} E_D^{MS-LBF} &= E_{in}^{MS-LBF} + E_{out}^{MS-LBF} \\ &= \int_{\Omega} M_{in}^{MS-LBF} H(\phi)\mathrm{d}x + \int_{\Omega} M_{out}^{MS-LBF}(1-H(\phi))\mathrm{d}x \end{aligned} \tag{16}$$

The energy functional of MS-LBF model is obtained by:

$$\begin{aligned} E &= E_D^{MS-LBF} + E_R \\ &= \int_{\Omega} M_{in}^{MS-LBF} H(\phi)\mathrm{d}x + \int_{\Omega} M_{out}^{MS-LBF}(1-H(\phi))\mathrm{d}x \\ &+ \mu \int_{\Omega} |\nabla H(\phi)|\mathrm{d}x + \upsilon \int_{\Omega} (\nabla\phi-1)^2\mathrm{d}x \end{aligned} \tag{17}$$

Finally, the energy functional (17) is minimized by gradient descend method. Keeping $LI_{k,in}^{MS-LBF}$ and $LI_{k,out}^{MS-LBF}$ fixed and minimizing the energy functional with respect to ϕ, the Euler-Lagrange equation for ϕ can be deduced. Parameterizing the descent direction with an artificial time t, the evolution equation of MS-LBF model can be written as:

$$\begin{aligned} \frac{\partial\phi}{\partial t} &= \delta(\phi)(M_{in}^{MS-LBF} - M_{out}^{MS-LBF}) \\ &+ \mu\delta(\phi)\cdot div(\frac{\nabla\phi}{|\nabla\phi|}) + \upsilon(\nabla^2\phi - div(\frac{\nabla\phi}{|\nabla\phi|})) \end{aligned} \tag{18}$$

3   Experimental Results

In this section, experiments on real and simulated data are carried out to evaluate the performance of the proposed method. We also compare with traditional local region-based level set methods, i.e. LBF model. The parameters are set as follows: $\upsilon = 1, \Delta t = 0.1$ (the time step), $\sigma_k = 2k+1, k \in [1,m]$. Here, m determines the range of local region scale. If m is too big, the computational burden at each iteration will be greatly increased since much statistical information needs to be calculated. If m is too small, the local region will be too narrow to cover adequate object and background pixels. Generally, m can be defined in the interval [8, 32]. In this paper, m is set as 16 for MS-LBF model.

Meanwhile, Jaccard similarity coefficient (JSC) is used as a quantitative measure to evaluate the segmentation results [17]. JSC is defined as:

$$J(O_m, O_t) = \frac{A(O_m \cap O_t)}{A(O_m \cup O_t)} \quad (19)$$

where O_m denotes the derived object region by the algorithm and O_t denotes its corresponding object region in the ground truth image. $A(*)$ represents the area of region. The Jaccard similarity coefficient is bounded in [0, 1], and the larger value implies better segmentation result.

3.1 Evolving Process and Visual Evaluation

In this experiment, the proposed MS-LBF models were applied to real images with intensity inhomogeneity. The evolving process and the segmentation results are shown in Fig. 3. We can see that the proposed method yields reasonable segmentation results. Meanwhile, the last column displays the final level set functions, which are smooth and steady, demonstrating the capability of the proposed methods to keep the level set function regular during the curve evolving.

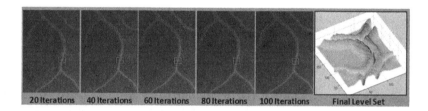

Fig. 3. The evolving process and the final level sets for MS-LBF model.

3.2 Robustness to Contour Initialization

We evaluated the influence of contour initialization on the final results with the MS-LBF. Real images with intensity inhomogeneity were used, and the initial contours (green polygons in Fig. 4.) were placed at different parts of the images. We also show the segmentation results with the LBF on the same image with same initial contours for comparison. The blue and red contours in Fig. 4 denote the final segmentation results of the LBF model and our method. It can be seen that the multi-scale model is robust to the contour initialization, and can obtain reasonable and almost same results, despite totally different initial contours. On the contrary, the LBF model cannot segment the image accurately by using the four different initial contours. This is because that the LMDD method can capture more boundary information, rather than be constrained by local region.

(a) LBF   (b) MS-LBF

Fig. 4. Influence of the contour initialization on the final segmentation results for (a) LBF model and (b) MS-LBF model. (Color figure online)

3.3 Comparison with LBF Model

In this experiment, the proposed MS-LBF model was compared with the traditional LBF model. When applying the LBF model, because there is no general guideline to choose suitable scale parameters, different scale values ranging from 1 to 16 were tested one by one. The segmentation results of LBF model are shown in Fig. 5(a). Among all the segmentation results of LBF model, the one with $\sigma = 3$ is best (enclosed by red rectangle in Fig. 5(a)). However, it still fails to segment the object accurately, especially in regions with severe intensity inhomogeneity (denoted by dotted blue circle in Fig. 5(a)). By comparison, Fig. 5(b) and (c) show the final scale map and the corresponding segmentation result generated by the proposed MS-LBF model. The optimal scale values in the image vary largely with different locations. Specifically, for pixels around the object boundary (denoted by yellow arrow in Fig. 5(b)) or regions with severe

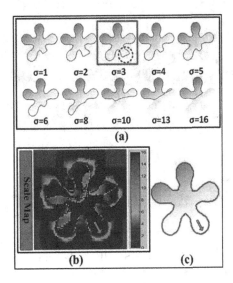

Fig. 5. (a) Segmentation results of LBF model with different scale parameter ranging from 1 to 16; (b) Final scale map obtained from LMDD feature; (c) Segmentation result of MS-LBF model. Yellow arrows in (b) and (c) point out boundary of the segmented object, where the scale values tend to be small. (Color figure online)

intensity inhomogeneity, the optimal scales for local regions tend to be small, so that the detailed information of the image can be captured. Whereas, for smooth regions, the optimal scale is big, and global information about intensity contrast is captured. In this way, the scale value can be adaptively determined to promote suitable local region descriptors to model the piecewise constant image, thus guiding the evolving contour toward desired boundary (Fig. 5(c)). Here, the parameter μ are set as 0.0001×255^2.

To further demonstrate the power of the proposed multi-scale method, two more experiments on real images are conducted and the results are shown in Fig. 6. It can been seen that MS-LBF model can generate reasonable segmentation results, while the LBF model fails in regions with severe intensity inhomogeneity (denoted by yellow arrows in Fig. 6), even various scale parameters are tried out. Here, the parameter μ are set as 0.001×255^2 and 0.01×255^2 for Fig. 6 (a) and (d), respectively.

To quantitatively evaluate the performance of the proposed method, the Jaccard similarity coefficient (JSC) between the ground truth and the segmentation results obtained by MS-LBF model and LBF model were calculated. The results are shown in Table 1. It is apparent that the JSC values of MS-LBF model are quite higher than that of LBF model, demonstrating better performance of the proposed MS-LBF model in comparison with LBF model for image with severe intensity inhomogeneity.

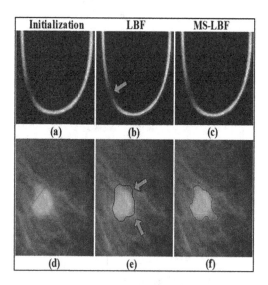

Fig. 6. Comparison of LBF model with MS-LBF model on two real medical images. (a) and (d) are the original images with initialized contours; (b) and (e) are the segmentation results of LBF model, where yellow arrows point to regions with segmentation obstacles; (e) and (f) show the segmentation results of MS-LBF model. (Color figure online)

Table 1. Jaccard similarity coefficients of LBF model and MS-LBF model for images in Figs. 5 and 6.

Model	Figure 5	Figure 6(a)	Figure 6(d)
LBF	0.9571	0.9316	0.7773
MS-LBF	0.9881	0.9985	0.9652

4 Discussion and Conclusion

Motivated by previous studies on multi-scale image processing and local region-based level set method, we propose a novel multi-scale local region-based level set method for segmentation of images with severe intensity inhomogeneity. By using the proposed LMDD feature, the optimal scale value of the local region for each image pixel is determined in an automatic, adaptive, and dynamic way. Then, the LMDD feature is incorporated into three classical local region-based level set model, such as LBF model, to complete the image segmentation. Experiments on synthetic and real images demonstrate better performance compared with the traditional local region-based level set models. It should be noted that since multi-scale or multi-layer structure is adopted in the proposed image segmentation method, the computational efficiency is suboptimal. Our future work will consider combining semantic information into the method, aiming to promote the computational efficiency of the proposed adaptive scale method.

Acknowledgments. This work was supported by the Major Science and Technology Program of Anhui Province (15czz02024), the National Natural Science Foundation of China (81401483), the Youth Innovation Promotion Association of CAS (2014290), Dean's Fund of Hefei Institute of Physical Science, CAS (YZJJ201525), the Natural Science Fund of Anhui Province (1708085MF141), Development Project of Foreign Expert Recruitment Program of Anhui Province, and John S. Dunn Research Foundation (STCW).

References

1. Brox, T., Cremers, D.: On local region models and a statistical interpretation of the piecewise smooth Mumford-Shah functional. Int. J. Comput. Vis. **84**(2), 184–193 (2009)
2. Darolti, C., Mertins, A., Bodensteiner, C., Hofmann, U.G.: Local region descriptors for active contours evolution. IEEE Trans. Image Process. **17**(12), 2275–2288 (2008)
3. Kim, D.H., Kim, H.K., Choi, K.S., Ko, S.J., et al.: Multi-scale level set based curve evolution for real-time non-rigid object contour tracking. In: TENCON 2009–2009 IEEE Region 10 Conference, pp. 1–5. IEEE (2009)
4. Lankton, S., Tannenbaum, A.: Localizing region-based active contours. IEEE Trans. Image Process. **17**(11), 2029–2039 (2008)
5. Li, C., Gore, J.C., Davatzikos, C.: Multiplicative intrinsic component optimization (mico) for MRI bias field estimation and tissue segmentation. Magn. Reson. Imaging **32**(7), 913–923 (2014)

6. Li, C., Huang, R., Ding, Z., Gatenby, J.C., Metaxas, D.N., Gore, J.C.: A level set method for image segmentation in the presence of intensity inhomogeneities with application to MRI. IEEE Trans. Image Process. **20**(7), 2007–2016 (2011)

7. Li, C., Kao, C., Gore, J.C., Ding, Z.: Implicit active contours driven by local binary fitting energy, pp. 1–7 (2007)

8. Li, C., Kao, C., Gore, J.C., Ding, Z.: Minimization of region-scalable fitting energy for image segmentation. IEEE Trans. Image Process. **17**(10), 1940–1949 (2008)

9. Lin, N., Yu, W., Duncan, J.S.: Combinative multi-scale level set framework for echocardiographic image segmentation. Med. Image Anal. **7**(4), 529–537 (2003)

10. Nercessian, S.C., Panetta, K.A., Agaian, S.S.: Non-linear direct multi-scale image enhancement based on the luminance and contrast masking characteristics of the human visual system. IEEE Trans. Image Process. **22**(9), 3549–3561 (2013)

11. Rajapakse, J.C., Kruggel, F.: Segmentation of mr images with intensity inhomogeneities. Image Vis. Comput. **16**(3), 165–180 (1998)

12. Sato, Y., Nakajima, S., Shiraga, N., Atsumi, H., Yoshida, S., Koller, T., Gerig, G., Kikinis, R.: Three-dimensional multi-scale line filter for segmentation and visualization of curvilinear structures in medical images. Med. Image Anal. **2**(2), 143–168 (1998)

13. Sui, H., Xu, C., Liu, J., Sun, K., Wen, C.: A novel multi-scale level set method for sar image segmentation based on a statistical model. Int. J. Remote Sens. **33**(17), 5600–5614 (2012)

14. Sum, K.W., Cheung, P.Y.S.: Vessel extraction under non-uniform illumination: a level set approach. IEEE Trans. Biomed. Eng. **55**(1), 358–360 (2008)

15. Vovk, U., Pernus, F., Likar, B.: A review of methods for correction of intensity inhomogeneity in MRI. IEEE Trans. Med. Imaging **26**(3), 405–421 (2007)

16. Wang, X., Huang, D., Xu, H.: An efficient local Chan-Vese model for image segmentation. Pattern Recognit. **43**(3), 603–618 (2010)

17. Zhang, H., Ye, X., Chen, Y.: An efficient algorithm for multiphase image segmentation with intensity bias correction. IEEE Trans. Image Process. **22**(10), 3842–3851 (2013)

18. Zhang, K., Liu, Q., Song, H., Li, X.: A variational approach to simultaneous image segmentation and bias correction. IEEE Trans. Cybern. **45**(8), 1426–1437 (2015)

19. Zhang, K., Song, H., Zhang, L.: Active contours driven by local image fitting energy. Pattern Recognit. **43**(4), 1199–1206 (2010)

20. Zhang, K., Zhang, L., Lam, K., Zhang, D.: A level set approach to image segmentation with intensity inhomogeneity. IEEE Trans. Syst. Man Cybern. **46**(2), 546–557 (2016)

Joint Supervision for Discriminative Feature Learning in Convolutional Neural Networks

Jianyuan Guo, Yuhui Yuan, and Chao Zhang[✉]

Key Laboratory of Machine Perception (MOE),
Peking University, Beijing 10087, China
{jyguo,yhyuan,c.zhang}@pku.edu.cn

Abstract. Convolutional Neural Networks have achieved excellent results in various tasks such as face verification and image classification. As a typical loss function in CNNs, the softmax loss is widely used as the supervision signal to train the model for multi-class classification, which can force the learned features to be separable. Unfortunately, these learned features aren't discriminative enough. In order to efficiently encourage intra-class compactness and inter-class separability of learned features, this paper proposes a H-contrastive loss based on contrastive loss for multi-class classification tasks. Jointly supervised by softmax loss, H-contrastive loss and center loss, we can train a robust CNN to enhance the discriminative power of the deeply learned features from different classes. It is encouraging to see that through our joint supervision, the results achieve the state-of-the-art accuracy on several multi-class classification datasets such as MNIST, CIFAR-10 and CIFAR-100.

Keywords: Convolutional neural networks · Joint supervision
H-contrastive loss

1 Introduction

Over the past several years, convolutional neural networks (CNNs) have efficiently boosted the state-of-the-art performance in many fields such as multi-class classification. The pipeline of multi-class classification can be summarized as: feature learning followed by classification. Firstly, the convolutional neural network uses the convolutional layers to learn the features from input images, then the inner-product layer outputs the score $z_i = \boldsymbol{w_i} \cdot \mathbf{f}$, where $\mathbf{f}$ represents the feature learned by the networks, $\boldsymbol{w_i}$ represents the weight vector belonging to the class i. Finally the last layer find out the highest score of feature $\mathbf{f}$, which means the input image belongs to the corresponding class. Many advanced network architectures [5–7] used softmax loss as the loss function in classification problem, which converges quickly in training and can be easily optimized by SGD (Stochastic Gradient Descent optimizer). If the features are separable in the feature embedding space after training, the tasks can be transformed to simply N classification problem while testing. Thus, it is crucial to learn separable features. Considering the

© Springer Nature Singapore Pte Ltd. 2017
J. Yang et al. (Eds.): CCCV 2017, Part II, CCIS 772, pp. 509–520, 2017.
https://doi.org/10.1007/978-981-10-7302-1_42

open-set protocol in face verification tasks, where the testing identities are usually disjoint from the training set, the deeply learned features need to be not only separable but discriminative enough. So the deep metric embedding needs to pull similar samples closer and push samples from different classes far away in embedding space. Inspired by the deep metric learning, we want to improve the separability of learned features, narrow the distance between features from the same class, and expand the distance between features from different classes simultaneously in both training and testing phases.

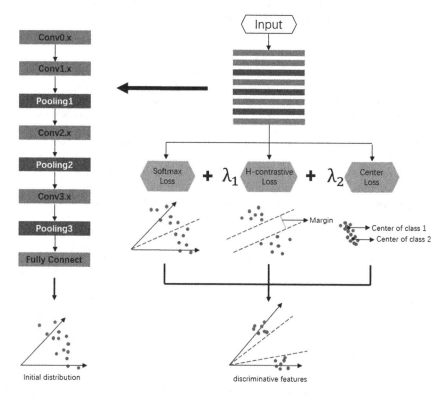

Fig. 1. Jointly supervised architecture for multi-class classification problem.

In tasks of classification and face recognition, pioneering works [5–7] learned features via the softmax loss, but softmax loss only learned separable features which are not discriminative enough. To improve this, some methods combined softmax loss with contrastive loss [2,3] or center loss [1] to learn more discriminative features, and [4] adopted triplet loss to supervise the embedding learning, leading to state-of-the-art face recognition results. However, center loss only decreases the intra-class distance while ignoring the inter-class separability. Both contrastive loss and triplet loss require carefully designed pair/triplet mining procedures because the results are sensitive to the mining hard samples, which are time-consuming and the results are determined by the quality of mining procedures. [11] proposed Hard-Aware Deeply Cascaded Embedding based on the contrastive loss to mine hard examples in deep metric embedding. Inspired by [11], we propose the H-contrastive loss function based on the contrastive loss to help efficiently enhance the discriminative power of the learned features in CNNs for classification problem, we will define it in detail in Sect. 3.1. While using the H-contrastive loss, we don't need to spend any time on the design of hard examples mining procedures. As shown in Fig. 1, the softmax loss learns separable features, the H-contrastive loss produces decision margin between features from different classes, the center loss further decreases the distance between features from the same class. With the joint supervision, not only the inter-class distances are enlarged, but also the intra-class distances are shrunk down.

Our major contributions can be summarized as follows:

1. We propose the H-contrastive loss for CNNs to enhance the discriminative power of learned features, which doesn't need to design pair mining procedures to pick hard samples.
2. We show that H-contrastive loss is robust enough and can be jointly supervised with other loss functions in CNNs easily. By using softmax loss, H-contrastive loss and center loss to jointly supervise the training, we achieved the state-of-the-art performance on several multi-class classification datasets, e.g. MNIST, CIFAR-10, and CIFAR-100.

2 Related Work

Center Loss. Wen et al. [1] proposed center loss to simultaneously learn a center for deep features of each class and penalize the distances between the deep features and their corresponding class centers. The softmax loss pushes deep features from different classes away from each other during training, while center loss forces deep features from the same class closer to the center of the class. The main idea of center loss is to shrink down the intra-class distance, then use the softmax loss to jointly supervise the training for expanding inter-class distance. Thus the ability in this joint supervision to force the features from different class is the same as the supervision conducted by softmax loss only. Different from the center loss, H-contrastive loss can accomplish these two tasks itself.

Large-Margin Softmax Loss. Liu et al. [9] proposed a loss function based on the softmax loss, called Large-Margin Softmax Loss(L-Softmax), to concentrate more on the angular decision margin between different classes by defining an adjustable margin parameter m. L-Softmax loss can replace the softmax loss in the training of CNNs, it firstly constrains L_2-normalization on both features $\mathbf{f}$ and weight vector $\mathbf{w}_i$, then enlarges the decision margin between different classes based on the cosine similarity.

Contrastive Loss/Triplet Loss. Because of the high intra-class and low inter-class variance, metric learning with CNNs [2–4] use contrastive loss and triplet loss to construct loss functions for image pairs and triplets. The goal is to use a

CNN to learn a feature embedding that captures the semantic similarity among images. Unlike deep learning, deep metric learning usually takes pairs or triplets of samples as input, and outputs the distance between them. The most widely used metric learning methods are contrastive loss and triplet loss, both of the two loss functions optimize the normalized Euclidean distance between feature pairs/triplets. But it is almost impossible to deal with all possible combinations during training, so sampling and mining procedures are necessary [4], and usually, these procedures are time-consuming. On the contrary, H-contrastive loss can jointly supervise the CNN training with other loss functions and be optimized by SGD easily without procedure carefully designed to mine hard examples.

3   The Proposed H-Contrastive Loss Function

In this section, we elaborate our approach, and use some toy examples to intuitively show the distributions of deeply learned features supervised by different loss functions.

3.1   Definition

First we give the notations that will be used to describe our method:

- $P = \{I_i^+, I_j^+\}$: all the positive input image pairs constructed from the mini-batch training set, where I_i^+ and I_j^+ are supposed to belong to the same class.
- $N = \{I_i^-, I_j^-\}$: all the negative input image pairs constructed from the mini-batch training set, where I_i^- and I_j^- are supposed to come from different classes.
- $\{f_i^+, f_j^+\}$: the computed feature vector for positive pairs $\{I_i^+, I_j^+\}$ after the transform function that transform the output of the computation block to a low dimensional feature vector for distance calculation.
- $\{f_i^-, f_j^-\}$: the computed feature vector for negative pairs $\{I_i^-, I_j^-\}$.

The H-contrastive loss is defined as:

$$L_H = \sum_{(i,j)\in P} L^+(i,j) + \sum_{(i,j)\in N} L^-(i,j) \qquad (1)$$

$$L^+(i,j) = D(f_{i,h}^+, f_{j,h}^+) \qquad (2)$$

$$L^-(i,j) = max\,\{0,\, M - D(f_{i,h}^-, f_{j,h}^-)\} \qquad (3)$$

where $D(f_{i,h}, f_{j,h})$ is the Euclidean distance between the two L_2-normalized feature vectors $f_{i,h}$ and $f_{j,h}$, M is the margin. It is difficult to predefine thresholds for hard sample selection as the loss distributions keep changing during training, so we use a simple way by ranking distances of all positive pairs in a mini-batch, and take top h percent samples as hard positive set, and similarly for hard negative example mining. In this way, we don't need to design the mining procedure and can still pick out hard samples. We use a hyperparameter h to control hard ratio in training, $(f_{i,h}, f_{j,h})$ means top h percent feature pairs in (f_i, f_j). The original softmax loss and center loss can be written as:

$$L_S = -\frac{1}{m} \sum_{i=1}^{m} \log \frac{e^{W_{y_i}^T f_i + b_{y_i}}}{\sum_{j=1}^{n} e^{W_j^T f_i + b_j}} \tag{4}$$

$$L_C = \sum_{i=1}^{m} \|x_i - c_{y_i}\|_2^2 \tag{5}$$

In Eq. (4), m is the batch size in training, n is the number of classes, x_i denotes the ith deep feature, y_i is the corresponding class label, W and b are weight and bias for the inner-product layer of CNNs. The c_{y_i} in Eq. (5) denotes the y_ith class center of learned features. We adopt the joint supervision of softmax loss, H-contrastive loss and center loss to train the CNNs to learn more discriminative features, the formulation is given in Eq. (6). λ_1 and λ_2 are used for balancing the three loss functions.

$$L = L_S + \lambda_1 L_H + \lambda_2 L_C \tag{6}$$

3.2 Toy Examples

In order to give an intuitive feeling about the distribution of deeply leaned features, we did some toy examples based on Wen et al.'s model in [1] except for some minor modifications on the MNIST [8] dataset, the CNN architecture we adopt is shown in Table 1: We reduce the output number of the fully connected (FC) layer to 2, which means the dimension of the deep features is 2, and then plot them by class, as shown in Fig. 2. We decrease the output number of the FC

Table 1. The CNN architectures for MNIST/CIFAR-10/CIFAR-100. Conv1.x denotes convolution units that may contain multiple convolution layers. E.g., [5 × 5, 32] × 2 denotes 2 cascaded convolution layers with 32 filters of size 5 × 5. All the pooling layers have the same pooling strides of 2.

Layer	MNIST-2D	MNIST	CIFAR-10	CIFAR-100
Conv0.x	N/A	[3 × 3, 64] × 1	[3 × 3, 64] × 1	[3 × 3, 128] × 1
Conv1.x	[5 × 5, 32] × 2	[3 × 3, 64] × 3	[3 × 3, 64] × 4	[3 × 3, 128] × 4
Pool1	2 × 2 Max	2 × 2 Max	2 × 2 Max	2 × 2 Max
Conv2.x	[5 × 5, 64] × 2	[3 × 3, 64] × 3	[3 × 3, 128] × 4	[3 × 3, 256] × 4
Pool2	2 × 2 Max	2 × 2 Max	2 × 2 Max	2 × 2 Max
Conv3.x	[5 × 5, 128] × 2	[3 × 3, 64] × 3	[3 × 3, 256] × 4	[3 × 3, 512] × 4
Pool3	2 × 2 Max	2 × 2 Max	2 × 2 Max	2 × 2 Max
FC	2	256	256	512

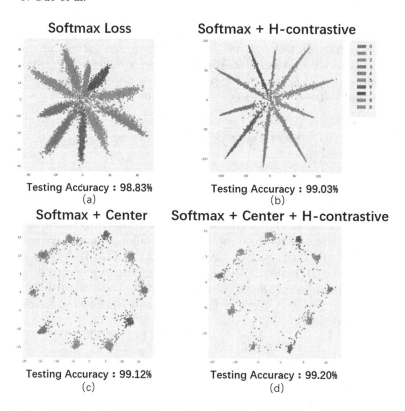

Fig. 2. 2-D feature distribution on MNIST dataset's test set. (a) Features learned in CNN supervised by softmax loss only. (b) Features learned in CNN jointly supervised by softmax loss and H-contrastive loss, it is obviously to see the distance between different classes is larger than (a). (c) Features learned in CNN jointly supervised by softmax loss and center loss, features from the same class are closer to their center. (d) Features learned in CNN jointly supervised by softmax loss, center loss and H-contrastive loss, H-contrastive loss helps to not only further expand the distance between different classes and narrow the intra-class distance, but also decrease the features scattered in the middle.

layer which affects the performance, that is the reason why the testing accuracy is not as good as Table. 2. We can find that under the supervision of softmax loss, the deeply learned features are separable, but not discriminative enough. Through the joint supervision, our H-contrastive loss enhances the discriminative power of features significantly.

3.3   Discussion

If we only use softmax loss to supervise the CNN, the features in test set would have short inter-class distances as well as long intra-class distances. After the joint supervision with center loss, features would have shorter intra-class distances, but the angle between different classes has no change, in other word, the cosine similarity between two different classes is the same as the features supervised by softmax loss only. The H-contrastive loss can help decrease the intra-class distance and enlarge angle between different classes efficiently.

4 Experiments

4.1 Experimental Settings

We evaluate the loss functions in three standard benchmark datasets: MNIST [8], CIFAR-10 [21] and CIFAR-100 [21]. In testing stage, we only use the softmax loss to classify the samples in all datasets. For convenience, we use HC to denote the H-contrastive loss, and the training on the same dataset supervised by different loss functions use the same CNN shown in Table 1.

General Settings: Our general framework to train and extract deeply learned features is illustrated in Fig. 1, while using softmax loss, center loss and H-contrastive loss to jointly supervise the training, we fix the λ_1 to 1 and the λ_2 to 0.05, the different combinations of λ_1 and λ_2 are analysed in Sect. 4.4 in detail, and set the margin M in H-contrastive loss function to 0.4. We implement the CNNs using the Caffe library [12] with our modifications. For experiments, we adopt the ReLU [10] as the activation function, a weight decay of 0.0005 and momentum of 0.9, and the batch size for all experiments is 256, in all convolution layers, the stride is set to 1. The weight initialization in [13] and batch normalization [14] are used in our networks to replace the dropout. For optimization, the SGD will work well, and during the training and testing, we don't adopt any data augmentation setup.

MNIST/CIFAR-10/CIFAR-100: We start with a learning rate of 0.1, divide it by 10 when the error plateaus, finally terminate training at 10k/30k/30k iterations in corresponding datasets. The training data and testing data is 60k/10k, 50k/10k and 50k/10k split following the standard settings for MNIST, CIFAR-10 and CIFAR-100.

4.2 Multi-class Classification Results

MNIST: The network architecture we adopt is shown in Table 1. It is obvious that our method boosts the performance efficiently, improves the softmax from 0.35% to 0.29%, improves the L-Softmax from 0.30% to 0.27% and improves the result supervised by softmax and center loss from 0.31% to 0.25%. Moreover, we use the same architecture with [9], and through joint supervision our method can achieve the sate-of-the-art results while training with less iterations on MNIST, and we believe that the improvement in relative error rate is more worthy of attention.

Table 2. Error rate (%) on MNIST/CIFAR-10/CIFAR-100.

Method	MNIST	CIFAR-10	CIFAR-100
CNN [15]	0.53	N/A	N/A
DropConnect [16]	0.57	9.41	N/A
Hinge Loss [9]	0.47	9.91	33.10
DSN [17]	0.39	9.69	34.57
R-CNN [18]	0.31	8.69	31.75
GenPool [19]	0.31	7.62	32.37
Softmax [9]	0.40	9.05	32.74
L-Softmax [9]	0.31	7.58	29.53
Softmax	0.35	8.59	31.80
L-Softmax	0.30	7.60	29.53
Softmax + HC	0.29	7.38	30.24
L-Softmax + HC	0.27	7.09	29.17
Softmax + Center	0.31	7.24	26.59
Softmax + Center + HC	**0.25**	**6.89**	**25.80**

CIFAR-10: Table 1 also shows the CNN architecture that we use to evaluate our method. Firstly, we reproduce the results following the same setting in [9], the L-Softmax has effectively improved the softmax loss and the performance of L-Softmax is already very high. The second column in Table 2 quantifies the effectiveness of our H-contrastive loss. Our H-contrastive loss improved the softmax loss from 8.59% to 7.38%, improved the L-Softmax loss from 7.60% to 7.09%, which is illustrated in Fig. 3(a). We achieve the best performance by jointly supervising the CNN with softmax loss, center loss and H-contrastive loss, which improve the error rate from 7.24% to 6.89%.

CIFAR-100: We also evaluate our method on more complicated dataset CIFAR-100 which has 10000 testing images belonging to 100 classes to further verify the effectiveness of H-contrastive loss and the necessity of joint supervision. The results are shown in the third column in Table 2, and are also illustrated in Fig. 3(b), the CNN architecture refers to Table 1. One can notice that the joint supervision outperforms the CNN with the other competitive methods. H-contrastive loss improved the softmax loss from 31.80% to 30.24%, improved the L-Softmax loss from 29.53% to 29.17% and beat the CNN jointly supervised by softmax loss and center loss for decreasing the error rate from 26.59% to 25.80%. We use the joint supervision to promote the performance for 6.00%, improving the performance from original 31.80% to 25.80%, and get the state-of-the-art result on the CIFAR-100.

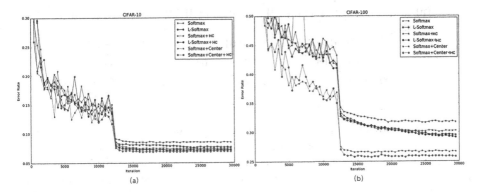

Fig. 3. Error rate vs. iteration with different loss functions on (a) CIFAR-10. (b) CIFAR-100.

4.3 Experiments on Parameter h

We also conduct experiments on CIFAR-10 and CIFAR-100 to investigate how the hard ratio h influences the result. Table 3 shows that different h lead to different results and we can see consistent improvement for all different choices of the h, in which S means softmax loss, C means center loss and HC means H-contrastive loss. With proper h, the performance of our method can be significantly boosted, which are illustrated in Fig. 4.

4.4 Experiments on Parameters λ_1 and λ_2

The hyperparameters λ_1 and λ_2 determine the inter-class separability and the intra-class variations in the joint supervision. Both of them are essential for the training, so we conduct two experiments to investigate the results in terms of various combination weights.

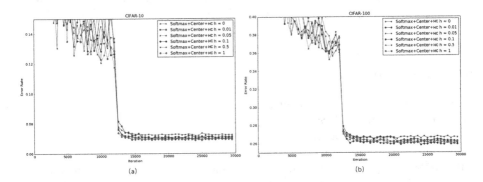

Fig. 4. Error rate vs. iteration with different hard ratios h on (a) CIFAR-10. (b) CIFAR-100.

Table 3. Error rate (%) on CIFAR-10 and CIFAR-100 with different h.

dataset	Method	$h=0$	$h=0.01$	$h=0.05$	$h=0.1$	$h=0.5$	$h=1$
CIFAR-10	S + C + HC	7.24	7.01	6.98	6.96	**6.89**	6.97
CIFAR-100	S + C + HC	26.59	26.27	**25.80**	25.98	26.19	26.09

In the first experiment, we fix λ_1 to 0.5, vary λ_2 from 0.0001 to 1 to supervise the training, the accuracies of the joint training on CIFAR-10 dataset are shown in Fig. 5(a). But the vanishing gradient problem occurs during the training when λ_2 is higher than 0.1, we think it is mainly caused by the initialization's instability of the center loss. Except that, it is very clear that using other loss functions to jointly train the CNN can obviously improve the results. In the second experiment, we fix λ_2 to 0.05 and vary λ_1 from 0.0001 to 1 to supervise the training, as shown in Fig. 5(b). Properly choosing the combination weights of λ_1 and λ_2 can train the network to learn more discriminative features and improve the accuracy on the multi-class classification dataset.

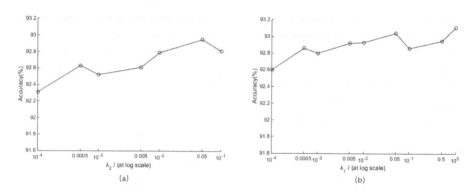

Fig. 5. Accuracies on CIFAR-10 dataset, which were respectively achieved by the same network settings except the values of hyperparameters λ_1 and λ_2. (a) results with different λ_2 while fixed $\lambda_1 = 0.5$, (b) results with different λ_1 while fixed $\lambda_2 = 0.05$.

5   Conclusion

In this paper, we propose H-contrastive loss based on the contrastive loss to help increase the inter-class separability and decrease the intra-class distance at the same time. We recommend that center loss and H-contrastive loss should jointly supervise the training. Extensive experiments on MNIST, CIFAR-10 and CIFAR-100 verify the effectiveness of our method and show clear advantages over current state-of-the-art CNNs and all compared baselines. In the future, we are going to evaluate our method on larger dataset like ImageNet and on other field such as face verification tasks.

Acknowledgments. This work is supported by the National Key Basic Research Project of China (973 Program) under Grant 2015CB352303 and the National Nature Science Foundation of China under Grant 61671027.

References

1. Wen, Y., Zhang, K., Li, Z., Qiao, Y.: A discriminative feature learning approach for deep face recognition. In: Leibe, B., Matas, J., Sebe, N., Welling, M. (eds.) ECCV 2016. LNCS, vol. 9911, pp. 499–515. Springer, Cham (2016). https://doi.org/10.1007/978-3-319-46478-7_31
2. Sun, Y., Chen, Y., Wang, X., et al.: Deep learning face representation by joint identification verification. In: Advances in Neural Information Processing Systems, pp. 1988–1996 (2014)
3. Sun, Y., Wang, X., Tang, X.: Sparsifying neural network connections for face recognition. In: Proceedings of the IEEE Conference on Computer Vision and Pattern Recognition, pp. 4856–4864 (2016)
4. Schroff, F., Kalenichenko, D, Philbin, J.: FaceNet: a unified embedding for face recognition and clustering. In: Proceedings of the IEEE Conference on Computer Vision and Pattern Recognition, pp. 815–823 (2015)
5. Sun Y, Wang X, Tang X.: Deep learning face representation from predicting 10,000 classes. In: Proceedings of the IEEE Conference on Computer Vision and Pattern Recognition, pp. 1891–1898 (2014)
6. Szegedy, C., Liu, W., Jia, Y., et al.: Going deeper with convolutions. In: Computer Vision and Pattern Recognition, pp. 1–9. IEEE (2015)
7. Simonyan, K., Zisserman, A.: Very deep convolutional networks for large-scale image recognition. In: Computer Science (2014)
8. LeCun, Y.: The MNIST database of handwritten digits (1998). http://yann.lecun.com/exdb/mnist/
9. Liu, W., Wen, Y., Yu, Z., et al.: Large-margin softmax loss for convolutional neural networks. In: Proceedings of The 33rd International Conference on Machine Learning, pp. 507–516 (2016)
10. Krizhevsky, A., Sutskever, I., Hinton, G.E.: ImageNet classification with deep convolutional neural networks. In: Advances in neural information processing systems, pp. 1097–1105 (2012)
11. Yuan, Y., Yang, K., Zhang, C.: Hard-aware deeply cascaded embedding. In: Proceedings of International Conference on Computer Vision (2017)
12. Jia, Y., Shelhamer, E., Donahue, J., et al.: Caffe: convolutional architecture for fast feature embedding. In: Proceedings of the 22nd ACM International Conference on Multimedia, pp. 675–678. ACM (2014)
13. He, K., Zhang, X., Ren, S., et al.: Delving deep into rectifiers: surpassing human-level performance on ImageNet classification. In: Proceedings of the IEEE International Conference on Computer Vision, pp. 1026–1034 (2015)
14. Ioffe, S., Szegedy, C.: Batch normalization: accelerating deep network training by reducing internal covariate shift. In: ICML (2015)
15. Jarrett, K., Kavukcuoglu, K., LeCun, Y.: What is the best multi-stage architecture for object recognition? In: 2009 IEEE 12th International Conference on Computer Vision, pp. 2146–2153. IEEE (2009)
16. Wan, L., Zeiler, M., Zhang, S., et al.: Regularization of neural networks using DropConnect. In: Proceedings of the 30th International Conference on Machine Learning (ICML-13), pp. 1058–1066 (2013)

17. Lee, C.Y., Xie, S., Gallagher, P., et al.: Deeply-supervised nets. In: Artificial Intelligence and Statistics, pp. 562–570 (2015)
18. Liang, M., Hu, X.: Recurrent convolutional neural network for object recognition. In: Proceedings of the IEEE Conference on Computer Vision and Pattern Recognition, pp. 3367–3375 (2015)
19. Lee, C.Y., Gallagher, P.W., Tu, Z.: Generalizing pooling functions in convolutional neural networks: mixed, gated, and tree. In: International Conference on Artificial Intelligence and Statistics (2016)
20. Cui, Y., Zhou, F., Lin, Y., et al.: Fine-grained categorization and dataset bootstrapping using deep metric learning with humans in the loop. In: Proceedings of the IEEE Conference on Computer Vision and Pattern Recognition, pp. 1153–1162 (2016)
21. Krizhevsky, A., Geoffrey, H.: Learning multiple layers of features from tiny images. Technical report (2009)

Instance Semantic Segmentation via Scale-Aware Patch Fusion Network

Jinfu Yang[1,2], Jingling Zhang[1,2(✉)], Mingai Li[1,2], and Meijie Wang[1,2]

[1] Faculty of Information Technology,
Beijing University of Technology, Beijing, China
13137730606@163.com
[2] Beijing Key Laboratory of Computational Intelligence
and Intelligent System, Beijing, China

Abstract. Instance semantic segmentation has already been a promising direction, but many leading approaches are lack of detailed structural information and unable to segment small size objects. In this paper, we present a novel segmentation framework, called Scale-aware Patch Fusion Network (SPF). Our unified end-to-end trainable network consists of three components, namely, multi-scale patch generator, semantic segmentation network and patch fusion algorithm. This patch-based method aggregates information from different scales of patches via fusing local segmentation prediction results. The proposed approach is thus more effective and simple. Experiments on VOC 2012 segmentation val, VOC 2012 SDS val, MS COCO datasets validate the effectiveness of our approach.

Keywords: Structural information · Scale-aware · SPF
Instance semantic segmentation

1 Introduction

Recently, instance semantic segmentation (i.e., simultaneous detection and segmentation, SDS [6]) has become an attractive object recognition goal. It combines elements from object detection and semantic segmentation. SDS is therefore much more challenging because it requires precise detection and correct segmentation of all objects in an image. SDS can be widely used in various fields, such as automatic driving, surveillance, visual question answering and robot obstacle avoidance, to name a few.

Instance semantic segmentation seeks to classify the semantic category of each pixel and relate each pixel to a physical instance. Related approaches are roughly divided into three streams. The first is based on segment-proposal. For example, Hariharan et al. [6] developed pioneer work. It started with category-independent bottom-up object proposals generated by MCG [10]. It then extracted features from both the bounding box of the region and the foreground of the region using convolution neural networks (CNNs). Finally, the

© Springer Nature Singapore Pte Ltd. 2017
J. Yang et al. (Eds.): CCCV 2017, Part II, CCIS 772, pp. 521–532, 2017.
https://doi.org/10.1007/978-981-10-7302-1_43

concatenated features were classified by SVM. The second is built upon Markov random field model (MRF). In [13], the authors used CNNs in conjunction with a global densely connected MRF to perform local object disambiguation and derive a globally consistent labels of the entire image. The last kind of methods [11,12] is to use recurrent neural network (RNN) for dense prediction. [11] presented an end-to-end RNN architecture with visual attention mechanism to perform instance segmentation. Typically, the segment-proposal based approaches have dominated the field of SDS.

However, those segment-proposal based methods have three drawbacks. First, the RoI (region of interest) pooling layer loses detailed spatial structure information because of feature pooling and resizing, which, however, is important to get fixed-size feature representation for fully-connected (fc) layers [3]. The object in an image may be mis-predicted or the pixels that belong to the same object may have inconsistent labels due to the fixed-size representation. Second, the size of object proposals greatly influences the performance of segmentation. The proposal-based approaches assume that instances are almost already in proposals and what they only need to do is to segment them out. Such characteristics cause the small instances not to be searched. Third, the proposals also contain much noise regarding other instances. The proposals not only contain the instances we are interested but also contain the other objects we are not interested.

In this paper, we present a novel segmentation framework, called Scale-aware Patch Fusion Network (SPF), as shown in Fig. 1. Studies of mid-level representation demonstrate that it is helpful to extract more structural information and model instance variation in local patches. Motivated by the spirit of mid-level representation and multi-scale orderless pooling [4], the proposed SPF accepts multiple scale patches as inputs, followed by a flexible patch fusion algorithm. Our system regards different patches as different semantic parts of the entire instance. Experiments on VOC 2012 segmentation val, VOC 2012 SDS val, MS COCO demonstrate excellent performance using the end-to-end training deep VGG-16 model (Titan X).

In addition, our framework to solve the SDS problem makes the following main contributions.

(1) We propose the strategy to generate the multi-scale patches for instance parsing.
(2) We develop an efficient algorithm to infer the segmentation mask for each instance by merging information from mid-level patches.
(3) We capture much more detail and discriminative information via different patches.

2   The Proposed Method

Figure 1 shows the overall architecture of the deep SPF network. The key components of the framework are multi-scale patch generator, semantic segmentation network and the patch fusion algorithm. First, multiple scales of patches are

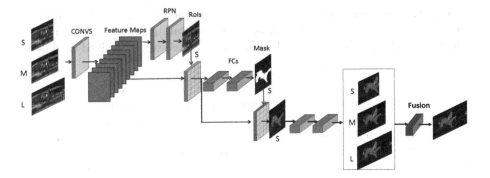

Fig. 1. The SPF framework for instance segmentation. S represents the small-level scale patch, M denotes the middle-level patch, and L is the large-level scale patch. We segment the different levels of scales patches, then merge the segmentation results via a new fusion algorithm. However, we just show the middle result of S scale in the framework.

generated, and then the local patches are segmented and classified via a multi-task segmentation network. Finally, the predicted results are fused by the new patch fusion algorithm.

2.1 Multi-scale Patch Generator

In this section, we describe how to generate multiple scales of patches from the original image. In this paper, we use three scales of patches, i.e., 64×64, 96×96, and 256×256. We first normalize all the images to the same scale of 256×256, then generate three scales of patches for each normalized image. The coarsest scale is the whole image with the global spatial information preserved. For the other two scales, we extract 64×64, 96×96 patches to capture more local and fine-grained information. To filter out redundant patches, we must reselect from the overlapping patches according to the following two constrains:

(1) The patch center is overlapped with the instance center.
(2) The area of P_i is half larger than that of I_i.

P_i denotes the i-th patch and I_i represents the i-th instance of the patch. We choose patches that satisfy the above two constrains, and store them in Γ_1. And represent each patch with a four-tuple (r, c, h, w), where h and w are height and width, respectively, while r and c are the coordinates of its top-left corner.

2.2 Semantic Segmentation Network

After reselecting patches, a cascade segmentation network is employed to get the corresponding patch segmentation results. Our cascade segmentation model contains three stages: differentiating instances, regressing mask-level instances, and categorizing instances.

Differentiating Instances. In the first stage, we use the Region Proposal Networks (RPNs) which has two sibling 1×1 convolutional layers for box regression and object classification. The loss function of this stage is defined as follows:

$$L_{RPN}(\Theta) = L_{cls}(B(\Theta)) + \lambda L_{reg}(B(\Theta)) \tag{1}$$

Here Θ and B separately denote the network parameters and the outputs of the first stage. The boxes list: $B = \{B_i\}$ and $B_i = \{x_i, y_i, w_i, h_i, p_i\}$, where B_i is a box indexed by i and (x_i, y_i) is the coordinate of its center. w_i and h_i represent the width and the height, respectively. p_i is the predicted objectness probability. In this paper, the balance weight λ is set to 1. The Eq. (1) indicates that the cost of stage-1 is the function of network parameters Θ.

Regressing Mask-Level Instances. In the second stage, the inputs are the shared convolutional features and the regressed bounding boxes. It outputs a pixel-level mask for each RoI. For generating fixed-size representation and being differentiable to the box coordinate, we perform the RoI pooling by a new strategy. Firstly, we use the RoI warping layer to crop a feature map region and warp it into a fixed-size (14×14) by bilinear interpolation. Then we perform standard max pooling after the warping operation. The RoI warping operation can be described as:

$$F_i^{RoI}(\Theta) = G(B_i(\Theta))F(\Theta) \tag{2}$$

Here $F(\Theta)$ represents the full-image feature map, which is reshaped as a m-d vector ($m = WH$) where $W \times H$ corresponds to the spatial resolution for the full-image feature map. G denotes the cropping and warping operations, and it is a $m' \times m'$ matrix with $m' = W'H'$ representing the RoI warping output. $F_i^{RoI}(\Theta)$ is a m'-d vector corresponding to the pre-defined warping output resolution $W' \times H'$. The computation in Eq. (2) has the following representation:

$$F_i^{RoI}(u', v'|\Theta) = \sum_{(u,v)}^{W \times H} G(u, v; u', v'|B_i(\Theta))F_{(u,v)}(\Theta)$$

$$= \sum_{(u,v)}^{W \times H} R(u, u'|x_i, w_i)R(v, v'|y_i, h_i)F_{(u,v)}(\Theta) \tag{3}$$

$$R(u, u'|x_i, w_i) = max(0, 1 - |(x_i + \frac{u'}{W'}w_i - u)|)$$

$$R(v, v'|y_i, h_i) = max(0, 1 - |(y_i + \frac{v'}{H'}h_i - u)|)$$

Here (u', v') is the pixel coordinate in the target $W' \times H'$ feature map, and (u, v) is defined similarly. The function G denotes transforming the bounding box size from $(x_i - w_i/2, x_i + w_i/2) \times (y_i - h_i/2, y_i + h_i/2)$ into $(-W'/2, W'/2) \times (-H'/2, H'/2)$, while R represents the bilinear interpolation function.

After the special RoI pooling, we utilize two fc layers to reduce dimension and regress mask. The feature dimension is reduced to 256 by the first fc layers.

The second fc layer has high dimension 784-way output for regressing a pixel-level mask Mi with a spatial resolution of $n \times n$ (we use $n = 28$). Then the mask is parameterized by an n^2-dimensional vector. The loss function of stage-2 is formally written as:

$$L_{mask}(\Theta) = L_{mask}(M(\Theta)|B(\Theta)) \tag{4}$$

As a related method, DeepMask also regresses discretized masks. DeepMask applies the regression layers to dense sliding windows (fully-convolutionally), but our method only regresses masks from a few proposed boxes and so reduces computational cost. Moreover, mask regression is only one stage in our network cascade that shares features among multiple stages, so the marginal cost of the mask regression layers is very small.

Categorizing Instance. Given the provided binary mask M_i and warped feature maps F_i^{RoI}, we can compute the masked feature maps F_i according to the following element-wise product:

$$F_i = F_i^{RoI}(\Theta) \bullet M_i(\Theta) \tag{5}$$

The loss term L_{mask} is described as:

$$L_{classify}(\Theta) = L_{classify}(C(\Theta)|B(\Theta), M(\Theta)) \tag{6}$$

where C is the output of stage-3, representing category prediction list: $C = \{C_i\}$.
Then the loss function of the whole network is defined as follows:

$$L(\Theta) = L_{RPN}(\Theta) + L_{mask}(M(\Theta)|B(\Theta)) + L_{classify}(C(\Theta)|B(\Theta), M(\Theta)) \tag{7}$$

where balance weights of 1 are implicitly used among the three terms. L is minimized w.r.t. the network parameters.

This loss function is unlike traditional multi-task learning, because the loss term of a later stage depends on the output of the earlier ones. Based on the chain rule of backpropagation, the gradient of L_{mask} involves the gradients w.r.t. B. The main technical challenge of applying the chain rule to Eq. (7) lies on the spatial transform of a predicted box B_i that determines RoI pooling. However, this can be solved by the RoI warping layer. And we finally train the model by Stochastic Gradient Descent (SGD) for the whole objective function.

2.3   Patch Fusion Algorithm

After network segmentation and classification, we get the predicted label y_i and semantic mask for each patch. We get the final results by fusing the semantic masks from nearby patches. In order to reduce the accumulative error, we fuse the patches with a pyramid polymerization method.

For each P_i, we compute the overlap score of semantic mask from neighboring patches which have the same predicted label y_i. We denote s_{mn} as the overlap score of P_m and P_n, which is defined by the intersection-over-union score (IoU). We search patches from two branches. The row search range includes the patches located on the left side of P_m, denoted as $C_l(P_i)$. The column search range includes the patches located on top of P_m. We denote patches in this range as $C_t(R_i)$. All the patches from row and column directions will be iterated. The overlapped scores s_{mn}, nearby patches P_m and P_n, are all stored in Γ_2, and we will merge patch pair which has the highest overlap score, until there is no patch pair with the overlap score higher than threshold τ.

3   Experiments

3.1   Implementation Details

Positive/Negative Samples. On the second stage, we compute the highest overlap score for each regressed box with respect to the ground truth mask. An RoI is considered positive and contributes to the mask branch if the box IoU is larger than 0.5, and negative otherwise. On stage 3, we adopt two sets of positive/negative samples. For the first set, if the box-level IoU between box and the nearest ground truth box $\geqslant 0.5$, the RoI is regarded as positive samples (the rest are the negative samples). For the second set, the positive samples are the objects that overlap with ground truth objects by box-level IoU $\geqslant 0.5$ and mask-level IoU $\geqslant 0.5$. The classification loss is only defined on positive RoIs. The stage 4 and stage 5 share the similar definition as stage 2 and stage 3.

Inference. For each scale of the original image, the RPN network generates $\sim 10^4$ RoIs on the first stage. Non-maximum suppression (NMS) with IoU ratio 0.7 is used to filter out redundant regressed boxes for stage 2. After that, the binary mask branch and classification branch are then applied to the top-ranked 300 RoIs boxes. For each box, we can archive corresponding binary mask (in probability) and classification scores by the two branches. Meanwhile, the RoIs are classified to categories with highest classification scores by the classification branch.

Training. We train and fine-tune our system based on the Caffe platform. We initialize the shared convolutional layers with the released pre-trained VGG-16 model. While for the extra layers, we randomly initialize them. We train the five stages end to end with convolutional features shared. For this cascade network, the later stage takes the outputs of the earlier stages as inputs. The initial learning ratio is set to 0.001 for 32k iterations, which is decreased by 10 at the 8k iterations. We use SGD optimization and train the model on a single Titan X GPU, with a weight decay rate of 0.0005 and a momentum value of 0.9. We use the image patches generated from Sect. 2.1, namely, 64×64, 96×96, and 256×256.

3.2 Experiments on PASCAL VOC 2012

We first conduct experiments on PASCAL VOC 2012. Following the protocols heavily used in [1–3,9], the models are trained on the VOC 2012 training set, and evaluated on the validation set. Note that we also use extra annotations from [5], which has 10,582 images for training, 1449 images for validating, and 1456 images for testing. All the approaches are evaluated by mask-level IoU between predicted segmentations and the ground-truth. We measure the mean Average Precision (mAP^r, the superscript r corresponds to the segmented region) using IoU threshold at 0.5 and 0.7. Our model is short for SPF$_S$, SPF$_M$, SPF$_L$ (input at different scales).

Comparisons with State-of-the-art Approaches. The proposed SPF method is compared with the state-of-the-art approaches on VOC 2012 segmentation val, as shown in Table 1. It is noteworthy that our method outperforms all the state-of-the-art models, including MNC [3] and FCIS [9] which are the winners of the COCO 2015 and 2016 segmentation challenges. [3] proposed a novel multi-task multi-stage cascade network that predicted segment proposals from RoIs, followed by a classification branch. Yet it is unable to capture detail spatial structural information. Li et al. [9] presented the first fully convolutional system for SDS. But this method creates artificial edge information. Contrarily, our method can provides more robust structural information and eliminate artificial edge information by taking the scale factor into account. In addition, the proposed approach can also handle challenging cases where the input image has multiple scales and where the instance category is similar to the background (e.g., Fig. 2). Table 1 shows that the SPF$_L$ is ∼4% higher mAP^r@0.5 than MNC and 1.7% higher mAP^r@0.5 than FCIS.

Influence of IoU Threshold. Table 2 explores the impact of different IoU thresholds by gradually increasing it from 0.6 to 0.9. We can find that the accuracy improves when the number of scales increases. And the accuracy is improved while the IoU threshold is decreased. Our method gets the best result when τ is set to 0.6. So we use the IoU threshold of 0.6 for all the other experiments.

Qualitative Results. Figure 3 depicts visualizations of several sample results on VOC 2012 validation set. We show the ground-truth label and our predicted results for each image. It is suggested that our approach can generate high-quality segmentation masks. Both the large instances and small instances are segmented out and classified.

3.3 Experiments on VOC 2012 SDS

We perform a thorough comparison of SPF to the leading approaches [2,6,7] on this dataset, as shown in Table 3. There are 5623 training images, and 5732 validation images in this subset. For evaluation, we use the metric of mAP^r

Table 1. Comparison of SDS on PASCAL VOC 2012 val set. MNC [3] and FCIS [9] are the winners of the COCO 2015 and COCO 2016 segmentation challenges.

Method	mAP^r@0.5	mAP^r@0.7
SDS [6]	49.7	25.3
Hypercolumn [7]	60.0	40.4
CFM [2]	60.7	39.6
MNC [3]	63.5	41.5
FCIS [9]	65.7	52.1
SPF$_S$	**64.0**	**42.2**
SPF$_M$	**66.5**	**48.6**
SPF$_L$	**67.4**	**53.0**

Table 2. Experimental results (in AP^r) on PASCAL VOC 2012 segmentation val with different IoU threshold τ.

IoU threshold	0.6	0.7	0.8	0.9
SPF$_S$	50.4	40.9	35.2	15.7
SPF$_M$	56.1	52.3	38.5	17.7
SPF$_L$	**67.4**	**55.2**	**40.3**	**20.1**

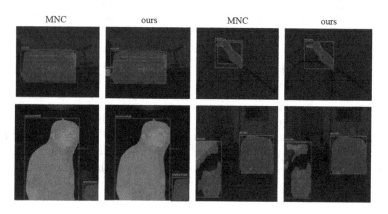

Fig. 2. Results produced by MNC and our SPF. We can generate fine segmentations compared to MNC, and handle challenging instance.

similar to [6] with the pre-trained VGG-16 model. We apply one scale or three scales as input(s) for our network during the course of inference.

Specifically, Hypercolumn [7] refined the mask using the independent detection model for better performance. It is evident that our method achieves large improvement, even without this effective strategy. Table 3 demonstrates that our model with the single-scale input (61.3% mAP^r) is 0.6 points higher than CFM

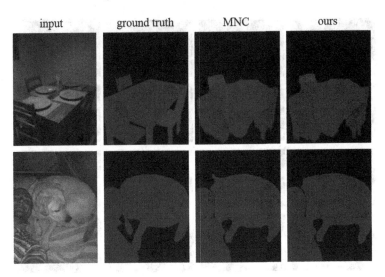

Fig. 3. Example results generated by our SPF network on PASCAL VOC 2012 validation set. For each image, we show the ground-truth label, MNC results and our segmentation results.

Table 3. More results on VOC 2012 SDS *val*. The best mAP^r is bold-faced.

Method	mAP^r
SDS [6]	49.7
Hypercolumn [7]	56.5
Hypercolumn-rescore [7]	60.0
CFM [2]	60.7
SPF$_S$	**61.3**
SPF$_L$	**66.5**

and further improves results using multiple scales inputs, with a margin of 5.8 points mAP^r over [2]. This again indicates that scale is essential for detection and segmentation. Some segmentation results are visualized in Fig. 4.

3.4   Experiments on MS COCO

We finally evaluate our approach on the MS COCO dataset. This dataset includes 80 object categories and numerous comprehensive images. Our network is trained on 80k + 40K trainval images, and the results are reported on the *test-std* and *test-dev* set. We evaluate the performance using three standard metrics: standard COCO evaluation metric (mAP^r@[0.5:0.95]), PASCAL VOC metric (mAP^r@0.5), as well as the traditional metric (mAP^r@0.75).

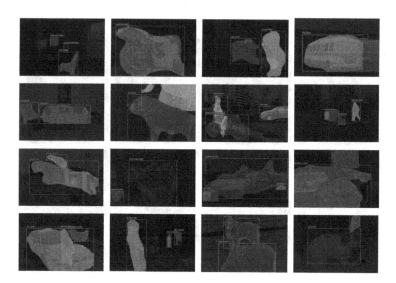

Fig. 4. Extra examples on VOC2012 SDS val set. Our approach can simultaneous segment the large and small instances out.

Comparison with MNC. We compare the SPF approach with MNC [3], the winner in 2015 COCO segmentation challenge. These two approaches share similar architecture and similar training/inference procedures. For fair comparison, the common implementation details are kept the same. Table 4 shows the results with VGG-16 model. The SPF_L archives $mAP^r@[0.5:0.95]$ score of 22.4% on COCO *test-dev* subset. Although without any effective strategies, the SPF_L is 2.9% absolutely higher than MNC. The experimental results demonstrate that the improved accuracy is more significant for small instances, suggesting that the SPF system can capture much more detail spatial structural information.

Ablation Study on MS COCO. We finally conduct a number of ablations with ResNet-101 model. The results are presented in Table 5, and analyzed in the following.

Table 4. Experimental results on Microsoft COCO *test-std* and *test-dev* set using the mAP^r metric at different thresholds using VGG-16 model.

Method	$mAP^r@[0.5:0.95]$	$mAP^r@0.5$	$mAP^r@0.75$
MNC [3]	-/19.5	-/39.7	-/-
SPF_S	14.5/14.3	31.6/31.7	16.8/17.0
SPF_M	**20.2/20.4**	34.5/34.8	21.2/20.9
SPF_L	**21.3/22.4**	**39.8/41.2**	23.2/24.6

Table 5. Results of instance semantic segmentation on COCO *test-dev* set using ResNet-101 model.

Method	mAP^r@[0.5:0.95]	mAP^r@0.5	mAP^r@0.75
FAIRCNN (2015)	25.1	45.8	-
MNC [3]	24.6	44.3	24.8
MNC+++ [3]	28.2	51.5	-
FCIS [9]	29.2	49.5	-
FCIS+++ [9]	37.6	59.9	-
SPF$_L$-101 baseline	**38.8**	52.9	**35.6**
+horizontal flip	**39.1**	54.1	**36.9**
+ensemble	**39.5**	57.3	**38.2**
+OHEM	**39.7**	60.0	**39.8**

SRF baseline: The baseline SPF approach obtains an mAP^r@[0.5:0.95] of 38.8%, which has already outperformed FCIS+++ [9] by 1.2%. That strongly confirms the effectiveness of SPF on segmenting instances.

Horizontal flip: Following [3,9], our SPF system is trained on both the original images and the horizontal flipped images. This preprocessing operation leads to a further improvement of 0.3%, verifying the translation-variant property of SPF.

Ensemble: Similar to [8], we only utilize two models with different depths to form the ensemble. The final performance is 39.5%, which is increased by 0.4%. We also apply OHEM to this method which helps improve the accuracy by 0.2%.

4 Conclusion

The presented scale-aware patch fusion network takes multiple scales patches as inputs and segments the patches via a multi-task network. Finally, we merge the patch segmentation result with fusion algorithm to get the whole result. This unified trainable network inherits all the merits of mid-level patches and we are provided much more detail information. We evaluate our approach on three datasets for semantic segmentation. Although our approach has achieved promising results, further research will be carried out to improve the efficiency of segmentation.

Acknowledgments. This work is partly supported by the National Natural Science Foundation of China under Grant nos. 61573351, and 81471770.

References

1. Dai, J., He, K., Li, Y., Ren, S., Sun, J.: Instance-sensitive fully convolutional networks. In: Leibe, B., Matas, J., Sebe, N., Welling, M. (eds.) ECCV 2016. LNCS, vol. 9910, pp. 534–549. Springer, Cham (2016). https://doi.org/10.1007/978-3-319-46466-4_32
2. Dai, J., He, K., Sun, J.: Convolutional feature masking for joint object and stuff segmentation. In: Proceedings of the IEEE Conference on Computer Vision and Pattern Recognition, pp. 3992–4000 (2015)
3. Dai, J., He, K., Sun, J.: Instance-aware semantic segmentation via multi-task network cascades. In: Proceedings of the IEEE Conference on Computer Vision and Pattern Recognition, pp. 3150–3158 (2016)
4. Gong, Y., Wang, L., Guo, R., Lazebnik, S.: Multi-scale orderless pooling of deep convolutional activation features. In: Fleet, D., Pajdla, T., Schiele, B., Tuytelaars, T. (eds.) ECCV 2014. LNCS, vol. 8695, pp. 392–407. Springer, Cham (2014). https://doi.org/10.1007/978-3-319-10584-0_26
5. Hariharan, B., Arbeláez, P., Bourdev, L., Maji, S., Malik, J.: Semantic contours from inverse detectors. In: 2011 IEEE International Conference on Computer Vision (ICCV), pp. 991–998. IEEE (2011)
6. Hariharan, B., Arbeláez, P., Girshick, R., Malik, J.: Simultaneous detection and segmentation. In: Fleet, D., Pajdla, T., Schiele, B., Tuytelaars, T. (eds.) ECCV 2014. LNCS, vol. 8695, pp. 297–312. Springer, Cham (2014). https://doi.org/10.1007/978-3-319-10584-0_20
7. Hariharan, B., Arbeláez, P., Girshick, R., Malik, J.: Hypercolumns for object segmentation and fine-grained localization. In: Proceedings of the IEEE Conference on Computer Vision and Pattern Recognition, pp. 447–456 (2015)
8. He, K., Zhang, X., Ren, S., Sun, J.: Deep residual learning for image recognition. In: Proceedings of the IEEE Conference on Computer Vision and Pattern Recognition, pp. 770–778 (2016)
9. Li, Y., Qi, H., Dai, J., Ji, X., Wei, Y.: Fully convolutional instance-aware semantic segmentation. arXiv preprint arXiv:1611.07709 (2016)
10. Pont-Tuset, J., Arbelaez, P., Barron, J.T., Marques, F., Malik, J.: Multiscale combinatorial grouping for image segmentation and object proposal generation. IEEE Trans. Pattern Anal. Mach. Intell. **39**(1), 128–140 (2017)
11. Ren, M., Zemel, R.S.: End-to-end instance segmentation and counting with recurrent attention. arXiv preprint arXiv:1605.09410 (2016)
12. Romeraparedes, B., Torr, P.H.S.: Recurrent instance segmentation. Computer Science (2016)
13. Zhang, Z., Fidler, S., Urtasun, R.: Instance-level segmentation for autonomous driving with deep densely connected MRFs. In: Proceedings of the IEEE Conference on Computer Vision and Pattern Recognition, pp. 669–677 (2016)

A Multi-size Kernels CNN with Eye Movement Guided Task-Specific Initialization for Aurora Image Classification

Bing Han[✉], Fuyue Chu, Xinbo Gao, and Yue Yan

School of Electronic Engineering, Xidian University, Xi'an 710071, China
bhan@xididan.edu.cn, heuchufuyue@163.com, xbgao@mail.xidian.edu.cn,
yuey8663@163.com

Abstract. Aurora is the ionosphere track generated by the interaction of solar wind and magnetosphere. Different aurora correspond to different dynamics process in the magnetosphere. So aurora image classification is the basis for aurora analysis. Deep networks available that pretrained with large scale natural image dataset are not suitable for aurora classification due to the characteristics of aurora images. In this paper, a novel multi-size kernels CNN (MSKCNN) with eye movement guided task-specific initialization is proposed to classify aurora images. First, according to the human visual and cognitive mechanism, the patches are extracted guided by eye movement information of space physics. Then, a task-specific initialization scheme is design by using the features learned from aurora image patches to initialize the first layer of our proposed network. In addition, the network contains three parallel streams to learn features under different scale. The comparison experimental results illustrate the effectiveness of the proposed method.

Keywords: Aurora image classification · Deep learning · Eye movement

1 Introduction

The aurora is a beautiful and natural phenomenon generated by the interaction between energetic charged particles from outer space and Earth's upper atmosphere. Different types of aurora are correlated with specific dynamic activities [1]. Aurora image classification is the basis for aurora semantic analysis. The aurora images can be divided into arc aurora, drapery corona aurora and radiation corona aurora. Figure 1 gives examples of these three kinds of aurora images.

Since Syrjasuo et al. firstly introduced computer vision technique to aurora image analysis in 2004 [2], there appears many automatic algorithms for aurora image classification [3–10]. Most available methods consist of two steps [10]. The features are extracted from raw input in the first step and a classifier is learned

© Springer Nature Singapore Pte Ltd. 2017
J. Yang et al. (Eds.): CCCV 2017, Part II, CCIS 772, pp. 533–544, 2017.
https://doi.org/10.1007/978-981-10-7302-1_44

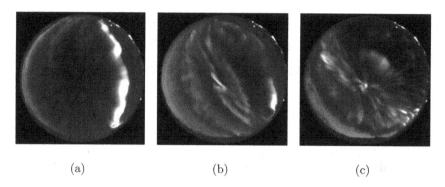

(a)   (b)   (c)

Fig. 1. Examples of three kinds of aurora images. (a) arc aurora; (b) drapery corona aurora; (c) radiation corona aurora.

based on the obtained features in the second step. However, the aurora image is gray image and has nonrigid structure. So most hand-crafted features designed for natural images cannot represent aurora images properly.

Recently, convolutional neural network (CNN) has been demonstrated as an effective model for many compute vision fields such as image classification [11], image segmentation [12] and object detection [13]. CNN has the ability of automatically learning complex pattern and representative features from data in a hierarchical stream. This motivates us to design an automatic aurora classification method based on CNN.

In most of the CNN models, initialization is crucial since poorly initialized networks are likely to find poor apparent local minimum [14]. Usually, networks are initialized with parameters that learned from large scale natural image dataset. But, as shown in Fig. 1, aurora images are essentially different from natural images. It's not reasonable to pretrain network for aurora image classification with natural images. To get a proper initialization for aurora image classification, a task-specific initialization scheme is designed in our method. The convolutional kernels of the first convolutional layer are initialized by the features learned from aurora image patches with auto-encoder.

Since there exist several black areas in the four corners of an aurora image and interferences such as cosmic ray tracks, dayglow contamination and system noise caused by equipment which can lead to mistakes on extracting the patches and labeling the aurora types [15, 16]. It is unreasonable for aurora images to extract patches randomly or by sliding window according zigzag order from up-left corner to down-right of an image as usual. Human have the ability to choose visual information of interest while looking at a static or dynamic scene [17]. This tremendous ability can help us to interact with complex environments by selecting relevant and important information to be proceeded in the brain. It has been claimed that visual attention is attracted to the most informative region [18]. That is to say, those areas where human attend to are usually the most informative areas in an image. So we employ a selection strategy to obtain valuable patches at a set of candidate locations from human fixation points captured by eye tracker.

Besides, most previous patch-based CNN methods took the same size patches extracted from image dataset [19]. Either global information or local details will be lost when the size of patches is inappropriate. To address this issue, we use different sizes of patches to learn features for initialization. A bio-inspired strategy based on eyemovement information is applied to determine the patch size. Correspondingly, a multi-size kernels CNN is constructed, which has a multi-scale architecture with three parallel streams in this paper. Those streams differ in the kernel size of their first convolutional layers. In this way, different streams get different receptive fields on the input image and learn multi-scale features. The stream with larger receptive field gets more global information, while the stream with smaller receptive field get more local details.

In conclusion, there are three major contributions in this paper: (1) A patch selection strategy guided by eye movement information for patch extraction is introduced. And a bio-inspired strategy that uses gaze duration recorded by eye tracker is applied to determine the patch size (2) A task-specific initialization scheme that initializes convolutional kernels with features learned from the patch is designed; (3) A new multi-size kernels CNN (MSKCNN) is designed to utilize multi-scale information.

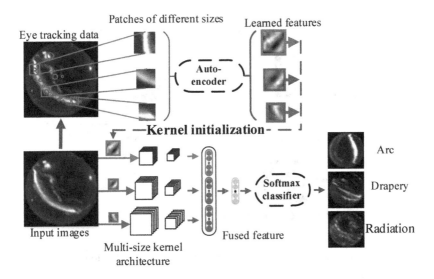

Fig. 2. Flow chart of the proposed model

2   The Proposed Model

The structure of proposed model is shown in Fig. 2, which is a multi-scale architecture containing single-layer auto-encoders in parallel and a multi-size kernels CNN. The details about our model are specified as follows.

2.1   Eye Tracking Data Collection

In this paper, EyeLink 1000 eye tracker with 2000 Hz sampling rate is used to record eye tracking data. A bite bar is used to minimize head movement. Subjects view the aurora images on a 19-inch monitor at 70 cm from human eyes.

Five subjects took part in the eye tracking experiment. They are all space physics experts who are very familiar with the aurora images. They were informed that they would be shown a series of aurora images in which they would give the type of the aurora image. Response is recorded by pressing button. The button '1', '2' and '3' represent for arc aurora, drapery corona aurora and radiation corona aurora respectively.

At the beginning of experiment, the eye tracker is calibrated using a 9-point calibration and validation procedure. As can be seen in Fig. 3, the fixation cross is shown for 200 ms before aurora image firstly. Each aurora image is displayed on the screen for 4 s followed by the fixation cross picture. To initiate the next aurora image, the subject has to fixate a cross centered on the screen for 200 ms again. According to our study purpose, the two measures, fixation point and gaze duration are used in this paper. As shown in Fig. 3, the center of the red circle indicates location of fixation points. And the radius of the circle is directly proportional to gaze duration.

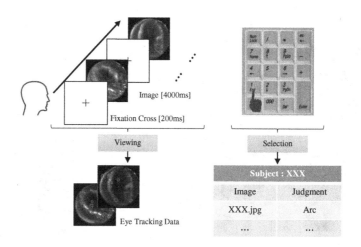

Fig. 3. Procedure of the eye tracking data collection

2.2   Patches Extraction Guided by Eye Movement Information

Attention selection and information abstraction of human beings are cognitive mechanisms employed by human for parsing, gazing, structuring and organizing perceptual stimuli [20]. A cognitive technique is adopted to obtain valuable patches at a set of candidate locations from the fixation points. And psychological research find that gaze duration corresponds to the duration of cognitive processing of the material located at fixation [21]. It means that an area is usually labeled as long gaze duration when the subject observes the details on the area carefully. So there should be an inverse correlation between gaze duration and the patch size.

We apply a linear function with negative slope for mapping the gaze duration in eye movement information to patch size. For an aurora image, the patch size k is assigned according to the information of fixation points as given in Eq. (1),

$$k = a - b * [floor(3 * \frac{t - t_{avemin}}{t_{avemax} - t_{avemin}}) + 1], t_{avemin} < t < t_{avemax}, \quad (1)$$

where t is gaze duration of a specific eye-fixation point, t_{avemin} and t_{avemax} are the average minimum and maximum gaze duration of all aurora images respectively. Those fixation points with gaze duration less than t_{avemin} or more than t_{avemax} are excluded as outliers. The patches are extracted from valid fixation points from all subjects, which guarantees that the features extracted from these valuable patches can reflect the generality of all subjects. The distribution of gaze duration is described in Fig. 4. a is a constant determined by experiments; b is the size difference between different patches which is empirically chosen as 2. So there are three kinds of patches with sizes $a - b$, a and $a + b$. Each kind of patch is used to train a single-layer auto-encoder.

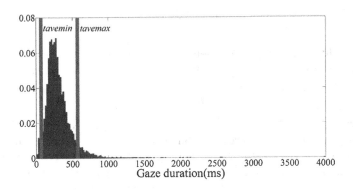

Fig. 4. Distribution of gaze duration

2.3   Task-Specific Initialization

This section will elaborate our task-specific initialization scheme which consists of three steps: data pre-processing, auto-encoder training and convolutional kernels constructing for CNN.

Data Pre-processing. We reshape each patch with size k to a column vector and combine all these column vectors into a matrix,

$$X^k_{original} \in \mathbb{R}^{k^2 * M}, \quad (2)$$

where M is the number of patches. Because the learning capability of the auto-encoder depends on the quality of input data [22], $X_{original}^k$ is preprocessed to reduce the redundancy of input. Firstly, we substract the mean value of each column and get $X_{zero\_mean}^k$,

$$X_{zero\_mean}^k = X_{original}^k - \frac{1}{k} \begin{bmatrix} sum(x_{original}^{k(1)}) & \cdots & sum(x_{original}^{k(M)}) \\ \vdots & \vdots & \vdots \\ sum(x_{original}^{k(1)}) & \cdots & sum(x_{original}^{k(M)}) \end{bmatrix}. \quad (3)$$

Then singular value decomposition (SVD) is performed on the covariance matrix Σ of $X_{zero\_mean}^k$:

$$U \Lambda U^T = svd(\Sigma), U, \Lambda, \Sigma \in \mathbb{R}^{M*M}. \quad (4)$$

The matrix U contains the eigenvectors of Σ, and $\Lambda = diag(\lambda_1, \lambda_2, \cdots, \lambda_M)$ contains the corresponding eigenvalues. Finally, we can get the preprocessed data of $X_{original}^k$ as:

$$X^k = U * diag(\frac{1}{\sqrt{\lambda_1 + \varepsilon}}, \frac{1}{\sqrt{\lambda_2 + \varepsilon}}, \cdots, \frac{1}{\sqrt{\lambda_M + \varepsilon}}) * U^T * X_{zero\_mean}^k. \quad (5)$$

where is ε a constant which is empirically chosen as $\varepsilon = 10^{-5}$.

Auto-Encoder Training. Auto-encoder model is an unsupervised learning algorithm that applies back propagation and tries to learn a function.

$$y = h_{W,b}(x) \approx x. \quad (6)$$

The architecture of the single-layer auto-encoder used in this work is shown in Fig. 5. By learning an approximation to identity function,. the auto-encoder can automatically learn representative features from unlabeled data. Fed with patches extracted from aurora image, the auto-encoder can get the task-specific features for aurora image. And these learned features will be used to initialize the convolutional kernels.

Suppose there is a training set:

$$(x^{k(1)}, y^1), (x^{k(2)}, y^2), \cdots, (x^{k(M)}, y^M), \quad (7)$$

the cost function of single-layer auto-encoder can be defined as,

$$J(W,b) = \frac{1}{M} \sum_{m=1}^{M} \frac{1}{2} ||h_{W,b}(x^{k(m)}))) - y^m||^2 + \frac{\lambda}{2} \sum_{l=1}^{s_2} \sum_{j=1}^{s_i} \sum_{i=1}^{s_{i+1}} (W_{ij}^l)^2, \quad (8)$$

where $h(x)$ is activate function; s_l is the number of units on the lth layer and W_{ij}^l is a weight between the jth unit on the lth layer and ith unit on the $(l+1)$th

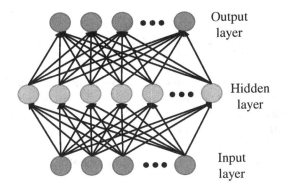

Fig. 5. Architecture of the single-layer auto-encoder.

layer. The first term in $J(W, b)$ is an average sum-of-squares error. The second term is a regular term against over-fitting.

A sparse constraint is imposed on the hidden layer to find representative features efficiently. Let $\hat{\rho}_j$ be the average activation of the hidden unit j:

$$\hat{\rho}_j = \frac{1}{M} \sum_{m=1}^{M} [a_j(x^{k(m)})]. \tag{9}$$

An extra penalty term is added to $J(W, b)$. This ensures that $\hat{\rho}_j$ is close to the sparse coefficient ρ. In this paper, penalty term is set as following:

$$\sum_{j=1}^{H} KL(\rho|\hat{\rho}_j), \tag{10}$$

where H is the number of hidden units and $KL(\rho|\hat{\rho}_j)$ is the Kullback-Leibler (KL) divergence between a Bernoulli random variable with mean ρ and a Bernoulli random variable with mean $\hat{\rho}_j$. Then the overall cost function of sparse auto-encoder is updated to:

$$J_{sparse} = J(W, b) + \beta \sum_{j=1}^{H} KL(\rho|\hat{\rho}_j), \tag{11}$$

where β is used to control the importance of the sparse constraint. To train our auto-encoder model, we can now repeatedly take steps of gradient descent to reduce the cost function.

Convolution Kernels Constructing for CNN. Patches that maximally activate hidden units of an auto-encoder are treated as the features learned by auto-encoder. These learned features are used to initialize convolution filters.

It has been proved that the input which maximally activates hidden unit i is given by setting value of each pixel [23],

$$C_i^k(j) = \frac{W_{ij}^l}{\sqrt{\sum_{j=1}^{k^2}(W_{ij}^l)^2}}(i = 1, 2, \cdots, H; j = 1, 2, \cdots, k^2), \qquad (12)$$

where H is the number of hidden units in auto-encoder and it is also the number of features map in the first convolutional layer. Then convolutional kernels in the first convolutional layer in our MSKCNN models are initialized with patches $C_i^k(i = 1, 2, \cdots, H)$ formed by these pixel values $C_i^k(j)(j = 1, 2, \cdots, k^2)$. As a result, the characteristic features of aurora images have been captured by the first convolutional layer. So that those parameters after the first convolutional layer can be initialized randomly.

2.4 Net Architecture of the Multi-size Kernels CNN

Our network starts from three streams. Each stream has the same architecture with the net designed by Alex et al. [11]. Differences between different streams are the kernel size of their first convolutional layers. So that different streams can get different receptive fields on the input image. As a result, multi-scale features can be learned with the proposed MSKCNN.

In our task-specific initialization, feature maps of first convolutional from each stream can be computed as,

$$I\_conved(row, col) = \sum_{\delta_i}^{k}\sum_{\delta_j}^{k}C_i^k(\delta_i, \delta_j) * X^k(s_c * row + \delta_i, s_c * col + \delta_j), \quad (13)$$

where s_c is the stride of convolution kernels.

The fusion of streams is done by linking the last fully connected layers of different streams end-to-end. Because the dimension of the last fully connected layer is 1024, the fused layer is a 1024*3 dimension feature vector. Then it is followed by an additional fully connected layer with 1024 dimension to reduce dimension.

3   Experiments and Analysis

3.1 Datasets and Experimental Setup

The experimental data used in this paper is obtained by the All-Sky imager in Arctic Yellow River Station. 9000 manual labeled images are used as our aurora dataset and each category contains 3000 images. All images have an original size of 512*512 and are resized to 256*256. We trained our model with a batch size of 100. Training and testing of the model are performed on a single GTX TITAN GPU with Caffe deep learning framework [24].

3.2 Experiment

Evaluation on Task-Specific Initialization. In this section, two comparison experiments are conducted to demonstrate the effectiveness of the proposed task-specific initialization.

In the first experiment, we investigate performance of the single stream model with different kernel sizes in the first convolutional layer under different kinds of initialization. The results are given in Table 1. The first row in Table 1 shows the performance with random initialization. And the second row shows the performance with the proposed task-specific initialization. The third row shows the performance with eye guided task-specific initialization. One-tenth of the images in our aurora image dataset are chosen randomly as the training data, while the others as the testing data. The average classification accuracy with our task-specific initialization is 61.3% while it is 56.1% with random initialization.

Table 1. Classification accuracy under different patch sizes

Method	Patch size								
	7	9	11	13	15	17	19	21	Ave
Random	0.564	0.566	0.539	**0.634**	0.591	0.530	0.523	0.539	0.561
Task-specific	**0.578**	**0.591**	**0.584**	0.602	**0.620**	**0.667**	**0.611**	**0.656**	**0.613**

In the second experiment, we test the performance of task-specific initialization under different proportion between training data and testing data. The experimental results are shown in Fig. 6. It can be concluded that our task-specific initialization outperforms random initialization consistently.

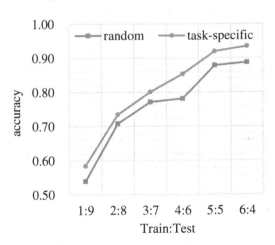

Fig. 6. Classification accuracy under different proportion of training data and testing data.

State-of-the-art Comparison. Performance under different parameter a in Eq. (1) is tested to determine the kernel sizes of the first convolutional layers in our model. As shown in Table 2, the best classification accuracy is 96.2% when a is 11. So, the kernel sizes of the first convolutional layers in the three streams of our proposed MSKCNN is set to be 9, 11 and 13 respectively.

Table 2. Classification accuracy under different a (train:test = 6:4)

a	7	9	11	13	15	17	19	21
Accuracy	0.954	0.940	**0.962**	0.885	0.930	0.921	0.886	0.879

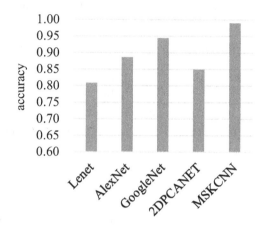

Fig. 7. Performance of different methods on our aurora image dataset. (train:test = 6:4)

We compare the proposed MSKCNN model with three convolutional neural networks for image classification: Lenet [25], GoogleNet [26], AlexNet [11] and the latest aurora image classification method based on deep learning: 2DPCANET [10]. The results are presented in Fig. 7. It shows that the classification accuracy increases from 88.7% with the original AlexNet to 96.2% with the proposed method. And our method outperforms the other classification method based on deep learning.

4   Conclusion

In this paper, we present an approach for aurora image classification. The proposed MSKCNN contains three different streams with different receptive field to utilize information flows with small to large contexts. Patch extraction is guided by eye fixation locations and gaze duration of space physics experts. So the patch extraction becomes more consistent with human visual and cognitive mechanism. Then, a task-specific initialization is designed. This novel initialization method learns features from aurora image patches and then initializes kernels of the first convolution layers with these learned features. Experiment results demonstrate the effectiveness of our proposed method on aurora datasets.

In the future, we will further extend the proposed work on other dataset.

Acknowledgments. This research was supported by the National Natural Science Foundation of China (41031064; 61572384; 61432014), China's postdoctoral fund first-class funding (2014M560752), Shaanxi province postdoctoral science fund, the central university basic scientific research business fee (JBG150225).

References

1. Feldstein, Y.I., Elphinstone, R.D.: Aurorae and the large-scale structure of the magnetosphere. Earth Planets Space **44**(12), 1159–1174 (1992)
2. Suo, M.T.S., Donovan, E.F.: Diurnal auroral occurrence statistics obtained via machine vision. Ann. Geophys. **22**(4), 1103–1113 (2004)
3. Donovan, E.F., Qin, X., Yang, Y.H.: Automatic classification of auroral images in substorm studies (2007)
4. Fu, R., Li, J., Gao, X., Jian, Y.: Automatic aurora images classification algorithm based on separated texture. In: IEEE International Conference on Robotics and Biomimetics, pp. 1331–1335 (2009)
5. Wang, Q., Liang, J., Hu, Z.J., Hu, H.H., Zhao, H., Hu, H.Q., Gao, X., Yang, H.: Spatial texture based automatic classification of dayside aurora in all-sky images. J. Atmos. Solar Terr. Phys. **72**(5), 498–508 (2010)
6. Han, S.M., Wu, Z.S., Wu, G.L., Tan, J.: Automatic classification of dayside aurora in all-sky images using a multi-level texture feature representation. Adv. Mater. Res. **341–342**, 158–162 (2012)
7. Syrjasuo, M., Partamies, N.: Numeric image features for detection of aurora. IEEE Geosci. Remote Sens. Lett. **9**(2), 176–179 (2012)
8. Han, B., Zhao, X., Li, X., Li, X., Hu, Z., Hu, H.: Dayside aurora classification via BIFs-based sparse representation using manifold learning. Int. J. Comput. Math. **91**(11), 2415–2426 (2014)
9. Rao, J., Partamies, N., Amariutei, O., Syrjäsuo, M., Sande, K.E.A.V.D.: Automatic auroral detection in color all-sky camera images. IEEE J. Sel. Topics Appl. Earth Observations Remote Sens. **7**(12), 4717–4725 (2014)
10. Jia, Z., Han, B., Gao, X.: 2DPCANet: Dayside Aurora Classification Based on Deep Learning. Springer, Heidelberg (2015)
11. Krizhevsky, A., Sutskever, I., Hinton, G.E.: ImageNet classification with deep convolutional neural networks. In: International Conference on Neural Information Processing Systems, pp. 1097–1105 (2012)
12. Long, J., Shelhamer, E., Darrell, T.: Fully convolutional networks for semantic segmentation. IEEE Trans. Patt. Anal. Mach. Intell. **39**(4), 640–651 (2014)
13. Girshick, R.: Fast R-CNN. Computer Science (2015)
14. Sutskever, I., Martens, J., Dahl, G., Hinton, G.: On the importance of initialization and momentum in deep learning. In: International Conference on International Conference on Machine Learning, pp. III–1139 (2013)

15. Han, B., Song, Y., Gao, X., Wang, X.: Dynamic aurora sequence recognition using volume local directional pattern with local and global features. Neurocomputing **184**, 168–175 (2016)

16. Li, X., Ramachandran, R., Movva, S., Graves, S.: Dayglow removal from FUV auroral images. In: 2004 IEEE International Geoscience and Remote Sensing Symposium, 2004, IGARSS 2004, Proceedings, vol. 6, pp. 3774–3777 (2004)

17. Xu, J., Jiang, M., Wang, S., Kankanhalli, M.S., Zhao, Q.: Predicting human gaze beyond pixels. J. Vis. **14**(1), 97–97 (2014)

18. Bruce, N.D.B.: Saliency, attention and visual search: an information theoretic approach. J. Vis. **9**(3), 5.1–24 (2009)

19. Zhang, W., Li, R., Deng, H., Wang, L., Lin, W., Ji, S., Shen, D.: Deep convolutional neural networks for multi-modality isointense infant brain image segmentation. Neuroimage **108**, 214 (2015)

20. Sevakula, R.K., Shah, A., Verma, N.K.: Data preprocessing methods for sparse auto-encoder based fuzzy rule classifier. In: Computational Intelligence: Theories, Applications and Future Directions, pp. 1–6 (2016)

21. Irwin, D.E.: Fixation location and fixation duration as indices of cognitive processing, pp. 105–134. Psychology Press (2004)

22. Evangelopoulos, G., Zlatintsi, A., Potamianos, A., Maragos, P., Rapantzikos, K., Skoumas, G., Avrithis, Y.: Multimodal saliency and fusion for movie summarization based on aural, visual, and textual attention. IEEE Trans. Multimedia **15**(7), 1553–1568 (2013)

23. Visualizing a trained autoencoder. http://ufldl.stanford.edu/wiki/index.php/Visualizing_a_Trained_Autoencoder

24. Jia, Y., Shelhamer, E., Donahue, J., Karayev, S., Long, J.: Caffe: convolutional architecture for fast feature embedding, pp. 675–678 (2014)

25. Lcun, Y., Bottou, L., Bengio, Y., Haffner, P.: Gradient-based learning applied to document recognition. Proc. IEEE **86**(11), 2278–2324 (1998)

26. Szegedy, C., Liu, W., Jia, Y., Sermanet, P., Reed, S., Anguelov, D., Erhan, D., Vanhoucke, V., Rabinovich, A.: Going deeper with convolutions. In: Computer Vision and Pattern Recognition, pp. 1–9 (2015)

Text Extraction for Historical Tibetan Document Images Based on Connected Component Analysis and Corner Point Detection

Xiqun Zhang[1,2], Lijuan Duan[1,3], Longlong Ma[4(✉)], and Jian Wu[4]

[1] Faculty of Information Technology, Beijing University of Technology, Beijing, China
zhangxiqun122@163.com, ljduan@bjut.edu.cn
[2] Beijing Key Laboratory of Trusted Computing, Beijing, China
[3] Beijing Key Laboratory on Integration and Analysis of Large-scale Stream Data,
Beijing, China
[4] Chinese Information Processing Laboratory, Institute of Software,
Chinese Academy of Sciences, Beijing, China
{longlong,wujian}@iscas.ac.cn

Abstract. In this paper, we present a text extraction method for historical Tibetan document images. The task of text extraction is considered as text area detection and location problem. Firstly, the historical Tibetan document image is preprocessed to correct imbalanced illumination, tilt and noises, then get the binary image. Secondly, the regions of interest in historical Tibetan documents are divided into three categories using connected components. The images are divided equally into grids and the grids are filtered by the information of the categories of CCs and corner point density. The remaining grids are used to compute vertical and horizontal grid projections. Thirdly, by analyzing the projections, the approximate location of the text area can be detected. Finally, the text area is extracted accurately by correcting the bounding box of the approximate text area. Experiments on the dataset of historical Tibetan document images demonstrate the effectiveness of the proposed method.

Keywords: Historical Tibetan document · Text extraction
Connected components · Corner point

1 Introduction

Nowadays, some historical Tibetan documents have been digitized and available to the public. Most of them are stored in the form of images and digitized by manually entering corresponding texts into the computer. The historical Tibetan documents stored in the form of images require a lot of storage space. The scanned image cannot be edited, so the research and the use of these images are restricted. Except of the relevant researchers, few people will read the

© Springer Nature Singapore Pte Ltd. 2017
J. Yang et al. (Eds.): CCCV 2017, Part II, CCIS 772, pp. 545–555, 2017.
https://doi.org/10.1007/978-981-10-7302-1_45

scanned image of historical documents. If there is an efficient document recognition method to digitize automatically historical Tibetan documents, it is very meaningful for the inheritance and protection of the Tibetan traditional culture.

Text extraction is an important initial step in the automatic digitization of historical documents. In the past decades, researchers have proposed some methods to extract texts from the document images. These methods rely on connected components analysis [7], corner point density [11,13], and feature extraction [1,2,10]. In addition, some researchers use edge detection [6] and similarity matching [5] methods. In the field of text extraction, it is usually impossible to use the same method to process document images with different layout structure and text features. Our goal is to develop a method to extract texts from historical Tibetan document images. The vast majority of Tibetan historical documents are written on Tibetan papers, which are handmade in traditional ways. The layout of historical Tibetan document is irregular and complex. The lines of some frames have double layers, even the same frame will be broken into several parts. Some text areas are surrounded by multiple layers of such frames. Due to the compositional characteristics of Tibetan, the consecutive text lines of Tibetan is often touching or overlapping. The text is also attached to non-text parts. All of the above mentioned document features bring difficulties for text extraction. Figure 1 gives one example of historical Tibetan document images.

Fig. 1. Historical Tibetan document image.

We transform the text extraction problem into text area detection and location problem. Firstly, the historical Tibetan document image is preprocessed to eliminate the effects of imbalanced illumination, tilt and noises. After this, the image is binarized using the Otsu algorithm. The seed-filling algorithm is used to detect the CCs, and CCs are divided into three categories (*text, frame, line*) according to the area threshold and the width-height or height-width ratio of CCs. Then, the image is divided equally into $N*N$ grids (non-overlapping) and the grids are filtered by the information of the categories of CCs and the corner point density. The remaining grids are used to compute vertical and horizontal grid projections. By analyzing the projections, we can obtain the approximate location of the text area. Finally, the text area is extracted accurately by correcting the bounding box of the approximate text area.

The rest of the paper is organized as follows. Section 2 presents an overview of the related work. Section 3 describes the proposed method in detail. Section 4 shows our experimental results. Finally, the conclusions are given in Sect. 5.

2   Related Work

There are some works related to text extraction tasks. Researchers used different methods, such as traditional methods, machine-learning methods, etc., to extract texts from different types of documents.

Traditional segmentation methods can be grouped into top-down, bottom-up and hybrid methods. In general, extracting text from the complex layout document is mainly based on the hybrid method. AGORA project [7] developed the user-driven approach based on the hybrid method to perform layout analysis for historical printed books. Their segmentation algorithm includes two maps: a shape map and a background map. The shape map is formed from the bounding box of the CCs in the page. The background map provides information about white areas corresponding to block separations in the page. Their method used the background map of the images to highlight the separation between blocks and the shape map. Then, they segment the image with the predefined rules and the information provided by these two representations. In addition, Winder [9] modified the RAST and Voronoi segmentation algorithms in OCRopus to process mixed content layouts at a variety of resolutions, and make the digitization of standard format historical documents by low-budget organizations feasible. Yu [12] improved the bottom-up page segmentation method based on the connected region of printed newspapers. In Chinese ancient book "Imperial Collection of Four" digitization project, Jiang [14] employed a hybrid method, associated with artificial correction, to analyze the document layout. Singh [8] analyzed the layout of Indian newspapers with the hybrid method.

Machine-learning method regards text extraction as a classification problem. These methods extract the features from different parts of document images to train the classifier, the classifier classifies image regions into text, illustration and so on. Chen [3] developed an unsupervised feature learning method for page segmentation of historical documents with color images. They used the convolutional autoencoder to learn features directly from pixel intensity values. Aiming at reducing the computation time, they present a superpixel-based method [2] to replace the original pixel-based method. Training a support vector machine to classify superpixels based on these features. Bukhari [1] extracted the relevant features in a connected-component level to train a multilayer perceptron. A voting scheme is then applied to refine the resulting segmentation and produce the final classification. Xiao [10] proposed a method to extract text from ancient Yi character documents. They combined edge and texture features to represent the characteristics of text in ancient Yi character documents accurately and adopt the GBDT (Gradient Boost Descent Tree) learning theory to design a classifier to classify text and non-text pixels.

Other methods, such as corner point detection, contour detection, can also extract texts from document images. Zeng [13] proposed a filtering method based on the Harris corner point detection. The algorithm can filter the background, which contains text image commendably of printed documents. Yadav [11] designed a very simple technique based on FAST key points to extract texts from document images. The image is divided into blocks and the blocks including more points are classified as text blocks. Then, the connectivity of blocks is checked to group and obtain complete text blocks. In order to segment text from degraded historical Indus script images, Kavitha [6] proposed a new combination of Sobel and Laplacian for enhancing degraded low contrast pixels to generate skeletons for text components. The component that gives a less number of branches is considered as a text cluster because text components usually has fewer branches compared to non-texts in historical Indus script images. Ha [4] proposed an adaptive over-split and merge algorithm for page segmentation of printed documents. Firstly, the document image is over-splitted into text blocks or text lines. Then, these text blocks or text lines are merged into text regions using a new adaptive threshold method. The local context analysis method used a set of text line separators to split homogeneous text regions with similar font size and further merged text blocks into paragraphs.

3   System Description

In order to extract texts accurately from historical Tibetan documents, we propose a text extraction method based on connected component analysis and corner point detection. Figure 2 gives the workflow of our method.

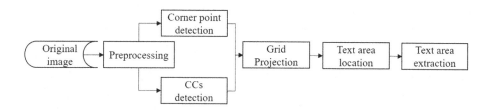

Fig. 2. Text extraction workflow.

3.1   Preprocessing

The preprocessing is to correct imbalanced illumination, tilt, noises. If these problems are not solved properly, will cause unnecessary trouble and affect the subsequent operations.

In order to eliminate the adverse effects mentioned above, the images are preprocessed by the following methods. In order to reduce the amount of calculation, we normalize the original image into a fixed size. The gamma correction algorithm is used to correct illumination. A tilt correction algorithm based on the Hough transform is used to correct the skew of historical Tibetan document images. Finally, the images are transformed into the binary image using the Otsu algorithm.

3.2   Grid Projection

The grid projections are used to locate the approximate text area and highlight the gap between the different parts in the historical Tibetan documents. From [11,13], we can know that text regions have more corner points than non-text regions, which is confirmed by our experiments on the historical Tibetan document images. Inspired by the use of key point density from the grid filter method [11], we combine the CCs classification information with the corner point density to filter the grid.

The CCs in the binary image are detected by the seed-filling algorithm. The CCs are classified using some priori information. By observing the document images, we can find that the frame is larger in area than the text and the width and height ratio or height and width ratio (w/h or h/w ratio) of the line has a greater difference from other categories. The CCs are classified according to the following rules. The area of the image is denoted as S.

- Rule 1: If the area of CC is bigger than $a*S$, where a is the threshold of frame, the CC is classified as frame.
- Rule 2: If the w/h or h/w ratio of CC is smaller than the ratio of R, the CC is classified as a line.
- Rule 3: If the CC does not match the above rules, it is classified as text.

All CC matches each rule in order. The classification labels of CCs are stored in an image named *labelImg*, where all pixel values of CC are equal to its label value. The Harris algorithm detects corner points in the binary image. The image is divided equally into $N*N$ grids, and the number of corner points in each grid is calculated. The maximum number of corner points ($Nmax$) in all grids is recorded. The grids are filtered with the following steps.

- Step 1: If the number of corner points of the grid is less than $c*Nmax$ (c is the threshold of grid filtering), the grid will be deleted directly. Otherwise, do the following steps.
- Step 2: If the grid contains more than two non-text classes in the corresponding location of *labelImg*, the grid is isolated, or the grid is an edge grid that only contains non-text class. The grid is also deleted. Because we thought that these grids are located in a corner-dense non-text area.
- Step 3: If the grid contains text and non-text classes in the corresponding location of *labelImg*, the text part of the grid will be preserved.

Filtered grids using the above rules, gaps between different parts are highlighted. The remaining grids are used to compute vertical and horizontal grid projections.

3.3   Text Area Location

The approximate text area will be located by analyzing the grid projections. The projections are analyzed with the following steps.

- Step 1: Search the change points from left to right according to the vertical grid projection. The two adjacent change points from zero to non-zero and from non-zero to zero are considered as the horizontal start and end positions of an area.
- Step 2: For the horizontal grid projection, the first non-zero point at both ends of the horizontal grid projection is considered as the vertical start and end positions.
- Step 3: The start and end positions of the horizontal and vertical directions are used to search for the first non-text points as the areas pixel point of the bounding box from the inside to the outside of the four edges in the corresponding area of *labelImg*. The pixel points of the bounding box near the broken frames, which are misclassified to texts, are filled with zero-to-nonzero change points. These points can be obtained by searching for the corresponding position of its nearby pixel point of the bounding box. The breakpoints of the broken frames are filled with its nearest bounding boxes pixel point.

Repeating the above steps until the end of vertical projection is searched. Approximate text areas are located, and their bounding boxes are also obtained.

3.4   Text Area Extraction

Text area are extracted accurately by correcting the bounding box of the approximate text area. The bounding box is corrected by the following strategy. Take the upper boundary as an example, the boundary points are selected to calculate the average of their horizontal ordinate (denoted as AO), and if any point's horizontal ordinate greater than AO and the difference between its horizontal ordinate and its previous points horizontal ordinate is greater than X, and its horizontal ordinate will be replaced by the previous points horizontal ordinate. Using the same method, continue to correct the horizontal ordinate of the bounding box. From the previous steps, we can know that some points from four corners of bounding box for the approximate text area are not found. These points are searched based on the existing detected neighbor points. After the above operations, the texts attached to the boundary are removed and the text areas are accurately extracted.

4   Experiments

This text extraction method is tested on manuscripts taken from *"The Complete Works of the Panchen Lama"*, provided by the Department of Computer Science, Qinghai Nationalities University. The layout structure of the collected images is irregular and complex. Due to the restriction of hardware, the images have serious imbalanced illumination and the quality of the images is poor. The color of documents is not uniform. The size of the images is not uniform, but within a certain range. We use the collected historical Tibetan document images to conduct the experiment, and the collected dataset contains 360 images.

4.1   Preprocessing

In order to obtain the preprocessed images from historical Tibetan documents, several methods of illumination equalization and binarization are experimented. We found that the combination of the gamma correction algorithm and the Otsu algorithm could obtain a better binarization result. Figure 1 shows the original image. The gamma value of gamma correction is set to 0.4. Figure 3(a) shows the result of elimination imbalanced illumination. Figure 3(a) is transformed into grayscale image, the skew of the image is corrected by a tilt correction algorithm based on the Hough transform. Figure 3(b) is the binary image of the grayscale image using the Otsu algorithm.

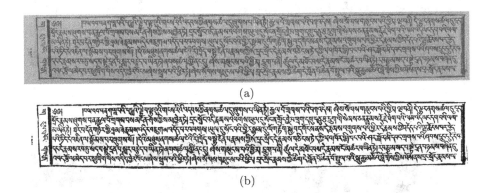

Fig. 3. Preprocessing. (a) Gamma correct result. (b) Binary image.

4.2   Grid Projection

In order to get the vertical and horizontal projections, we perform the operations according to Sect. 3.2. Firstly, the threshold of a, R are set to 1/4 and 0.05, respectively. The classification results of CCs are shown in Fig. 4(a). The regions of frame, line and text are labeled separately using red, green and black color. We can see that the broken parts of the frame are misclassified as other classes. Some sticky texts were misclassified as non-text classes. These misclassifications of sticky texts will be corrected in the following operations. Figure 4(b) shows the corner point detection results. If the grids are filtered using the method of [11], the text area could not be located accurately from historical Tibetan documents. Figure 4(c) shows the results of grid filtering, which are filtered according to the method described in Sect. 3.2. It is not difficult to find that the grids of Fig. 4(c) can not only locate the text area, but also highlight the gap between the different text areas.

The vertical and horizontal grid projections are computed based on the remaining grids. Figure 5(a) and (b) shows the vertical and horizontal projections, respectively. It is possible to determine the approximate text area by analyzing the grid projections.

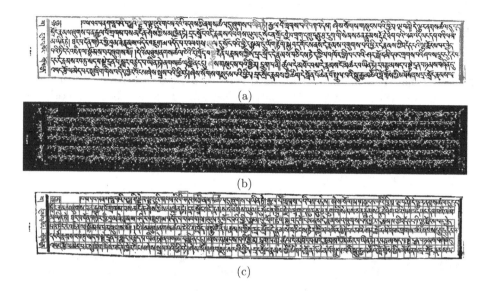

(a)

(b)

(c)

Fig. 4. CCs classification and grid filter. (a) CCs classification results. (b) Corner point detection result. (c) The result of filtered grids. (Color figure online)

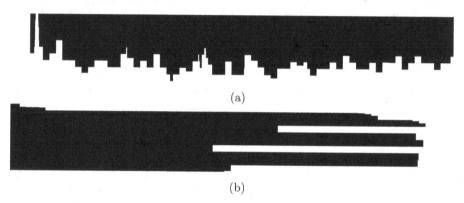

(a)

(b)

Fig. 5. Grid projections. (a) Vertical projection. (b) Horizontal projection.

4.3 Text Area Location

As we can see from the projections, there is a clear gap between different parts. So the approximate text area can be obtained by the method described in Sect. 3.3. Figure 6(a) shows the result of approximate text area location. The red bounding box is the boundary of the text area.

4.4 Text Area Extraction

Based on the result of previous step, the strategy, described in Sect. 3.4, is used to correct the bounding box of the approximate text area. The threshold of X

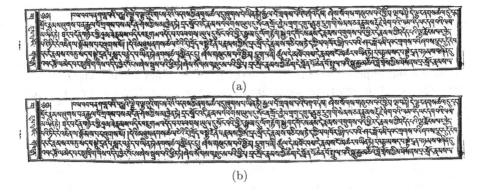

Fig. 6. The bounding box and final result of text area. (a) The bounding box of the approximate text area. (b) The result of text extraction. (Color figure online)

is set to 3. The results are shown in Fig. 6(b). The points of the extracted text areas bounding box are labeled as red color.

4.5 Experimental Results

As we all know, the text extraction algorithm has a strong correlation with the specific document layout. Usually a layout type of document corresponds to a specific text extraction algorithm. In fact, we have also tried some of the methods in other paper, most of the methods are not suitable for historical Tibetan documents. And there are no papers about extracting text from historical Tibetan document or similar layout historical document. Therefore, only the method proposed in this paper is evaluated. We adopt the F-measure metric, which combines precision with recall values into a single scalar representative, to evaluate the text extraction accuracy. Precision and recall are estimated according to Eqs. 1 and 2, respectively.

$$Precision = \frac{TP}{TP + FP} \tag{1}$$

$$Recall = \frac{TP}{TP + FN} \tag{2}$$

where True-Positive(TP), False-Positive(FP) and False-Negative(FN) with respect to text area, are defined as following:

TP: Number of the extracted text area correctly
FP: Number of the misclassified text area
FN: Number of the undetected text area

Once we have the precision and recall values. F-measure is calculated according to Eq. 3.

$$F - Measure = \frac{\left(1 + \beta^2\right) \cdot Precision \cdot Recall}{\left(\beta^2 \cdot Recall\right) + Precision} \tag{3}$$

The β is set to 1, it means that precision and recall are equally important in the F-measure estimation.

In order to compare the rigorous evaluation of the experimental results, we labeled ground truth of text areas on our dataset, and the use of pixel accuracy to evaluate our experimental results. If the pixel accuracy of the extracted area bigger than the threshold acc, we thought it is a text area. The acc for larger text areas and smaller text areas are set to 95% and 90%, respectively. From Table 1, we can see that the F-measure reaches 85.60%, but the recall reaches 98.58%, while the precision is only 75.64%. By analyzing the experimental results, we found that main reasons of lower precision are led by the complexity of the layout and the document image quality. In our dataset, lines of the frame are rough and irregular. Due to the quality of the image and other reasons, these lines are usually broken into shorter lines or even dot sequences, so these broken lines and dots are often extracted as the text and effects the precision of text extraction.

Table 1. Performance evaluation

Precision(%)	Recall(%)	F-Measure(%)
75.64	98.58	85.60

5   Conclusion

In this paper, we presented an efficient method to extract text areas from historical Tibetan documents. It makes a classification of CCs using the priori rules. Corner points are detected in binary images, and images are equally divided into grids. Based on the classification information of CCs and corner points, the grids are filtered by the predefined rules. By analyzing vertical and horizontal projections of remaining grids, the approximate text area is located, and the bounding box of the text area is searched. The text area is extracted accurately in the approximate area through the bounding box correction strategies. The experimental results verify the effectiveness of our method.

Our future work will focus on improving the adaptability of the algorithm. As we all know, different historical documents have different layout structures and characteristics. Our method needs to adapt to historical Tibetan documents with different layout structures. Enhancing the performance of the text extraction is also the direction of our future efforts.

Acknowledgments.. This work was supported by the Science and Technology Project of Qinghai Province (no. 2016-ZJ-Y04) and the Basic Research Project of Qinghai Province (no. 2016-ZJ-740). The authors would like to thank Qilong Sun, the Department of Computer Science, Qinghai Nationalities University for providing the experimental dataset of historical Tibetan document images.

References

1. Bukhari, S.S., Breuel, T.M., Asi, A., El-Sana, J.: Layout analysis for arabic historical document images using machine learning. In: 2012 International Conference on Frontiers in Handwriting Recognition (ICFHR), pp. 639–644. IEEE (2012)
2. Chen, K., Liu, C.L., Seuret, M., Liwicki, M., Hennebert, J., Ingold, R.: Page segmentation for historical document images based on superpixel classification with unsupervised feature learning. In: 2016 12th IAPR Workshop on Document Analysis Systems (DAS), pp. 299–304. IEEE (2016)
3. Chen, K., Seuret, M., Liwicki, M., Hennebert, J., Ingold, R.: Page segmentation of historical document images with convolutional autoencoders. In: 2015 13th International Conference on Document Analysis and Recognition (ICDAR), pp. 1011–1015. IEEE (2015)
4. Dai-Ton, H., Duc-Dung, N., Duc-Hieu, L.: An adaptive over-split and merge algorithm for page segmentation. Pattern Recognit. Lett. **80**, 137–143 (2016)
5. Fu, H., Liu, X., Jia, Y.: Text extraction based on maximum-minimum similarity training method (in Chinese). J. Softw. **19**(3), 621–629 (2008)
6. Kavitha, A., Shivakumara, P., Kumar, G., Lu, T.: Text segmentation in degraded historical document images. Egypt. Inf. J. **17**(2), 189–197 (2016)
7. Ramel, J.Y., Leriche, S., Demonet, M., Busson, S.: User-driven page layout analysis of historical printed books. Int. J. Doc. Anal. Recognit. **9**(2), 243–261 (2007)
8. Singh, V., Kumar, B.: Document layout analysis for indian newspapers using contour based symbiotic approach. In: 2014 International Conference on Computer Communication and Informatics (ICCCI), pp. 1–4. IEEE (2014)
9. Winder, A., Andersen, T., Smith, E.H.B.: Extending page segmentation algorithms for mixed-layout document processing. In: 2011 International Conference on Document Analysis and Recognition (ICDAR), pp. 1245–1249. IEEE (2011)
10. Xiao, R.: Research on the Method of Extracting Ancient Yi Text from Complex Background (in Chinese). Ph.D. thesis, South-Center University for Nationalities (2011)
11. Yadav, V., Ragot, N.: Text extraction in document images: highlight on using corner points. In: 2016 12th IAPR Workshop on Document Analysis Systems (DAS), pp. 281–286. IEEE (2016)
12. Yu, M., Guo, J., Wang, D., Yu, Y.: Improved page segmentation method based on connected domain (in Chinese). Comput. Eng. Appl. **49**(17), 195–198 (2013)
13. Zeng, F., Zhang, G., Jiang, J.: Text image with complex background filtering method based on Harris corner-point detection. J. Softw. **8**(8), 1827–1834 (2013)
14. Jiang, S.P.Z., Ma, Y.X.: Automatic document layout analysis system for the large scale Chinese antient books 'imperial collection of four' (in Chinese). J. Chin. Inf. Process. **17**(2), 14–20 (2000)

Image-Text Dual Model for Small-Sample Image Classification

Fangyi Zhu[1], Xiaoxu Li[1,2], Zhanyu Ma[1(✉)], Guang Chen[1], Pai Peng[3], Xiaowei Guo[3], Jen-Tzung Chien[4], and Jun Guo[1]

[1] Pattern Recognition and Intelligent System Lab,
Beijing University of Posts and Telecommunications, Beijing, China
mazhanyu@bupt.edu.cn

[2] School of Computer and Communication, Lanzhou University of Technology,
Lanzhou, China

[3] Youtu Lab, Tecent Technology, Shanghai, China

[4] Department of Electrical and Computer Engineering,
National Chiao Tung University, Hsinchu City, Taiwan

Abstract. Small-sample classification is a challenging problem in computer vision and has many applications. In this paper, we propose an image-text dual model to improve the classification performance on small-sample dataset. The proposed dual model consists of two sub-models, an image classification model and a text classification model. After training the sub-models respectively, we design a novel method to fuse the two sub-models rather than simply combining the two models' results. Our image-text dual model aims to utilize the text information to overcome the problem of training deep models on small-sample datasets. To demonstrate the effectiveness of the proposed dual model, we conduct extensive experiments on LabelMe and UIUC-Sports. Experimental results show that our model is superior to other models. In conclusion, our proposed model can achieve the highest image classification accuracy among all the referred models on LabelMe and UIUC-Sports.

Keywords: Small-sample image classification · Ensemble learning
Deep convolutional neural network

1 Introduction

With the wide use of the internet, the image data on the network is increasing dramatically. How to retrieve and understand the image data correctly is a hot and difficult problem in the current research of computer vision. In recent years, with the development of deep learning, learning and extracting semantic information of massive images using convolutional neural network provides an effective solution for image understanding. However, we usually consider too much about the ideal situation but ignore the actual cases. In many applications, there is no sufficiently large amount of data but only small amount of labeled data, e.g., LabelMe dataset [12] and UIUC-Sports dataset [8].

© Springer Nature Singapore Pte Ltd. 2017
J. Yang et al. (Eds.): CCCV 2017, Part II, CCIS 772, pp. 556–565, 2017.
https://doi.org/10.1007/978-981-10-7302-1_46

The two datasets consist of 8 classes. Each class has more than 130 and less than 326 images. The total numbers of images in the two datasets are much smaller than that of other image datasets, such as Flickr [3] and MS COCO [10], which are frequently used in many image classification tasks. However, a large number of labeled samples are required in supervised learning. Adding annotations to data will cost a lot of manpower. On the other hand, the efficiency of artificially image tagging is very unstable and the results can easily be affected. Generally speaking, manual annotation is not only laborious but also easy to get subjective and individual tagging errors. Therefore, how to achieve similar performance on the small-sample dataset as on the large dataset is an important problem in the field of computer vision. Moreover, how to learn semantic information from small amount of labeled samples, among others, is one striking challenge. Motivated by this observation, this paper aims to study image semantic learning on small-sample datasets. The category of the images can be regarded as a general description. Hence, the image classification task is one of the fundamental tasks.

To tackle the challenges, we decompose the problem of image classification into two manageable sub-problems. An image classification model is trained on the given images, and a text classification model is built on the annotations. On top of the two classification models, we introduce a fusion process to learn the connection between the two sub-models.

The main contribution of our paper is to present an image-text dual model. Comparing with some existing models for image classification on small-sample datasets, our model achieves the best performance in terms of classification accuracy on LabelMe dataset and UIUC-Sports dataset [8, 12, 16, 17]. Our model can also save computational resources significantly.

2   Related Work

The method based on topic model to learn semantic features is the direction that researchers have been focused on recently [2, 4, 6, 14, 16, 17]. A topic model refers to a statistical model that discovers or learns abstract topics of documents, which origins from natural language processing (NLP) [2, 4]. In recent years, with the fast developing of neural network research, the research about neural topic model, which is the topic model based on neural network [6], and image classification based on neural topic model has been started [2, 4, 6, 14, 16, 17].

Larochelle et al. proposed a model at the 2012's NIPS conference, the model named Document Neural Autoregressive Distribution Estimator (DocNADE) [6], can obtain good topic features. The model assumes that the generation of each word is only associated with the words that generated before it, and the document is the product of a set of conditional probabilities, which is generated by a feedforward neural network. The advantage is that the method models the relationship between words and the calculation of latent variables does not require complex approximate reasoning as other probabilities generation models. Zheng et al. presented the SupDocNADE, a shallow model based on the DocNADE [16, 17]. The model got 83.43% accuracy on LabelMe and 77.29%

accuracy on UIUC-Sports [8,12,16,17]. Recently, there has been remarkable progress in the direction under big data with the development of deep learning, however there is still room for improvement in this area. AlexNet trained on ImageNet obtained perfect performance for image classification [5]. Zhou et al. [18] proposed a method of extracting features trained on Places dataset. The model is similar to AlexNet, and got perfect performance on several datasets. Zisserman et al. proposed VGG-Net [15], which achieved the first place on the localization task of ILSVR and the second place on the classification task of ILSVR. However, the models mentioned above are all deep models. Since deep models have a large number of parameters needed to be trained, the models can not be trained adequately on small-sample datasets, therefore, the performance will be constrained. To tackle this problem, we utilize the semantic of annotations to learn images' deep semantic features and improve the performance of image classification on small-sample dataset.

In the previous methods, there were two ways to learn the connections between image and the corresponding annotations. One was to input the joint features of images and annotations into a classification model, the other was to input the features of images and annotations to the classification models separately. Summarizing the existing methods, the classification models on small-sample dataset mostly utilize traditional methods not deep models. This is because the deep models can not be trained reliably on small-sample dataset. Hence, we propose an image-text dual model in order to utilize the annotations' semantic information to overcome the insufficient training problem. By comparing with some recently proposed methods, we find that our semantic information can achieve significant improvement in terms of image classification accuracy.

The main contribution of our paper is to propose an image-text dual model. The model utilizes two models to respectively learn image and text features and fuse the two models' results in the end. Our model can utilize annotations' semantic information to improve the performance of image classification model. The proposed dual model achieves 97.75% classification accuracy on LabelMe dataset [12] and 99.51% classification accuracy on UIUC-Sports dataset [6].

3   Image-Text Dual Model

In order to utilize annotations' semantic information to improve the performance of the image classification model, we propose a simple yet effective image-text dual model. It decomposes the traditional image classification model into two models.

Image model: It is an end-to-end network fine-tuned by VGG16 [15], which gives the image classification results.

Text model: It is an end-to-end neural network as well, it gives the annotations classification results.

Ultimately, we propose a method to fuse the image classification results and the annotations classification results to predict the final class that the input

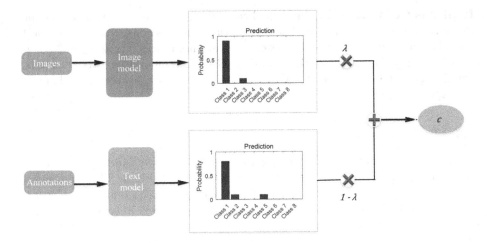

Fig. 1. Illustration of the image-text dual model. Images and annotations are trained respectively. On top of the two classification models, we add another fusion processing to merge the two models' results. To learn the connection between the two models, we propose a method in the fusion process.

image belongs to. The architecture of the proposed image-text dual model is shown in Fig. 1.

3.1 Image Model

Instead of simply jointing the image features and annotations features as input of one model, we train two models respectively. We use transfer learning [13] to build the model. Transfer learning allows to utilize known relevant task data to solve new unknown tasks. We use VGG16 [15] as pre-trained model and fine-tune the model on our pre-processed datasets [8,12]. The structure of our image model is shown in Fig. 2.

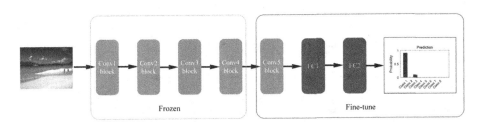

Fig. 2. Illustration of image classification model. We take only the convolutional layers of VGG16 and drop the FC layers. On top of the convolutional layers, we add two FC layers. When fine-tune the model, we freeze the first four convolution blocks and fine-tune the fifth convolution block and new FC layers

Implementation details. The model consists of five convolutional blocks and two full-connected(FC) layers. The convolutional blocks consist of 3×3 stride 1 convolutions and ReLU. Between two convolutional blocks, there is 2×2 stride 2 maxpooling. The stack of convolutional layers is followed by two FC layers: the first contains 512 channels, the second is a soft-max layer with 8 channels (one for each class).

3.2 Text Model

To learn the semantic information of the annotations, we build a text classification model consisted of three FC layers as shown in Fig. 3. There are many ways to build word embedding. We applied the word2vec [7, 11] and bag of words model to build the text vectors. The two methods work well in text classification tasks. And we further attempt to use PCA [1] to reduce word vecotrs to lower dimensions.

Implementation details. The model consists of three FC layers: the first contains 64 channels, the second contains 512 channels, the third is a soft-max layer with 8 channels (one for each class).

3.3 Fusion Models

We get the soft-max layer's output $P_{img}(c)$ of the image model and the soft-max layer's output $P_{text}(c)$ of the text model. $P_{img}(c)$ and $P_{text}(c)$ are all of eight dimensions. Each of the dimensions represents the probability of the sample belongs to a specific class. The final result c_{final} is predicted by merging the two models' results. Simple way of fusion is to add the two models' results together. However this fusion strategy can not get good results. We propose a method to combine the results of the two models as shown in (1). λ is a regularization parameter that controls the balance between the two sub-models. The range of values of λ is between 0 to 1.

$$c_{final} = max_c(\lambda P_{img}(c) + (1 - \lambda)P_{text}(c)). \tag{1}$$

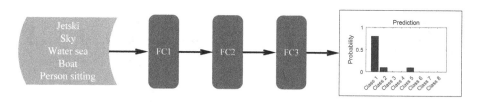

Fig. 3. Illustration of text classification model.

4   Experiment Results and Discussions

To evaluate the proposed image-text dual model, we conducted extensive quantitative and qualitative evaluations on LabelMe dataset and UIUC-Sports dataset [8,12]. The two datasets contain images' annotations and are popular classification benchmarks. We provided a quantitative comparison between SupDocNADE, Fu-L, Mv-sLDA [9,16,17]. Following Li et al. [9], we randomly extracted five subsets of LabelMe dataset [12] and five subsets of UIUC-Sports dataset [8]. We used the mean classification accuracy of the five subsets to measure the performance of image classification.

4.1   Datasets Description

The LabelMe dataset [12] collected by Aude Oliva, Antonio Torralba. LabelMe dataset [12] has eight classes: coast, forest, highway, inside city, open country, street, tall building. For each class in one of our subsets, 200 images were randomly selected and split evenly in the training and test sets, yielding 1600 images.

The UIUC-Sports dataset [8] collected by L. Li and F. Li. It contains 1792 images, classified into eight classes. Each subset we constructed consists of 1720 images: badminton (300 images), bocce (130 images), croquet (300 images), polo (190 images), rockclimbing (190 images), rowing (250 images), sailing (180 images), snowboarding (190 images). The images are randomly selected and split evenly in the training and test sets.

4.2   Performance Analysis

In this section, we analyze our proposed dual model on LabelMe dataset [12] and UIUC-Sports dataset [8] with baseline models shown in Fig. 4.

The mean image classification accuracy of our single image model on LabelMe [12] is 80.9% and on UIUC-Sports [8] is 70.7%. Our single image model can get better performance than single image classification model in Fu-L and Mv-sLDA [9].

We use word2vec [7,11] and bag of words to build text vectors. However, word2vec [7,11] in our task does not have good performance. The reason is that word2vec fits for sentences but not for separate words. Furthermore, only less than 1000 words in total can not train by neural network sufficiently to get a good word2vec embedding. In our work, we choose the bag of words model to build the text vectors. We further attempted to use PCA [1] to reduce word vectors to lower dimensions. The mean text classification accuracy on LabelMe dataset without PCA [1] is 95.13%. The accuracy is 94.73% after reducing to 480 dimensions with PCA. The accuracy is 93.78% after reducing to 240 dimensions. The model without PCA is always higher than the model with PCA [1]. This is due to the fact that the dimension of the original vectors is not high for the neural model to train. Therefore, without dimension reduction, better result can be obtained. The mean text classification accuracy on UIUC-Sports dataset [8] without PCA [1] is 99.35%.

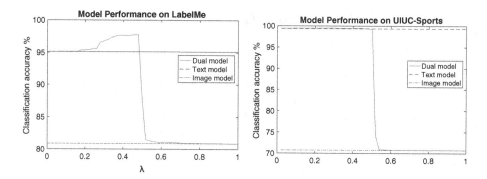

Fig. 4. The vertical axis presents the classification accuracy and the horizontal axis represents the value of the weight λ in (1).

The aforementioned text classification model also can get better performance than the text classification models in Fu-L, Mv-sLDA [9].

Figure 4 shows the trend of image classification accuracy with different values of λ. When λ is 0, the model is the text classification model. When λ is 1, the model is the image classification model. Good performance can be obtained when λ falls in [0.4, 0.5] on LabelMe and can be obtained when λ falls in [0.2, 0.3] on UIUC-Sports. As illustrated in Fig. 4, the dual model achieves significant improvement than the single models. It exhibits that our image-text dual model reliably effective. These experimental results demonstrate that the proposed dual model can overcome the insufficient training of deep models on small-sample datasets. The idea that utilizing annotations' semantic information to improve image classification model's capability is proved to be feasible. As reason that the probability values of classes from image classification model's soft-max layer are much bigger than the values of text classification model's, so the dual model's accuracy drops rapidly when λ is equal or greater than 0.5.

Table 1. Performance comparison of dual model between different variants on LabelMe.

Model	LabelMe accuracy (mean)%	LabelMe accuracy (max)%
Dual model ($\lambda = 0.40$)	97.65	98.25
Dual model ($\lambda = 0.42$)	97.63	98.25
Dual model ($\lambda = 0.44$)	97.65	98.25
Dual model ($\lambda = 0.46$)	**97.75**	**98.25**
Dual model ($\lambda = 0.48$)	97.75	98.13
Dual model ($\lambda = 0.50$)	88.35	90.63

Table 2. Performance comparison of dual model between different variants on UIUC-Sports.

Model	UIUC-Sports accuracy (mean)%	UIUC-Sports accuracy (max)%
Dual model ($\lambda = 0.20$)	99.48	99.76
Dual model ($\lambda = 0.22$)	99.48	99.76
Dual model ($\lambda = 0.24$)	99.48	99.76
Dual model ($\lambda = 0.26$)	**99.51**	**99.76**
Dual model ($\lambda = 0.28$)	99.48	99.76
Dual model ($\lambda = 0.30$)	99.46	99.76

As shown in Tables 1 and 2, we can conclude that the proposed image-text dual model can get the highest image classification accuracy on LabelMe dataset when λ is 0.46 and on UIUC-Sports dataset when λ is 0.26.

4.3   Experimental Results

In this section, we describe quantitative comparison between our dual model and other methods. The classification results are illustrated in Table 3. Our dual model obtains an accuracy of 97.75% on the LabelMe [12] dataset and 99.51% on UIUC-Sports dataset. This is significantly superior to other models' performance on the two datasets as shown in Table 3.

We compare the proposed image-text dual model with the SupDocNADE [16,17], which uses the image-text joint embedding during training and inputs images only during test stage. Since using the code published by the author can not get the accuracy in paper, we compare directly with the results found in the corresponding paper [16,17]. The reported accuracy is 83.43% on LabelMe and 77.29% on UIUC-Sports for SupDocNADE [16,17], which is lower than our image-text dual model as shown in Table 3. Therefore, we can conclude that training the image and text separately can yield better results. In addition, the max iteration number for training SupDocNADE [16,17] is 3000. Our image model only needs 350 epochs to train at most, 100 epochs to fine-tune and the text model also only needs 100 epochs at most, as our models' parameters are less than SupDocNADE'. For the reason that we decompose one model into two sub-models and use fine-tune during training the image model. What' more, the two sub-models can be trained parallelly at the same time. Our single image model's accuracy is lower than SupDocNADE because deep models have the problems of insufficient training and the single image model don't utilize the annotations. The final result shows that our model can learn the semantic information of images' annotations better.

We also compare the proposed model requiring both images and annotations during test stage, i.e., Fu-L and Mv-sLDA [9]. Fu-L and Mv-sLDA all utilize traditional methods but not deep learning [9]. The structure of Fu-L is similar to

Table 3. Performance comparison of different models.

Datasets	SupDocNADE	Fu-L	Mv-sLDA	Dual model
LabelMe	83.43%	82.3%	92.2%	**97.75%**
UIUC-Sports	77.29%	84.1%	99.0%	**99.51%**

our model. Fu-L builds two separate traditional models on image modal and text modal. Then Fu-L utilizes the third model to fuse the two models. Mv-sLDA builds a traditional model can classify both image modal and image modal. After training separately, it fuses the results of image model and text model. As shown in Table 3, our model's image classification accuracy achieves the best performance for the reason that we introduce the deep models to the small-sample datasets.

5   Conclusions

In this paper, we proposed an image-text dual model for the task of small-sample image classification. The proposed method decomposes the image classification model into two manageable ones, i.e., an image classification model and a text classification model. Furthermore, we propose a method to fuse the two sub-models. Extensive quantitative and qualitative results demonstrate the effectiveness of our proposed model. Compared to some recently proposed models, our method can better incorporate the semantic information of the annotations. Therefore, we can get higher image classification accuracy. Moreover, our image-text dual model needs few epochs to train because it contains the few parameters. In our work, the two sub-models can be trained at the same time also contributes to saving the computational efficiency. In addition, the structure of our dual model can be extended to other modalities, e.g. image-sketch, image-video, text-video.

Acknowledgement. This work was supported in part by the National Natural Science Foundation of China (NSFC) under Grant 61773071, Grant 61628301, Grant 61402047 and Grant 61563030, in part by the Beijing Nova Program Grant Z171100001117049, in part by the Beijing Natural Science Foundation (BNSF) under Grant 4162044, and in part by the CCF-Tencent Open Research Fund.

References

1. Bishop, C.M.: Pattern Recognition and Machine Learning. Springer, New York (2006)
2. Blei, D.M., Ng, A.Y., Jordan, M.I.: Latent dirichlet allocation. JMLR **3**(1), 993–1022 (2003)
3. Hare, J.S., Lewis, P.H.: Automatically annotating the MIR Flickr dataset: experimental protocols, openly available data and semantic spaces. In: ACM MIR, pp. 547–556. ACM (2010)

4. Hofmann, T.: Probabilistic latent semantic indexing. In: ACM SIGIR, pp. 50–57. ACM (1999)
5. Krizhevsky, A., Sutskever, I., Hinton, G.E.: Imagenet classification with deep convolutional neural networks. In: NIPS, pp. 1097–1105 (2012)
6. Larochelle, H., Lauly, S.: A neural autoregressive topic model. In: NIPS, pp. 2708–2716 (2012)
7. Le, Q., Mikolov, T.: Distributed representations of sentences and documents. In: ICML, pp. 1188–1196 (2014)
8. Li, L., Li, F.: What, where and who? classifying events by scene and object recognition. In: IEEE ICCV, pp. 1–8. IEEE (2007)
9. Li, X., Li, R., Feng, F., Cao, J., Wang, X.: Multi-view supervised latent dirichlet allocation. Acta Electron. Sin. **42**(10), 2040–2044 (2014)
10. Lin, T.-Y., Maire, M., Belongie, S., Hays, J., Perona, P., Ramanan, D., Dollár, P., Zitnick, C.L.: Microsoft COCO: common objects in context. In: Fleet, D., Pajdla, T., Schiele, B., Tuytelaars, T. (eds.) ECCV 2014. LNCS, vol. 8693, pp. 740–755. Springer, Cham (2014). https://doi.org/10.1007/978-3-319-10602-1_48
11. Mikolov, T., Chen, K., Corrado, G., Dean, J.: Efficient estimation of word representations in vector space. arXiv preprint arXiv:1301.3781 (2013)
12. Oliva, A., Torralba, A.: Modeling the shape of the scene: a holistic representation of the spatial envelope. IJCV **42**(3), 145–175 (2001)
13. Pan, S.J., Yang, Q.: A survey on transfer learning. IEEE TKDE **22**(10), 1345–1359 (2010)
14. Putthividhya, D., Attias, H.T., Nagarajan, S.S.: Topic regression multi-modal latent dirichlet allocation for image annotation. In: IEEE CVPR, pp. 3408–3415. IEEE (2010)
15. Simonyan, K., Zisserman, A.: Very deep convolutional networks for large-scale image recognition. arXiv preprint arXiv:1409.1556 (2014)
16. Zheng, Y., Zhang, Y., Larochelle, H.: Topic modeling of multimodal data: an autoregressive approach. In: IEEE CVPR, pp. 1370–1377 (2014)
17. Zheng, Y., Zhang, Y., Larochelle, H.: A deep and autoregressive approach for topic modeling of multimodal data. IEEE TPAMI **38**(6), 1056–1069 (2016)
18. Zhou, B., Lapedriza, A., Xiao, J., Torralba, A., Oliva, A.: Learning deep features for scene recognition using places database. In: NIPS, pp. 487–495 (2014)

Learning Deep Feature Fusion for Group Images Classification

Wenting Zhao[1], Yunhong Wang[1], Xunxun Chen[2], Yuanyan Tang[3],
and Qingjie Liu[1(✉)]

[1] The State Key Laboratory of Virtual Reality Technology and Systems,
School of Computer Science and Engineering, Beihang University,
Beijing 100191, China
{wtzhao,qingjie.liu}@buaa.edu.cn

[2] National Computer Network Emergency Response Technical Team/Coordination
Center of China, Beijing 100029, China

[3] The Department of Computer and information Science,
Faculty of Science and Technology, University of Macau, Taipa 853, Macau

Abstract. With the rapid development of social media, people tend to post multiple images under the same message. These images, we call it group images, may have very different contents, however are highly correlated in semantic space, which refers to the same theme that can be understood by a reader, easily. Understanding images present in one group has potential applications such as recommendation, user analysis, etc. In this paper, we propose a new research topic beyond the traditional image classification that aims at classifying a group of images in social media into corresponding classes. To this end, we design an end-to-end network which accepts variable number of images as input and fuses features extracted from them for classification. The method are tested on two newly collected datasets from Microblog and compared with a baseline method. The experiment demonstrates the effectiveness of our method.

Keywords: Group images understanding · Social activity analysis
Image classification · Social media

1 Introduction

Image classification is a fundamental task in computer vision which aims at classifying images into corresponding categories in term of its scene or objects. Recent researches on this topic mainly focus on classifying a single image into its scene or object categories, such as works in [2,8,19,21]. Recently, with the rapid development of social media and mobile devices, more and more people are

---

Wenting Zhao is the first author of this paper and is a graduate student who is pursuing her master degree in School of Computer Science and Engineering, Beihang University.

J. Yang et al. (Eds.): CCCV 2017, Part II, CCIS 772, pp. 566–576, 2017.
https://doi.org/10.1007/978-981-10-7302-1_47

willing to share ideas and experiences on the social media platforms, e.g. Twitter, Flicker, Microblog. Most of these messages posted by users contain not only text information but also some images which express emotions, opinions and social activities, etc. Obtaining information from these data is quite important for a wide range of applications, including advertisements, recommendation, etc. Most researches about these social messages now focus on analyzing text or extracting information from a single image. However, the fact is that one post or message usually consists of multiple images and sometimes there is no text messages. Therefore, lots of important information will be lost if only text or single image is used. We believe that images within one message or post should be processed in their entirety as they show similar semantic information and most of these have the same theme. Therefore, analyzing images in a group manner may help to better understand the message. Thus, how to simply use the relationship between a group of images to analyze the user's behavior state or the semantic theme of the group images is a very interesting and challenging subject. To solve this, we do a lot of researches on multiple images processing.

Aside from the widely studied image or scene classification, one similar topic to ours is action recognition from video sequences [5, 7, 16], in which adjacent frames referring to the same action in a video clip can be assigned with a label indicating a specific action category. Another similar research topics are image co-segmentation and co-saliency detection, which aim at detecting foreground objects or salient regions from multiple images containing similar of same objects over simple background [3, 6, 9, 12, 14, 15]. All these researches as well as our's take multiple images as input. However, our research is fundamentally different from that of the aforementioned researches in that multiple images posted under the same message show very complex contents. For example, one user post a group of images when he/she is attending a soccer game. The images may comprise of playground, audiences and soccer players, etc. They hardly contain the same objects. An example is shown in Fig. 1, from which we can see the images in the soccer games including playground, scoreboard and players, etc. It is unlikely that the same object appears in different images. Even if the same object appears in more than one images, it is still hard to understand the relationship between these images due to scale variations caused by the different angle or the distance while taking photos. Furthermore, the shooting time span is extensive and discontinuous. From Fig. 1 we can see that both the space and times of photoing these images within the same group are discrete.

In this paper, we put forward a two-step framework to classify the group images in the social media platforms. Firstly, a VGG-16 model [17] pre-trained on ImageNet is employed to extract high level semantic features from images. Then, we propose a feature map fusion strategy to combine all features extracted from different images together to perform classification. We elevate our method on two newly collected datasets GUD-5 and GUD-12, and compare it with a baseline method. Experimental results demonstrate that a group of images under the same message show strong correlation in semantic space and it is feasible to use deep learning method to classify them.

Fig. 1. Examples of group images collected from microblog. The images are retrieved from Sina Microblog using key words such as "birthday", "cycling" etc. We collect two datasets, one contains 5 classes (we name it GUD-5 as abbreviation of group image understanding dataset), the other one is an extension of GUD-5, which has 12 classes and each of them consists of 100 groups. Because the maximum number of images under one message is 9, in our datasets, each group may contain 2 to 9 images. From these examples, we can see the contents of these images varies greatly and the backgrounds are complex, making it very difficulty to be classified.

2   Group Image Classification

One image can show very rich information. People may tell different stories from the same image. However, we believe that, although having very different content, images under the same post or message in social media platforms are highly relevant in semantic space. They narrow down the semantic space so that people can easily understand what is the meaning of this message or post. Here, we demonstrate that, it is possible to classify multiple images in the same message into different categories. In this section, we describe the details of our method to classify the group images. The brief architecture of our method is shown in Fig. 2.

2.1 Feature Extraction

We use a pre-trained CNN on ImageNet to extract features from each image. In this work, VGG-16 [17] is employed. Let $\mathcal{I} = \{I_i\}_{i=1}^{n}$ be a set of images contained in one group, n is the number of images ranging from 2 to 9. After feature extraction, we have a set of high level features to represent semantic information of each image $f_i = \text{VGG}(I_i)$. It should be noted the number of images are not fixed in each group, thus the size of features varies with the number of images in each group. Because we need fixed-sized features to be classified, these features for each group should be further processed to have the same size. One way is using bag of visual words method [13] to encode features extracted from different images. Inspired by the spatial pooling method that used within the same feature map, in this paper, we proposed to use feature map pooling that can fuse feature maps of different images into a fixed-sized feature map.

2.2 Feature Map Fusion

One straightforward way to tackle varying size features is using bag of visual words method to fuse features extracted from multiple images into one single fix-sized vector [13]. Another method is using recurrent neural networks to encode information across frames [11]. Inspired by the spatial pooling used in CNN and the observation that strong activations in high level feature maps corresponding to object level information, we proposed to use feature map pooling to aggregate features extracted from different images.

After through VGG-16, feature maps for each image can be obtained. Similar to spatial pooling, there are three ways to aggregate these features, max pooling, mean pooling and subtraction pooling.

$$v(k, h, w) = \max_{i=1,...,n} f_i(k, h, w) \tag{1}$$

where $v(k, h, w)$ is the value of kth fused feature map at position (h, w), and $f_i(k, h, w)$ is value of kth feature map of image i at (h, w). Finally, arbitrary number feature maps are merged into one.

Mean pooling is also a commonly used aggregation strategy. It is used here to produce a single feature maps averaged over all the extracted feature maps, as follows:

$$v(k, h, w) = \frac{1}{n} \sum_{i=1,...,n} f_i(k, h, w) \tag{2}$$

Subtraction pooling can be achieved similarly as follows:

$$v(k, h, w) = |f_i(k, h, w) - f_{i-1}(k, h, w)| \qquad (1 <= i <= n) \qquad (3)$$

In this paper, we test all pooling methods.

2.3   Group Images Classification

After extraction of high level semantic information using a VGG-16 network, we adopt three different fusion methods as described in Sect. 2.2 to encode the features from multiple images into one representation. After fusing these feature maps, a following five-layer convolution and three-layer fully connected layers with softmax classifier is applied to perform the classification.

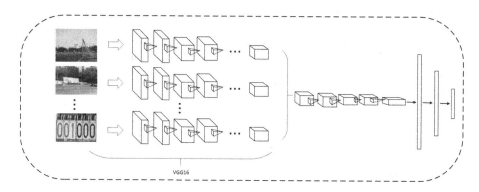

Fig. 2. The network of classifying group images. Firstly, we utilize the VGG-16 network pre-trained on ImageNet to extract the feature maps. Then, three different fusion strategies are applied to encode all the feature maps to a fix-sized features. Finally, we design a classification network with 5 convolutional layers and 3 fully-connected layers to perform classification.

3   Experiments

3.1   Dataset

The existing image classification methods mainly focus on classifying a single image into different categories in term of the scene or objects it contains. The popular datasets for image classification are ImageNet [4], Places [20, 21], etc. To the best of our knowledge, there is no public dataset that can be used for testing our method. To evaluate our method and facilitate future researches on this topic, we build two datasets named GUD-5 and GUD-12 (GUD is a acronym for Group image Understanding Dataset), respectively. GUD-5 contains 5 classes, and GUD-12 is an extension of GUD-5 to have 12 classes. Each class in these two datasets is comprised of 100 groups and each group contains 2 to 9 images. The images are downloaded from

Sina Microblog[1] by searching the social activity key words such as skiing, birthday party, etc. We only select the results which contain multiple images. Then, 9 students are asked to filter out the results by answering the question "Does this group of images show skiing?". Some examples are showned in Fig. 1. 10% of these samples are used as testing set, and the remaining 90% are as training set.

3.2 Training

We use the VGG-16 network [17] pre-trained on ImageNet to extract features from each image in the group. Then we select high level feature maps as representation of this image and fuse them using three different strategies as described in Sect. 2.2. The following is a network with 5 convolutional and 3 fully-connected layers that used for group images classification. The network can be trained in an end-to-end manner. It was implemented in Tensorflow [1] and trained on a Nvidia GeoForce GTX 1080 GPU with a batch size of 128 and a momentum of 0.9 using SGD optimization method. The learning rate was set starting from 0.01 and decreased by every 50 epochs. The training will converge in 300 epochs.

3.3 Impacts of Fusion Layers

There are 13 convolutional layers and 3 full-connected layers in VGG-16. Which layer should be selected to fuse is undermined. We test performances of different layers to select the suitable layer to fuse. The simplest method is choosing the last fully-connected layer of the VGG-16 to represent the image. However, the experiments demonstrate a poor performance, as shown in Figs. 3 and 4. And compared with lower level features, we get the best results when the feature map sized 28×28 that is the 10th or 11th layer features. We can see the results are very poor when we fuse the low level features such as 5th and 6th layer features. This is because it is hard to get the global information when only extract texture features or color features from the group images. However, it is also difficult to fuse the sematic features if we extract too high level features. Therefore, good classification results can be obtained if we select the 10th or 11th layer features.

3.4 Impacts of Fusion Strategies

We evaluate performances of three fusion strategies, as shown in Figs. 3 and 4. It can be seen that, compared with the maximum fusion method, average fusion works most stable. Results obtained by subtracting fusion in GUD-5 and GUD-12 are very different. The reason behind this may be that average fusion can better balance the global information, subtracting can better capture the major difference information between group images. Global information can get better performance when the category is small. However, due to the interference of similar categories, average fusion will be difficult to obtain better result for large category. In this work, the difference comparison between images will be more important. Therefore, subtracting fusion can achieve better results for 12 classes.

---

[1] weibo.com.

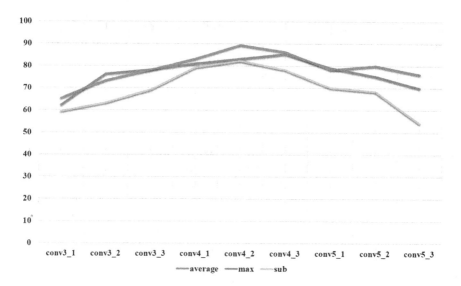

Fig. 3. Performances of our method by using different fusion strategies and fusing different layers on GUD-5 dataset.

3.5   Comparison with Baseline Method

To further demonstrate the effectiveness of the proposed method, we compare it with a baseline method i.e. bag of visual word (BOW). We use SIFT descriptor [10] to extract feature from each image and cluster SIFT features of all the training images into 1000 clusters. Using BOW model, each group can be represented as a 1000 dimensional vector. A linear SVM [18] is adopted to classify groups into corresponding categories. The experimental settings are the same to that of the proposed method, i.e. 10% of the dataset are used for testing and 90% are used as training set. We compare it with our method on the both datasets. The results are shown in Table 1. We can see that, comparing with the baseline, our method obtains a much higher results both on the two datasets. This proves that compared with the traditional processing method, CNN network has more powerful feature extraction ability and can better learn more high level semantic information. In this paper, fusing the high level features can make better use of the relationship between images than simply extracting low-level features from each image for classification.

3.6   Results Analysis

There may exist noise images in one group that do not agree with other images to have the same or similar theme. How to select the most relevant images to the theme of the group images and filter out irrelevant parts from the group images is a very important problem. In order to solve this, we train a classification network to remove these images and compare the results with that of the original one on

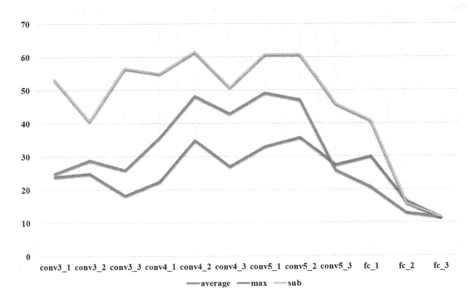

Fig. 4. Performances of our method by using different fusion strategies and fusing different layers on GUD-12 dataset.

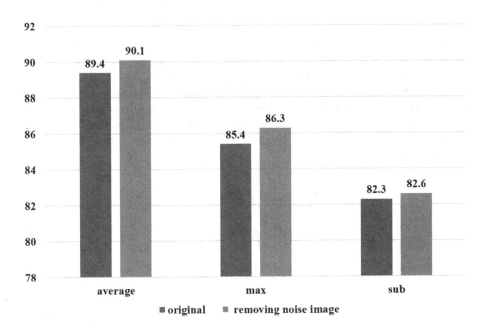

Fig. 5. Influence of with and without the noise images for classification. It can be seen that although with much noise, our method can still classify group images with a high accuracy.

Table 1. Comparisons between the proposed and BOW method on the two datasets

Method	GUD-5	GUD-12
Proposed (mean)	89.4%	49.16%
Proposed (max)	87.2%	35%
Proposed (sub)	82.3%	61.6%
BOW	45%	25%

the GDU-5 dataset. We use five-layer convolution and three-layer fully connected layers with softmax classifier to classify these images. We give each image in the same group the same label which is the theme of the group images and try to classify them. If it is difficult for the classified network to judge which topic the image should belong to, we will think this image has nothing to do with it's group images' theme and delete it from this group. Through the filter processing, we can get the images making largest contribution to the theme of the group and filter out irrelevant ones from the group. Figure 5 shows the results. As it can be seen, we can get better results after the filter processing. Even with these noisy images, our method can still have promising performance (Fig. 6).

Compared with the results on GUD-5 dataset, our method have poor results on GUD-12 dataset. This maybe because some classes in GUD-12 are very hard to be discriminated. Figure 7 shows the confusion matrix on GUD-12. From which we can see the classification results are very good on "soccer games" or "amusement park". This may because these two classes have distinct theme and salient objects such as football, ferris wheel and so on. However for activities such as "barbecue" and "visiting aquarium", it is hard to classify them because these activities are all with dense population and have few salient objects. At the same time, these subjects may contain more kinds of objects which are not very similar such as varieties of foods and water animals. These could also lead to the bad experimental result. Increasing training data may help to alleviate this problem.

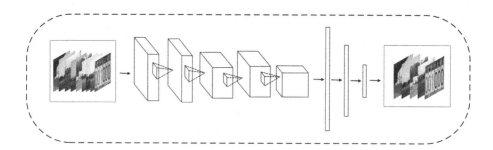

Fig. 6. The network of filtering out irrelevant parts from the group images. We use a classification network with 5 convolutional layers and 3 fully-connected layers to get the parts which make largest contribution to the theme of the group images.

	soccer games	cycling	birthday	barbecue	wedding	Flag raising	skiing	zoo	gallery	aquarium	auto show	amusement
soccer games	1	0	0	0	0	0	0	0	0	0	0	0
cycling	0	0.5	0	0	0	0	0	0	0.5	0	0	0
birthday party	0	0	0.5	0	0.5	0	0	0	0	0	0	0
barbecue	0	0	0	0.3	0	0	0.7	0	0	0	0	0
wedding	0.3	0	0	0	0.7	0	0	0	0	0	0	0
Flag raising ceremony	0	0	0	0	0	0.5	0	0	0	0.5	0	0
skiing	0	0	0	0	0	0	0.4	0	0	0	0	0.6
visiting the zoo	0	0	0.2	0	0	0	0	0.8	0	0	0	0
visiting the gallery	0	0	0	0	0	0	0	0.3	0.7	0	0	0
visiting the aquarium	0	0	0	0.7	0	0	0	0	0	0.3	0	0
auto show	0	0	0	0	0	0.3	0	0	0	0	0.7	0
amusement park	0	0	0	0	0	0	0	0	0	0	0	1

Fig. 7. Confusion matrix on GUD-12 dataset.

4 Conclusion

We have proposed a new architecture for group images classification that may be useful for understanding images in social media platforms. We believe considering images under the same message in a entirety manner is better than processing them one by one, for images may show a strong correlated theme even they have very different contents. To this end, we propose a fusion method to fuse different features extracted from each image within the same group, and design a deep fusion network to perform classification. We test our method on two datasets collected from Chinese biggest microblog platform Sina Microblog. Experiments demonstrate that it is challenging yet possible to classify a group of images. In the future work, we shall work on generating the theme of a group images in an unsupervised manner.

Acknowledgement. This work was supported in part by the Hong Kong, Macao, and Taiwan Science and Technology Cooperation Program of China under Grant L2015TGA9004. This work is also supported by research grant 008/2014/AMJ from Maco.

References

1. Abadi, M., Agarwal, A., Barham, P., Brevdo, E., Chen, Z., Citro, C., Corrado, G.S., Davis, A., Dean, J., Devin, M., et al.: Tensorflow: large-scale machine learning on heterogeneous distributed systems. arXiv preprint arXiv:1603.04467 (2016)
2. Antonio Torralba, R.F., Freeman, W.T.: 80 million tiny images: a large dataset for non-parametric object and scene recognition. IEEE Trans. Pattern Anal. Mach. Intell. **30**, 1958–1970 (2008)

3. Chang, K.Y., Liu, T.L., Lai, S.H.: From co-saliency to co-segmentation: an efficient and fully unsupervised energy minimization model. In: Computer Vision and Pattern Recognition, pp. 2129–2136 (2011)
4. Deng, J., Dong, W., Socher, R., Li, L.J., Li, K., Fei-Fei, L.: Imagenet: a large-scale hierarchical image database. In: Computer Vision and Pattern Recognition, pp. 248–255. IEEE (2009)
5. Feichtenhofer, C., Pinz, A., Zisserman, A.: Convolutional two-stream network fusion for video action recognition. In: Computer Vision and Pattern Recognition, pp. 1933–1941 (2016)
6. Joulin, A., Bach, F., Ponce, J.: Discriminative clustering for image co-segmentation. In: Computer Vision and Pattern Recognition, pp. 1943–1950 (2010)
7. Karpathy, A., Toderici, G., Shetty, S., Leung, T., Sukthankar, R., Fei-Fei, L.: Large-scale video classification with convolutional neural networks. In: Computer Vision and Pattern Recognition, pp. 1725–1732 (2014)
8. Krizhevsky, A., Sutskever, I., Hinton, G.E.: Imagenet classification with deep convolutional neural networks. In: Advances in Neural Information Processing Systems, pp. 1097–1105 (2012)
9. Li, H., Ngan, K.N.: A co-saliency model of image pairs. IEEE Trans. Image Process. **20**, 3365–3375 (2011)
10. Liao, K., Liu, G., Hui, Y.: An improvement to the sift descriptor for image representation and matching. Pattern Recogn. Lett. **34**, 1211–1220 (2013)
11. McLaughlin, N., Martinez del Rincon, J., Miller, P.: Recurrent convolutional network for video-based person re-identification. In: Computer Vision and Pattern Recognition, pp. 1325–1334 (2016)
12. Mukherjee, L., Singh, V., Dyer, C.R.: Half-integrality based algorithms for cosegmentation of images. In: Computer Vision and Pattern Recognition, pp. 2028–2035 (2009)
13. Peng, X., Wang, L., Wang, X., Qiao, Y.: Bag of visual words and fusion methods for action recognition: comprehensive study and good practice. Comput. Vis. Image Underst. **150**, 109–125 (2016)
14. Rother, C., Kolmogorov, V., Minka, T., Blake, A.: Cosegmentation of image pairs by histogram matching - incorporating a global constraint into MRFs. In: Computer Vision and Pattern Recognition, pp. 993–1000 (2006)
15. Rubinstein, M., Joulin, A., Kopf, J., Liu, C.: Unsupervised joint object discovery and segmentation in internet images. In: Computer Vision and Pattern Recognition, pp. 1939–1946 (2013)
16. Simonyan, K., Zisserman, A.: Two-stream convolutional networks for action recognition in videos. In: Advances in Neural Information Processing Systems, pp. 568–576 (2014)
17. Simonyan, K., Zisserman, A.: Very deep convolutional networks for large-scale image recognition. arXiv preprint arXiv:1409.1556 (2014)
18. Tsang, I.W., Kwok, J.T., Cheung, P.M.: Core vector machines: fast SVM training on very large data sets. J. Mach. Learn. Res. **6**, 363–392 (2005)
19. Xiao, J., Ehinger, K.A., Hays, J., Torralba, A., Oliva, A.: Sun database: exploring a large collection of scene categories. Int. J. Comput. Vision **119**(1), 3–22 (2016)
20. Zhou, B., Khosla, A., Lapedriza, A., Torralba, A., Oliva, A.: Places: an image database for deep scene understanding. arXiv preprint arXiv:1610.02055 (2016)
21. Zhou, B., Lapedriza, A., Xiao, J., Torralba, A., Oliva, A.: Learning deep features for scene recognition using places database. In: Advances in Neural Information Processing Systems, pp. 487–495 (2014)

Discriminative Region Guided Deep Neural Network Towards Food Image Classification

Yali Chen[✉], Yanping Yang, Qing Fang, and Xiaoyu Yao

School of Electronic Engineering,
University of Electronic Science and Technology of China,
Chengdu 611731, China
chenyalicqu@163.com,
{YYanping94,fountain_uestc,yaoxiaoyulzh}@outlook.com

Abstract. Food image classification plays an important role in smart health management, such as, diet analysis and food recommendation. Due to the similar appearance and shape between different foods, it is quite challenging to distinguish various food categories from their images. To address this issue, we propose a discriminative region guided deep neural network to classify the food images. More specifically, a saliency map based pooling strategy is applied to the input image to preserve the category aware discriminative regions. Meanwhile, the multi-scale fusion scheme is employed in our deep neural network to describe the discriminative regions across different resolutions. Experimental results on a large-scale Chinese food database show that, the average accuracy the proposed method is as high as 91.18%, and outperforms the baseline by 2.58%.

Keywords: Image classification · CNN · Deep learning
Discriminative regions

1 Introduction

Accurately classifying food category is an important step for healthy diet management [11,23]. Although the consulting service from experts could offer professional dietary analysis and advice, the high price and time costs fairly limit its popularization to the public [20]. As an alternative, it is much more desirable to classify and manage personal diet using the easily accessible food images, which could be captured by various smart devices, such as, cell phone, tablet and so on [6,8].

In recent years, the deep convolutional neural network (CNN) based algorithms have achieved great success in image classification [15,21,22], He *et al.* [10] presented deep residual network and made a breakthrough on generic image classification. Deep CNN was also applied to food image recognition in [24]. Zhang *et al.* [28] proposed to pick deep filter responses for fine-grained image classification. The methods mentioned above used whole image as input, which easily involves

© Springer Nature Singapore Pte Ltd. 2017
J. Yang et al. (Eds.): CCCV 2017, Part II, CCIS 772, pp. 577–587, 2017.
https://doi.org/10.1007/978-981-10-7302-1_48

the background noise and interferences the recognition for foreground objects. To address this issue, some region based algorithms are proposed to classify the image from its discriminative areas. Huang *et al.* [12] proposed polygon-based classifier to detect the discriminative regions for fine-grained classification. Yang *et al.* [25] utilized statistics of pairwise local features for food recognition. In [2,12,13,25], some manually annotated landmarks are also employed to locate the discriminative areas, such as, the beaks and eyes of the birds, which efficiently improves the classification accuracy for bird species.

Specific to the food image, the discriminative regions could also guide us to focus on the category related subtle details. More specifically, Chinese food categories vary widely and their ingredients change a lot. With its high similarity between categories and great difference among intra categories, it is very challenging to recognise food images in computer vision [3]. As discussed in many fine-grained image classification literatures [2,12,13], the discriminative regions play an important role in distinguishing object categories with subtle difference. The similar observations could be found in food images. As shown in Fig. 1, it is difficult to distinguish these two dishes between *Hongshaorou* and *Hongshao-niurou*. By removing the useless round plate and only focusing on the details in the center of the dishes, we can differentiate them from the details of containing or not containing fat.

Fig. 1. Examples of food images that have high similarities between categories and great difference among intra categories. The first row shows *Hongshaorou* images and the second row shows the similar dish *Hongshaoniurou*. It is difficult to distinguish these two kinds of dishes if we do not pay attention to the fat of pork marked by the green window in the first row and fat-free of beef marked by the red window in the second row. (Color figure online)

Inspired by the aforementioned observations, we propose a discriminative region guided deep neural network, which follows a two-step scheme to implement food image classification. At first, we utilize the normalized average saliency map to extract the most discriminative regions which are applied to all input images. The proposed method explores the common property of multiple saliency maps and significantly reduces the computational complexity and hardware overhead in training phase. Secondly, the multi-scale strategy is applied to the pre-processed input image, which describes the discriminative regions from different resolutions. More specifically, the input image and the feature map. We resize the input image to 224×224 and 448×448 on VGG-16 to obtain a base model separately. In addition, inspired by [26], we fuse the feature maps generated from lower and intermediate layers to capture the multi-scale information, where lower layers describe color and edge features, and intermediate layers describe texture and contextual information. Experimental results on a large scale database confirm that the proposed method could efficiently improve the food image classification accuracy.

In comparison with previous works, the contributions of this paper could be summarized into three folds. Firstly, we utilize the normalized average saliency map to derive the most discriminative region of each image. Secondly, the multi-scale strategy is proposed which combines multi-scale input image and multi-scale feature fusion. Thirdly, we build a challenging large-scale Chinese food image dataset CF90 which includes 135,000 images.

The remainder of the paper is organized as follows. We describe our proposed normalized average saliency map strategy and multi-scale method in Sect. 2. Then the dataset and experimental results are evaluated in Sect. 3. We make a conclusion in Sect. 4.

2   Our Approach

With a two-step scheme method in our approach, at first we apply a normalized average saliency map based pooling strategy to the input image to preserve the category aware discriminative regions. Secondly, we apply multi-scale strategy on VGG-16, which includes multi-scale input image and multi-scale feature fusion.

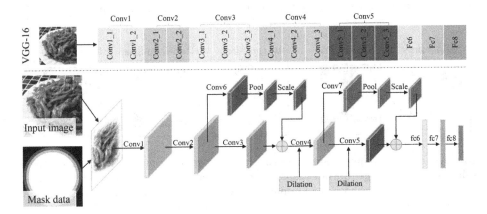

Fig. 2. Overview of the network architecture. The first row shows VGG-16 framework. Mask data is generated by the normalized saliency map to locate the discriminative regions in the image. After operating the multi-scale input image on VGG-16, the resolution of the input image is 448×448, we use *dilation* operation in convolution layers to match the dimension of the fc6. We continue with fusing the multi-scale feature map in the feature extraction stage through cascading the output of conv2_2 to conv3_3, conv4_3 to conv5_3.

2.1   Network Architecture

The network architecture is shown in Fig. 2. We implement our experiment on AlexNet [15] and VGG-16 [21], and select VGG-16 as our base network because of its better performance. We compute the saliency map of each image and generate a normalized average saliency map. The mask-layer add the normalized average saliency map on the input image and extract the most discriminative regions. We resize the image to 448×448 which changes the default input resolution of 224×224 of VGG-16. In order to match the input dimension of fc6-layer and utilize the pre-trained model parameters, we adopt *dilation* operation which will be discussed later.

2.2   Normalized Average Saliency Map

Visual saliency is the most attention-catching parts in images [9]. The information a saliency map contains reveals the discriminative region and the core massage that the image wants to transfer. Deep CNN was adopted to find saliency map in [16,29], Zhang *et al.* [27] proposed an effective method at 80 FPS based on minimum barrier salient object. Deep network based saliency map methods are always not so efficient as traditional ones, hence we choose Zhang's method to obtain saliency map in this paper.

Instead of only extracting the saliency map to implement the classification task, we use the normalized average saliency map to maintain the useful information and enough noises in the food image. Figure 3 shows some original saliency maps, average saliency maps, and normalize average saliency maps in our dataset. We transfer the normalized average saliency map to a mask, add it on the input image to preserve the most critical information of the image.

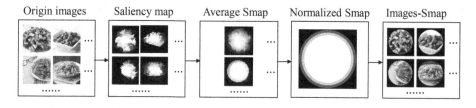

Fig. 3. The first column shows two categories of original images, followed by their saliency maps, average saliency maps, normalize average saliency map, and the images after operation. We choose the normalized average saliency map as a universal mask and operate it on the input image.

2.3   Multi-scale Input Image

Multi-scale input image contains different details cause its various resolution [18]. In this paper, we choose VGG-16 as our base model which has an input image resolution of 224×224 and obtain an average accuracy at 89%. CF90 dataset has

W_{11}	W_{12}	W_{13}
W_{21}	W_{22}	W_{23}
W_{31}	W_{32}	W_{33}

W_{11}	0	W_{12}	0	W_{13}
0	0	0	0	0
W_{21}	0	W_{22}	0	W_{23}
0	0	0	0	0
W_{31}	0	W_{32}	0	W_{33}

Fig. 4. The examples of different resolutions after resizing the image. The first row shows the image of resolution of 224×224, the second row shows the image of resolution of 448×448.

Fig. 5. The first column shows the original convolutional kernel, the second column shows the dilated convolutional kernel.

images resolution more than 280×280, we adopt multi-scale input image method and resize the image to 448×448. Through the improvement of input resolution, the extracted feature will be more precise. There are a number of resize methods available in Opencv such as *NEAREST, LINEAR*. We choose *LANCZOS* in order to gain a better resizing results. Figure 4 shows an example of different resolution. We can see the details are well defined in the image after resizing operation. Full connecting layers in the deep learning network always perform a significant role to gain the high-level features [17]. In the pre-trained caffemodel of VGG-16, the parameters of fc6, fc7-layer consists the most important part of the model [19]. We must match the input dimension if we want to use these parameters. After operating 448×448 input image on the network, the conv5_3-layer's output dimension changes to 14×14. In order to match the fc6-layer's dimension of 7×7, we utilize the *dilation* operation as shown in Fig. 5. *Dilation* operation means we add 0 to the kernels when doing convolutional computation and expand the kernel_size. The relation between kernel_size α, dilation β, and kernel_extend γ is shown in Eq. 1. We set the dilation parameter β to 2 in layer conv4_3, conv5_1, conv5_2, and conv5_3. After 4 times of *dilation*, we down sample the feature map to a size of 7×7 and match the fc6-layer well.

$$\gamma = \beta * (\alpha - 1) + 1 \tag{1}$$

2.4 Multi-scale Feature Fusion

Deep CNN shows a good performance on feature extraction and gives a reference to the classification or detection tasks for further process [3]. With the deepening of the network, the high-level feature can be learned and the details lost at the same time. For food image classification, the low-level features such as the edge, color and texture will be important for the ultimate classification stage [7]. In this paper, we fuse the feature maps generated from lower and intermediate layers to cascade multi-scale information.

As shown in Fig. 2, we combine the output of con2_2 with conv3_3, conv4_3 with conv5_3. After pooling the feature map into a suitable size, we add a scale layer to automatical learn the weights of the adding feature, where λ is a learnable parameter. The adding feature as an auxiliary information, we call it μ, and name the origin feature as ν, then we get the fused feature ξ,

$$\xi = \nu + \lambda * (0.0001\mu) \tag{2}$$

where λ means the learnable parameters in the network. We set the initial weight of μ to 0.0001, which is small enough to balance the most important origin feature and the auxiliary information.

3   Experiment

This section introduces CF90 dataset and evaluates our approaches on food image classification based on CF90. We randomly choose 1,200 images to train and 300 to test in each category on a single NVIDIA Titan GPU. The measure standard is average accuracy which display the proportion of test images that are correctly classified. The experiment will be discussed in 2 aspects: food image classification baseline and improvement. In the improvement stage, we first implement normalized average saliency map on the input image, then apply the multi-scale method based on the best model in the experiment. With a two-step scheme training method, we wish the approaches improve performance on our dateset.

3.1   CF90 Dataset

In this section, we will discuss and introduce CF90 dataset. To the best of our knowledge, the publicly available food image dataset till now is Recipe1M [1], UNIMIB2016 [5], and FOOD-101 [4]. The Recipe1M dataset consists of 1m cooking recipe and related 800k food images. UNIMIB2016 contains 73 categories of dishes. The images of FOOD-101 dataset was collected from foodspotting.com which are upload by people in real life. All above are Western food images, hence we collect images from www.xiachufang.com, image.baidu.com and images.google.com to build the largest Chinese food image dataset: CF90.

Our dataset contains 90 categories of common Chinese food and the most popular dishes on the menu website. Each category contains 1,500 images and we randomly select 1,200 images for training and the rest for testing. The resolution of the images are more than 280×280. Cause the images are taken by people from different conditions without artificial data clean up, the dataset keep a good redundancy to train. With 135,000 images in total, CF90 contains the popular food classes such as *Yuxiangrousi, Qingjiaorousi, Chaotudousi, Hongshaoqiezi, Fanqiechaodan, Hongshaorou, Pidanshourouzhou, Mantou, Shuizhuyu, Qingzhengyu, Paiguluobotang, Paiguyumitang, Nanguazhou, Baozi,* etc., as shown in Fig. 6. The images diverse from the light, shooting angle, and satisfy the semantic and visual richness. Email to scharlie92@outlook.com and reach the dataset.

Fig. 6. Examples of CF90 dataset.

3.2   Food Image Classification Baseline

In the first stage, we implement our experiment on AlexNet and VGG-16 based on the public Caffe platform [14]. For each of them, we fine-tune the model from the pre-trained ImageNet caffemodel as its good result. We resize the images to a fixed 256×256 resolution, from which randomly crop 224×224 when training on VGG-16 and 227×227 when training on AlexNet separately. We evaluate the average accuracy during testing stage.

We compare the average accuracy of AlexNet and VGG-16, as shown in Fig. 7, VGG-16 has a better performance of accuracy at 89% cause the deepening of the network. We exhibit the best 5 classification results and worst 5 results with their wrong results in VGG-16 by the way. The top accuracy reaches 98% and the worst is 32%. We choose VGG-16 to be the base model and implement the following experiment on it.

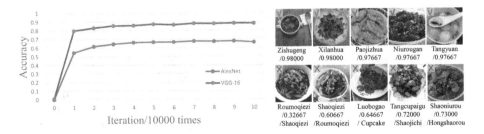

Fig. 7. The accuracy curve when training VGG-16 and AlexNet. VGG-16 obtains the best accuracy of 89%, AlexNet obtains the best accuracy at 67%.

3.3 Food Image Classification Improvement

In the second stage, we test our method one-by-one, firstly is the normalized average feature map strategy, then the multi-scale network including multi-scale input and multi-scale feature fusion.

Normalized Average Saliency Map. We perform the normalized average saliency map method on base model and use batch size of 5, learning rate of 0.0001, weight decay of 0.0005 and momentum of 0.9. We terminate training at 100k iterations, which is determined on a 108k\27k train\val split. Figure 8 shows the accuracy curve when training and Table 1 shows the specific accuracy.

Multi-scale Network. This section is followed by the multi-scale input image and the multi-scale feature fusion to improve the classification performance. The multi-scale input image strategy means we train the model with 448 × 448 images unlikely in Sect. 3.2 we use the input of 224 × 224. As mentioned in Fig. 5, we adopt the *dilation* operation to match the dimension requirement of fc6-layer. Following is the multi-scale feature fusion strategy which means the combination of low/mid/high-level feature in the network as shown in Fig. 2. Own to this multi-scale strategy, we improve the accuracy by 1%. Figure 8 shows the final accuracy curve and our proposed approach achieve 2.58% improvement from VGG-16 baseline overall.

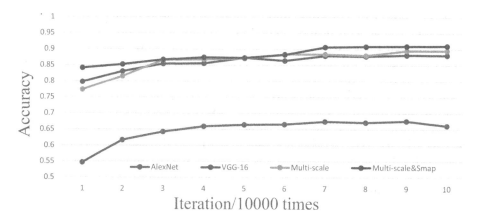

Fig. 8. Accuracy curves of proposed methods in this paper.

We give an example of top-10 error and top-10 right results on CF90 dataset in Fig. 9. After 4.6 epochs, the top accuracy is 99.667% and the top error is 1.667%. The texts below the first row images show their label names and their classification accuracy. The texts below the second row images show their label names, their classification accuracy, and the wrong categories they are classified. The best 10 categories have discriminative appearances to be classified and the worst 10 categories have high similarity with the wrongly classified categories, some have the same ingredients, the similar color and shape.

Table 1. Top-1 accuracy and top-5 accuracy of our approaches. With a two-step scheme training method, we improve the accuracy 2.585% by the baseline.

Network\Method	AlexNet	VGG-16	Multi-sclae	Smap+Multi-scale
Top-1(%)	67.629	88.5966	89.567	**91.1816**
Top-5(%)	89.663	97.7851	98.1518	**98.3443**

| Tangyuan /0.99667 | Pidanzhou /0.99333 | Chaocaixin /0.98000 | Zicaijuan /0.98000 | Yinertang /0.97667 | Zhengxie /0.97333 | Xilanhua /0.97333 | Paigutang /0.96667 | Mapotoufu /0.96667 | Paojizhua /0.96333 |

| Roumoqiezi /0.01667 /Shaoqiezi | Shaojitui /0.65000 /Shaozhuti | Jianbing /0.71333 /Shuizhuyu | Shaojichi /0.72000 /Shaojitui | Zhajiangmian /0.76667 /Chaomian | Huiguorou /0.78333 /Shaoniurou | Laziji /0.79000 /Shaozhuti | Hulobogao /0.79607 /Cupcake | Tudoushaorou /0.81000 /Hongshaorou | Hongshaorou /0.81667 /Tudoushaorou |

Fig. 9. The best 10 classification results and the worst 10 classification results.

4   Conclusion

This paper focus on food image classification, we propose normalized average saliency map for discriminative regions and multi-scale method to improve the performance. With a two-step scheme training methods and fine-tune the ImageNet pre-trained VGG-16 network, we improve the accuracy 2.58% by baseline. Our method does not rely on the pairwised saliency map of each image and is more universal to match the image that the object located in the center. At the same time, we contribute a novel Chinese food image dataset: CF90. We future demonstrate optimizing the dataset and the efficiency of the proposed method on CF90.

References

1. Salvador, A., Hynes, N., Aytar, Y., Marin, J., Ofli, F., Weber, I., Torralba, A.: Learning cross-modal embeddings for cooking recipes and food images. In: CVPR (2017)
2. Angelova, A., Zhu, S.: Efficient object detection and segmentation for fine-grained recognition. In: CVPR (2013)
3. Beijbom, O., Joshi, N., Morris, D., Saponas, T.S., Khullar, S.: Menu-match: Restaurant-specific food logging from images. In: Computer Vision (2015)
4. Bossard, L., Guillaumin, M., Gool, L.V.: Food-101: mining discriminative components with random forests. In: ECCV (2014)
5. Ciocca, G., Napoletano, P., Schettini, R.: Food recognition: a new dataset, experiments and results. IEEE J. Biomed. Health Inform. **21**(3), 588–598 (2016)

6. Cordeiro, F., Bales, E., Cherry, E., Fogarty, J.: Rethinking the mobile food journal: exploring opportunities for lightweight photo-based capture. In: ACM Conference on Human Factors in Computing Systems (2015)

7. Fan, J., Gao, Y., Luo, H., Jain, R.: Mining multilevel image semantics via hierarchical classification. IEEE Trans. Multimedia **10**(2), 167–184 (2008)

8. Farooq, M., Sazonov, E.: A novel wearable device for food intake and physical activity recognition. Sensors **16**(7), 1067 (2016)

9. Girshick, R., Donahue, J., Darrell, T., Malik, J.: Rich feature hierarchies for accurate object detection and semantic segmentation. In: CVPR (2014)

10. He, K., Zhang, X., Ren, S., Sun, J.: Deep residual learning for image recognition. In: CVPR (2016)

11. Herranz, L., Jiang, S., Xu, R.: Modeling restaurant context for food recognition. IEEE Trans. Multimedia **19**(2), 430–440 (2017)

12. Huang, C., Li, H., Xie, Y., Wu, Q., Luo, B.: PBC: polygon-based classifier for fine-grained categorization. IEEE Trans. Multimedia **19**(4), 673–684 (2016)

13. Huang, C., Meng, F., Luo, W., Zhu, S.: Bird breed classification and annotation using saliency based graphical model. J. VCIR **25**(6), 1299–1307 (2014)

14. Jia, Y., Shelhamer, E., Donahue, J., Karayev, S., Long, J.: Caffe: convolutional architecture for fast feature embedding. Eprint arXiv (2014)

15. Krizhevsky, A., Sutskever, I., Hinton, G.E.: Imagenet classification with deep convolutional neural networks. In: NIPS (2012)

16. Lee, G., Tai, Y.W., Kim, J.: Deep saliency with encoded low level distance map and high level features. In: CVPR (2016)

17. Lin, M., Chen, Q., Yan, S.: Network in network. Comput. Sci. (2013)

18. Lin, Z., Hua, G., Davis, L.S.: Multi-scale shared features for cascade object detection. In: ICIP (2013)

19. Long, J., Shelhamer, E., Darrell, T.: Fully convolutional networks for semantic segmentation. In: CVPR (2015)

20. Martin, C.K., Correa, J.B., Han, H., Allen, H.R., Rood, J.C., Champagne, C.M., Gunturk, B.K., Bray, G.A.: Validity of the remote food photography method (rfpm) for estimating energy and nutrient intake in near real time. Obesity **20**(4), 891–899 (2012)

21. Simonyan, K., Zisserman, A.: Very deep convolutional networks for large-scale image recognition. In: ICLR (2015)

22. Szegedy, C., Liu, W., Jia, Y., Sermanet, P., Reed, S., Anguelov, D., Erhan, D., Vanhoucke, V., Rabinovich, A.: Going deeper with convolutions. In: CVPR (2015)

23. Tammachat, N., Pantuwong, N.: Calories analysis of food intake using image recognition. In: ICITEE (2014)

24. Yanai, K., Kawano, Y.: Food image recognition using deep convolutional network with pre-training and fine-tuning. In: ICME (2015)

25. Yang, S., Chen, M., Pomerleau, D., Sukthankar, R.: Food recognition using statistics of pairwise local features. In: CVPR (2010)

26. Zeiler, M.D., Fergus, R.: Visualizing and understanding convolutional networks. In: Fleet, D., Pajdla, T., Schiele, B., Tuytelaars, T. (eds.) ECCV 2014. LNCS, vol. 8689, pp. 818–833. Springer, Cham (2014). https://doi.org/10.1007/978-3-319-10590-1_53

27. Zhang, J., Sclaroff, S., Lin, Z., Shen, X., Price, B., Mech, R.: Minimum barrier salient object detection at 80 fps. In: ICCV (2016)
28. Zhang, X., Xiong, H., Zhou, W., Lin, W., Tian, Q.: Picking deep filter responses for fine-grained image recognition. In: CVPR (2016)
29. Zhou, B., Khosla, A., Lapedriza, A., Oliva, A., Torralba, A.: Learning deep features for discriminative localization. In: CVPR (2016)

Image-Based Modeling

ARSAC: Robust Model Estimation via Adaptively Ranked Sample Consensus

Rui Li[1], Jinqiu Sun[2(✉)], Yu Zhu[1], Haisen Li[1], and Yanning Zhang[1]

[1] School of Computer Science, Northwestern Polytechnical University, Xi'an, China
lirui.david@gmail.com
[2] School of Astronautics, Northwestern Polytechnical University, Xi'an, China
sunjinqiu@nwpu.edu.cn

Abstract. RANSAC is a popular model estimation algorithm in various of computer vision applications. However, it easily gets slow as the inlier rate of the measurements declines. In this paper, a novel Adaptively Ranked Sample Consensus (ARSAC) algorithm is presented to boost the speed and robustness of RANSAC. Our algorithm adopts non-uniform sampling based on the ranked measurements. We propose an adaptive scheme which updates the ranking of the measurements on each trial, to incorporate high quality measurement into sample at high priority. We also design a geometric constraint during sampling process, which could alleviate degenerate cases caused by non-uniform sampling in epipolar geometry. Experiments on real-world data demonstrate the effectiveness and robustness of the proposed method compared to the state-of-the-art methods.

Keywords: Model estimation · Efficiency
Adaptively ranked measurements · Non-uniform sampling
Geometric constraint

1 Introduction

Model estimation is a crucial step in various computer vision problems, including structure from motion (SFM), dim targets detection, image retrieval and SLAM, *etc.* One of the main challenges during model estimation is that the measurements (often points or correspondences) are unavoidably contaminated with outliers due to the imperfection of current measurements acquisition algorithms. These contaminated measurements will probably lead to an arbitrary bad model which could bring a catastrophic impact on the final result in real world applications.

In order to verify and exclude the outliers, a variety of robust model fitting techniques [3,10,14,15,19,25–27] have been studied for years and can be roughly categorized into two kinds. One kind of methods [4,17,27,28] analyze

---

R. Li—Student first author.

© Springer Nature Singapore Pte Ltd. 2017
J. Yang et al. (Eds.): CCCV 2017, Part II, CCIS 772, pp. 591–602, 2017.
https://doi.org/10.1007/978-981-10-7302-1_49

each measurement by its residuals of different model hypotheses, then distinguish inliers from the outliers. This kind of method are usually used in multi-model fitting problems and limited by the requirement of abundant model hypotheses. The other kind of algorithms generate model hypotheses by sampling in the measurements, then find the best model via optimizing a well-designed cost function. The cost function can be solved by random sampling [2,14,19,25,26] or branch-and-bound (BnB) algorithm which guarantees a optimal solution with much more time consuming [3,15,15,29]. In real-world applications, these algorithms are carefully selected in terms of different usages. Among these methods, random sample consensus (RANSAC) algorithm [10] is one of the most popular algorithms for model estimation, which is broadly known by its simplicity and effectiveness. The algorithm follows a vote-and-verify paradigm. It repeatedly draw minimal measurement sets by random sampling, and calculate model parameters through each measurement set to form a model hypothesis. For every model hypothesis, a verification step is also needed. All the measurements are incorporated to calculate their residuals with current model parameters. Then, measurements which residuals below a certain threshold will be grouped as the consensus set of that model hypothesis. After sufficient trials of model generation and verification, RANSAC selects the model hypothesis with the largest consensus set as the optimal hypothesis. And the final model parameters are calculated from the consensus set of the optimal hypothesis.

However, RANSAC still needs improvement for better efficiency. In order to get sufficient trials in RANSAC, we must draw k trials to get at least one outlier-free sample with confidence η_0,

$$k \geq \frac{log(1 - \eta_0)}{log(1 - \epsilon^m)}, \tag{1}$$

where ϵ is the fraction of inliers in the dataset, m denotes the size of the minimal sampling set. It's obvious that as ϵ declines, k grows dramatically which brings heavy computational burden on the algorithm. This phenomenon results from the uniform sampling in RANSAC, that the algorithm samples every measurement at equivalent probability. Therefore some valuable information indicating true inliers will be overlooked, which could have been used to cut down the running time of the algorithm.

In this paper, we mainly address on this problem and present ARSAC, which aims to boosts the efficiency of the algorithm. Our contributions are as follows:

- We propose a method which adopts non-uniform sampling based on adaptively ranked measurements. It updates the ranking of measurements at each trial, and incorporate most prominent measurement into sample, which is experimentally proved to achieve high efficiency.
- We designed a mild geometric constraint, which samples measurements with broad spatial distribution. It largely alleviates degeneracy in epipolar geometry estimation and improves the overall robustness of the algorithm.
- We develop ARSAC, an efficient algorithm which is able to achieve better robustness compared to other efficiency-improving algorithms.

2   Related Work

Due to the inefficiency of RANSAC, a number of schemes have been proposed to improve its performance. Basic ideas focus efficiency improving on sampling and model verification. For efficiency on sampling, non-uniform sampling is conducted by leveraging prior knowledge of the measurements. NAPSAC [20] selects measurements that are sampled within a hypersphere in high dimensional space of radius r. It can effectively handle the problem of poor sampling in higher dimensional space, but may result in degeneration for the measurements may be too close to each other. GroupSAC [21] divides the measurements into different groups, and samples from the most prominent group. However, it relies on the performance of the group function which is supposed to make sense in the algorithm. PROSAC algorithm [7] ranks the measurements by their quality, then draw non-uniform sampling with measurements in descending quality order. However, the fixed quality order of measurements could not be sufficient to precisely reflect the probability of true inliers, especially for the scenes with unapparent features. PROSAC also needs to be improved in case of degeneracy [23] during non-uniform sampling.

For the efficiency on model verification, a subset of all measurements is often chosen for verification. Preemptive methods [1,6] follows the paradigm that a very small number of measurements is randomly selected for verification to filter out extremely invalid models. SPRT test [8,18] uses Wald's Theory of sequential testing [13] on model verification with adaptive likelihood ratio to be a criterion. However, these methods may lead to misjudgments towards false outliers, that the conclusion is usually drawn before all measurements are verified.

For the purpose of robust estimation, some methods [5,24] extended the inlier set by conducting extra RANSAC trials based on the current best consensus set, which explores potential inliers that is not strictly consistent with current best model. In order to handle with degeneracy, QDEGSAC [11] uses several runs of RANSAC to provide the most constraining model, which leads to the least probability of degeneracy. However, these methods requires further runs of RANSAC which helps little for efficiency.

Our method adopts non-uniform sampling for efficiency by maintaining an ordered measurements set. Different from the methods above, we adaptively update the ranking of measurements to combine prior knowledge and current model information together. At the mean time, in order to alleviate degeneracy which usually caused by non-uniform sampling, we constrain the spatial distribution of measurements during the sampling process, which requires no further trials.

3   Adaptively Ranked Sampling Consensus Algorithm

In order to improve the efficiency of RANSAC, ARSAC adopts non-uniform sampling that measurements with high quality are sampled in advance. Different from methods with fixed ranking of measurements [7], in ARSAC, the ranking

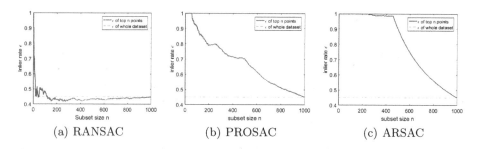

Fig. 1. The inlier rate ϵ out of top n measurements for different algorithms from synthetic dataset. (a) denotes RANSAC which draws uniform sampling, (b) denotes PROSAC that draws non-uniform sampling with fixed ranking and (c) denotes ARSAC which adopts adaptively ranked non-uniform sampling. The blue dashed denotes the inlier rate of the whole dataset, which is 0.45. (Color figure online)

of measurements is iteratively updated, which offers a better description of the measurements with high quality. In the experiment of synthetic data shown in Fig. 1, the inlier rate of RANSAC in top n measurements fluctuate around that of whole dataset, for RANSAC does not rank the measurements. Both PROSAC and ARSAC keep high inlier rate in the top ranked measurements. However, the inlier rate of PROSAC falls down quickly as n grows and there are ups and downs in its inlier rate curve, which means the ranking strategy that PROSAC implemented could not effectively choose real inliers. While ARSAC finds the largest number of top measurements which keep very high inlier rate. That means in limited sampling trials, ARSAC reaches the highest probability to get all-inlier samples, which strongly contributes the convergence of the algorithm. At the same time, a mild geometric constraint is proposed to constrain the sampling of ARSAC. Note that $\mathcal{X} = \{x_i\}_{i=1}^{N}$ is the dataset with N measurements, where i is the index of every measurement. $\mathcal{M}_j$ is the minimal sample set of size m, j denotes the index of the trials. θ_j is the model hypothesis generated by $\mathcal{M}_j$ and its consensus set is noted as I_j. $\mathcal{U}$ is the ranked measurements, where $\mathcal{U}_n$ denotes top n measurements of it. The whole procedures of ARSAC are described in Algorithm 1.

3.1   Adaptively Ranked Progressive Sampling

In ARSAC, we adopts non-uniform sampling and improve the scheme by adaptively updating the ranking of measurements. We first show the sampling process when a ranked measurements set is given, then introduce the updating strategy for ranked measurements with posterior information and geometric constraint respectively.

Non-uniform Sampling with Ranked Measurements. When given a set of ranked measurements $\mathcal{U}$, it's intuitively difficult to ensure which measurements should be selected or how many times should a measurement be selected.

Algorithm 1. Adaptively Ranked Sample Consensus Algorithm

Input: Ranked measurement set $\mathcal{U}$, minimal sample length m, length of the measurement set N;
Output: The best model parameter θ^*;
1: Initialize $j \leftarrow 0$, $n \leftarrow m$, $\mathcal{U}_n \leftarrow \mathcal{U}(1:n)$, $\alpha_n \leftarrow 0$;
2: **while** Stopping conditions are not achieved **do**
3:     $j \leftarrow j + 1$;
4:     Update the ranking of $\mathcal{U}$ and select current sample M_j (see Sect. 3.1);
5:     Compute current model parameter θ_j and its consensus set I_j;
6:     Set θ^* to the model parameter with the largest consensus set;
7:     Set the stopping criteria of current iteration (see Sect. 3.2);
8: return θ^*.

We thus use a progressive scheme which was inspired by [7], that a subset $\mathcal{U}_n$ containing n top-ranked measurements is chosen to draw a minimal sized sample. As the sampling process continues, $\mathcal{U}_n$ will gradually increase. This aims to draw the same samples as RANSAC does even with a non-uniform sampling strategy. Note that S_N is the total samples drawn in standard RANSAC, let S_n denotes the average number of samples that only contains measurements from $\mathcal{U}_n$

$$S_n = S_N \frac{\binom{n}{m}}{\binom{N}{m}} = S_N \prod_{i=0}^{m-1} \frac{n-i}{N-i} . \tag{2}$$

Considering that there is overlap between samples in S_n and S_{n-1}, so the number of samples S'_n that need to be drawn for current $\mathcal{U}_n$ will be

$$S'_n = \lceil S_n - S_{n-1} \rceil, \tag{3}$$

where $\lceil \bullet \rceil$ denotes the operation of getting upper bound. It's worth noticing that new sample drawn from $\mathcal{U}_n$ follows the principle that the nth measurement in $\mathcal{U}_n$ must be selected and the rest $m-1$ measurements are randomly chosen from $\mathcal{U}_{n-1}$. When S'_n samples is drawn from current $\mathcal{U}_n$, n will increase by 1, and $\mathcal{U}_n$ will in turn expand by including the best inlier from the remaining measurements $\mathcal{U} \backslash \mathcal{U}_n$.

Measurements Updating. Measurements updating in ARSAC is related to the sampling process. The ranking of measurements is initialized by the evaluation of isolated measurements [16], then is iteratively updated whenever $\mathcal{U}_n$ is about to expand (see Sect. 3.1). We assume that the current best model θ_j^* is more reliable to evaluate the quality of remaining measurements $\mathcal{U} \backslash \mathcal{U}_n$ than the initial ranking. In ARSAC, the best inlier x_i^* of θ_j^* is calculated by the residuals of the model. Then the location of $x_{\mathcal{U}_n}^*$ in current quality ranking is leveraged to judge the reliability of current ranking. Once x_i^* does not locate in the top β fraction of $\mathcal{U} \backslash \mathcal{U}_n$, the existing ranking will be regarded as unreliable and $x_{\mathcal{U}_n}^*$ will be inserted after $\mathcal{U}_n$, which also results in an updated $\mathcal{U}$ that offers a better description of the latent inliers. The default value of β is set to 0.05.

We enhance the sampling strategy by a simple constraint to alleviate degeneracy. Inspired by situations illustrated in [9], We assume that measurements leading to degeneracy are liable to gather in limited areas in the image. In ARSAC, a constraint circle $C_{\mathcal{U}_n}$ is presented which contains all measurements in $\mathcal{U}_n$. The center of the circle $c_{\mathcal{U}_n}$ lies on the center of measurements in $\mathcal{U}_n$, and the radius $r_{\mathcal{U}_n}$ is set to be

$$r_{\mathcal{U}_n} = \max \|x_i - c_{\mathcal{U}_n}\|_2 + \lambda, \quad x_i \in \mathcal{U}_n \tag{4}$$

where λ is a step factor to decide the extra expansion of the circle. New measurement to be included in $\mathcal{U}_n$ should meet with the constraint that the position of it should be out of the range of $C_{\mathcal{U}_n}$ in the image, in order to avoid degeneracy which result from the clustering nature of degenerated measurements.

3.2   Stopping Criteria

In order to get an optimal model after a number of trials, The basic stopping criteria of ARSAC are designed by three constraints: maximality constraint, non-randomness constraint and geometry constraint.

The maximality constraint guarantees that after k_η trails, the probability of exsiting another model which has a larger consensus set falls below a certain threshold η

$$(1 - \epsilon_0^m)^{k_\eta} \leq \eta, \tag{5}$$

where ϵ_0 denotes the inlier rate of the measurements. The trials of our algorithm k must be larger than k_η before being terminated.

The non-randomness constraint is designed to make the probability of the situation below a certain threshold Ψ, that a bad model is calculated and supported by a consensus set by chance. The cardinal g of the consensus set for a wrong model follows the binomial distribution B (n, β), that

$$P^n(g) = \beta^{g-m}(1 - \beta)^{n-g+m}\binom{n-m}{g-m}, \tag{6}$$

where β is the probability of a measurement in $\mathcal{U}_n$ to be misclassified as an inlier by a wrong model. Thus for each n, the minimal size of consensus set L_{min}^n is

$$L_{min}^n = \min\{j : \sum_{g=j}^{n} P^n(g) < \Psi\}. \tag{7}$$

For the concern of time saving, we further assume the size of the subset $\mathcal{U}_n$ is big enough that the distribution can be approximated by Gaussian distribution according to central limit theorem

$$B(n, \beta) \sim N(\mu, \sigma), \tag{8}$$

where $\mu = n\beta$, and $\sigma = \sqrt{n\beta(1 - \beta)}$. Then Eq. (7) could be demonstrated by Chi-square distribution. So the minimal size of consensus should be

$$L_{min}^n = \lceil m + \mu + \sigma * \sqrt{Chi^2}\rceil, \tag{9}$$

where Chi^2 is determined by the threshold Ψ. The trails of ARSAC will not stop if the size of the best consensus set is not bigger than L_{min}^n.

Geometric constraint is used to judge whether the measurements in $\mathcal{U}_n$ distribute in an adequate range. Note that $D_{\mathcal{U}_n}$ is the bounding box of measurements in $\mathcal{U}_n$, and D_{total} is the bounding box of all measurements in the image. We terminate the trails when the condition is met

$$\frac{D_{\mathcal{U}_n}}{D_{total}} > r_{range}, \tag{10}$$

where r_{range} is the acceptable ratio of two bounding boxes.

4   Experiments

In this section, we conduct experiments on real-world data to verify the efficiency and robustness of ARSAC. We evaluate different algorithms on two well-known model estimation problems: fundamental matrix (**F**) estimation and homography (**H**) estimation. Both model estimation problems require effective outlier removal process to guarantee the accuracy of the final solution. Fundamental matrix constrains the 3D spatial relationship between two views and homography describes the transformation between two plane objects. Solutions to both problems can be get by solving least-square problems. The minimal sample size for fundamental matrix is 7 while that for homography is 4. As discussed in Sect. 2, there has been some algorithms for better efficiency during measurements sampling. So in the experiments, except for RANSAC we further compare the proposed ARSAC with the following state-of-the-art algorithms: NAPSAC, PROSAC, GroupSAC. We implement these algorithms in Matlab. All the evaluations are performed on an Intel i7 CPU with 32 GB RAM.

We tested ARSAC's performance on real-world images for model estimation problems. The datasets is provided by [22], each dataset presents various of challenges in terms of low inlier rate and degeneracy. In the experiments, dataset **A~D** is selected to estimate the fundamental matrix, while dataset **E~H** are used to perform homography estimation. Since the ground truth of inlier is unknown on real-world images, we approximate it by performing 10^6 trials of random sampling. The baseline comparison is shown in Table 1. For each algorithm, the table lists the found inliers (I), the number of trails (k) and the total runtime (time) measured by millisecond. The error is computed by Sampson error [12] and return the mean value after 500 executions for each algorithm. As shown in Table 1, in most cases, ARSAC performs the best in terms of average trials and time than other algorithms. What's more, the average error of ARSAC keeps at the lowest level compared with other efficient-driven algorithms. It is worth noticing that RANSAC could find more inliers and estimate model with lower error, but this performance builds on significant number of trials which is the case we are trying to avoid.

In order to compare the efficiency of different methods intuitively, we display the least trials which are needed for different methods to fall below a certain

Table 1. Baseline comparison of ARSAC with other algorithms on real-world data.

		RANSAC	NAPSAC	PROSAC	GroupSAC	ARSAC
A: $\epsilon = 0.48, N = 3154$	I	1429.15	1476.43	610.43	1433.95	595.35
	k	1234.50	868.10	4.91	55.26	4.35
	time	12230.63	10351.39	563.47	4213.75	447.13
	error	0.54	1.25	1.46	0.96	0.73
B: $\epsilon = 0.22, N = 1516$	I	275.20	254.7	291.45	255.85	283.55
	k	50000	5023	36.85	32.45	4.1
	time	140720.86	14300	248.20	746.98	76.61
	error	2.85	20.35	14.59	6.14	3.59
C: $\epsilon = 0.39, N = 422$	I	147.05	142.37	140.70	138.60	155.3
	k	6501.05	3886.47	283.70	20.95	11.30
	time	8093.21	7203.83	352.24	262.92	139.28
	error	4.19	9.89	10.51	2.15	0.54
D: $\epsilon = 0.92, N = 786$	I	687.55	688.2	604.00	605.10	604.00
	k	12.35	11.83	2.75	4.3	1.86
	time	153.21	251.22	57.32	294.07	50.27
	error	0.71	4.35	5.58	12.56	3.19
E: $\epsilon = 0.52, N = 2540$	I	1281.10	1196.13	1189.24	1162.70	1134.20
	k	93.55	63.90	18.27	8.45	8.00
	time	545.29	1500.35	169.73	289.89	91.5
	error	0.73	2.37	3.23	0.69	0.74
F: $\epsilon = 0.15, N = 514$	I	73.30	68.14	68.70	66.73	67.90
	k	12266.65	6847.9	15.38	7.46	13.15
	time	16474.05	7951.13	44.98	53.67	40.5
	error	0.40	9.39	8.41	1.63	0.40
G: $\epsilon = 0.34, N = 1967$	I	632.10	603.53	426.03	409.62	448.5
	k	482.05	313.10	12.18	8.9	11.25
	time	2218.73	2401.05	72.97	127.32	74.27
	error	0.80	11.54	3.34	1.65	1.34
H: $\epsilon = 0.12, N = 979$	I	107.90	106.40	100.95	97.40	93.38
	k	32585.20	10114	861.06	177.15	159.30
	time	73441.53	14825.28	2220.70	282.08	211.83
	error	0.47	6.53	1.27	1.47	0.46

error threshold. Considering that ARSAC, PROSAC and GroupSAC shows far better performance in efficiency compared with NAPSAC, we compare those three algorithms as shown in Fig. 2. As is illustrated in the diagrams, for some cases, all methods provide satisfactory results for efficiency, but for some specific cases, e.g. low inlier rate or degeneracy, ARSAC shows its superiority in fastest convergence compared with other methods which the adaptively ranked scheme is not implemented.

We further compare the robustness of ARSAC with other algorithms. Table 2 shows the number of degenerate cases (k_{Deg}) among total trials (k) in fundamental matrix estimation on real-world dataset **A**~**D**. Each algorithm is executed 500 times. The result shows that ARSAC performs the best with very few degenerate

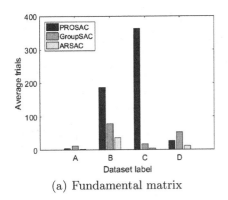
(a) Fundamental matrix

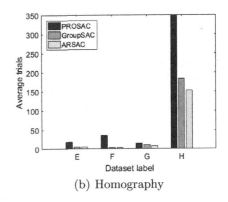
(b) Homography

Fig. 2. Average trials by each algorithm to reach predefined error threshold δ. The label on X-axis correspond to datasets in Table 1, Y-axis denotes the minimal trials of each algorithm to make the error of model falls below δ, where δ is set to 3.0. The plots represent average value of 500 runs.

Table 2. Degenerate cases for fundamental matrix estimation.

	NAPSAC		PROSAC		GroupSAC		ARSAC	
	k_{Deg}	k	k_{Deg}	k	k_{Deg}	k	k_{Deg}	k
A	261.75	868.10	1.31	4.91	3.22	75.20	0.06	4.30
B	23.41	5023	0.67	36.85	1.26	32.45	0	4.10
C	869.51	3886.47	231.80	334.60	3.54	20.95	0.91	11.30
D	4.27	11.83	0.93	2.75	3.34	4.3	0.52	1.86

cases for every dataset, thanks to its geometric constraint which samples measurements with broad spatial distribution. The true inlier rate for different algorithms' consensus set also demonstrate the robustness of ARSAC, as the optimal consensus set determines the final result of the model. We compare them as shown in Fig. 3, it can be seen that ARSAC prevails over all other non-uniform sampling algorithms. At the mean time, the true inlier rate of ARSAC could reach the same level as that of RANSAC, with a significant improvement of efficiency.

It could be observed that in the context of efficiency-driven robust model fitting problems, algorithms like PROSAC and NAPSAC improve the efficiency a lot, but the error of estimated model seems too large to be acceptable. GroupSAC performs well in many cases, but it could suffer from degeneracy in fundamental matrix estimation for scenes with dominant planes. Among all the algorithms, ARSAC presents the best performance in terms of both efficiency and robustness, which provide a better solution for robust model fitting problems.

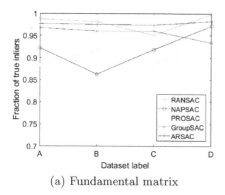 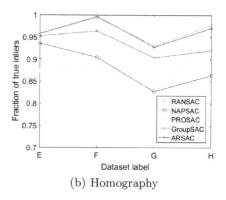

(a) Fundamental matrix   (b) Homography

Fig. 3. Fraction of true inliers returned by each algorithm for fundamental matrix estimation and homography estimation problems. The label on X-axis correspond to datasets in Table 1, the Y-axis denotes the inlier rate of each dataset estimated by different algorithms. The plots represent average value of 500 runs.

5   Conclusion

We present ARSAC, a novel variant of RANSAC that draws non-uniform sampling with adaptively ranked measurements. At each trial of ARSAC, the algorithm selects measurement with highest quality into the new sample set. We also propose a mild geometric constraint in ARSAC to alleviate degeneracy. Our algorithm is capable of handling low inlier rate measurements in model estimation and is shown to be more efficient and robust compared with state-of-the-art algorithm. Though proved to be effective, our algorithm may be bothered by user-provided parameter settings. In future work, we plan to explore a parameter-free strategy for simplicity with the performance being guaranteed.

Acknowledgments. This work is supported by the National Natural Science Foundation of China (No. 61231016), the National 863 Program (No. 2015AA016402), Seed Foundation of Innovation and Creation for Graduate Students in Northwestern Polytechnical University (Z2017184).

References

1. Capel, D.P.: An effective bail-out test for ransac consensus scoring. In: BMVC (2005)
2. Chen, H., Meer, P.: Robust regression with projection based m-estimators. In: 2003 Proceedings of IEEE International Conference on Computer Vision, pp. 878–885, vol. 2 (2003)
3. Chin, T.J., Yang, H.K., Eriksson, A., Neumann, F.: Guaranteed outlier removal with mixed integer linear programs. In: IEEE Conference on Computer Vision and Pattern Recognition, pp. 5858–5866 (2016)
4. Chin, T.J., Yu, J., Suter, D.: Accelerated hypothesis generation for multi-structure robust fitting. In: European Conference on Computer Vision, pp. 533–546 (2010)

5. Chum, O., Matas, J., Kittler, J.: Locally optimized RANSAC. In: Joint Pattern Recognition Symposium, pp. 236–243 (2003)

6. Chum, O., Matas, J.: Randomized ransac with $T_{d,d}$ test. In: Proceedings of the British Machine Vision Conference, vol. 2, pp. 448–457 (2002)

7. Chum, O., Matas, J.: Matching with PROSAC-progressive sample consensus. In: IEEE Computer Society Conference on Computer Vision and Pattern Recognition, CVPR 2005, vol. 1, pp. 220–226. IEEE (2005)

8. Chum, O., Matas, J.: Optimal randomized RANSAC. IEEE Trans. Pattern Anal. Mach. Intell. **30**(8), 1472–1482 (2008)

9. Chum, O., Werner, T., Matas, J.: Two-view geometry estimation unaffected by a dominant plane. In: 2005 IEEE Computer Society Conference on Computer Vision and Pattern Recognition, CVPR 2005, vol. 1, pp. 772–779. IEEE (2005)

10. Fischler, M.A., Bolles, R.C.: Random sample consensus: a paradigm for model fitting with applications to image analysis and automated cartography. Commun. ACM **24**(6), 381–395 (1981)

11. Frahm, J.M., Pollefeys, M.: RANSAC for (quasi-) degenerate data (QDEGSAC). In: 2006 IEEE Computer Society Conference on Computer Vision and Pattern Recognition, vol. 1, pp. 453–460. IEEE (2006)

12. Hartley, R., Zisserman, A.: Multiple View Geometry in Computer Vision, 2nd edn. Cambridge University Press, Cambridge (2000)

13. Lai, T.L.: Sequential analysis. Wiley Online Library (2001)

14. Lee, K.M., Meer, P., Park, R.H.: Robust adaptive segmentation of range images. IEEE Computer Society (1998)

15. Li, H.: Consensus set maximization with guaranteed global optimality for robust geometry estimation. In: IEEE International Conference on Computer Vision, pp. 1074–1080 (2009)

16. Lowe, D.G.: Distinctive image features from scale-invariant keypoints. Kluwer Academic Publishers (2004)

17. Magri, L., Fusiello, A.: T-linkage: a continuous relaxation of J-linkage for multi-model fitting. In: Computer Vision and Pattern Recognition, pp. 3954–3961 (2014)

18. Matas, J., Chum, O.: Randomized RANSAC with sequential probability ratio test. In: 2005 Tenth IEEE International Conference on Computer Vision, ICCV 2005, vol. 2, pp. 1727–1732. IEEE (2005)

19. Miller, J.V., Stewart, C.V.: Muse: robust surface fitting using unbiased scale estimates. In: Conference on Computer Vision and Pattern Recognition, p. 300 (1996)

20. Nasuto, D., Craddock, J.B.R.: NAPSAC: high noise, high dimensional robust estimation-its in the bag (2002)

21. Ni, K., Jin, H., Dellaert, F.: GroupSAC: efficient consensus in the presence of groupings. In: 2009 IEEE 12th International Conference on Computer Vision, pp. 2193–2200. IEEE (2009)

22. Raguram, R., Chum, O., Pollefeys, M., Matas, J., Frahm, J.M.: USAC: a universal framework for random sample consensus. IEEE Trans. Pattern Anal. Mach. Intell. **35**(8), 2022–2038 (2013)

23. Raguram, R., Frahm, J.-M., Pollefeys, M.: A comparative analysis of RANSAC techniques leading to adaptive real-time random sample consensus. In: Forsyth, D., Torr, P., Zisserman, A. (eds.) ECCV 2008. LNCS, vol. 5303, pp. 500–513. Springer, Heidelberg (2008). https://doi.org/10.1007/978-3-540-88688-4_37

24. Raguram, R., Frahm, J.M., Pollefeys, M.: Exploiting uncertainty in random sample consensus. In: IEEE International Conference on Computer Vision, pp. 2074–2081 (2010)

25. Rousseeuw, P.J.: Least median of squares regression. J. Am. Stat. Assoc. **79**(388), 871–880 (1984)
26. Stewart, C.V.: MINPRAN: a new robust estimator for computer vision. IEEE Trans. Pattern Anal. Mach. Intell. **17**(10), 925–938 (1995)
27. Toldo, R., Fusiello, A.: Robust multiple structures estimation with J-linkage. In: European Conference on Computer Vision, pp. 537–547 (2008)
28. Zhang, W., Kosecka, J.: A new inlier identification procedure for robust estimation problems. In: Robotics: Science and Systems (2006)
29. Zheng, Y., Sugimoto, S., Okutomi, M.: Deterministically maximizing feasible subsystem for robust model fitting with unit norm constraint. In: Computer Vision and Pattern Recognition, pp. 1825–1832 (2011)

Sparse Softmax Vector Coding Based Deep Cascade Model

Ji Liu and Lei Zhang$^{(\boxtimes)}$

College of Communication Engineering, Chongqing University,
No. 174 Shazheng street, Shapingba district, Chongqing 400044, China
{jiliu,leizhang}@cqu.edu.cn

Abstract. Recently, many sparse coding techniques like sparse representation based classification (SRC) have been proposed to deal with face recognition (FR) problem. In SRC, a testing image is linearly coded by the training images to calculate the sparse coefficients by l_1-norm minimization. Then, SRC needs to compute the representation error of each category when classifying the testing image. The corresponding category of the testing image would have the minimum representation error. In other words, representation errors of all classes show class discrimination. In this paper, we take advantage of this distinct representation errors that are transformed into softmax vector and find that the sub-pattern of the whole image is sometimes more discriminative than the whole image. Sparse softmax vector coding based deep cascade model (SSVD) is proposed to improve the pattern classification performance. The experiments demonstrate that the proposed model is much more effective than state-of-the-art methods.

Keywords: Sparse coding · Softmax vector · Spatial pyramid
Deep cascade model

1 Introduction

Face recognition can be viewed as one of the most popular and challenging topic in computer vision and pattern recognition. In the past 20 years, substantial face recognition methods [1–13] have been developed by numerous researchers. Among these methods, sparse coding and discriminative methods have yielded significant results.

Nassem et al. [1] proposed the linear regression classifier (LRC) for face recognition. The main idea of LRC is representing a testing face by a suitable way and classifying it to one class, which can represent it better than other classes. One after another, l_1-norm regularization term is imposed upon the LRC model to avoid over-fitting by Wright et al. [2] who proposed a sparse representation based classification (SRC) framework to solve FR problems. In SRC, a testing image is coded by a sparse linear combination of training samples via the l_1-norm minimization. SRC classifies the testing image through estimating which class of training

© Springer Nature Singapore Pte Ltd. 2017
J. Yang et al. (Eds.): CCCV 2017, Part II, CCIS 772, pp. 603–614, 2017.
https://doi.org/10.1007/978-981-10-7302-1_50

samples could generate the smallest reconstruction error of it with the corresponding class coding coefficients. Zhang et al. [3] illustrate that not only l_1-norm but also l_2-norm could achieve parallel results on coding coefficients and proposed the collaborative representation classifier (CRC) scheme. Among the above models, the fidelity terms are measured by the l_2-norm or l_1-norm, which follows the assumption that the pixels of error obey Gaussian or Laplacian distribution independently. Nevertheless, if there were some illumination variation, occlusion, or disguise in the images, the above assumption might be unconscionable.

Subsequently, several scholars enhanced the sparse coding based models and proposed some new methods. Typically, to obtain more robustness, Yang et al. [4] proposed a robust sparse coding (RSC) model for FR, in which the residual of the testing image and the estimated one is assumed independently and identically distributed according to some probability density function (PDF), where the parameter characterizes the distribution. Then, RSC finds an maximum likelyhood estimation solution of the sparse coding, which can be viewed as a weighted LASSO problem. He et al. [5] took advantage of the correntropy induced robust error metric and proposed the correntropy based sparse representation (CESR) model. What is interesting is that RSC and CESR can be viewed similar work of M-estimator with different kernel size. Recently, He et al. [6] proposed a new model of using different half-quadratic functions to measure the error image, which combines the ideas of SRC, CESR and RSC. In addition, to make the LRC more robust to random pixel disguise, occlusion, or illumination, Nassem et al. [7] extended the LRC to the robust linear regression classification (RLRC) by making use of Huber estimator. Zhou et al. [8] borrowed the markov random field model into the sparse coding scheme and proposed sparse error correction with MRF model. Jia et al. [9] utilized structured sparsity-inducing norm into the SRC model and presented a structured sparse representing classifier (SSRC).

To improve the recognition rate of sparse coding methods, we propose a deep cascade model based on sparse softmax vector coding (SSVD) in this paper, inspired by [23]. The main contributions of our work are as follows. (1) The use of discriminative softmax vector. SRC codes a testing image by sparse linear combination of all training images and classifies it to the class which has minimum representation error. In other words, representation errors of all classes show class discrimination. Most existing sparse coding based methods only focus on the original or extracted image feature. To further explore the effectiveness of sparse coding method on the discriminative representation errors, we propose the SSVD method, in which the softmax vectors transformed by representation errors are used to do sparse representation repeatedly. (2) Three-level spatial pyramid structure is used to enhance class discrimination. Most of the sparse coding methods are based on the whole images, which ignores the local information of the subregion. Because the subregions of the whole image show more detailed local information and more discriminative than the whole image, SSVD combines the whole image and its subregions to obtain softmax vectors by using three-level spatial pyramid structure as shown in the image coding part of Fig. 1.

(3) Deep cascade model based on concatenated softmax vectors is proposed. As the cascade model goes deep, the concatenated softmax vectors obtain more class discrimination, which is in favour of classification. Our extensive experiments in benchmark databases show that the proposed deep model achieves better performance than many existing sparse coding methods.

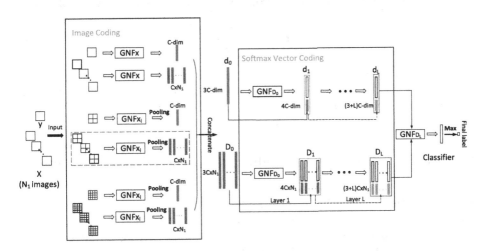

Fig. 1. An example is given to illustrate how the deep model works when classifying a test image **y** under all training images **X**. (Color figure online)

The rest of this paper is organized as follows. Section 2 presents the proposed deep model. Section 3 presents the solving algorithm of sparse representation. The experiment results are shown in Sect. 4. Section 5 concludes this paper.

2   The Proposed Approach

In this section, we illustrate how we classify the testing image by giving all training images. First, we define a procedure getting new feature in the first part. Then, in the second part, we present a detailed illustration that how the deep cascade model goes as shown in Fig. 1

2.1   Getting New Feature

According to SRC, suppose that we have C classes of subjects and define that $\mathbf{d}$ represents one of testing sample and $\mathbf{D} = [\mathbf{D}_1, \mathbf{D}_2, \cdots, \mathbf{D}_C]$ represents the dictionary. The representation model can be transformed into following problem [15]:

$$\min_{\alpha} \| \mathbf{d} - \mathbf{D}\alpha \|_2^2 + \lambda \| \alpha \|_1 \tag{1}$$

where λ is a scalar constant. After solving the above function, we compute the representation error of each class as follow:

$$r_c = \| \mathbf{d} - \mathbf{D}^c \boldsymbol{\alpha}^c \|_2^2 \tag{2}$$

where $\mathbf{D}^c$ is the c-th class samples, and $\boldsymbol{\alpha}^c$ is the coefficient vector associated with c-th class. Softmax vector $\mathbf{r}$ is computed by softmax function as follow:

$$\mathbf{r} = \frac{e^{-r_c}}{\sum_{c=1}^{C} e^{-r_c}} \tag{3}$$

where $\mathbf{r} = [r_1, r_2, \cdots, r_C] \in \mathbb{R}^C$. If the testing sample $\mathbf{d}$ belonged to class $i (\leq C)$, r_i should be bigger than other atoms in softmax vector $\mathbf{r}$, which is called class discrimination. The above process of obtaining softmax vector $\mathbf{r}$ is named as Getting New Feature on dictionary $\mathbf{D}$ (GNF_D)

2.2   Sparse Softmax Vector Coding Based Deep Cascade Model

Without loss of generality, we let $\mathbf{X}$ represents the training images and $\mathbf{Y}$ represents the testing images. The class number is C. The numbers of training images and testing images are N_1 and N_2. For each image, a three-level spatial pyramid is used to compute the softmax vector. We take one testing image $\mathbf{y}$ and all training images $\mathbf{X}$ as an example to explain how to obtain the softmax vectors and classify the testing image $\mathbf{y}$ as shown in Fig. 1.

There are 3 parallel channels that are designed to process the input images. In the first channel, the original testing image $\mathbf{y}$ is represented by all training images $\mathbf{X}$ and goes through the GNF_X procedure to get a softmax vector. Similarly, a softmax vector set of training images will be obtained after each training image goes through GNF_X procedure. In the second channel, all the input images are equally divided into 4 subregions. Let $\mathbf{y}_i$ denote the i-$th (i = 1, \cdots, 4)$ subregion of test image $\mathbf{y}$ and $\mathbf{X}_i$ denote the i-$th (i = 1, \cdots, 4)$ subregion set of all the training images $\mathbf{X}$. Similar to the first channel, $\mathbf{y}_i$ goes through GNF_{X_i} procedure, then 4 softmax vectors will be generated. Those 4 softmax vectors are transformed into one vector after max pooling or average pooling. Like the testing image, each subregion of per training image goes through the corresponding GNF_{X_i} procedure, and 4 softmax vectors will be generated. After the max pooling or average pooling, the 4 softmax vectors are transformed into 1 vector. Then the transformed vector of each image is parallel integrated into one matrix as shown in Fig. 2 that is an instance presented in red dashed part of the second channel in Fig. 1 to illustrate max pooling and average pooling. In the third channel, each input images are equally divided into 16 subregions. Using the same approach in the second channel, a transformed softmax vector of testing image $\mathbf{y}$ and a transformed softmax vector set of training images $\mathbf{X}$ will be generated.

After those 3 channels, 3 softmax vectors (tinted with blue) of testing image are concatenated into one vector $\mathbf{d}_0$ and 3 softmax vector sets (tinted with red)

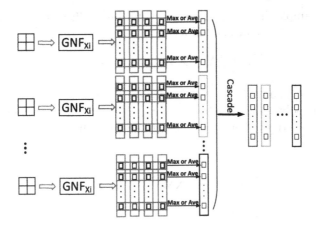

Fig. 2. Illustration of the max pooling and average pooling. 4 softmax vectors will be obtained after each subregion of per image goes through GNF_{X_i} procedure. Then, we compute the maximum value or average of the 4 values in the corresponding dimension to construct a new vector. Finally, the new vector of each image is parallel integrated into one matrix.

are concatenated into one vector set $\mathbf{D_0}$. Then, $\mathbf{d_0}$ goes through GNF_{D_0} procedure to compute the softmax vector that is concatenated with $\mathbf{d_0}$ to construct input sample $\mathbf{d_1}$ of second layer. Similarly, each column in $\mathbf{D_0}$ also goes through GNF_{D_0} procedure to compute the softmax set that is concatenated with $\mathbf{D_0}$ to construct input dictionary $\mathbf{D_1}$ of second layer. Using the same way, we can obtain the testing sample $\mathbf{d_L}$ and dictionary $\mathbf{D_L}$ of level L. Finally, $\mathbf{d_L}$ goes through GNF_{D_L} to get the softmax vector. The prediction will be obtained by taking the class with the maximum value in softmax vector.

3   Solving Algorithm of Sparse Representation

In recent years, many algorithms have been proposed for sparse representation. In particular, the alternating direction method of multipliers (ADMM), first proposed in 1970s [14], has drawn a lot of attention. Yang and Zhang [15] integrated the proximal methods into ADMM when solving l_1-norm minimization problems.

In this paper, we also use ADMM method to solve sparse representation problem. In the SSVD model, sparse representation problems need to be solved in different stages. We just take the first channel in Fig. 1 for an instance to illustrate how we solve the sparse coding coefficients of testing samples based on the dictionary $\mathbf{X}$. Let $\mathbf{X} = [\mathbf{x_1}, \mathbf{x_2}, \cdots, \mathbf{x_{N_1}}] \in \mathbb{R}^{d \times N_1}$ denote training samples and $\mathbf{Y} = [\mathbf{y_1}, \mathbf{y_2}, \cdots, \mathbf{y_{N_2}}] \in \mathbb{R}^{d \times N_2}$ denote testing samples. Each column represents a sample. To learn the representation coefficients, a general sparse representation model is formulated as

$$\min_{\mathbf{W}} \| \mathbf{Y} - \mathbf{XW} \|_F^2 + \lambda \| \mathbf{W} \|_1 \tag{4}$$

where λ is the regularization parameter for balancing respective term. We introduce $\mathbf{Z} = \mathbf{W}$ to solve model (4) by using augmented Lagrangian function according to ADMM method. The augmented Lagrangian function of problem (4) is formulated as

$$\begin{aligned} \mathbf{L}_\mu(\mathbf{W}, \mathbf{Z}, \mathbf{\Lambda}) = \min_{\mathbf{W}, \mathbf{Z}, \mathbf{\Lambda}} & \| \mathbf{Y} - \mathbf{XW} \|_F^2 + \lambda \| \mathbf{Z} \|_1 + <\mathbf{\Lambda}, \mathbf{W} - \mathbf{Z}> \\ & + \frac{\mu}{2} \| \mathbf{W} - \mathbf{Z} \|_F^2 \end{aligned} \tag{5}$$

where $<\mathbf{P}, \mathbf{Q}> = tr(\mathbf{P}^\mathbf{T}\mathbf{Q})$, $\mathbf{\Lambda}$ is a Lagrange multiplier and μ is a scalar constant. The augmented Lagrangian is minimized alone one coordinate direction at each iteration. ADMM consists of the following iterations.

(i) Given $\mathbf{Z} = \mathbf{Z}^t, \mathbf{\Lambda} = \mathbf{\Lambda}^t$, updating $\mathbf{W}$ by

$$\mathbf{W}^{t+1} = arg\min_{\mathbf{W}} L_\mu(\mathbf{W}, \mathbf{Z}, \mathbf{\Lambda}) \tag{6}$$

(ii) Given $\mathbf{W} = \mathbf{W}^{t+1}, \mathbf{\Lambda} = \mathbf{\Lambda}^k$, updating $\mathbf{Z}$ by

$$\mathbf{Z}^{t+1} = arg\min_{\mathbf{Z}} L_\mu(\mathbf{W}, \mathbf{Z}, \mathbf{\Lambda}) \tag{7}$$

(iii) Given $\mathbf{W} = \mathbf{W}^{t+1}, \mathbf{Z} = \mathbf{Z}^{t+1}$, updating $\mathbf{\Lambda}$ by

$$\mathbf{\Lambda}^{t+1} = \mathbf{\Lambda}^t + \mu(\mathbf{W}^{t+1} + \mathbf{Z}^{t+1}) \tag{8}$$

The key steps are to solve the optimization problems in Eqs.(6) and (7). Based on the augmented Lagrangian function in Eqs.(5) and (6) can be expressed as

$$\mathbf{W}^{t+1} = arg\min_{\mathbf{W}}(\| \mathbf{Y} - \mathbf{XW} \|_F^2 + <\mathbf{\Lambda}, \mathbf{W} - \mathbf{Z}> + \frac{\mu}{2} \| \mathbf{W} - \mathbf{Z} \|_F^2) \tag{9}$$

Since Eq. (9) is a standard regression model, we can get its closed-form solution as follows

$$\mathbf{W}^{t+1} = (\mathbf{X}^T\mathbf{X} + \mu\mathbf{I})^{-1}(\mathbf{X}^T\mathbf{Y} - \mathbf{\Lambda}^t + \mu\mathbf{Z}^t) \tag{10}$$

where $\mathbf{I}$ is a identity matrix. Based on the augmented Lagrangian function in Eqs. (5) and (7) can be rewritten as

$$\mathbf{Z}^{t+1} = arg\min_{\mathbf{Z}}(\lambda \| \mathbf{Z} \|_1 + <\mathbf{\Lambda}, \mathbf{W} - \mathbf{Z}> + \frac{\mu}{2} \| \mathbf{W} - \mathbf{Z} \|_F^2) \tag{11}$$

Because l_1-norm problem is indifferentiable, the shrinkage technique [15] is used to solve this problem. The optimal solution presents as

$$\mathbf{W}^{t+1} = shrinkage_{\frac{\lambda}{\mu}}(\mathbf{W}^{t+1} + \frac{\Lambda^t}{\mu}) \tag{12}$$

According to ADMM algorithm, the objective function value will be convergence until certain optimality conditions and stopping criteria are satisfied. In this paper, to simplify this problem, we set a max iteration instead. The detailed process for solving problem (4) is summarized in Algorithm 1.

Algorithm 1. The proposed SSVD

Input: Training samples $\mathbf{X}$ and testing samples $\mathbf{Y}$ normalized with l_2-norm,
  parameters $\lambda = 10^{-4}, \mu = 10^{-1}$, identity matrix $\mathbf{I}$
Output: $\mathbf{W}, \mathbf{Z}, \Lambda$
 1: **Initialize:** $\mathbf{W}^0 = \mathbf{Z}^0 = \Lambda^0 = 0$
 2: **repeat**
 3:    update $\mathbf{W}$: $\mathbf{W}^{t+1} = (\mathbf{X}^T\mathbf{X} + \mu\mathbf{I})^{-1}(\mathbf{X}^T\mathbf{Y} - \Lambda^t + \mu\mathbf{Z}^t)$
 4:    update $\mathbf{Z}$: $\mathbf{Z}^{t+1} = shrinkage_{\frac{\lambda}{\mu}}(\mathbf{W}^{t+1} + \frac{\Lambda^t}{\mu})$
 5:    update Λ: $\Lambda^{t+1} = \Lambda^t + \mu(\mathbf{W}^{t+1} - \mathbf{Z}^{t+1})$
 6: **until** convergence

---

4   Experimental Results

In this section, we present the experimental results of our proposed SSVD method on publicly available databases, following the same experimental settings in [16]. We randomly split the databases into two part. To avoid special case, all the experiments are run 10 times, and the average recognition rates are reported. Different from [16], we just validate our proposed framework on three face databases (Extended Yale B [17], CMU PIE [18], AR [19]) and one object database (COIL-100 [20]). We compare the proposed method with the popular methods such as LLC, LRC, CRC, SRC, SVM [21] and three methods (ENLR, DENLR, MENLR) proposed in [16]. Our (Max) and Our (Ave) respectively represent the methods to obtain the final softmax vectors in the Image Coding part by using max pooling and average pooling.

In the experiments, we reshape each image into one vector or extract the random feature of image. The l_2-normalization is used for all the samples. The experimental results shows that our method can achieve more significant results than many compared methods especially on face databases. The bold numbers represent the best recognition rate. In the following experiments, we let λ_1, μ_1 represent the parameters in image coding part and λ_2, μ_2 represent the parameters in softmax vector coding part in Fig. 1. The number of layers is set as 10 on all database.

(1) Extended Yale B Database: The Extended Yale B database contains 2414 frontal face images of 38 individuals each of them has around 64 near frontal

images under different illuminations. We randomly select 15, 20, 25, 30 images per person for training, and the rest for testing. We set $\lambda_1 = 10^{-4}$, $\mu_1 = 10^{-1}$, $\lambda_2 = 10^{-4}$, and $\mu_2 = 1.7$. The recognition rates of different methods on this database are summarized in Table 1. Note that the mean recognition rate are reported, and the bold numbers represent the best recognition rates. It is worth noting that our method can achieve the best recognition rates. Typically, when the number of training samples is 15, the recognition rate of our method is 4 percent higher than MENLR that achieves the best result among the compared methods. Besides, it means that our method can achieve good recognition rate when there are less training samples on this database.

(2) CMU PIE Database: The CMU PIE face database contains 41,368 face images from 68 subjects as a whole. The images under five near frontal poses (C05, C07, C09, C27 and C29) are used in our experiment. We randomly select 15, 20, 25, 30 images from each subject as training samples and the remaining images as test samples. We set $\lambda_1 = 10^{-4}$, $\mu_1 = 10^{-1}$, $\lambda_2 = 10^{-4}$, and $\mu_2 = 10^{-2}$. The classification rates of different methods are summarized in Table 2. It is clear that our method outperforms the compared methods in different cases.

Table 1. Recognition rates (%) on Extended Yale B database with different number of training samples

Alg.	15	20	25	30
LLC	88.63	91.52	94.20	95.21
LRC	89.47	92.52	93.50	94.62
CRC	91.39	94.26	95.91	97.04
SRC	91.72	93.71	95.56	96.37
SVM	89.35	92.74	95.07	96.20
ENLR	92.18	94.28	95.70	96.80
DENLR	94.34	96.66	97.70	98.51
MENLR	94.76	97.27	97.68	98.74
Our (Max)	**98.87**	**99.51**	**99.63**	**99.79**
Our (Ave)	98.68	99.44	99.62	99.73

Table 2. Recognition rates (%) on CMU PIE database with different number of training samples

Alg.	15	20	25	30
LLC	84.62	90.90	93.27	94.46
LRC	85.61	90.17	92.65	94.01
CRC	89.76	92.42	93.80	94.61
SRC	88.97	91.14	92.62	93.71
SVM	86.66	90.70	92.66	93.06
ENLR	90.47	92.82	93.94	94.67
DENLR	92.25	94.06	95.61	95.86
MENLR	93.21	94.88	95.74	96.18
Our (Max)	91.44	93.73	94.95	95.66
Our (Ave)	**93.79**	**95.59**	**96.37**	**96.84**

(3) AR Database: The AR face database contains about 4,000 color face images of 126 subject, which consist of the frontal faces with different facial expressions, illuminations and disguises. In this experiment, we select a subset including 2600 images from 50 female and 50 male subjects. We randomly select 8, 11, 14, 17 images for each subject as training samples and the rest of images as test samples. Following the experiment in [22], each image and its subregion are projected onto a 540-dimensional feature vector with a randomly generated matrix from a zero-mean normal distribution. We set $\lambda_1 = 10^{-5}$, $\mu_1 = 2$, $\lambda_2 = 10^{-5}$, and $\mu_2 = 10^{-3}$.

The recognition rates of different methods on this database are summarized in Table 3. From the table, we can see that our method achieves the best recognition rates.

(4) COIL-100 Database: Columbia Object Image Library (COIL-100) database contains various views of 100 objects (72 images per object) with different lighting conditions. In our experiments, the images are converted to gray-scale images with the 32×32 pixels. We randomly select 15, 20, 25, 30 images per object to construct the training set, and the test set contains the rest of the images. We set $\lambda_1 = 10^{-2}$, $\mu_1 = 1$, $\lambda_2 = 10^{-4}$, and $\mu_2 = 10^{-2}$. The recognition rates of different methods on this database are summarized in Table 4. We can see that our method is inferior to the best MENLR, but still better than other methods.

Table 3. Recognition rates (%) on AR database with different number of training samples

Alg.	8	11	14	17
LLC	54.26	60.87	66.88	71.58
LRC	63.87	76.87	85.20	90.88
CRC	86.53	91.66	94.06	95.74
SRC	84.08	89.45	92.20	95.14
SVM	75.74	86.19	91.99	95.08
ENLR	90.42	93.80	95.41	96.31
DENLR	91.94	95.69	97.30	98.21
MENLR	92.61	95.63	97.16	98.56
Our (Max)	92.72	96.66	97.65	98.27
Our (Ave)	**95.15**	**97.31**	**98.13**	**98.70**

Table 4. Recognition rates (%) on COIL-100 database with different number of training samples

Alg.	15	20	25	30
LLC	86.93	90.25	92.50	93.84
LRC	85.33	88.79	91.09	92.63
CRC	81.36	84.33	86.33	87.72
SRC	86.10	89.47	91.99	93.91
SVM	84.89	88.10	90.80	92.44
ENLR	88.40	91.28	93.37	94.66
DENLR	91.92	94.36	95.80	96.87
MENLR	**92.75**	**94.88**	**96.34**	**97.36**
Our (Max)	89.51	92.77	94.55	95.90
Our (Ave)	91.09	93.95	95.48	96.89

In summary, the proposed SSVD model can achieve remarkable results on face databases. It is also worth noting that SSVD (Max) outperforms SSVD (Ave) on Extended Yale B database and is inferior to SSVD (Ave) on CMU PIE, AR and COIL-100 database. The important advantage of SSVD model is that each image is divided into 4 or 16 subregions, which means that one image can be represented 4 or 16 times. It is useful to amend the misclassified image. We take an example to explain the effect of max pooling and average pooling. Suppose that there is a four categories image set split into two parts training set and testing set. Given a misclassified testing image that is actually from class 1, we will obtain its softmax vector $r = [0.25 \quad 0.40 \quad 0.15 \quad 0.20]^T$. As for its subregions, there are two cases. (1) There exist one subregion (first subregion we suppose) which shows much more discriminative than the whole image and other subregions. We let $r_1 = [0.60 \quad 0.20 \quad 0.10 \quad 0.10]^T$, $r_2 = [0.30 \quad 0.45 \quad 0.10 \quad 0.15]^T$,

$r_3 = [0.25 \quad 0.50 \quad 0.10 \quad 0.15]^T$, and $r_4 = [0.30 \quad 0.35 \quad 0.25 \quad 0.10]^T$ respectively represent the softmax vectors of the 4 subregions. After the max pooling, we will obtain the final softmax vector $r' = [0.60 \quad 0.50 \quad 0.25 \quad 0.15]^T$ which can amend the misclassified image. (2) The above case is unusual in reality. Instead, there more likely exist most subregions which show a lot discriminative than other subregions. The misclassified image and its softmax vector are the same as case (1). We let $r_1 = [0.35 \quad 0.25 \quad 0.15 \quad 0.25]^T$, $r_2 = [0.40 \quad 0.20 \quad 0.30 \quad 0.10]^T$, $r_3 = [0.20 \quad 0.50 \quad 0.10 \quad 0.20]^T$, and $r_4 = [0.45 \quad 0.15 \quad 0.20 \quad 0.20]^T$ respectively represent the softmax vectors of the 4 subregions. After the average pooling, we will obtain the final softmax vector $r' = [0.35 \quad 0.28 \quad 0.19 \quad 0.19]^T$ which can also amend the misclassified image.

Discussion of Layers: To better illustrate our methods, we give the sample curves (presented in Fig. 3) that shows the recognition rates with different layers in the deep model for each database. It is clear that as the number of layers increases, the recognition rate represents a rising tendency, which demonstrates the effectiveness of deep cascade model.

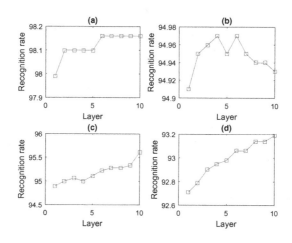

Fig. 3. The recognition rates with different layers on different database: (a) Extended Yale B database, (b) CMU PIE Database, (c) AR database, (d) COIL-100 database.

Convergence: To illustrate the effectiveness of our solving algorithm for problem (4), we show the objective function values with the varying iteration number (presented in Fig. 4) on the Extended Yale B database by using the Algorithm 1 to solve problem (4). It is easy to find that the objective function values present a convergence trend, which demonstrates the effectiveness of the Algorithm 1.

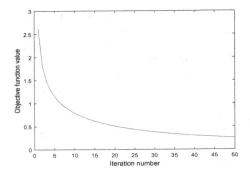

Fig. 4. The convergence of Algorithm 1.

5 Conclusion

This paper presented a novel sparse softmax vector coding based deep cascade model (SSVD). One important advantage of this model is using the class discrimination softmax vector. Besides, some sub-patterns show more discriminative than the whole image, which can amend the misclassified image by using max-polling or average-polling. We also explored the effectiveness of the concatenated softmax vector. The extensive experimental results clearly demonstrated that the proposed method outperforms significantly previous methods.

Acknowledgements. This work was supported by the National Science Fund of China under Grants (61771079, 61401048) and the Fundamental Research Funds for the Central Universities (No. 106112017CDJQJ168819).

References

1. Naseem, I., Togneri, R., Bennamounand, M.: Linear regression for face recognition. IEEE Trans. PAMI **32**(11), 2106–2112 (2010)
2. Wright, J., Yang, A.Y., Ganesh, A., Sastry, S.S., Ma, Y.: Robust face recognition via sparse representation. IEEE Transa. PAMI **31**(2), 210–227 (2009)
3. Zhang, L., Yang, M., Feng, X.: Sparse representation or collaborative representation: which helps face recognition? In: ICCV, pp. 471–478. IEEE (2011)
4. Yang, M., Zhang, D., Yang, J., Zhang, D.: Robust sparse coding for face recognition. In: CVPR, pp. 625–632. IEEE (2011)
5. He, R., Zheng, W.-S., Hu, B.-G.: Maximum correntropy criterion for robust face recognition. IEEE Trans. PAMI **33**(8), 1561–1576 (2011)
6. He, R., Zheng, W.-S., Tan, T., Sun, Z.: Half-quadratic-based iterative minimization for robust sparse representation. IEEE Trans. PAMI **36**(2), 261–275 (2014)
7. Naseem, I., Togneri, R., Bennamoun, M.: Robust regression for face recognition. Pattern Recogn. **45**(1), 104–118 (2012)
8. Zhou, Z., Wagner, A., Mobahi, H., Wright, J., Ma, Y.: Face recognition with contiguous occlusion using markov random fields. In: ICCV, pp. 1050–1057. IEEE (2009)

9. Jia, K., Chan, T.-H., Ma, Y.: Robust and practical face recognition via structured sparsity. In: Fitzgibbon, A., Lazebnik, S., Perona, P., Sato, Y., Schmid, C. (eds.) ECCV 2012. LNCS, vol. 7575, pp. 331–344. Springer, Heidelberg (2012). https://doi.org/10.1007/978-3-642-33765-9_24

10. Ren, C., Dai, D., Yan, H.: Coupled kernel embedding for low-resolution face image recognition. IEEE Trans. Image Process. **21**(8), 3770–3783 (2012)

11. Xu, Y., Li, X., Yang, J., Lai, Z., Zhang, D.: Integrating conventional and inverse representation for face recognition. IEEE Trans. Cybern. **44**(10), 1738–1746 (2014)

12. Cands, E.J., Li, X., Ma, Y., Wright, J.: Robust principal component analysis? J. ACM **58**(3), 11 (2011)

13. Li, Y., Liu, J., Lu, H., Ma, S.: Learning robust face representation with class-wise block-diagonal structure. IEEE Trans. Inf. Forensics Secur. **9**(12), 2051–2062 (2014)

14. Gabay, D., Mercier, B.: A dual algorithm for the solution of nonlinear variational problems via finite element approximations. IEEE Trans. Image Process. **22**(1), 17–40 (1976)

15. Yang, J., Zhang, Y.: Alternating direction algorithms for l_1-problems in compressive sensing. SIAM J. Sci. Comput. **33**(1), 250–278 (2011)

16. Zhang, Z., Lai, Z., Xu, Y., Shao, L., Xu, J., Xie, G.-S.: Discriminative elastic-net regularized linear regression. IEEE Trans. Image Process. **26**(3), 1466–1481 (2016)

17. Georghiades, A.S., Belhumeur, P.N., Kriegman, D.: From few to many: illumination cone models for face recognition under variable lighting and pose. IEEE Trans. PAMI **23**(6), 643–660 (2001)

18. Sim, T., Baker, S., Bsat, M.: The CMU pose, illumination, and expression (PIE) database. In: Proceedings of the 5th IEEE International Conference on Automatic Face Gesture Recognition, pp. 46–51 (2002)

19. Martinez, A.M., Benavente, R.: The AR Face Database. CVC Technical report, 24 June 1998

20. Nene, S.A., Nayar, S.K., Murase, H.: Columbia object image library (COIL-100). Technical report CUCS-006-96 (1996)

21. Chang, C.-C., Lin, C.-J.: LIBSVM: a library for support vector machines. ACM Trans. Intell. Syst. Technol. **2**(3), 1–27 (2011)

22. Jiang, Z., Lin, Z., Davis, L.S.: Label consistent K-SVD: learning a discriminative dictionary for recognition. IEEE Trans. PAMI **35**(11), 2651–2664 (2013)

23. Zhou, Z.-H., Feng, J.: Deep forest: toward an alternative to deep neural network. In: IJCAI (2017)

Author Index

Printed in the United States
By Bookmasters